THE ROLE OF BIOINFORMATICS IN AGRICULTURE

THE ROLE OF BIOINFORMATICS IN AGRICULTURE

Edited by

Santosh Kumar, PhD

Apple Academic Press

TORONTO NEW JERSEY

Apple Academic Press Inc.
3333 Mistwell Crescent
Oakville, ON L6L 0A2
Canada

Apple Academic Press Inc.
9 Spinnaker Way
Waretown, NJ 08758
USA

International Standard Book Number-13: 978-1-77188-003-9 (Hardcover)

Library of Congress Control Number: 2013958424

Library and Archives Canada Cataloguing in Publication

The role of bioinformatics in agriculture/edited by Santosh Kumar, PhD.

Includes bibliographical references and index.
ISBN 978-1-77188-003-9 (bound)
1. Agriculture--Data processing. 2. Bioinformatics.
3. Agricultural informatics. I. Kumar, Santosh, 1974-, editor of compilation

S494.5.D3R64 2013 630.285 C2013-907167-9

Apple Academic Press also publishes its books in a variety of electronic formats. Some content that appears in print may not be available in electronic format. For information about Apple Academic Press products, visit our website at **www.appleacademicpress.com** and the CRC Press website at **www.crcpress.com**

ABOUT THE EDITOR

Santosh Kumar, PhD
Research Associate in Plant Science, University of Manitoba, Canada

Dr. Santosh Kumar obtained his BSc Agriculture degree in plant breeding from Punjab Agricultural University, Ludhiana, India. He obtained his MSc working on wheat phytosiderophores from the Indian Agricultural Research Institute, New Delhi, India, and PhD working on control of flowering and germination in barley from University of Manitoba, Canada. At present, Dr. Kumar is stationed at the Cereal Research Centre, Manitoba, Canada, working for the University of Manitoba. Dr. Kumar has studied various disciplines in plant sciences and received many awards throughout his academic career. His current interests include crop genomics and bioinformatics. Dr. Kumar is member of various plant science societies in Canada and the US. Dr. Kumar has publications that include research articles, a review paper, and a book chapter in areas of plant physiology, genetics, and bioinformatics.

CONTENTS

Acknowledgment and How to Cite .... *xi*

List of Contributors .... *xi*

Introduction .... *xix*

Part I: Genome Data Mining Through Modern Sequencing Technologies

1. **Comparison of Next-Generation Sequencing Systems** **1**
   Lin Liu, Yinhu Li, Siliang Li, Ni Hu, Yimin He, Ray Pong, Danni Lin, Lihua Lu, and Maggie Law

2. **Why Assembling Plant Genome Sequences Is So Challenging** **27**
   Manuel Gonzalo Claros, Rocío Bautista, Darío Guerrero-Fernández, Hicham Benzerki, Pedro Seoane, and Noé Fernández-Pozo

3. **Genome Walking by Next Generation Sequencing Approaches** **55**
   Mariateresa Volpicella, Claudia Leoni, Alessandra Costanza, Immacolata Fanizza, Antonio Placido, and Luigi R. Ceci

4. **Molecular Tools for Exploring Polyploid Genomes in Plants** **73**
   Riccardo Aversano, Maria Raffaella Ercolano, Immacolata Caruso, Carlo Fasano, Daniele Rosellini, and Domenico Carputo

5. **Wheat Genomics: Present Status and Future Prospects** **99**
   P. K. Gupta, R. R. Mir, A. Mohan, and J. Kumar

6. **Development in Rice Genome Research Based on Accurate Genome Sequence** **167**
   Takashi Matsumoto, Jianzhong Wu, Baltazar A. Antonio, and Takuji Sasaki

7. **Genomic Databases for Crop Improvement** **189**
   Kaitao Lai, Michał T. Lorenc, and David Edwards

Part II: Going Beyond DNA Variations to Understand Environmental Adaptations

8. **Uncovering the Complexity of Transcriptomes with RNA-Seq 207**
Valerio Costa, Claudia Angelini, Italia De Feis, and Alfredo Ciccodicola

9. **Efficient Experimental Design and Analysis Strategies for the Detection of Differential Expression Using RNA-Sequencing........... 249**
José A. Robles, Sumaira E. Qureshi, Stuart J. Stephen, Susan R. Wilson, Conrad J. Burden, and Jennifer M. Taylor

10. **Analyzing the microRNA Transcriptome in Plants Using Deep Sequencing Data 281**
Xiaozeng Yang and Lei Li

11. **Maize (*Zea mays* L.) Genome Diversity as Revealed by RNA-Sequencing.. 299**
Candice N. Hansey, Brieanne Vaillancourt, Rajandeep S. Sekhon, Natalia de Leon, Shawn M. Kaeppler, and C. Robin Buell

12. **Molecular Mechanisms of Epigenetic Variation in Plants 327**
Ryo Fujimoto, Taku Sasaki, Ryo Ishikawa, Kenji Osabe, Takahiro Kawanabe, and Elizabeth S. Dennis

13. **Metagenomics: A Guide from Sampling to Data Analysis.................. 357**
Torsten Thomas, Jack Gilbert, and Folker Meyer

Part III: Single Nucleotide Polymorphism Is the Ideal Marker

14. **SNP Markers and Their Impact on Plant Breeding............................ 387**
Jafar Mammadov, Rajat Aggarwal, Ramesh Buyyarapu, and Siva Kumpatla

15. **Genotyping-by-Sequencing in Plants... 415**
Stéphane Deschamps , Victor Llaca, and Gregory D. May

Author Notes 447

Index 453

ACKNOWLEDGMENT AND HOW TO CITE

The chapters in this book were previously published in various places and in various formats. By bringing them together here in one place, we offer the reader a comprehensive perspective on recent investigations into the role of bioinformatics in agriculture. Each chapter is added to and enriched by being placed within the context of the larger investigative landscape. Specifically:

- Chapter 1 provides a description of hands-on experience with functioning, accuracy, and cost of sequencing from second generation sequencers. The authors also discuss the applications of data generated through high throughput second generation sequencers.
- Chapter 2 focuses on the typical challenges of plant genomes that explains why plant genomics is less developed than animal genomics. It provides a discussion on factors hindering accurate plant genome assembly.
- Chapter 3 sheds light on the development of several protocols for the application of next generation sequencing technologies to genome walking.
- Chapter 4 focuses on technologies and strategies that may allow an in-depth analysis of polyploid genomes. Origin and genetics of polyploids as well as the main tools available for genome and gene expression analysis are discussed. The implications of next generation sequencing in study of polyploids is also discussed.
- Chapter 5 discusses the advances in wheat genomics and also describes the available resources which can be used for future genomics research; this review is important because wheat has a large genome and is a relatively difficult for genomics research.
- Chapter 6 focuses on producing accurate genome sequences of rice and studying its functional and evolutionary implications for comparative genomics
- Chapter 7 is a concise list of plant genome databases, resources for individual crop improvement and potential course of action for future improvement of genome databases.
- Chapter 8 surveys the RNA-Seq methods and the bioinformatics challenges that follow in analyzing the sequence data.
- Chapter 9 provides insights into power and accuracy of differential gene estimation based on different experimental designs.

- Chapter 10 sheds light on the computational tools that are required to understand the large amount of complex small RNA data generated through high throughput sequencing.
- Chapter 11 focuses on evaluation of genome diversity in maize by analyzing expression profile of genes. Maize is one of the most important cereals used as human and animal feed throughout the world. For this reason, maize has evolved to be genetically and phenotypically diverse. Not only genes but gene expression products also contribute to genetic and phenotypic plasticity.
- Chapter 12 discusses the heritability of epigenetic changes and its potential to drive natural variation. The epigenetic modifications alter gene expression without any change in the nucleotide sequence but by changing structural conformation of chromatin. They are key to adaptations occurring in short time spans.
- Chapter 13 summarizes the current protocols in metagenomics, experimental designs, guidance on sample processing, sequencing technology and sequence assembly, annotation, statistical analysis, data management and dissemination.
- Chapter 14 discusses the historical perspective of SNP development and the current experimental designs for SNP discovery and validation using specific examples. Molecular markers are regarded as foundation of genetic analysis and are indispensable for efficient breeding. The single nucleotide polymorphic sites are the ideal markers as they can be identified on the genome wide scale by a single experiment. The next generation sequencing has simplified SNP identification and validation.
- Chapter 15 summarizes an approach where sequencing data is used simultaneously to discover and validate SNPs thereby bypassing the entire marker development stages. With more and more genomes being sequenced through next generation sequencing, this is a timely review that summarizes the current state of genotyping-by-sequencing on genome wide scale.

We wish to thank the authors who made their research available for this book, whether by granting permission individually or by releasing their research as open source articles. When citing information contained within this book, please do the authors the courtesy of attributing them by name, referring back to their original articles, using the credits provided at the beginning of each chapter.

LIST OF CONTRIBUTORS

Rajat Aggarwal
Department of Trait Genetics and Technologies, Dow AgroSciences LLC, 9330 Zionsville Road, Indianapolis, IN 46268, USA

Claudia Angelini
Istituto per le Applicazioni del Calcolo "Mauro Picone", IAC-CNR, 80131 Naples, Italy

Baltazar A. Antonio
Division of Genome and Biodiversity Research, National Institute of Agrobiological Sciences, 2-1-2, Kannondai, Tsukuba, Ibaraki 3058602, Japan

Riccardo Aversano
Department of Soil, Plant, Environmental and Animal Production Sciences, University of Naples Federico II, Via Università 100, Portici 80055, Italy

Rocío Bautista
Bioinformatics Andalusian Platform, Bio-innovation Building, University of Malaga, 29590 Málaga, Spain

Hicham Benzerki
Department of Molecular Biology and Biochemistry, Faculty of Sciences, University of Malaga, 29071 Málaga, Spain and Bioinformatics Andalusian Platform, Bio-innovation Building, University of Malaga, 29590 Málaga, Spain

C. Robin Buell
Department of Plant Biology, Michigan State University, East Lansing, Michigan, United States of America and Department of Energy Great Lakes Bioenergy Research Center, Michigan State University, East Lansing, Michigan, United States of America

Conrad J. Burden
Mathematical Sciences Institute, Australian National University, Canberra, Australia

Ramesh Buyyarapu
Department of Trait Genetics and Technologies, Dow AgroSciences LLC, 9330 Zionsville Road, Indianapolis, IN 46268, USA

Domenico Carputo
Department of Soil, Plant, Environmental and Animal Production Sciences, University of Naples Federico II, Via Università 100, Portici 80055, Italy

Immacolata Caruso
Department of Soil, Plant, Environmental and Animal Production Sciences, University of Naples Federico II, Via Università 100, Portici 80055, Italy

Alfredo Ciccodicola
Institute of Genetics and Biophysics "A. Buzzati-Traverso", IGB-CNR, 80131 Naples, Italy

Luigi R. Ceci
Institute of Biomembranes and Bioenergetics, Italian National Research Council (CNR), Via Amendola 165/A, 70126 Bari, Italy

Valerio Costa
Institute of Genetics and Biophysics "A. Buzzati-Traverso", IGB-CNR, 80131 Naples, Italy

Alessandra Costanza
Department of Biosciences, Biotechnology and Pharmacological Sciences, University of Bari, Via Amendola 165/A, 70126 Bari, Italy

Elizabeth S. Dennis
Commonwealth Scientific and Industrial Research Organisation (CSIRO) Plant Industry, Canberra ACT 2601, Australia

Stéphane Deschamps
DuPont Agricultural Biotechnology, Experimental Station, PO Box 80353, 200 Powder Mill Road, Wilmington, DE 19880-0353, USA

Italia De Feis
Istituto per le Applicazioni del Calcolo "Mauro Picone", IAC-CNR, 80131 Naples, Italy

Natalia de Leon
Department of Agronomy, University of Wisconsin-Madison, Madison, Wisconsin, United States of America and Department of Energy Great Lakes Bioenergy Research Center, University of Wisconsin-Madison, Madison, Wisconsin, United States of America

David Edwards
School of Agriculture and Food Sciences, University of Queensland, Brisbane, QLD 4072, Australia and Australian Centre for Plant Functional Genomics, University of Queensland, Brisbane, QLD 4072, Australia

Maria Raffaella Ercolano
Department of Soil, Plant, Environmental and Animal Production Sciences, University of Naples Federico II, Via Università 100, Portici 80055, Italy

Immacolata Fanizza
Department of Biosciences, Biotechnology and Pharmacological Sciences, University of Bari, Via Amendola 165/A, 70126 Bari, Italy

Carlo Fasano
Department of Soil, Plant, Environmental and Animal Production Sciences, University of Naples Federico II, Via Università 100, Portici 80055, Italy

Noé Fernández-Pozo
Department of Molecular Biology and Biochemistry, Faculty of Sciences, University of Malaga, 29071 Málaga, Spain

Ryo Fujimoto
Graduate School of Science and Technology, Niigata University, Nishi-ku, Niigata 950-2181, Japan

Jack Gilbert
Argonne National Laboratory, 9700 South Cass Avenue, Argonne, IL 60439, USA and Department of Ecology and Evolution, University of Chicago, 5640 South Ellis Avenue, Chicago, IL 60637, USA

Manuel Gonzalo Claros
Department of Molecular Biology and Biochemistry, Faculty of Sciences, University of Malaga, 29071 Málaga, Spain and Bioinformatics Andalusian Platform, Bio-innovation Building, University of Malaga, 29590 Málaga, Spain

Darío Guerrero-Fernández
Bioinformatics Andalusian Platform, Bio-innovation Building, University of Malaga, 29590 Málaga, Spain

P. K. Gupta
Molecular Biology Laboratory, Department of Genetics and Plant Breeding, Ch. Charan Singh University, Meerut 250 004, India

Candice N. Hansey
Department of Plant Biology, Michigan State University, East Lansing, Michigan, United States of America and Department of Energy Great Lakes Bioenergy Research Center, Michigan State University, East Lansing, Michigan, United States of America

Yimin He
NGS Sequencing Department, Beijing Genomics Institute (BGI), 4th Floor, Building 11, Beishan Industrial Zone, Yantian District, Guangdong, Shenzhen 518083, China

Ni Hu
NGS Sequencing Department, Beijing Genomics Institute (BGI), 4th Floor, Building 11, Beishan Industrial Zone, Yantian District, Guangdong, Shenzhen 518083, China

Ryo Ishikawa
Laboratory of Plant Breeding, Graduate School of Agricultural Science, Kobe University, Nada, Kobe 657-8510, Japan and Cell and Developmental Biology, John Innes Centre, Norwich Research Park, Colney, Norwich NR4 7UH, UK

Shawn M. Kaeppler
Department of Agronomy, University of Wisconsin-Madison, Madison, Wisconsin, United States of America and Department of Energy Great Lakes Bioenergy Research Center, University of Wisconsin-Madison, Madison, Wisconsin, United States of America

Takahiro Kawanabe
Watanabe Seed Co., Ltd, Machiyashiki, Misato-cho, Miyagi 987-8607, Japan

J. Kumar
Molecular Biology Laboratory, Department of Genetics and Plant Breeding, Ch. Charan Singh University, Meerut 250 004, India

Siva Kumpatla
Department of Biotechnology Regulatory Sciences, Dow AgroSciences LLC, 9330 Zionsville Road, Indianapolis, IN 46268, USA

Kaitao Lai
School of Agriculture and Food Sciences, University of Queensland, Brisbane, QLD 4072, Australia and Australian Centre for Plant Functional Genomics, University of Queensland, Brisbane, QLD 4072, Australia

Maggie Law
NGS Sequencing Department, Beijing Genomics Institute (BGI), 4th Floor, Building 11, Beishan Industrial Zone, Yantian District, Guangdong, Shenzhen 518083, China

Claudia Leoni
Department of Biosciences, Biotechnology and Pharmacological Sciences, University of Bari, Via Amendola 165/A, 70126 Bari, Italy

Lei Li
Department of Biology, University of Virginia, Charlottesville VA 22904, USA

Siliang Li
NGS Sequencing Department, Beijing Genomics Institute (BGI), 4th Floor, Building 11, Beishan Industrial Zone, Yantian District, Guangdong, Shenzhen 518083, China

Yinhu Li
NGS Sequencing Department, Beijing Genomics Institute (BGI), 4th Floor, Building 11, Beishan Industrial Zone, Yantian District, Guangdong, Shenzhen 518083, China

Danni Lin
NGS Sequencing Department, Beijing Genomics Institute (BGI), 4th Floor, Building 11, Beishan Industrial Zone, Yantian District, Guangdong, Shenzhen 518083, China

Lin Liu
NGS Sequencing Department, Beijing Genomics Institute (BGI), 4th Floor, Building 11, Beishan Industrial Zone, Yantian District, Guangdong, Shenzhen 518083, China

Victor Llaca
DuPont Agricultural Biotechnology, Experimental Station, PO Box 80353, 200 Powder Mill Road, Wilmington, DE 19880-0353, USA

Michał T. Lorenc
School of Agriculture and Food Sciences, University of Queensland, Brisbane, QLD 4072, Australia

Lihua Lu
NGS Sequencing Department, Beijing Genomics Institute (BGI), 4th Floor, Building 11, Beishan Industrial Zone, Yantian District, Guangdong, Shenzhen 518083, China

Jafar Mammadov
Department of Trait Genetics and Technologies, Dow AgroSciences LLC, 9330 Zionsville Road, Indianapolis, IN 46268, USA

Takashi Matsumoto
Division of Genome and Biodiversity Research, National Institute of Agrobiological Sciences, 2-1-2, Kannondai, Tsukuba, Ibaraki 3058602, Japan

Gregory D. May
DuPont Pioneer, 7300 NW 62nd Ave., P.O. Box 1004, Johnston, IA 50131-1004, USA

Folker Meyer
Argonne National Laboratory, 9700 South Cass Avenue, Argonne, IL 60439, USA and Computation Institute, University of Chicago, 5640 South Ellis Avenue, Chicago, IL 60637, USA

R. R. Mir
Molecular Biology Laboratory, Department of Genetics and Plant Breeding, Ch. Charan Singh University, Meerut 250 004, India

A. Mohan
Molecular Biology Laboratory, Department of Genetics and Plant Breeding, Ch. Charan Singh University, Meerut 250 004, India

Kenji Osabe
Commonwealth Scientific and Industrial Research Organisation (CSIRO) Plant Industry, Canberra ACT 2601, Australia

Antonio Placido
Department of Biosciences, Biotechnology and Pharmacological Sciences, University of Bari, Via Amendola 165/A, 70126 Bari, Italy

Ray Pong
NGS Sequencing Department, Beijing Genomics Institute (BGI), 4th Floor, Building 11, Beishan Industrial Zone, Yantian District, Guangdong, Shenzhen 518083, China

Sumaira E. Qureshi
Mathematical Sciences Institute, Australian National University, Canberra, Australia

José A. Robles
CSIRO Plant Industry, Black Mountain Laboratories, Canberra, Australia

Daniele Rosellini
Department of Applied Biology, University of Perugia, Borgo XX Giugno 74, Perugia 06121, Italy

Taku Sasaki
Gregor Mendel Institute of Molecular Plant Biology, Austrian Academy of Sciences, Dr. Bohrgasse 3, Vienna 1030, Austria

Takuji Sasaki
Division of Genome and Biodiversity Research, National Institute of Agrobiological Sciences, 2-1-2, Kannondai, Tsukuba, Ibaraki 3058602, Japan

Rajandeep S. Sekhon
Department of Agronomy, University of Wisconsin-Madison, Madison, Wisconsin, United States of America and Department of Energy Great Lakes Bioenergy Research Center, University of Wisconsin-Madison, Madison, Wisconsin, United States of America

Pedro Seoane
Department of Molecular Biology and Biochemistry, Faculty of Sciences, University of Malaga, 29071 Málaga, Spain

Stuart J. Stephen
CSIRO Plant Industry, Black Mountain Laboratories, Canberra, Australia

Jennifer M. Taylor
CSIRO Plant Industry, Black Mountain Laboratories, Canberra, Australia

Torsten Thomas
School of Biotechnology and Biomolecular Sciences & Centre for Marine Bio-Innovation, The University of New South Wales, Sydney, NSW 2052, Australia

Brieanne Vaillancourt
Department of Plant Biology, Michigan State University, East Lansing, Michigan, United States of America and Department of Energy Great Lakes Bioenergy Research Center, Michigan State University, East Lansing, Michigan, United States of America

Mariateresa Volpicella
Department of Biosciences, Biotechnology and Pharmacological Sciences, University of Bari, Via Amendola 165/A, 70126 Bari, Italy

Susan R. Wilson
Mathematical Sciences Institute, Australian National University, Canberra, Australia and Prince of Wales Clinical School and School of Mathematics and Statistics, University of New South Wales, Sydney, Australia

Jianzhong Wu
Division of Genome and Biodiversity Research, National Institute of Agrobiological Sciences, 2-1-2, Kannondai, Tsukuba, Ibaraki 3058602, Japan

Xiaozeng Yang
Department of Biology, University of Virginia, Charlottesville VA 22904, USA

INTRODUCTION

The use of computing resources to manage, analyse and interpret biological data led to inception of an interdisciplinary branch of science termed Bioinformatics. The advances in information technology and next generation sequencing have propelled the use of bioinformatics in agriculture especially in the area of crop improvement. An extremely large amount of genomics data is available from plants and animals due to tremendous improvements in second and third generation sequencing technologies. Presently, fourty-one plant genome sequences are publicly available and many more are being sequenced. The computing hardware has become increasingly powerful yet at a declining cost. Various genomic resources such as reference genome sequence, transcriptome sequence and single nucleotide polymorphic markers that were available only for model plant species a decade ago are now available for majority of economically important plant species. The greatest challenge that we currently face is to make sense and then use the wealth of data generated by genome, transcriptome and methylome sequencing projects. Bioinformatics in agriculture, compared to human studies, faces additional challenges due to the genomic diversity in different plant and animal species that needs to be factored in prior to start of any bioinformatics study. In addition, presence of repetitive elements, transposons, gene duplications and variations in polyploidy levels pose significant constrains on developing bioinformatics tools that can be used universally on all plant species. Therefore, there is a great need for bioinformatics professionals in all areas of applied agriculture.

This book intends to acquaint the readers in state-of-the-art sequencing technologies, recent developments in computing algorithms and certain biological perspectives that influence development of bioinformatics tools by giving specific examples from model plant species. I sincerely hope that this book will immensely help the young and aspiring professionals who intend to hone their computing skills and biological knowledge to advance the promising field of Bioinformatics.

With fast development and wide applications of next-generation sequencing (NGS) technologies, genomic sequence information is within reach to aid the achievement of goals to decode life mysteries, make better crops, detect pathogens, and improve life qualities. NGS systems are typically represented by SOLiD/Ion Torrent PGM from Life Sciences, Genome Analyzer/HiSeq 2000/MiSeq from Illumina, and GS FLX Titanium/GS Junior from Roche. Beijing Genomics Institute (BGI), which possesses the world's biggest sequencing capacity, has multiple NGS systems including 137 HiSeq 2000, 27 SOLiD, one Ion Torrent PGM, one MiSeq, and one 454 sequencer. In Chapter 1, Liu and colleagues have accumulated extensive experience in sample handling, sequencing, and bioinformatics analysis. In this paper, technologies of these systems are reviewed, and first-hand data from extensive experience is summarized and analyzed to discuss the advantages and specifics associated with each sequencing system. At last, applications of NGS are summarized.

Gonzalo Claros and colleagues argue that in spite of the biological and economic importance of plants, relatively few plant species have been sequenced. Only the genome sequence of plants with relatively small genomes, most of them angiosperms, in particular eudicots, has been determined. The arrival of next-generation sequencing technologies has allowed the rapid and efficient development of new genomic resources for non-model or orphan plant species. But the sequencing pace of plants is far from that of animals and microorganisms. Chapter 2 focuses on the typical challenges of plant genomes that can explain why plant genomics is less developed than animal genomics. Explanations about the impact of some confounding factors emerging from the nature of plant genomes are given. As a result of these challenges and confounding factors, the correct assembly and annotation of plant genomes is hindered, genome drafts are produced, and advances in plant genomics are delayed.

Chapter 3, by Volpicella and colleagues, explores Genome Walking methods. Genome Walking (GW) comprises a number of PCR-based methods for the identification of nucleotide sequences flanking known regions. The different methods have been used for several purposes: from de novo sequencing, useful for the identification of unknown regions, to the characterization of insertion sites for viruses and transposons. In the latter cases Genome Walking methods have been recently boosted by coupling to Next Generation Sequencing technologies. This review will focus on the development of several protocols for the application of Next Generation Sequencing (NGS) technologies to GW, which have been developed in the course of analysis of insertional libraries. These analyses find broad application in protocols for functional genomics and gene therapy. Thanks to the application of NGS technologies, the original vision of GW as a procedure for walking along an unknown genome is now changing into the possibility of observing the parallel marching of hundreds of thousands of primers across the borders of inserted DNA molecules in host genomes.

Polyploidy is a very common phenomenon in the plant kingdom, where even diploid species are often described as paleopolyploids. The polyploid condition may bring about several advantages compared to the diploid state. Polyploids often show phenotypes that are not present in their diploid progenitors or exceed the range of the contributing species. Some of these traits may play a role in heterosis or could favor adaptation to new ecological niches. Advances in genomics and sequencing technology may create unprecedented opportunities for discovering and monitoring the molecular effects of polyploidization. In chapter 4, Aversano and colleagues provide an overview of technologies and strategies that may allow an in-depth analysis of polyploid genomes. After introducing some basic aspects on the origin and genetics of polyploids, we highlight the main tools available for genome and gene expression analysis and summarize major findings. In the last part of this review, the implications of next generation sequencing are briefly discussed. The accumulation of knowledge on polyploid formation, maintenance, and divergence at whole-genome and subgenome levels will not only help plant biologists to understand how plants have evolved and diversified, but also assist plant breeders in designing new strategies for crop improvement.

Gupta and colleagues explore the implications for using genomics research to study wheat in chapter 5. Wheat (*Triticum aestivum L.*), with a large genome (16000 Mb) and high proportion (80%) of repetitive sequences, has been a difficult crop for genomics research. However, the availability of extensive cytogenetics stocks has been an asset, which facilitated significant progress in wheat genomic research in recent years. For instance, fairly dense molecular maps (both genetic and physical maps) and a large set of ESTs allowed genome-wide identification of gene-rich and gene-poor regions as well as QTL including eQTL. The availability of markers associated with major economic traits also allowed development of major programs on marker-assisted selection (MAS) in some countries, and facilitated map-based cloning of a number of genes/QTL. Resources for functional genomics including TILLING and RNA interference (RNAi) along with some new approaches like epigenetics and association mapping are also being successfully used for wheat genomics research. BAC/BIBAC libraries for the subgenome D and some individual chromosomes have also been prepared to facilitate sequencing of gene space. In this brief review, the authors discuss all these advances in some detail, and also describe briefly the available resources, which can be used for future genomics research in this important crop.

Chapter 6, by Matsumoto and colleagues, focuses on another prominant crop; rice is one of the most important crops in the world. Although genetic improvement is a key technology for the acceleration of rice breeding, a lack of genome information had restricted efforts in molecular-based breeding until the completion of the high-quality rice genome sequence, which opened new opportunities for research in various areas of genomics. The syntenic relationship of the rice genome to other cereal genomes makes the rice genome invaluable for understanding how cereal genomes function. Producing an accurate genome sequence is not an easy task, and it is becoming more important as sequence deviations among, and even within, species highlight functional or evolutionary implications for comparative genomics.

Genomics is playing an increasing role in plant breeding and this is accelerating with the rapid advances in genome technology. Translating the vast abundance of data being produced by genome technologies requires the development of custom bioinformatics tools and advanced databases. These range from large generic databases which hold specific data types for a broad range of species, to carefully integrated and curated databases which act as a resource for the improvement of specific crops. In chapter 7, Lai and colleagues outline some of the features of plant genome databases, identify specific resources for the improvement of individual crops and comment on the potential future direction of crop genome databases.

In chapter 8, Costa and colleagues show that in recent years, the introduction of massively parallel sequencing platforms for Next Generation

Sequencing (NGS) protocols, able to simultaneously sequence hundred thousand DNA fragments, dramatically changed the landscape of the genetics studies. RNA-Seq for transcriptome studies, Chip-Seq for DNA-proteins interaction, CNV-Seq for large genome nucleotide variations are only some of the intriguing new applications supported by these innovative platforms. Among them RNA-Seq is perhaps the most complex NGS application. Expression levels of specific genes, differential splicing, allele-specific expression of transcripts can be accurately determined by RNA-Seq experiments to address many biological-related issues. All these attributes are not readily achievable from previously widespread hybridization-based or tag sequence-based approaches. However, the unprecedented level of sensitivity and the large amount of available data produced by NGS platforms provide clear advantages as well as new challenges and issues. This technology brings the great power to make several new biological observations and discoveries, it also requires a considerable effort in the development of new bioinformatics tools to deal with these massive data files. The paper aims to give a survey of the RNA-Seq methodology, particularly focusing on the challenges that this application presents both from a biological and a bioinformatics point of view.

Chapter 9, by Robles and colleagues, demonstrates that RNA sequencing (RNA-Seq) has emerged as a powerful approach for the detection of differential gene expression with both high-throughput and high resolution capabilities possible depending upon the experimental design chosen. Multiplex experimental designs are now readily available, these can be utilised to increase the numbers of samples or replicates profiled at the cost of decreased sequencing depth generated per sample. These strategies impact on the power of the approach to accurately identify differential expression. This study presents a detailed analysis of the power to detect differential expression in a range of scenarios including simulated null and differential expression distributions with varying numbers of biological or technical replicates, sequencing depths and analysis methods. Differential and non-differential expression datasets were simulated using a combination of negative binomial and exponential distributions derived from real RNA-Seq data. These datasets were used to evaluate the performance of three commonly used differential expression analysis algorithms and to quantify the changes in power with respect to true and false positive rates when simulating variations in sequencing depth, biological replication and multiplex experimental design choices. The work in this chapter quantitatively explores comparisons between contemporary analysis tools and experimental design choices for the detection of differential expression using RNA-Seq. The authors found that the DESeq algorithm performs more conservatively than edgeR and NBPSeq. With regard to testing of various experimental designs, this work strongly suggests that greater power is gained through the use of biological replicates relative to library (technical) replicates and sequencing depth. Strikingly, sequencing depth could be reduced as low as 15% without substantial impacts on false positive or true positive rates.

In chapter 10, Yang and Li explore the role of MicroRNAs: 20- to 24-nucleotide endogenous small RNA molecules emerging as an important class of sequence-specific, trans-acting regulators for modulating gene expression at the post-transcription level. There has been a surge of interest in the past decade in identifying miRNAs and profiling their expression pattern using various experimental approaches. In particular, ultra-deep sampling of specifically prepared low-molecular-weight RNA libraries based on next-generation sequencing technologies has been used successfully in diverse species. The challenge now is to effectively deconvolute the complex sequencing data to provide comprehensive and reliable information on the miRNAs, miRNA precursors, and expression profile of miRNA genes. Here, the authors review the recently developed computational tools and their applications in profiling the miRNA transcriptomes, with an emphasis on the model plant Arabidopsis thaliana. Highlighted is also progress and insight into miRNA biology derived from analyzing available deep sequencing data.

Hansey and colleagues demonstrate the genetic diversity of maize in chapter 11. Maize is rich in genetic and phenotypic diversity. Understanding the sequence, structural, and expression variation that contributes to phenotypic diversity would facilitate more efficient varietal improvement. RNA based sequencing (RNA-seq) is a powerful approach for transcriptional analysis, assessing sequence variation, and identifying novel transcript sequences, particularly in large, complex, repetitive genomes such as maize. In this study, the authors sequenced RNA from whole seedlings of 21 maize inbred lines representing diverse North American and exotic germplasm. Single nucleotide polymorphism (SNP) detection identified 351,710 polymorphic loci distributed throughout the genome covering 22,830 annotated genes. Tight clustering of two distinct heterotic groups and exotic lines was evident using these SNPs as genetic markers. Transcript abundance analysis revealed minimal variation in the total number of genes expressed across these 21 lines (57.1% to 66.0%). However, the transcribed gene set among the 21 lines varied, with 48.7% expressed in all of the lines, 27.9% expressed in one to 20 lines, and 23.4% expressed in none of the lines. De novo assembly of RNA-seq reads that did not map to the reference B73 genome sequence revealed 1,321 high confidence novel transcripts, of which, 564 loci were present in all 21 lines, including B73, and 757 loci were restricted to a subset of the lines. RT-PCR validation demonstrated 87.5% concordance with the computational prediction of these expressed novel transcripts. Intriguingly, 145 of the novel de novo assembled loci were present in lines from only one of the two heterotic groups consistent with the hypothesis that, in addition to sequence polymorphisms and transcript abundance, transcript presence/absence variation is present and, thereby, may be a mechanism contributing to the genetic basis of heterosis.

Natural variation is defined as the phenotypic variation caused by spontaneous mutations. In general, mutations are associated with changes of nucleotide sequence, and many mutations in genes that can cause changes in plant development have been identified. Epigenetic change, which does not involve alteration to the nucleotide sequence, can also cause changes in gene activity by changing the structure of chromatin through DNA methylation or histone modifications. In chapter 12, Fujimoto and colleagues show that now there is evidence based on induced or spontaneous mutants that epigenetic changes can cause altering plant phenotypes. Epigenetic changes have occurred frequently in plants, and some are heritable or metastable causing variation in epigenetic status within or between species. Therefore, heritable epigenetic variation as well as genetic variation has the potential to drive natural variation.

In chapter 13, Thomas and colleagues explore the field of metagenomics. Metagenomics applies a suite of genomic technologies and bioinformatics tools to directly access the genetic content of entire communities of organisms. The field of metagenomics has been responsible for substantial advances in microbial ecology, evolution, and diversity over the past 5 to 10 years, and many research laboratories are actively engaged in it now. With the growing numbers of activities also comes a plethora of methodological knowledge and expertise that should guide future developments in the field. This review summarizes the current opinions in metagenomics, and provides practical guidance and advice on sample processing, sequencing technology, assembly, binning, annotation, experimental design, statistical analysis, data storage, and data sharing. As more metagenomic datasets are generated, the availability of standardized procedures and shared data storage and analysis becomes increasingly important to ensure that output of individual projects can be assessed and compared.

The use of molecular markers has revolutionized the pace and precision of plant genetic analysis which in turn facilitated the implementation of molecular breeding of crops. The last three decades have seen tremendous advances in the evolution of marker systems and the respective detection platforms. Markers based on single nucleotide polymorphisms (SNPs) have rapidly gained the center stage of molecular genetics during the recent years due to their abundance in the genomes and their amenability for high-throughput detection formats and platforms. Computational approaches dominate SNP discovery methods due to the ever-increasing sequence information in public databases; however, complex genomes pose special challenges in the identification of informative SNPs warranting alternative strategies in those crops. Many genotyping platforms and chemistries have become available making the use of SNPs even more attractive and efficient. Mammadov and colleagues provide a review of historical and current efforts in the development, validation, and application of SNP markers in QTL/gene discovery and plant breeding by discussing key experimental strategies and cases exemplifying their impact in chapter 14.

The final chapter, by Deschamps and colleagues, argues that the advent of next-generation DNA sequencing (NGS) technologies has led to the development of rapid genome-wide Single Nucleotide Polymorphism (SNP) detection applications in various plant species. Recent improvements in sequencing throughput combined with an overall decrease in costs per gigabase of sequence is allowing NGS to be applied to not only the evaluation of small subsets of parental inbred lines, but also the mapping and characterization of traits of interest in much larger populations. Such an approach, where sequences are used simultaneously to detect and score SNPs, therefore bypassing the entire marker assay development stage, is known as genotyping-by-sequencing (GBS). This review will summarize the current state of GBS in plants and the promises it holds as a genome-wide genotyping application.

PART I

GENOME DATA MINING THROUGH MODERN SEQUENCING TECHNOLOGIES

CHAPTER 1

COMPARISON OF NEXT-GENERATION SEQUENCING SYSTEMS

LIN LIU, YINHU LI, SILIANG LI, NI HU, YIMIN HE, RAY PONG, DANNI LIN, LIHUA LU, and MAGGIE LAW

1.1 INTRODUCTION

(Deoxyribonucleic acid) DNA was demonstrated as the genetic material by Oswald Theodore Avery in 1944. Its double helical strand structure composed of four bases was determined by James D. Watson and Francis Crick in 1953, leading to the central dogma of molecular biology. In most cases, genomic DNA defined the species and individuals, which makes the DNA sequence fundamental to the research on the structures and functions of cells and the decoding of life mysteries [1]. DNA sequencing technologies could help biologists and health care providers in a broad range of applications such as molecular cloning, breeding, finding pathogenic genes, and comparative and evolution studies. DNA sequencing technologies ideally should be fast, accurate, easy-to-operate, and cheap. In the past thirty years, DNA sequencing technologies and applications have undergone tremendous development and act as the engine of the genome era which is characterized by vast amount of genome data and subsequently broad range of research areas and multiple applications. It is necessary to look

This chapter was originally published under the Creative Commons Attribution License. Liu L, Li Y, Li S, Hu N, He Y, Pong R, Lin D, Lu L, and Law M. Comparison of Next-Generation Sequencing Systems. Journal of Biomedicine and Biotechnology. ***2012** (2012), 11 pages. doi:10.1155/2012/251364.*

back on the history of sequencing technology development to review the NGS systems (454, GA/HiSeq, and SOLiD), to compare their advantages and disadvantages, to discuss the various applications, and to evaluate the recently introduced PGM (personal genome machines) and third-generation sequencing technologies and applications. All of these aspects will be described in this paper. Most data and conclusions are from independent users who have extensive first-hand experience in these typical NGS systems in BGI (Beijing Genomics Institute).

Before talking about the NGS systems, we would like to review the history of DNA sequencing briefly. In 1977, Frederick Sanger developed DNA sequencing technology which was based on chain-termination method (also known as Sanger sequencing), and Walter Gilbert developed another sequencing technology based on chemical modification of DNA and subsequent cleavage at specific bases. Because of its high efficiency and low radioactivity, Sanger sequencing was adopted as the primary technology in the "first generation" of laboratory and commercial sequencing applications [2]. At that time, DNA sequencing was laborious and radioactive materials were required. After years of improvement, Applied Biosystems introduced the first automatic sequencing machine (namely AB370) in 1987, adopting capillary electrophoresis which made the sequencing faster and more accurate. AB370 could detect 96 bases one time, 500 K bases a day, and the read length could reach 600 bases. The current model AB3730xl can output 2.88 M bases per day and read length could reach 900 bases since 1995. Emerged in 1998, the automatic sequencing instruments and associated software using the capillary sequencing machines and Sanger sequencing technology became the main tools for the completion of human genome project in 2001 [3]. This project greatly stimulated the development of powerful novel sequencing instrument to increase speed and accuracy, while simultaneously reducing cost and manpower. Not only this, X-prize also accelerated the development of next-generation sequencing (NGS) [4]. The NGS technologies are different from the Sanger method in aspects of massively parallel analysis, high throughput, and reduced cost. Although NGS makes genome sequences handy, the followed data analysis and biological explanations are still the bottle-neck in understanding genomes.

Following the human genome project, 454 was launched by 454 in 2005, and Solexa released Genome Analyzer the next year, followed by (Sequencing by Oligo Ligation Detection) SOLiD provided from Agencourt, which are three most typical massively parallel sequencing systems in the next-generation sequencing (NGS) that shared good performance on throughput, accuracy, and cost compared with Sanger sequencing (shown in Table 1(a)). These founder companies were then purchased by other companies: in 2006 Agencourt was purchased by Applied Biosystems, and in 2007, 454 was purchased by Roche, while Solexa was purchased by Illumina. After years of evolution, these three systems exhibit better performance and their own advantages in terms of read length, accuracy, applications, consumables, man power requirement and informatics infrastructure, and so forth. The comparison of these three systems will be focused and discussed in the later part of this paper (also see Tables 1(a), 1(b), and 1(c)).

TABLE 1: (a) Advantage and mechanism of sequencers. (b) Components and cost of sequencers. (c) Application of sequencers.

(A)

Sequencer	454 GS FLX	HiSeq 2000	SOLiDv4	Sanger 3730xl
Sequencing mechanism	Pyrosequencing	Sequencing by synthesis	Ligation and two-base coding	Dideoxy chain termination
Read length	700 bp	50SE, 50PE, 101PE	50 + 35 bp or 50 + 50 bp	400 ~ 900 bp
Accuracy	99.9%*	98%, (100PE)	99.94% *raw data	99.999%
Reads	1 M	3 G	1200~1400 M	—
Output data/run	0.7 Gb	600 Gb	120 Gb	1.9~84 Kb
Time/run	24 Hours	3~10 Days	7 Days for SE	
14 Days for PE	20 Mins~3 Hours			
Advantage	Read length, fast	High throughput	Accuracy	High quality, long read length
Disadvantage	Error rate with polybase more than 6, high cost, low throughput	Short read assembly	Short read assembly	High cost low throughput

TABLE 1: *Cont.*

(B)

Sequencers	454 GS FLX	HiSeq 2000	SOLiDv4	3730xl
Instrument price	Instrument $500,000, $7000 per run	Instrument $690,000, $6000/(30x) human genome	Instrument $495,000, $15,000/100 Gb	Instrument $95,000, about $4 per 800 bp reaction
CPU	2* Intel Xeon X5675	2* Intel Xeon X5560	8* processor 2.0 GHz	Pentium IV 3.0 GHz
Memory	48 GB	48 GB	16 GB	1 GB
Hard disk	1.1 TB	3 TB	10 TB	280 GB
Automation in library preparation	Yes	Yes	Yes	No
Other required device	REM e system	cBot system	EZ beads system	No
Cost/million bases	$10	$0.07	$0.13	$2400

(C)

Sequencers	454 GS FLX	HiSeq 2000	SOLiDv4	3730xl
Resequencing		Yes	Yes	
De novo	Yes	Yes		Yes
Cancer	Yes	Yes	Yes	
Array	Yes	Yes	Yes	Yes
High GC sample	Yes	Yes	Yes	
Bacterial	Yes	Yes	Yes	
Large genome	Yes	Yes		
Mutation detection	Yes	Yes	Yes	Yes

(1) All the data is taken from daily average performance runs in BGI. The average daily sequence data output is about 8 Tb in BGI when about 80% sequencers (mainly HiSeq 2000) are running.
(2) The reagent cost of 454 GS FLX Titanium is calculated based on the sequencing of 400 bp; the reagent cost of HiSeq 2000 is calculated based on the sequencing of 200 bp; the reagent cost of SOLiDv4 is calculated based on the sequencing of 85 bp.
(3) HiSeq 2000 is more flexible in sequencing types like 50SE, 50PE, or 101PE.
(4) SOLiD has high accuracy especially when coverage is more than 30x, so it is widely used in detecting variations in resequencing, targeted resequencing, and transcriptome sequencing. Lanes can be independently run to reduce cost.

1.2 ROCHE 454 SYSTEM

Roche 454 was the first commercially successful next generation system. This sequencer uses pyrosequencing technology [5]. Instead of using dideoxynucleotides to terminate the chain amplification, pyrosequencing technology relies on the detection of pyrophosphate released during nucleotide incorporation. The library DNAs with 454-specific adaptors are denatured into single strand and captured by amplification beads followed by emulsion PCR [6]. Then on a picotiter plate, one of dNTP (dATP, dGTP, dCTP, dTTP) will complement to the bases of the template strand with the help of ATP sulfurylase, luciferase, luciferin, DNA polymerase, and adenosine 5′ phosphosulfate (APS) and release pyrophosphate (PPi) which equals the amount of incorporated nucleotide. The ATP transformed from PPi drives the luciferin into oxyluciferin and generates visible light [7]. At the same time, the unmatched bases are degraded by apyrase [8]. Then another dNTP is added into the reaction system and the pyrosequencing reaction is repeated.

The read length of Roche 454 was initially 100–150 bp in 2005, 200000+ reads, and could output 20 Mb per run [9, 10]. In 2008 454 GS FLX Titanium system was launched; through upgrading, its read length could reach 700 bp with accuracy 99.9% after filter and output 0.7 G data per run within 24 hours. In late 2009 Roche combined the GS Junior a bench top system into the 454 sequencing system which simplified the library preparation and data processing, and output was also upgraded to 14 G per run [11, 12]. The most outstanding advantage of Roche is its speed: it takes only 10 hours from sequencing start till completion. The read length is also a distinguished character compared with other NGS systems (described in the later part of this paper). But the high cost of reagents remains a challenge for Roche 454. It is about $\$12.56 \times 10^{-6}$ per base (counting reagent use only). One of the shortcomings is that it has relatively high error rate in terms of poly-bases longer than 6 bp. But its library construction can be automated, and the emulsion PCR can be semiautomated which could reduce the manpower in a great extent. Other informatics infrastructure and sequencing advantages are listed and compared with HiSeq 2000 and SOLiD systems in Tables 1(a), 1(b), and 1(c).

1.2.1 454 GS FLX TITANIUM SOFTWARE

GS RunProcessor is the main part of the GS FLX Titanium system. The software is in charge of picture background normalization, signal location correction, cross-talk correction, signals conversion, and sequencing data generation. GS RunProcessor would produce a series of files including SFF (standard flowgram format) files each time after run. SFF files contain the basecalled sequences and corresponding quality scores for all individual, high-quality reads (filtered reads). And it could be viewed directly from the screen of GS FLX Titanium system. Using GS De Novo Assembler, GS Reference Mapper and GS Amplicon Variant Analyzer provided by GS FLX Titanium system, SFF files can be applied in multiaspects and converted into fastq format for further data analyzing.

1.3 AB SOLID SYSTEM

(Sequencing by Oligo Ligation Detection) SOLiD was purchased by Applied Biosystems in 2006. The sequencer adopts the technology of two-base sequencing based on ligation sequencing. On a SOLiD flowcell, the libraries can be sequenced by 8 base-probe ligation which contains ligation site (the first base), cleavage site (the fifth base), and 4 different fluorescent dyes (linked to the last base) [10]. The fluorescent signal will be recorded during the probes complementary to the template strand and vanished by the cleavage of probes' last 3 bases. And the sequence of the fragment can be deduced after 5 round of sequencing using ladder primer sets.

The read length of SOLiD was initially 35 bp reads and the output was 3 G data per run. Owing to two-base sequencing method, SOLiD could reach a high accuracy of 99.85% after filtering. At the end of 2007, ABI released the first SOLiD system. In late 2010, the SOLiD 5500xl sequencing system was released. From SOLiD to SOLiD 5500xl, five upgrades were released by ABI in just three years. The SOLiD 5500xl realized improved read length, accuracy, and data output of 85 bp, 99.99%, and 30 G per run, respectively. A complete run could be finished within 7 days. The sequencing cost is about $\$40 \times 10^{-9}$ per base estimated from reagent use only by BGI users. But the short read length and resequencing only in applications is still its major shortcoming [13]. Application of SOLiD includes whole genome resequencing, targeted resequencing, transcriptome research (including gene expression profiling, small RNA analysis, and whole transcriptome analysis), and epigenome (like ChIP-Seq and methylation). Like other NGS systems, SOLiD's computational infrastructure is expensive and not trivial to use; it requires an air-conditioned data center, computing cluster, skilled personnel in computing, distributed memory cluster, fast networks, and batch queue system. Operating system used by most researchers is GNU/LINUX. Each solid sequencer run takes 7 days and generates around 4 TB of raw data. More data will be generated after bioinformatics analysis. This information is listed and compared with other NGS systems in Tables 1(a), 1(b), and 1(c). Automation can be used in library preparations, for example, Tecan system which integrated a Covaris A and Roche 454 REM e system [14].

1.3.1 SOLID SOFTWARE

After the sequencing with SOLiD, the original sequence of color coding will be accumulated. According to double-base coding matrix, the original color sequence can be decoded to get the base sequence if we knew the base types for one of any position in the sequence. Because of a kind of color corresponding four base pair, the color coding of the base will directly influence the decoding of its following base. It said that a wrong color coding will cause a chain decoding mistakes. BioScope is SOLiD data analysis package which provides a validated, single framework for resequencing, ChIP-Seq, and whole transcriptome analysis. It depends on reference for the follow-up data analysis. First, the software converts the base sequences of references into color coding sequence. Second, the color-coding sequence of references is compared with the original sequence of color-coding to get the information of mapping with newly developed mapping algorithm MaxMapper.

1.4 ILLUMINA GA/HISEQ SYSTEM

In 2006, Solexa released the Genome Analyzer (GA), and in 2007 the company was purchased by Illumina. The sequencer adopts the technology of sequencing by synthesis (SBS). The library with fixed adaptors is denatured to single strands and grafted to the flowcell, followed by bridge amplification to form clusters which contains clonal DNA fragments. Before sequencing, the library splices into single strands with the help of linearization enzyme [10], and then four kinds of nucleotides (ddATP, ddGTP, ddCTP, ddTTP) which contain different cleavable fluorescent dye and a removable blocking group would complement the template one base at a time, and the signal could be captured by a (charge-coupled device) CCD.

At first, solexa GA output was 1 G/run. Through improvements in polymerase, buffer, flowcell, and software, in 2009 the output of GA increased to 20 G/run in August (75PE), 30 G/run in October (100PE), and 50 G/run in December (Truseq V3, 150PE), and the latest GAIIx series can attain 85 G/run. In early 2010, Illumina launched HiSeq 2000, which adopts the same sequencing strategy with GA, and BGI was among the first globally to adopt the HiSeq system. Its output was 200 G per run initially, improved to 600 G per run currently which could be finished in 8 days. In the foreseeable future, it could reach 1 T/run when a personal genome cost could drop below $1 K. The error rate of 100PE could be below 2% in average after filtering (BGI's data). Compared with 454 and SOLiD, HiSeq 2000 is the cheapest in sequencing with $0.02/million bases (reagent counted only by BGI). With multiplexing incorporated in P5/P7 primers and adapters, it could handle thousands of samples simultaneously. HiSeq 2000 needs (HiSeq control software) HCS for program control, (real-time analyzer software) RTA to do on-instrument base-calling, and CASAVA for secondary analysis. There is a 3 TB hard disk in HiSeq 2000. With the aid of Truseq v3 reagents and associated softwares, HiSeq 2000 has improved much on high GC sequencing. MiSeq, a bench top sequencer launched in 2011 which shared most technologies with HiSeq, is especially convenient for amplicon and bacterial sample sequencing. It could sequence 150PE and generate 1.5 G/run in about 10 hrs including sample and library preparation time. Library preparation and their concentration measurement can both be automated with compatible systems like Agilent Bravo, Hamilton Banadu, Tecan, and Apricot Designs.

1.4.1 HISEQ SOFTWARE

HiSeq control system (HCS) and real-time analyzer (RTA) are adopted by HiSeq 2000. These two softwares could calculate the number and position of clusters based on their first 20 bases, so the first 20 bases of each sequencing would decide each sequencing's output and quality. HiSeq 2000 uses two lasers and four filters to detect four types of nucleotide (A, T, G, and C). The emission spectra of these four kinds of nucleotides have cross-talk, so the images of four nucleotides are not independent and the distribution of bases would affect the quality of sequencing. The standard sequencing output files of the HiSeq 2000 consist of *bcl files, which contain the base calls and quality scores in each cycle. And then it is converted into *_qseq.txt files by BCL Converter. The ELAND program of CASAVA (offline software provided by Illumina) is used to match a large number of reads against a genome.

In conclusion, of the three NGS systems described before, the Illumina HiSeq 2000 features the biggest output and lowest reagent cost, the SOLiD system has the highest accuracy [11], and the Roche 454 system has the longest read length. Details of three sequencing system are list in Tables 1(a), 1(b), and 1(c).

1.5 COMPACT PGM SEQUENCERS

Ion Personal Genome Machine (PGM) and MiSeq were launched by Ion Torrent and Illumina. They are both small in size and feature fast turnover rates but limited data throughput. They are targeted to clinical applications and small labs.

1.5.1 ION PGM FROM ION TORRENT

Ion PGM was released by Ion Torrent at the end of 2010. PGM uses semiconductor sequencing technology. When a nucleotide is incorporated into the DNA molecules by the polymerase, a proton is released. By detecting the change in pH, PGM recognized whether the nucleotide is added or not. Each time the chip was flooded with one nucleotide after another, if it is not the correct nucleotide, no voltage will be found; if there is 2 nucleotides added, there is double voltage detected [15]. PGM is the first commercial sequencing machine that does not require fluorescence and camera scanning, resulting in higher speed, lower cost, and smaller instrument size. Currently, it enables 200 bp reads in 2 hours and the sample preparation time is less than 6 hours for 8 samples in parallel.

An exemplary application of the Ion Torrent PGM sequencer is the identification of microbial pathogens. In May and June of 2011, an ongoing outbreak of exceptionally virulent Shiga-toxin- (Stx) producing *Escherichia coli* O104:H4 centered in Germany [16, 17], there were more than 3000 people infected. The whole genome sequencing on Ion Torrent PGM sequencer and HiSeq 2000 helped the scientists to identify the type of *E. coli* which would directly apply the clue to find the antibiotic resistance. The strain appeared to be a hybrid of two *E. coli* strains—entero aggregative *E. coli* and entero hemorrhagic *E. coli*—which may help explain why it has been particularly pathogenic. From the sequencing result of *E. coli* TY2482 [18], PGM shows the potential of having a fast, but limited throughput sequencer when there is an outbreak of new disease.

In order to study the sequencing quality, mapping rate, and GC depth distribution of Ion Torrent and compare with HiSeq 2000, a high GC Rhodobacter sample with high GC content (66%) and 4.2 Mb genome was sequenced in these two different sequencers (Table 2). In another experiment, E. coli K12 DH10B (NC_010473.1) with GC 50.78% was sequenced by Ion Torrent for analysis of quality value, read length, position accuracies, and GC distribution (Figure 1).

TABLE 2: Comparison in alignment between Ion Torrent and HiSeq 2000.

	Ion Torrent[a]	HiSeq 2000[b]
Total reads num	165518	205683
Total bases num	18574086	18511470
Max read length	201	90
Min read length	15	90
Map reads num	157258	157511
Map rate	95%	76.57%
Covered rate	96.50%	93.11%
Total map length	15800258	14176420
Total mismatch base	53475	142425
Total insertion base	109550	1397
Total insertion num	95740	1332
Total deletion base	152495	431
Total deletion num	139264	238
Ave mismatch rate	0.338%	1.004%
Ave insertion rate	0.693%	0.009%
Ave deletion rate	0.965%	0.003%

a: use TMAP to align; b: use SOAP2 to align.

1.5.1.1 SEQUENCING QUALITY

The quality of Ion Torrent is more stable, while the quality of HiSeq 2000 decreases noticeably after 50 cycles, which may be caused by the decay of fluorescent signal with increasing the read length (shown in Figure 1).

1.5.1.2 MAPPING

The insert size of library of Rhodobacter was 350 bp, and 0.5 Gb data was obtained from HiSeq. The sequencing depth was over 100x, and the contig and scaffold N50 were 39530 bp and 194344 bp, respectively. Based on the

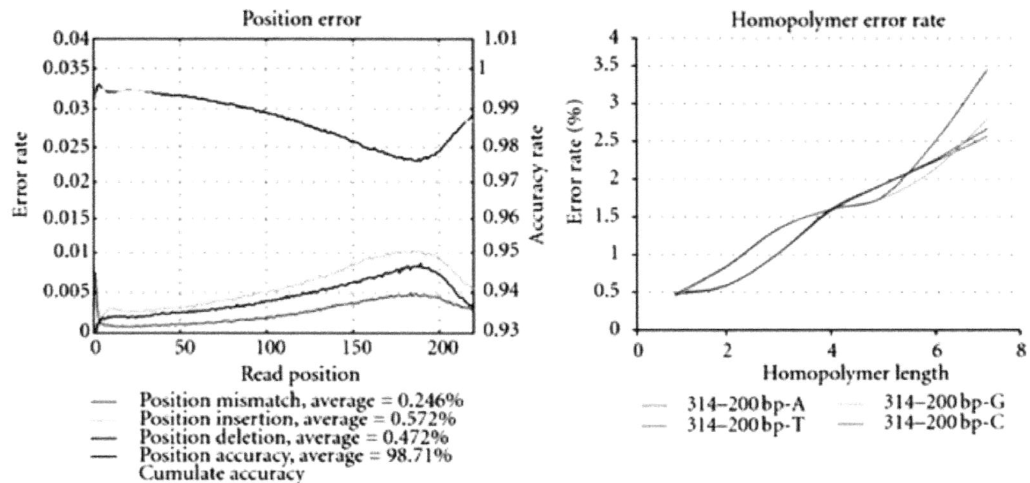

FIGURE 1: Ion Torrent sequencing quality. E. coli K12 DH10B (NC_010473.1) with GC 50.78% was used for this experiment. (a) is 314–200 bp from Ion Torrent. The left figure is quality value: pink range represents quality minimum and maximum values each position has. Green area represents the top and bottom quarter (1/4) reads of quality. Red line represents the average quality value in the position. The right figure is read length analysis: colored histogram represents the real read length. The black line represents the mapped length, and because it allows 3′ soft clipping, the length is different from the real read length. (b) is accuracy analysis. In each position, accuracy type including mismatch, insertion, and deletion is shown on the left y-axis. The average accuracy is shown the right y-axis. Accuracy of 200 bp sequencing could reach 99%. (c) is base composition along reads (left) and GC distribution analysis (right). The left figure is base composition in each position of reads. Base line splits after about 95 cycles indicating an inaccurate sequencing. The right one uses 500 bp window and the GC distribution is quite even. The data using high GC samples also indicates a good performance in Ion Torrent (data not shown).

assembly result, we used 33 Mb which is obtained from ion torrent with 314 chip to analyze the map rate. The alignment comparison is Table 2.

The map rate of Ion Torrent is higher than HiSeq 2000, but it is incomparable because of the different alignment methods used in different sequencers. Besides the significant difference on data including mismatch rate, insertion rate, and deletion rate, HiSeq 2000 and Ion Torrent were still incomparable because of the different sequencing principles. For example, the polynucleotide site could not be indentified easily in Ion Torrent. But it is shown that Ion Torrent has a stable quality along sequencing reads and a good performance on mismatch accuracies, but rather a bias in detection of indels. Different types of accuracy are analyzed and shown in Figure 1.

1.5.1.3 GC DEPTH DISTRIBUTION

The GC depth distribution is better in Ion Torrent from Figure 1. In Ion Torrent, the sequencing depth is similar while the GC content is from 63% to 73%. However in HiSeq 2000, the average sequencing depth is 4x when the GC content is 60%, while it is 3x with 70% GC content.

Ion Torrent has already released Ion 314 and 316 and planned to launch Ion 318 chips in late 2011. The chips are different in the number of wells resulting in higher production within the same sequencing time. The Ion 318 chip enables the production of >1 Gb data in 2 hours. Read length is expected to increase to >400 bp in 2012.

1.5.2 MISEQ FROM ILLUMINA

MiSeq which still uses SBS technology was launched by Illumina. It integrates the functions of cluster generation, SBS, and data analysis in a single instrument and can go from sample to answer (analyzed data) within a single day (as few as 8 hours). The Nextera, TruSeq, and Illumina's reversible terminator-based sequencing by synthesis chemistry was used in this innovative engineering. The highest integrity data and broader range of application, including amplicon sequencing, clone checking, ChIP-Seq, and small genome sequencing, are the outstanding parts of MiSeq. It is also flexible to perform single 36 bp reads (120 MB output) up to 2 × 150 paired-end reads (1–1.5 GB output) in MiSeq. Due to its significant improvement in read length, the resulting data performs better in contig assembly compared with HiSeq (data not shown). The related sequencing result of MiSeq is shown in Table 3. We also compared PGM with MiSeq in Table 4.

TABLE 3: MiSeq 150PE data.

Sample	GC	Q20	Q30
Human HPV	33.57; 33.62	98.26; 95.52	93.64; 88.52
Bacteria	61.33; 61.43	90.84; 83.86	78.46; 69.04

(1) The data in the table includes both read 1 and read 2 from paired-end sequencing.
(2) GC represents the GC content of libraries.
(3) Q20 value is the average Q20 of all bases in a read, which represents the ratio of bases with probability of containing no more than one error in 100 bases. Q30 value is the average Q30 of all bases in a read, which represents the ratio of bases with probability of containing no more than one error in 1,000 bases.

TABLE 4: The comparison between PGM and MiSeq.

	PGM	MiSeq
Output	10 MB–100 MB	120 MB–1.5 GB
Read length	~200 bp	Up to 2 × 150 bp
Sequencing time	2 hours for 1 × 200 bp	3 hours for 1 × 36 single read
		27 hours for 2 × 150 bp pair end read
Sample preparation time	8 samples in parallel, less than 6 hrs	As fast as 2 hrs, with 15 minutes hand on time
Sequencing method	semiconductor technology with a simple sequencing chemistry	Sequencing by synthesis (SBS)
Potential for development	Various parameters (read length, cycle time, accuracy, etc.)	Limited factors, major concentrate in flowcell surface size, insert sizes, and how to pack cluster in tighter
Input amount	μg	Ng (Nextera)
Data analysis	Off instrument	On instrument

1.5.3 COMPLETE GENOMICS

Complete genomics has its own sequencer based on Polonator G.007, which is ligation-based sequencer. The owner of Polonator G.007, Dover, collaborated with the Church Laboratory of Harvard Medical School, which is the same team as SOLiD system, and introduced this cheap open system. The Polonator could combine a high-performance instrument at very low price and the freely downloadable, open-source software and protocols in this sequencing system. The Polonator G.007 is ligation detection sequencing, which decodes the base by the single-base probe in nonanucleotides (nonamers), not by dual-base coding [19]. The fluorophore-tagged nonamers will be degenerated by selectively ligate onto a series of anchor primers, whose four components are labeled with one of four fluorophores with the help of T4 DNA ligase, which correspond to the base type at the query position. In the ligation progress, T4 DNA ligase is particularly sensitive to mismatches on 3′-side of the gap which is benefit to improve the accuracy of sequencing. After imaging, the Polonator chemically strips the array of annealed primer-fluorescent probe complex; the anchor primer is replaced and the new mixture are fluorescently tagged nonamers is introduced to sequence the adjacent base [20]. There are two updates compared with Polonator G.007, DNA nanoball (DNB) arrays, and combinatorial probe-anchor ligation (cPAL). Compared with DNA cluster or microsphere, DNA nanoball arrays obtain higher density of DNA cluster on the surface of a silicon chip. As the seven 5-base segments are discontinuous, so the system of hybridization-ligation-detection cycle has higher fault-tolerant ability compared with SOLiD. Complete genomics claim to have 99.999% accuracy with 40x depth and could analyze SNP, indel, and CNV with price 5500$–9500$. But Illumina reported a better performance of HiSeq 2000 use only 30x data (Illumina Genome Network). Recently some researchers compared CG's human genome sequencing data with Illumina system [21], and there are notable differences in detecting SNVs, indels, and system-specific detections in variants.

1.5.4 THE THIRD GENERATION SEQUENCER

While the increasing usage and new modification in next generation sequencing, the third generation sequencing is coming out with new insight in the sequencing. Third-generation sequencing has two main characteristics. First, PCR is not needed before sequencing, which shortens DNA preparation time for sequencing. Second, the signal is captured in real time, which means that the signal, no matter whether it is fluorescent (Pacbio) or electric current (Nanopore), is monitored during the enzymatic reaction of adding nucleotide in the complementary strand.

Single-molecule real-time (SMRT) is the third-generation sequencing method developed by Pacific Bioscience (Menlo Park, CA, USA), which made use of modified enzyme and direct observation of the enzymatic reaction in real time. SMRT cell consists of millions of zero-mode waveguides (ZMWs), embedded with only one set of enzymes and DNA template that can be detected during the whole process. During the reaction, the enzyme will incorporate the nucleotide into the complementary strand and cleave off the fluorescent dye previously linked with the nucleotide. Then the camera inside the machine will capture signal in a movie format in real-time observation [19]. This will give out not only the fluorescent signal but also the signal difference along time, which may be useful for the prediction of structural variance in the sequence, especially useful in epigenetic studies such as DNA methlyation [22].

Comparing to second generation, PacBio RS (the first sequencer launched by PacBio) has several advantages. First the sample preparation is very fast; it takes 4 to 6 hours instead of days. Also it does not need PCR step in the preparation step, which reduces bias and error caused by PCR. Second, the turnover rate is quite fast; runs are finished within a day. Third, the average read length is 1300 bp, which is longer than that of any second-generation sequencing technology. Although the throughput of the PacBioRS is lower than second-generation sequencer, this technology is quite useful for clinical laboratories, especially for microbiology research. A paper has been published using PacBio RS on the Haitian cholera outbreak [19].

	Prefilter	Post-QC filter*
Number of bases	84, 110, 272 bp	22, 373, 400 bp
Number of reads	46, 861	6, 754
Mean read length	513 bp	2, 566 bp
Mean read score	0.144	0.819

* MinRL = 50, MinRS = 0.75

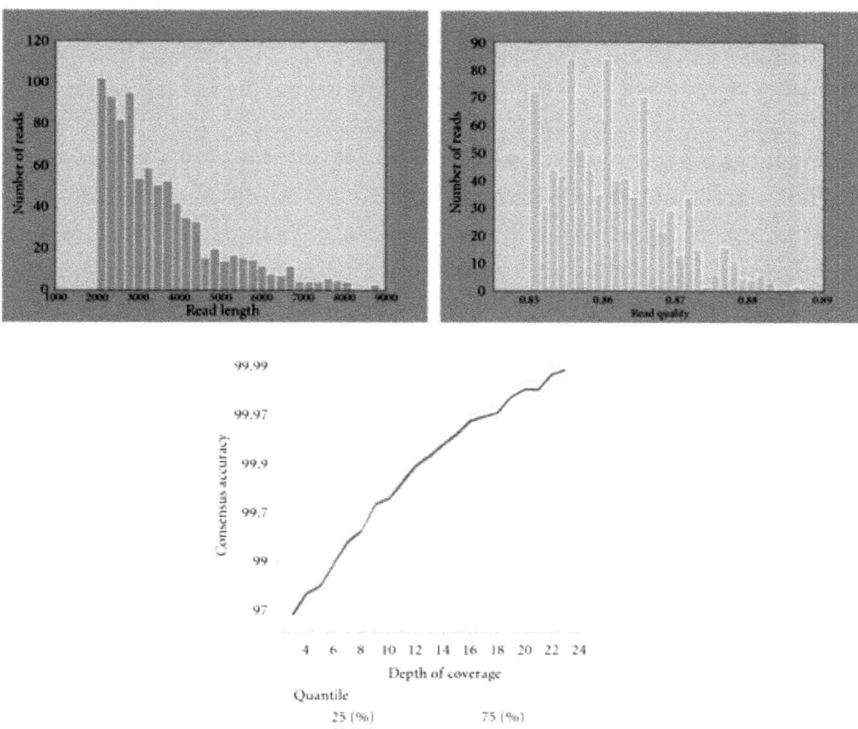

FIGURE 2: Sequencing of a fosmid DNA using Pacific Biosciences sequencer. With coverage, the accuracy could be above 97%. The figure was constructed by BGI's own data.

We have run a de novo assembly of DNA fosmid sample from Oyster with PacBio RS in standard sequencing mode (using LPR chemistry and SMRTcells instead of the new version FCR chemistry and SMRTcells). An SMRT belt template with mean insert size of 7500 kb is made and run in one SMRT cell and a 120-minute movie is taken. After Post-QC filter, 22,373,400 bp reads in 6754 reads (average 2,566 bp) were sequenced with the average Read Score of 0.819. The Coverage is 324x with mean read score of 0.861 and high accuracy (~99.95). The result is exhibited in Figure 2.

Nanopore sequencing is another method of the third generation sequencing. Nanopore is a tiny biopore with diameter in nanoscale [23], which can be found in protein channel embedded on lipid bilayer which facilitates ion exchange. Because of the biological role of nanopore, any particle movement can disrupt the voltage across the channel. The core concept of nanopore sequencing involves putting a thread of single-stranded DNA across α-haemolysin (αHL) pore. αHL, a 33 kD protein isolated from *Staphylococcus aureus* [20], undergoes self-assembly to form a heptameric transmembrane channel [23]. It can tolerate extraordinary voltage up to 100 mV with current 100 pA [20]. This unique property supports its role as building block of nanopore. In nanopore sequencing, an ionic flow is applied continuously. Current disruption is simply detected by standard electrophysiological technique. Readout is relied on the size difference between all deoxyribonucleoside monophosphate (dNMP). Thus, for given dNMP, characteristic current modulation is shown for discrimination. Ionic current is resumed after trapped nucleotide entirely squeezing out.

Nanopore sequencing possesses a number of fruitful advantages over existing commercialized next-generation sequencing technologies. Firstly, it potentially reaches long read length >5 kbp with speed 1 bp/ns [19]. Moreover, detection of bases is fluorescent tag-free. Thirdly, except the use of exonuclease for holding up ssDNA and nucleotide cleavage [24], involvement of enzyme is remarkably obviated in nanopore sequencing [22]. This implies that nanopore sequencing is less sensitive to temperature throughout the sequencing reaction and reliable outcome can be maintained. Fourthly, instead of sequencing DNA during polymerization, single DNA strands are sequenced through nanopore by means of DNA strand depolymerization. Hence, hand-on time for sample preparation such as cloning and amplification steps can be shortened significantly.

1.6 DISCUSSION OF NGS APPLICATIONS

Fast progress in DNA sequencing technology has made for a substantial reduction in costs and a substantial increase in throughput and accuracy. With more and more organisms being sequenced, a flood of genetic data is inundating the world every day. Progress in genomics has been moving steadily forward due to a revolution in sequencing technology. Additionally, other of types-large scale studies in exomics, metagenomics, epigenomics, and transcriptomics all become reality. Not only do these studies provide the knowledge for basic research, but also they afford immediate application benefits. Scientists across many fields are utilizing these data for the development of better-thriving crops and crop yields and livestock and improved diagnostics, prognostics, and therapies for cancer and other complex diseases.

BGI is on the cutting edge of translating genomics research into molecular breeding and disease association studies with belief that agriculture, medicine, drug development, and clinical treatment will eventually enter a new stage for more detailed understanding of the genetic components of all the organisms. BGI is primarily focused on three projects. (1) The Million Species/Varieties Genomes Project, aims to sequence a million economically and scientifically important plants, animals, and model organisms, including different breeds, varieties, and strains. This project is best represented by our sequencing of the genomes of the Giant panda, potato, macaca, and others, along with multiple resequencing projects. (2) The Million Human Genomes Project focuses on large-scale population and association studies that use whole-genome or whole-exome sequencing strategies. (3) The Million Eco-System Genomes Project has the objective of sequencing the metagenome and cultured microbiome of several different environments, including microenvironments within the human body [25]. Together they are called 3 M project.

In the following part, each of the following aspects of applications including de novo sequencing, mate-pair, whole genome or target-region resequencing, small RNA, transcriptome, RNA seq, epigenomics, and metagenomics, is briefly summarized.

In DNA de novo sequencing, the library with insert size below 800 bp is defined as DNA short fragment library, and it is usually applied in de novo and resequencing research. Skovgaard et al. [26] have applied a combination method of WGS (whole-genome sequencing) and genome copy number analysis to identify the mutations which could suppress the growth deficiency imposed by excessive initiations from the E. coli origin of replication, oriC.

Mate-pair library sequencing is significant beneficial for de novo sequencing, because the method could decrease gap region and extend scaffold length. Reinhardt et al. [27] developed a novel method for de novo genome assembly by analyzing sequencing data from high-throughput short read sequencing technology. They assembled genomes into large scaffolds at a fraction of the traditional cost and without using reference sequence. The assembly of one sample yielded an N50 scaffold size of 531,821 bp with >75% of the predicted genome covered by scaffolds over 100,000 bp.

Whole genome resequencing sequenced the complete DNA sequence of an organism's genome including the whole chromosomal DNA at a single time and alignment with the reference sequence. Mills et al. [28] constructed a map of unbalanced SVs (genomic structural variants) based on whole genome DNA sequencing data from 185 human genomes with SOLiD platform; the map encompassed 22,025 deletions and 6,000 additional SVs, including insertions and tandem duplications [28]. Most SVs (53%) were mapped to nucleotide resolution, which facilitated analyzing their origin and functional impact [28].

The whole genome resequencing is an effective way to study the functional gene, but the high cost and massive data are the main problem for most researchers. Target region sequencing is a solution to solve it. Microarray capture is a popular way of target region sequencing, which uses hybridization to arrays containing synthetic oligo-nucleotides matching the target DNA sequencing. Gnirke et al. [29] developed a captured method that uses an RNA "baits" to capture target DNA fragments from the "pond" and then uses the Illumina platform to read out the sequence.

About 90% of uniquely aligning bases fell on or near bait sequence; up to 50% lay on exons proper [29].

Fehniger et al. used two platforms, Illumina GA and ABI SOLiD, to define the miRNA transcriptomes of resting and cytokine-activated primary murine NK (natural killer) cells [30]. The identified 302 known and 21 novel mature miRNAs were analyzed by unique bioinformatics pipeline from small RNA libraries of NK cell. These miRNAs are overexpressed in broad range and exhibit isomiR complexity, and a subset is differentially expressed following cytokine activation, which were the clue to identify the identification of miRNAs by the Illumina GA and SOLiD instruments [30].

The transcriptome is the set of all RNA molecules, including mRNA, rRNA, tRNA, and other noncoding RNA produced in one or a population of cells. In these years, next-generation sequencing technology is used to study the transcriptome compares with DNA microarray technology in the past. The *S. mediterranea* transcriptome could be sequenced by an efficient sequencing strategy which designed by Adamidi et al. [31]. The catalog of assembled transcripts and the identified peptides in this study dramatically expand and refine planarian gene annotation, which is demonstrated by validation of several previously unknown transcripts with stem cell-dependent expression patterns.

RNA-seq is a new method in RNA sequencing to study mRNA expression. It is similar to transcriptome sequencing in sample preparation, except the enzyme. In order to estimate the technical variance, Marioni et al. [32] analyzed a kidney RNA samples on both Illumina platform and Affymetrix arrays. The additional analyses such as low-expressed genes, alternative splice variants, and novel transcripts were found on Illumina platform. Bradford et al. [33] compared the data of RNA-seq library on the SOLiD platform and Affymetrix Exon 1.0ST arrays and found a high degree of correspondence between the two platforms in terms of exon-level fold changes and detection. And the greatest detection correspondence was seen when the background error rate is extremely low in RNA-seq. The difference between RNA-seq and transcriptome on SOLiD is not so obvious as Illumina.

There are two kinds of application of epigenetic, Chromatin immunoprecipitation and methylation analysis. Chromatin immunoprecipitation

(ChIP) is an immunoprecipitation technique which is used to study the interaction between protein and DNA in a cell, and the histone modifies would be found by the specific location in genome. Based on next-generation sequencing technology, Johnson et al. [34] developed a large-scale chromatin immunoprecipitation assay to identify motif, especially noncanonical NRSF-binding motif. The data displays sharp resolution of binding position (±50 bp), which is important to infer new candidate interaction for the high sensitivity and specificity (ROC (receiver operator characteristic) area ≥0.96) and statistical confidence ($P < 10^{-4}$). Another important application in epigenetic is DNA methylation analysis. DNA methylation exists typically in vertebrates at CpG sites; the methylation caused the conversion of the cytosine to 5-methylcytosine. Chung presented a whole methylome sequencing to study the difference between two kinds of bisulfite conversion methods (in solution versus in gel) by SOLiD platform [35].

The world class genome projects include the 1000 genome project, and the human ENCODE project, the human Microbiome (HMP) project, to name a few. BGI takes an active role in these and many more ongoing projects like 1000 Animal and Plant Genome project, the MetaHIT project, Yanhuang project, LUCAMP (Diabetes-associated Genes and Variations Study), ICGC (international cancer genome project), Ancient human genome, 1000 Mendelian Disorders Project, Genome 10K Project, and so forth [25]. These internationally collaborated genome projects greatly enhanced genomics study and applications in healthcare and other fields.

To manage multiple projects including large and complex ones with up to tens of thousands of samples, a superior and sophisticated project management system is required handling information processing from the very beginning of sample labeling and storage to library construction, multiplexing, sequencing, and informatics analysis. Research-oriented bioinformatics analysis and followup experiment processed are not included. Although automation techniques' adoption has greatly simplified bioexperiment human interferences, all other procedures carried out by human power have to be managed. BGI has developed BMS system and Cloud service for efficient information exchange and project management. The behavior management mainly follows Japan 5S onsite model. Additionally, BGI has passed ISO9001 and CSPro (authorized by Illumina) QC system and is currently taking (Clinical Laboratory Improvement Amendments) CLIA and (American Society for Histocompatibility and Immunogenetics) AShI tests. Quick, standard, and open reflection system guarantees an efficient troubleshooting pathway and high performance, for example, instrument design failure of Truseq v3 flowcell resulting in bubble appearance (which is defined as "bottom-middle-swatch" phenomenon by Illumina) and random N in reads. This potentially hazards sequencing quality, GC composition as well as throughput. It not only effects a small area where the bubble locates resulting in reading N but also effects the focus of the place nearby, including the whole swatch, and the adjacent swatch. Filtering parameters have to be determined to ensure quality raw data for bioinformatics processing. Lead by the NGS tech group, joint meetings were called for analyzing and troubleshooting this problem, to discuss strategies to best minimize effect in terms of cost and project time, to construct communication channel, to statistically summarize compensation, in order to provide best project management strategies in this time. Some reagent QC examples are summaried in Liu et al. [36].

BGI is establishing their cloud services. Combined with advanced NGS technologies with multiple choices, a plug-and-run informatics service is handy and affordable. A series of softwares are available including BLAST, SOAP, and SOAP SNP for sequence alignment and pipelines for RNAseq data. Also SNP calling programs such as Hecate and Gaea are about to be released. Big-data studies from the whole spectrum of life and biomedical sciences now can be shared and published on a new journal GigaSicence cofounded by BGI and Biomed Central. It has a novel publication format: each piece of data links to a standard manuscript publication with an extensive database which hosts all associated data, data analysis tools, and cloud-computing resources. The scope covers not just omic type data and the fields of high-throughput biology currently serviced by large public repositories but also the growing range of more difficult-to-access data, such as imaging, neuroscience, ecology, cohort data, systems biology, and other new types of large-scale sharable data.

REFERENCES

1. G. M. Church and W. Gilbert, "Genomic sequencing," Proceedings of the National Academy of Sciences of the United States of America, vol. 81, no. 7, pp. 1991–1995, 1984.
2. http://en.wikipedia.org/wiki/DNA_sequencing/.
3. F. S. Collins, M. Morgan, and A. Patrinos, "The Human Genome Project: lessons from large-scale biology," Science, vol. 300, no. 5617, pp. 286–290, 2003.
4. http://genomics.xprize.org/.
5. http://my454.com/products/technology.asp.
6. J. Berka, Y. J. Chen, J. H. Leamon, et al., "Bead emulsion nucleic acid amplification," U.S. Patent Application, 2005.
7. T. Foehlich, et al., "High-throughput nucleic acid analysis," U.S. Patent, 2010.
8. http://www.pyrosequencing.com/DynPage.aspx.
9. http://www.roche-applied-science.com/.
10. E. R. Mardis, "The impact of next-generation sequencing technology on genetics," Trends in Genetics, vol. 24, no. 3, pp. 133–141, 2008.
11. S. M. Huse, J. A. Huber, H. G. Morrison, M. L. Sogin, and D. M. Welch, "Accuracy and quality of massively parallel DNA pyrosequencing," Genome Biology, vol. 8, no. 7, article R143, 2007.
12. "The new GS junior sequencer," http://www.gsjunior.com/instrument-workflow.php.
13. "SOLiD system accuray," http://www.appliedbiosystems.com/absite/us/en/home/applications-technologies/solid-next-generation-sequencing.html.
14. http://www.tecan.com/platform/apps/product/index.asp?MenuID=3465&ID=7191&Menu=1&Item=33.52.2.
15. B. A. Flusberg, D. R. Webster, J. H. Lee et al., "Direct detection of DNA methylation during single-molecule, real-time sequencing," Nature Methods, vol. 7, no. 6, pp. 461–465, 2010.
16. A. Mellmann, D. Harmsen, C. A. Cummings et al., "Prospective genomic characterization of the german enterohemorrhagic Escherichia coli O104:H4 outbreak by rapid next generation sequencing technology," PLoS ONE, vol. 6, no. 7, Article ID e22751, 2011.
17. H. Rohde, J. Qin, Y. Cui, et al., "Open-source genomic analysis of Shiga-toxin-producing E. coli O104:H4," New England Journal of Medicine, vol. 365, no. 8, pp. 718–724, 2011.
18. C. S. Chin, J. Sorenson, J. B. Harris et al., "The origin of the Haitian cholera outbreak strain," New England Journal of Medicine, vol. 364, no. 1, pp. 33–42, 2011.
19. W. Timp, U. M. Mirsaidov, D. Wang, J. Comer, A. Aksimentiev, and G. Timp, "Nanopore sequencing: electrical measurements of the code of life," IEEE Transactions on Nanotechnology, vol. 9, no. 3, pp. 281–294, 2010.
20. D. W. Deamer and M. Akeson, "Nanopores and nucleic acids: prospects for ultrarapid sequencing," Trends in Biotechnology, vol. 18, no. 4, pp. 147–151, 2000.
21. "Performance comparison of whole-genome sequencing systems," Nature Biotechnology, vol. 30, pp. 78–82, 2012.

22. D. Branton, D. W. Deamer, A. Marziali, et al., "The potential and challenges of nanopore sequencing," Nature Biotechnology, vol. 26, no. 10, pp. 1146–1153, 2008.
23. L. Song, M. R. Hobaugh, C. Shustak, S. Cheley, H. Bayley, and J. E. Gouaux, "Structure of staphylococcal α-hemolysin, a heptameric transmembrane pore," Science, vol. 274, no. 5294, pp. 1859–1866, 1996.
24. J. Clarke, H. C. Wu, L. Jayasinghe, A. Patel, S. Reid, and H. Bayley, "Continuous base identification for single-molecule nanopore DNA sequencing," Nature Nanotechnology, vol. 4, no. 4, pp. 265–270, 2009.
25. Website of BGI, http://www.genomics.org.cn.
26. O. Skovgaard, M. Bak, A. Løbner-Olesen, et al., "Genome-wide detection of chromosomal rearrangements, indels, and mutations in circular chromosomes by short read sequencing," Genome Research, vol. 21, no. 8, pp. 1388–1393, 2011.
27. J. A. Reinhardt, D. A. Baltrus, M. T. Nishimura, W. R. Jeck, C. D. Jones, and J. L. Dangl, "De novo assembly using low-coverage short read sequence data from the rice pathogen Pseudomonas syringae pv. oryzae," Genome Research, vol. 19, no. 2, pp. 294–305, 2009.
28. R. E. Mills, K. Walter, C. Stewart et al., "Mapping copy number variation by population-scale genome sequencing," Nature, vol. 470, no. 7332, pp. 59–65, 2011.
29. A. Gnirke, A. Melnikov, J. Maguire et al., "Solution hybrid selection with ultra-long oligonucleotides for massively parallel targeted sequencing," Nature Biotechnology, vol. 27, no. 2, pp. 182–189, 2009.
30. T. A. Fehniger, T. Wylie, E. Germino et al., "Next-generation sequencing identifies the natural killer cell microRNA transcriptome," Genome Research, vol. 20, no. 11, pp. 1590–1604, 2010.
31. C. Adamidi, Y. Wang, D. Gruen et al., "De novo assembly and validation of planaria transcriptome by massive parallel sequencing and shotgun proteomics," Genome Research, vol. 21, no. 7, pp. 1193–1200, 2011.
32. J. C. Marioni, C. E. Mason, S. M. Mane, M. Stephens, and Y. Gilad, "RNA-seq: an assessment of technical reproducibility and comparison with gene expression arrays," Genome Research, vol. 18, no. 9, pp. 1509–1517, 2008.
33. J. R. Bradford, Y. Hey, T. Yates, Y. Li, S. D. Pepper, and C. J. Miller, "A comparison of massively parallel nucleotide sequencing with oligonucleotide microarrays for global transcription profiling," BMC Genomics, vol. 11, no. 1, article 282, 2010.
34. D. S. Johnson, A. Mortazavi, R. M. Myers, and B. Wold, "Genome-wide mapping of in vivo protein-DNA interactions," Science, vol. 316, no. 5830, pp. 1497–1502, 2007.
35. H. Gu, Z. D. Smith, C. Bock, P. Boyle, A. Gnirke, and A. Meissner, "Preparation of reduced representation bisulfite sequencing libraries for genome-scale DNA methylation profiling," Nature Protocols, vol. 6, no. 4, pp. 468–481, 2011.
36. L. Liu, N. Hu, B. Wang, et al., "A brief utilization report on the Illumina HiSeq 2000 sequencer," Mycology, vol. 2, no. 3, pp. 169–191, 2011.

CHAPTER 2

WHY ASSEMBLING PLANT GENOME SEQUENCES IS SO CHALLENGING

MANUEL GONZALO CLAROS, ROCÍO BAUTISTA, DARÍO GUERRERO-FERNÁNDEZ, HICHAM BENZERKI, PEDRO SEOANE, and NOÉ FERNÁNDEZ-POZO

2.1 INTRODUCTION

Higher plants are the Earth's dominant vegetation in nearly all ecosystems. They sustain living beings (including humans) by providing oxygen, food, fiber, fuel, medicines, spirits, erosion defense, flooding control, soil regeneration, (bio)remediation, urban cooling, green spaces (including gardens) and CO_2 lowering, and contributing to the control of global warming [1]. Higher plants also exhibit a wide range of forms, with individuals ranging in size from floating Wolffia plants of 1 mm in length to trees of more than 100 m in height or with a trunk diameter exceeding 10 m (such as the angiosperm *Eucaliptus regnans* and the gymnosperms *Sequoia sempervirens* and *Taxodium mucronatum*). Plants also contain the longest-living organisms (with *Pinus longaeva*, *Taxus baccata* and *Picea abies* individuals living on Earth for nearly 5,000–8,000 years). Moreover, plants are stuck in place and cannot escape enemies or uncomfortable conditions and need to develop strategies that improve their chances of survival due to sessility. So, plants have evolved tens of thousands of chemical compounds which they use to ward off competition from other plants, to fight infections,

This chapter was originally published under the Creative Commons Attribution License. Gonzalo Claros MG, Bautista R, Guerrero-Fernández D, Benzerki H, Seoane P, and Fernández-Pozo N. Why Assembling Plant Genome Sequences is so Challenging. Biology. ***1** (2012), 439–459. doi:10.3390/biology1020439.*

and to respond generally to the environment [2]. In consequence, plant species have larger and more complex genome sizes and structures than animal species and exhibit tremendous diversity in both size and structure [3]. Therefore, plants seem to be an important source of biological knowledge and economic profit, but relatively few plant species have been sequenced. In fact, in a world with >370,000 known plant species (with probably many thousands more still unclassified), only ~80,000 species have at least one single sequence in GenBank.

The publication of the first plant genome sequence of *Arabidopsis thaliana* [4] provided and improved the genetic landscape for studying all plants and has paved the way for sequencing several other plant genomes. It has also transformed the methods and tools for plant research and crop improvement [5]. *Arabidopsis*, and later *Oryza sativa* (rice) [6], *Carica papaya* (papaya) [7] and *Zea mays* (maize) [8] were sequenced using the classical Sanger method. The arrival of next-generation sequencing (NGS) technologies has allowed the rapid and efficient development of genomic resources for non-model or orphan plant species [9–13]. However, only *Arabidopsis* and rice—sequenced by Sanger's method using a BAC-by-BAC approach—have been really finished to date, the rest being drafts in a greater or lesser stage of completion. Unfortunately, even the complete or gold standard genomes contain gaps in their sequences corresponding to highly repetitive sequences, which are recalcitrant to sequencing and assembly methods [14]. A summary of all published plant genome sequences to date can be found in Table 3 in [14] and in Table 3 in [5].

Since there is no central focus in the scientific plant world, the choice of plant genomes for sequencing has been driven mainly by cost efficiency and the avoidance of complexity, and hence only plants with relatively small genomes (median size of 466 Mbp) were selected for sequencing in the first instance, although the most important crops have a median size of 766 Mbp [5]. In fact, *Arabidopsis thaliana* proves to be an outlier amongst plants because its genome has undergone a 30% reduction in genome size and at least nine rearrangements in the short time since its divergence to *Arabidopsis lyrata* [1,15]. In many plant species, it is now clear that a single genome sequence does not necessarily reflect the entire genetic complement [16,17], opening a new branch in the study of pan-genomes and core genomes [18].

Most plant sequencing efforts have been dedicated to angiosperms, mainly the eudicots, under which the most economically important crops are classified [19,20]. But sequencing efforts should be expanded beyond the traditional commodity crops and include other non-commodity crops and non-model species (e.g., conifers, ferns and other bryophytes). We present here the current state of the art of challenges and confounding factors that explain why plant genomics is less developed than animal genomics and remains so focused on small genomes. We also discuss why challenges are not overcome by the arrival of NGS.

2.2 FROM SANGER TECHNOLOGY TO NGS: GETTING PLANTS OFF THE GROUND

While extremely successful in the past, Sanger sequencing [21] does present the following drawbacks for actual sequencing projects: (1) requirement of nucleic acid subcloning, (2) clone amplification in hosts, (3) low throughput, (4) slow sequencing speed, and (5) high costs (both in terms of consumables and salaries, averaging $1,330 per Mbp [22]). This is the reason why sequencing projects with Sanger technology have always been carried out by international consortia [4,8,23,24].

NGS strategies allow a single template molecule to be directly used to generate millions of bases at low cost with a less cumbersome laboratory protocol. There are three NGS platforms widely used nowadays that are considered to be second-generation sequencing: (1) the Genome Sequencer FLX+/454 from Roche which is capable of producing over a million reads of up to 800 bases per 10 hour run, yielding a total of 0.7–1 Gbp at a price of approximately $90 per megabase; (2) the Genome Analizer from Illumina, of which the latest version, HiSeq2000, yields 100 Gbp of bases per day (26–150 bp read length) at a cost of $4 per megabase; and (3) the Applied Biosystems SOLiD (Sequencing by Oligo Ligation and Detection) that produces 10–300 Gbp of short reads (up to 75 bp) per run at a similar cost. The three platforms offer the paired-end sequencing technique. As a result, even large plant genomes can count on relatively inexpensive deep coverage with reads of 100 bp and paired-end libraries from 1 to 5 kbp (we will see that deep coverage does not allow for complete plant sequencing). A detailed description is beyond the scope of this article, and several reviews illustrate the rapid evolution of these and the newest NGS technologies (to cite a few, [25–31]). While 454 FLX+ and Sanger technologies are considered to produce long reads (600–800 pb in average), the other two produce short reads (<150 bp in average). Short-read technologies compensate the shortness of the sequences with a high coverage, so that bacteria can be successfully sequenced with a 40×–50× coverage, but as the genome increases in complexity, coverage of 100× may still be inefficient [32–34]. In contrast, long-read technologies do not need such deep coverage, with 20×–30× being enough for a good compromise between costs and assembly quality [32].

NGS is becoming the new sequencing standard for the following reasons: (1) simplification of the sequencing process (DNA cloning is not required); (2) miniaturization and parallelization (low cost); and (3) good adaptation to a broad range of biological phenomena (genetic variation, RNA expression, protein-DNA interactions, gene capture, methylation, etc.). But not everything about NGS is an advantage [25]: (i) the base calls are at least tenfold less accurate than Sanger sequencing base calls; (ii) the sequence length is shorter than in Sanger technology and requires dedicated assembly algorithms; and (iii) the quality of the NGS assemblies is also lower than Sanger assemblies. As a result, most plant genomes sequenced by NGS produce "drafts" that are suitable for (1) establishing gene catalogues, (2) deciphering the repeat content, (3) glimpsing evolutionary mechanisms, and (4) performing early studies on comparative genomics and phylogeny. Unfortunately, drafts (i) hinder the progress of capturing accurately the information embedded in the repetitive fraction of the genome; (ii) make it difficult to distinguish genes from pseudogenes; and (iii) make it difficult to differentiate between alleles and even paralogues [35]. If only draft genomes are produced in the short future, plant genomics may face a crisis since, although the complex genomes of many more species are now accessible, the portion of each genome that can be reliably accessed has diminished substantially (<80%). The expertise and motivation for sequencing plant genomes to a high quality is disappearing, pushed by the rapid publication of a new draft genome lacking up to 20% of the genome [33].

Widespread adoption of NGS technology is tightly bound to bioinformatics. Integration of the many complex and rich sequencing datasets has yielded cohesive views of cellular activities and dynamics (for example, see [36–38]). The increase in plant sequence data has also prompted the development of dedicated repositories, such as the general purpose Phytozome [13], the comparative plant genomics resource PLAZA [39], plant family databases such as TreeGenes for forest tree genome data [40], or species specific databases (e.g., EuroPineDB for maritime pine [41], Eucawood for Eucalyptus [42], or MeloGen for *Cucumis melo* [43]). It is worth mentioning the iPlant project [44], which emerged with the aim of creating an innovative, comprehensive and foundational cyber infrastructure to support plant biology research, the VirtualPlant platform [45], integrating genome-wide data on the known and predicted relationships among genes, proteins, and molecules in order to enable scientists to visualize, integrate, and analyze genomic data from a systems biology perspective or the Plantagora platform [34], which addresses the gap between having the technical tools for plant genome sequencing and knowing precisely the best way to use them.

NGS can be said to have accelerated biological research in plants by enabling the comprehensive analysis of genomes, transcriptomes and interactomes. Moreover, translational research has been spurred by NGS, the most successful case being the application of a gene from *A. thaliana* to improve abiotic stress tolerance traits in crops [5]. But if NGS only produces draft genomes, it could drive plant functional genomics into a dead end in the near future.

2.3 CHALLENGING FEATURES OF PLANT GENOMES

Genome size, duplications and repeat content are factors to be considered for all genomes to be sequenced. In particular, plant genomes usually appear as gene islands among the background of high copy repeats (usually >80%), where 95% coverage of the genes is assumed, based on comparisons with cDNA databases. This discouraging situation can be explained by several plant features that hinder the sequence assembly and annotation, and severely limiting genomics research productivity.

2.3.1 SAMPLING

The main drawback of plant sequencing is that it is often very hard to extract large quantities of high-quality DNA from plant material, making it difficult to prepare proper libraries for sequencing. Additionally, although any genome sequencing project is afforded with samples from a single plant, the situation is completely different in transcriptome sequencing, where the traditional approach was to use a variety of tissues and conditions from different multiple accessions by different researchers, resulting in many extremely similar unigenes representing the same gene [41–43]. When such a heterogeneous transcriptome is studied using long reads, the presence of multiple alleles does not significantly hamper the unigene assembly [22], but when the transcriptome is studied with NGS technologies providing reads <100 bp, alleles and paralogues really do impair the assembly result.

2.3.2 GENOME SIZE AND COMPLEXITY

Plant-specific needs are sustained by new genes that may arise from gene duplications, alternative gene splicing, ploidy or gene retention following genome duplication, making plant genomes large and complex, as pointed out in the introduction. In fact, genome sizes across land plants range over two to three orders of magnitude, with an average around 6 Gbp, which is one order of magnitude larger than the average size of genomes sequenced so far [3]. Current sequencing technologies can manage large, complex genomes, such as wheat (*Triticum aestivum* with 16 Gbp split in 21 chromosomes) or pines (22–33 Gbp split in 12 chromosomes), so the genome size is not an unassailable issue anymore. The real problem is not the genome size per se but the complexity of the genomes, since the number of genes does not vary to the same extent as the genome size. The length of single-copy regions (always flanked by repeated sequences [12]) varies widely among plant species. In general, two types of arrangements are recognized: (1) short period interspersion (single copy sequences of 300–1,200 bp interspersed as islands among short lengths (50–2,000 nt) of repeat sequences); and (2) long period interspersion (single copy sequence islands of 2,000–6,000 bp interspersed among long repeat sequences). Genome size appears to be related to the type of interspersion: Plant species with small genomes, such as *Arabidopsis*, have long period interspersion and longer lengths of non repetitive sequences; on the contrary, plant species with large genomes, such as wheat, rye or maize, have short period interspersion and shorter non-repetitive sequences [46]. This confirms the intuitive notion that small genomes are therefore less difficult to assemble than larger genomes. The different factors that can contribute to the large variation of genome size and complexity in plants are discussed below.

2.3.3 TRANSPOSABLE ELEMENTS

During evolution, transposons have introduced profound changes in genome size, structure and function between species and within species [18], accounting for the major force in reshaping genomes [47]. This could explain why Chromosomes 1 and 2 of *A. thaliana* are a fusion of Chromosomes 1 and 2, and 3 and 4, respectively, of *A. lyrata* [15,47]. Transposable elements are by far the most highly represented repetitive sequence in plant genomics: due to the replicative nature of the retrotransposition process, Class I transposons (including retrotransposons) can account for up to 90% of all the transposons, while Class II elements are much less abundant [48]. Small-genome plants like *Arabidopsis* and rice are sparsely populated by transposons, containing 5.6% and 17% respectively. In contrast, the transposon-derived fraction of medium/large genomes may reach 85% in maize and >70% in barley [8,49,50]. Owing to their abundance and repetitive nature, transposable elements complicate genome assembly, particularly when short-read technologies are used [51].

2.3.4 HETEROZYGOSITY

Most plants are heterozygous, particularly those that have not been domesticated in laboratories [52]. Since it is a kind of redundancy, which is always a challenging factor in assemblies, only euchromatic regions of the genomes can be assembled, and a high percentage of NGS reads remain unassembled (15% in poplar [*Populus trichocarpa*] [53]). This happens even if a hierarchical clustering guided by a physical map is used to guide the sequence assembly. As a result, the poplar genome seems to contain a duplicated gene content since most loci present both possible alleles. The relative incompleteness of both heterozygous genomes demonstrates the difficulty of producing high-quality genome sequences for a natural, heterozygous cultivar with current sequencing technologies. As a consequence, some plant-sequencing projects tend to focus on homozygous derivatives, even if they are not commercially or agronomically important. This was the case, for example, for the highly homozygous genotype of *Vitis vinifera* (grape) in 2007 [54]. Another problem introduced by heterozygosity is the creation of false segmental duplications in assemblies that occur when heterozygous sequences from two haplotypes are assembled into separate contigs and are scaffolded adjacent to each other rather than being merged [55]. In conclusion, only the use of longer reads would improve the ability to assemble separate haplotypes within a genome (see "Polyploidy" section below)

2.3.5 POLYPLOIDY

Polyploidy is the result of the fusion of two or more genomes within the same nucleus. It originates from either whole-genome doubling (autopolyploidy) or by interspecific or intergeneric hybridizations followed by chromosome doubling (allopolyploidy). Genome duplication has the following potential advantages: (i) it is a source of genes with new functions and new phenotypes, (ii) some polyploids appear to be better adapted as a consequence of genome plasticity [56], and (iii) others lose their self-incompatibility, gain asexual reproduction, and produce higher levels of heterozygosity; this may explain the widespread occurrence of polyploids in plants [57]. Polyploidization is therefore one of the major driving forces in plant evolution and is extremely relevant to speciation and diversity [1,58]. An ancestral triplication affecting most (or perhaps all) dicots was followed by two additional whole-genome duplications [1,15]. Every plant lineage shows traces of additional, independent and more recent whole genome duplications somewhere between 50 and 70 million years ago [15]. Some genes have been repeatedly restored to single-copy status following many different genome duplications [59], with the degree of gene retention differing substantially in the different taxa. Therefore, the resulting assembly of a plant genome is dependent on whether the species is an autopolyploid, an allopolyploid, or on the age of the ploidization event. Sequencing of recent polyploids will be especially complex depending on the divergence of the duplicated genes, particularly in the case of many important crops that are true polyploids (banana, potato, cotton, wheat or sugarcane). The redundancy created by the presence of two or more sets of genes within a nucleus can affect the accuracy of the assembly, and the need to differentiate between homologues could influence the final utility of the obtained sequence. Indeed, contigs can break at polymorphic regions or misassembliets can be obtained between large-scale duplications.

The ploidy issue has been 'resolved' in different ways. For example, since most cultivated potatoes are tetraploids, the Potato Genome Sequencing Consortium decided to use as reference a doubled monoploid that was homozygous for a single set of the 12 chromosomes [60]. The authors found that the two haplotypes within a heterozygous diploid were more divergent from each other than from the single haplotype used as reference. In the case of the cultivated strawberry, which is allo-octaploid, the diploid species *Fragaria vesca* (woodland strawberry) was sequenced to bypass the difficulties of polyploidy [61]. For hexaploid wheat, the Wheat Genome Initiative has decided to follow another strategy: a flow cytometry separation of the 10 chromosomes one by one or in groups, the construction of a tiling BAC physical map, and subsequent sequencing of each chromosome using a BAC-by-BAC strategy [8].

2.3.6 GENE CONTENT AND GENE FAMILIES

The gene content in plants can be very complex, as shown by the presence of large gene families and abundant pseudogenes derived from recent genome duplication events and transposon activity (see above and [8]). For example, there are remnants of chloroplast and mitochondrial genes in the nuclear genome that skew coverage levels [7], such as ~270 kbp of the mitochondrial genome inserted into Chromosome 2 of *Arabidopsis* [62]. But gene duplication is regarded as a major force in the origin of new genes and genetic functions. By way of example, the appearance of C4 photosynthesis has evolved from the C3 pathway and has appeared independently on at least 50 occasions during plant evolution [63]. Other examples of gene duplication are the striking increase in the number of starch-associated genes in papaya (39) with respect to *Arabidopsis* (20), or the expanded number of kinase family members, cytochromes P450 and the enzymes engaged in plant secondary metabolism [64]. However, recent comparisons of *Arabidopsis*, poplar, grapevine, papaya and rice genomes estimated that the angiosperm ancestor should contain between 12,000 and 14,000 genes [15]. As a result, more than half of plant genes are really a gene family, 45% of them with the same function but different expression patterns [65]. Specific strategies are required to distinguish alleles from paralogues when sequencing natural heterozygous isolates, although this is not expected to have a very promising success in the near future [59]. Moreover, the presence of out-paralogues produced by duplication prior to the divergence of two lineages and in-paralogues produced in each lineage, together with the multiple rounds of polyploidy in plant lineages, accentuate these problems as divergence between paralogues occurs at different paces.

A curious finding in virtually all eukaryotic genomes sequenced to date (including plants) is the existence of lineage-specific genes for which an orthologue cannot be discerned in closely related species. Lineage-specific genes are a tantalizing target for functional studies since they should distinguish closely related taxa, but unfortunately, these apparently 'lineage-specific genes' could simply be the result of misassemblies [1]. A tention should be paid to these genes before a promising theory can be proposed. Bioinformatic efforts should be made to distinguish real, new genes from misassembled sequences, since we suspect that apparently new genes in sequences <150 bp in length correspond to misassemblies [66]. This also explains the fact that gene sequences may not always be correct, since nearly identical gene families are notoriously difficult to assemble and may collapse into a mosaic sequence without necessarily representing any member of the family [67].

Finally, gene movements can affect plant genome assembly. Gene movement studies found that many gene categories in *Arabidopsis*, papaya and grape were recently transposed at a basal frequency of 5%. The most striking result was that some gene families exhibited very high movement frequencies (50%–90%) [1,68]. This should not be a problem for any assembly procedure since jumping usually occurred a long time ago and the sequences have diverged, but the real drawback is that the region around the transposed gene is enriched with authentic transposons, phantom transposons and pseudogenes [69]. This situation directly impinges on the problem of assembly of repeated sequences and can cause gene loss in the assembly due to collapse of the repetitive surroundings.

2.3.7 NON-CODING RNAS

Non-coding RNAs (ncRNAs) were first described in plants in 1993 [70] and since then, they have provided new insights into gene regulation in plant and animal systems. The advent of NGS has produced a profound impact on the discovery of new ncRNAs. There are small ncRNAs with mature lengths below 30 bp, such as microRNA (miRNA), small interfering RNA (siRNA) and Piwi-interacting RNAs (piRNAs, usually found in animals). Long ncRNAs (200 bp long or more) are another subset of ncRNAs that contain many signatures of mRNAs, including 5' capping, splicing and poly-adenylation, but have little or no open reading frame [71]. Genomic sequences within ncRNAs are often shared within a number of different coding and non-coding transcripts in the sense and antisense directions giving rise to a complex hierarchy of overlapping isoforms. To add even more complexity to ncRNAs, a high proportion of them are variants of protein-coding cDNAs. When using short-read NGS strategies, sequence complexity frustrates the assembly of ncRNA precursors due to their repetitive nature since most ncRNAs contain fragments that are complementary to one or more genes, which causes the collapse of assemblers at the exon or, primarily, at the ncRNA gene [72]. Only long read-based strategies could cover both mature ncRNAs and ncRNA precursors provided that long ncRNAs are not longer than the read lengths.

2.3.8 WIDELY DISTRIBUTED REPETITIVE SEQUENCES (LOW COMPLEXITY SEQUENCES)

Plants share with other organisms a common source of general repetitive sequences [73] that are a source of low complexity regions, which are always a problem for assemblies. The main sources of repeats are the following:

- Repetitions among chromosomes: Duplications occurring both within chromosomes (e.g., ~250 tandem duplications each of ~10 kbp on Chromosome 2 of Arabidopsis) and between chromosomes (e.g., ~4 Mbp long regions between Chromosomes 2 and 4, or 700 Mbp long regions between Chromosomes 1 and 2 in Arabidopsis; ~3 Mbp at the termini of the short arms of Chromosomes 11 and 12 in rice, as well as Chromosomes 5 and 8 in sorghum) [62,74].
- rDNA units: These contain the rRNA genes, which are presented as hundreds of copies. Each unit is typically 10 kbp in plants and as a whole they represent up to 10% of the genome (for example, 8% in *Arabidopsis* [75]). They have not been resolved by any sequencing technology.
- Satellites: These are arrays of many tens or even thousands of identical or nearly identical copies of a repeated unit. They are abundant at centromeres and constitutive heterochromatin. For example, a total of 3% of the *Arabidopsis* genome consists of the 180 bp centromeric repeat [76]. As a result of microsatellites, most sequenced chromosomes are split into two sequences, the right arm and the left arm, since the repetitive, centromeric sequence is unknown.
- Microsatellites or SSRs (simple sequence repeats): These are short tandem repeats (in the range of kbp) of short motifs (1–5 bp) repeated a few hundred times or less, with different microsatellites having different motifs. They are often highly polymorphic with regard to the number of repeat units in a repeat [77]. Microsatellites are mainly located at the subtelomeric region that forms a border between distally positioned structural genes and telomeres, but they can also be found elsewhere.
- Telomeric sequences: These consist of a short repeat of a sequence motif similar to TTTAGGG in tandem arrays many hundreds of units long at the physical end of each chromosome arm. The number of telomeric repeats is a species-specific characteristic ranging from 2–5 kbp in *Arabidopsis* to 60–160 kbp in tobacco [62]. Moreover, the number of copies of the repeat motif also differs among the chromosome arms for the same genome, and

may even vary from cell to cell and tissue to tissue [78]. They are usually still unknown at the sequence level in most species sequenced to date since they are nearly impossible to assemble.

2.4 CONFOUNDING FACTORS FOR PLANT GENOME ASSEMBLY

The apparent disconnection between the limitations of sequencing technologies (several hundreds of base pairs per read in the better cases) and their successful application to genome projects (several hundreds of megabase pairs for small-genome plants) can be explained by the clever combination of sequencing and computation. The resulting reads of a sequencing run must be combined into a reconstruction of the original genome using a computer program called 'assembler.' The assembler tries to construct a 'superstring' that contains all reads as 'substrings.' It must be understood that different assemblers are needed for de novo genome assembly, transcriptome assembly, or genome resequencing (the different rationales for assemblers are beyond the scope of this article), so no assembler is suitable for all approaches. Assembly and analysis of raw sequence data requires substantial bioinformatic effort and expertise [79]. In spite of the fact that different sequencing goals will require different assemblers, the confounding factors emerging from the nature of plant genomes, which are discussed in the following sections, complicate any assembly of plant reads.

2.4.1 REPETITIVE NATURE OF PLANT GENOMES

Most of the challenging features of plant genomes discussed above produce some kind of repeats in DNA. Repeat sequences are difficult to assemble since high-identity reads could come from different portions of the genome, generating gaps, ambiguities and collapses in alignment and assembly, which, in turn, can produce biases and errors when interpreting results. Simply ignoring repeats is not an option, as this creates problems of its own and may mean that important biological phenomena are missed [50]. Repeats would be easily resolved if a single read could span a repeat instance with sufficient unique sequence on either side of the repeat. But repeats longer than the read length specifically create gaps in the assembly and can only be resolved if there are paired-ends that span the repeat instance. Nearly identical tandem repeats (>97% identity) are often collapsed into fewer copies, and it is difficult for an assembler to determine the true copy number since genomic regions that share the same repeats can be indistinguishable, especially if the repeats are longer than the reads [50]. Inexact repeats (<95% identity) can be separated using high-stringency parameters. Repeats were not so critical in Sanger sequencing in which misassemblies and collapses occurred for only ~8% of the genome when duplications or repeats exceeded 95% sequence identity. Consequently, it is expected that repeats longer than 800 bp will suffer from the read-length methodology, regardless of whether it is NGS or Sanger [33]. It can be speculated that NGS short reads will have less power to resolve genomic repeats and require higher coverages to increase the chance of spanning short repeats. As a consequence, the most recent genome assemblies are much more fragmented than assemblies from a few years ago [51].

Repeat separation is assisted by high coverage but confounded by high sequencing error frequency: error tolerance leads to false positive joints that can induce chimeric assemblies, and this becomes especially problematic with reads from inexact (polymorphic) repeats. As a result, depletion of repeated sequences in assemblies becomes acute when the sequence identity exceeds 85%, resulting in the loss of ~16% of the genome [33], or ~5% of the genome being misassembled or missing [5]. The presence of duplicated and repetitive sequences in introns (a frequent event for genes in regions with >50% repetitive content) complicates complete gene assembly and annotation, leading to genes being broken among multiple sequence scaffolds: the more repetitive the region, the more scaffolds are obtained for the gene. After an assembly, nearly 70% of the genes are usually contained in single scaffolds [33], although exon shuffling is an artifact present in ~0.2% of those genes.

The current and most robust methods for overcoming the repeat issues when assembling shotgun reads are: (1) increasing the read length (in fact, nowadays, a compromise solution is to combine short reads with long reads), (2) producing paired-end reads longer than the repeated regions [12], and (3) correlating contigs with genetic maps and/or FISH. This can be easily seen with recently assembled potato [60], tomato [80] and melon [81] genomes. In conclusion, the day that sequencing platforms generate error-free reads at high coverage and assembly software can operate at 100% stringency, repeats would be resolved and a single superstring solution would be obtained. However, advances in the newer technologies based on single-molecule sequencing are giving longer reads (2,000–5,000 bp by now), which will clearly help in the resolution of long repetitive DNAs.

2.4.2 DNA CONTAMINATION

Plant nuclear DNA extractions are always contaminated with mitochondrial and chloroplast DNAs that can confound further assemblies since there always are homologous genes between organelle and nucleus DNA. Moreover, samples from, for example, plant roots where the rhizosphere is not easily removed, are usually highly contaminated with cells from other organisms; and these contaminating cells contain their own DNA, which is usually not of interest in the sequencing goals. Also, contamination can be introduced during laboratory manipulation (adaptors, vector, linkers, poly-A, etc.). Unfortunately contamination is especially difficult to discern when sequencing is based on short reads. In fact, it has been found that contaminating sequences are usually present in the targeted, species-specific sequences, mainly in those that do not match with any homologous sequence in databases [33]. Therefore, in order to obtain a reliable assembly of genomes or transcriptomes, any possible contamination or artifact-prone sequence must be removed with pre-processing software (better than manual or in-house scripting methods), such as SeqTrimNext [82] (an evolution of SeqTrim fully prepared for NGS [83]). It must be taken into account, particularly in the case of genome assembly, that the phraes 'garbage in, garbage out' holds 100%, and that it can even be converted to 'garbage in, nothing out.' Reads devoid of any contamination are less cumbersome to assemble and less prone to misassembling, and produce more reliable consensus [84].

2.4.3 SEQUENCING ERRORS

If sequencing datasets were completely error-free, every read (substring) should be contained within a superstring. But real biological sequences are more complicated since error rates may be as high as 1–4% per nucleotide, implying that many reads contain mismatches with respect to the solution superstring [85]. For example, it has been reported that the Illumina sequencers result in sequence-specific miscalls, GC biased errors [86,87], and more substitution-type miscalls than indel-type miscalls [88]. Roche/454 sequencers produce more indel-type miscalls than substitution-type miscalls due to well-known homopolymer length inaccuracy concerns [89]. The newer technologies based on single-molecule sequencing have been reported to have a 5–15% error rate [90]. Error frequencies can explain the sequence coverage variability and the unfavorable bias observed in reads [91]. In practice, tolerance for sequencing errors makes it difficult to resolve a wide range of genomic phenomena, ranging from polymorphisms to paralogues.

2.4.4 READ LENGTH

Shorter reads are inherent to NGS technologies and deliver less information per read, thus confounding the computational problem of assembly by hindering the detection of contamination, repeats or polymorphisms/errors. Short reads cannot be assembled using any typical overlap-lay-outconsensus algorithm [92] because the repetitive sequences are usually longer than the reads, so many reads cannot be unambiguously assigned, resulting in very short sequence contigs. This prompted the development of new bioinformatic approaches such as de Bruijn graphs combined with Eulerian paths [93,94], and the over-sampling of the target genome from random positions. Assemblies constructed from short-read datasets are highly fragmented and require long reads to increase their contiguity

[60,80]. The assemblers mostly recommended for short reads are ALL-PATHS-LG, SOAPdenovo and SGA, each one with its own pros and cons with respect to assembly length and consensus errors [95]. The advent of technologies based on single-molecule sequencing are now giving reads of 2,000–5,000 bp in length [90], which could simplify the assembling process in the near future.

2.4.5 QUALITY VALUES

The quality value (QV) of each called base was widely used for Sanger sequences assembling [96]. Since its use greatly increases CPU and RAM requirements, QVs are used only by a small set of NGS assemblers [92]. Consequently, to save time and computational resources, most current assemblers assume that base calls are reliable. The presence of low-quality reads will reduce the effective coverage and obscure true overlaps between sequencing reads, thus fragmenting the assembly and risking the collapse of more repeats. This reinforces the need for a good pre-processing of NGS reads (e.g., using SeqTrimNext as explained before) to discard low QV fragments before assembly in order to avoid the assembling of inexistent sequences. For example, a 30 Gbp file of mate-pairs from HiSeq2000 could not be assembled within one week due to the presence of low quality nucleotides in the sequencer output; but this assembly was finished in four days in the same mainframe when reads were filtered for QV20 nucleotides [97].

2.4.6 NUMBER OF READS AND COVERAGE

Assembly is confounded by locations in which there are not enough overlapping reads to extend the sequence with confidence. It is easy to deduce that shorter read lengths will produce a larger number of gaps. The Lander-Waterman model offers a theoretical prediction of the minimum coverage needed to assemble large contigs depending on the read length [98]. For example, a three-fold (3×) coverage is sufficient when using Sanger technology, but a minimum of 15× coverage is required to assemble 100 bp reads into large contigs. However, considering the challenges depicted in the previous section, a minimum coverage of 7×–10× can work for Sanger technology, while 80×–100× is recommended in practice for short reads [32,33]. This high coverage will not resolve the concern about repeats but it is required to compensate the effective shorter length and sequencing errors of NGS technologies, which increase assembly complexity and intensify computational issues related to large datasets.

Short-read NGS technologies nowadays provide terabyte-sized data files, so coverage does not seem to be an issue, and previously intractable plant genomes (for example, pine genomes, which are seven- to 10-fold longer than the human genome and probably contain >95% repetitive sequences) are now feasible, at least in theory. Variation in coverage is introduced by chance, by variation of the copy number within DNA (i.e., repeats), and by the technology per se. But when coverage is homogeneous along the genome, local biases can be interpreted as follows: Gaps are a consequence of very low coverage, and high-coverage is a diagnosis of an over-collapsed repeat. Unfortunately, coverage variability is the rule and undermines the coverage-based diagnostics. It can be speculated that the sequencing itself needs to be improved to reduce the biases, for example from GC composition and PCR, so that the coverage along the genome will be uniform and complete [99].

The overwhelming throughput of NGS raises a collateral issue related to data overload on a laboratory, institutional and community scale. In fact, the infrastructure costs for data storage, processing and handling are becoming more worrying than the costs of generating the reads. Since sequencing throughput is expected to increase in coming years, data storage and handling are becoming a real concern [14]. A more critical issue is computation: The comparison of each read with others required by the overlap-layout-consensus algorithms as well as the resolution of the Eulerian paths for de Bruijn graphs are the most time-consuming part of the assembling process. Therefore, the task could become never-ending or result in a faulty execution when temporary data do not fit in available RAM. The situation could arise that the right data and the right algorithm are available, but the right computer or software to hold calculations and memory are not. The most recent assemblers are focused on distributing among CPUs the processing load that cannot be managed with current serial algorithms. The de Bruijn graphs methods for assembly have the advantage of avoiding the all-versus-all comparisons, but their use is limited when there are too many errors or there is too low coverage, since they lead to infinite loops in the Eulerian paths that produce erroneous 'superstrings' [100]. In conclusion, the type of choices to be made for plant sequencing using NGS remain the same: The importance of assembly size should be balanced against the cost of sequencing, the bioinformatics resources available, and the time the research team has to devote to the project (as in Heisenberg's uncertainty principle, less costs and time in sequencing, more costs and time in assembling).

2.5 SEEKING FOR THE BEST ASSEMBLY

When discussing plant genome assembly, it is important to distinguish between de novo approaches (where the aim is to reconstruct a new genome or transcriptome) and comparative approaches (referred to as mapping since the assembly uses a genome or transcriptome reference, or both). Mathematically, de novo assembly is such a difficult problem that, as yet, there is no efficient computational solution; in contrast, mapping is a much easier task. Neither approach is exclusive since after resequencing (mapping), there are always regions that differ significantly from the reference that can only be reconstructed through de novo assembly. Since de novo assemblies constructed from NGS technologies are highly fragmented, it has been proposed that a good genome assembly would have $N50_{contigs} > 30$ kbp, $N50_{scaffolds} > 250$ kbp, $N50_{super\text{-}scaffolds} > 1$ Mbp, >90% of the genes represented (as measured by previous transcriptomics analyses), and >90% coverage of full-length cDNAs [14]. For now, it should be evident that the ability to assemble plant genome data is constrained by the absence of bioinformatics tools designed to cope with the challenging features present in all plant genomes. Hence, genome assembly is far from being a resolved problem, and the worst consequence is the probably unexpected, artifactual explosion of apparent lineage-specific genes leading to gross incongruities [1]. It is a fact that different transcriptomics projects contain 20–40% unigenes that do not have an orthologue in another plant (e.g., [41,42]). Besides the species-specific genes, the most part of these unigenes may represent 'garbage sequences' generated by errors within the amplification and/or sequencing technology. The percentage of this garbage will be known more precisely as more and more transcriptomes and genomes are reported. In the meantime, we have developed the bioinformatics tool Full-LengtherNext [101] that can inform which unigenes may be garbage or putative species-specific unigenes [66].

Many assemblers designed to handle Sanger reads were found to be impractical when dealing with NGS data. The response was to develop new assemblers employing qualitatively new approaches that seemed to be appropriate for assembly from human to zgenomes (to cite a few, CA-BOG, Newbler, ABySS, SOAPdenovo or ALLPATHS), although their true success depends largely on the sophistication of their heuristics for real reads to solve the existing issues [12]. They generally require servers or clusters with >500 gigabytes of RAM and many terabytes of available disk space. The decrease in cost of servers, the emergence of supercomputing centers, and the development of cloud computing, mean that they are available at a negligible cost. But new sequencing projects such as loblolly pine [102] or maritime pine [103] with 22–30 Gbp genomes, are increasing the computational demands by nearly another order of magnitude, and no proven technology is available to resolve this assembly. Assembler performance was evaluated last year in a competitive framework with both simulated and real datasets of small, simple genomes. Results confirmed that the final sequences were highly dependent on the assembler and pipeline used [95], although it can be said that assemblers for long reads produce longer contigs and scaffolds with more indels and underrepresentation of repeats, while the de Bruijn-based assemblers include shorter contigs and scaffolds, more mismatches and the highest representation for repeat regions [34]. Most assemblies nowadays rely on one single assembler, but as different assemblers use different underlying algorithms, combining different optimal assemblies from different programs can give a more credible final assembly [104]. The combination usually increased the N50 and median contig size, mapped more original reads, and diminished the final number of contigs/scaffolds. This strategy is currently used for transcriptomes, and CAP3 [96] or Minimus [105] are good candidates for the second assembly process [106–108]. In the case of genome assembly, mammalian genomes have recently been assembled using this combined strategy [109], running SOAPdenovo and ABySS separately, and then combining the assembly with GAP5 to generate the final consensus sequences.

As the choices made at the beginning of any study will determine the degree of success of the sequencing project, it can be concluded that there is a strong need to develop plant-specific assemblers that can overcome the challenges of these genomes; moreover, new software should expend efforts in producing user-friendly interfaces since most bioinformatics projects are developing software tailored to their needs, which leads to the same software being reinvented over and over again by different research groups [79]. Researchers have to decide which plant genome will be sequenced, which NGS technology will be applied, and which assembling approach should be used. In Plantagora [34], researchers can find a substantial body of information for comparing different approaches to sequencing a plant genome, providing a platform of metrics and tools for studying the process of sequencing and assembling that can aid in the critical decision-making required for planning a plant-sequencing project.

2.6 CONCLUDING REMARKS

Plant genome sequencing is a long way away from automatic sequencing and assembly providing a completely finished genome at low cost. At the moment, we are able to afford the reconstruction of complex plant genomes into highly useful drafts. The need remains for an assembler that can deal with the plant genome features that challenge sequencing and assembly, i.e., mainly large, repetitive genomes; moreover, incremental algorithms that can update the assembly as new data become available are also desirable. To circumvent the bioinformatics bottleneck in the near future, efforts should be invested in (1) parallelization of the assembly process, which has been shyly approached with ABySS [110] and ALLPATHS-LG [109]; (2) processing speed and storage capacity of computers; and (3) developing a new sequencing platform that can provide longer reads with unbiased coverage that can overcome the complex repeats. This last point refers to the so-called third-generation sequencing based on single-molecule sequencing, which is very promising with reads of 2,000–5,000 nt [90]. However, these technologies are relatively immature for immediate widespread application to plant genomes since to date an error rate of 5–15% has been reported.

REFERENCES

1. Paterson, A.H.; Freeling, M.; Tang, H.; Wang, X. Insights from the comparison of plant genome sequences. Annu. Rev. Plant Biol. 2010, 61, 349–372.
2. Sterck, L.; Rombauts, S.; Vandepoele, K.; Rouze, P.; van de Peer, Y. How many genes are there in plants (... and why are they there)? Curr. Opin. Plant Biol. 2007, 10, 199–203.
3. Gregory, T.R. The C-value enigma in plants and animals: A review of parallels and an appeal for partnership. Ann. Bot. 2005, 95, 133–146.
4. Arabidopsis Genome, I. Analysis of the genome sequence of the flowering plant Arabidopsis thaliana. Nature 2000, 408, 796–815.
5. Feuillet, C.; Leach, J.E.; Rogers, J.; Schnable, P.S.; Eversole, K. Crop genome sequencing: Lessons and rationales. Trends Plant Sci. 2011, 16, 77–88.
6. International Rice Genome Sequencing, P. The map-based sequence of the rice genome. Nature 2005, 436, 793–800.
7. Ming, R.; Hou, S.; Feng, Y.; Yu, Q.; Dionne-Laporte, A.; Saw, J.H.; Senin, P.; Wang, W.; Ly, B.V.; Lewis, K.L.; et al. The draft genome of the transgenic tropical fruit tree papaya (Carica papaya Linnaeus). Nature 2008, 452, 991–996.
8. Schnable, P.S.; Ware, D.; Fulton, R.S.; Stein, J.C.; Wei, F.; Pasternak, S.; Liang, C.; Zhang, J.; Fulton, L.; Graves, T.A.; et al. The B73 maize genome: Complexity, diversity, and dynamics. Science 2009, 326, 1112–1115.
9. Duvick, J.; Fu, A.; Muppirala, U.; Sabharwal, M.; Wilkerson, M.D.; Lawrence, C.J.; Lushbough, C.; Brendel, V. PlantGDB: A resource for comparative plant genomics. Nucleic Acids Res. 2008, 36, D959–D965.
10. Varshney, R.K.; Close, T.J.; Singh, N.K.; Hoisington, D.A.; Cook, D.R. Orphan legume crops enter the genomics era! Curr. Opin. Plant Biol. 2009, 12, 202–210.
11. Armstead, I.; Huang, L.; Ravagnani, A.; Robson, P.; Ougham, H. Bioinformatics in the orphan crops. Brief. Bioinform. 2009, 10, 645–653.
12. Imelfort, M.; Edwards, D. De novo sequencing of plant genomes using second-generation technologies. Brief. Bioinform. 2009, 10, 609–618.
13. Goodstein, D.M.; Shu, S.; Howson, R.; Neupane, R.; Hayes, R.D.; Fazo, J.; Mitros, T.; Dirks, W.; Hellsten, U.; Putnam, N.; et al. Phytozome: A comparative platform for green plant genomics. Nucleic Acids Res. 2012, 40, D1178–D1186.
14. Hamilton, J.P.; Buell, C.R. Advances in plant genome sequencing. Plant J. 2012, 70, 177–190.
15. Proost, S.; Pattyn, P.; Gerats, T.; van de Peer, Y. Journey through the past: 150 million years of plant genome evolution. Plant J. 2011, 66, 58–65

16. Ossowski, S.; Schneeberger, K.; Clark, R.M.; Lanz, C.; Warthmann, N.; Weigel, D. Sequencing of natural strains of Arabidopsis thaliana with short reads. Genome Res. 2008, 18, 2024–2033.
17. Springer, N.M.; Ying, K.; Fu, Y.; Ji, T.; Yeh, C.T.; Jia, Y.; Wu, W.; Richmond, T.; Kitzman, J.; Rosenbaum, H.; et al. Maize inbreds exhibit high levels of copy number variation (CNV) and presence/absence variation (PAV) in genome content. PLoS Genet. 2009, 5, e1000734.
18. Morgante, M.; de Paoli, E.; Radovic, S. Transposable elements and the plant pan-genomes. Curr. Opin. Plant Biol. 2007, 10, 149–155.
19. Plant Genomes Central. Available online: http://www.ncbi.nlm.nih.gov/genomes/PLANTS/ PlantList.html (accessed on 14 September 2012).
20. List of Sequenced Plant Genomes. Available online: http://en.wikipedia.org/wiki/List_of_sequenced_plant_genomes (accessed on 14 September 2012).
21. Sanger, F.; Nicklen, S.; Coulson, A.R. DNA sequencing with chain-terminating inhibitors. Proc. Natl. Acad. Sci. USA 1977, 74, 5463–5467.
22. Bräutigam, A.; Gowik, U. What can next generation sequencing do for you? Next generation sequencing as a valuable tool in plant research. Plant Biol. (Stuttg) 2010, 12, 831–841.
23. Goff, S.A.; Ricke, D.; Lan, T.H.; Presting, G.; Wang, R.; Dunn, M.; Glazebrook, J.; Sessions, A.; Oeller, P.; Varma, H.; et al. A draft sequence of the rice genome (Oryza sativa L. ssp. japonica). Science 2002, 296, 92–100.
24. Yu, J.; Hu, S.; Wang, J.; Wong, G.K.; Li, S.; Liu, B.; Deng, Y.; Dai, L.; Zhou, Y.; Zhang, X.; et al. A draft sequence of the rice genome (Oryza sativa L. ssp. indica). Science 2002, 296, 79–92.
25. Shendure, J.; Ji, H. Next-generation DNA sequencing. Nat. Biotechnol. 2008, 26, 1135–1145.
26. Mardis, E.R. Next-generation DNA sequencing methods. Annu. Rev. Genomics Hum. Genet. 2008, 9, 387–402.
27. Ansorge, W.J. Next-generation DNA sequencing techniques. N. Biotechnol. 2009, 25, 195–203.
28. Kircher, M.; Kelso, J. High-throughput DNA sequencing□Concepts and limitations. Bioessays 2010, 32, 524–536.
29. Zhou, X.; Ren, L.; Meng, Q.; Li, Y.; Yu, Y.; Yu, J. The next-generation sequencing technology and application. Protein Cell 2010, 1, 520–536.
30. Niedringhaus, T.P.; Milanova, D.; Kerby, M.B.; Snyder, M.P.; Barron, A.E. Landscape of next-generation sequencing technologies. Anal. Chem. 2011, 83, 4327–4341.
31. Pareek, C.S.; Smoczynski, R.; Tretyn, A. Sequencing technologies and genome sequencing. J. Appl. Genet. 2011, 52, 413–435.
32. Finotello, F.; Lavezzo, E.; Fontana, P.; Peruzzo, D.; Albiero, A.; Barzon, L.; Falda, M.; di Camillo, B.; Toppo, S. Comparative analysis of algorithms for whole-genome assembly of pyrosequencing data. Brief. Bioinform. 2012, 13, 269–280.
33. Alkan, C.; Sajjadian, S.; Eichler, E.E. Limitations of next-generation genome sequence assembly. Nat. Methods 2011, 8, 61–65.

34. Barthelson, R.; McFarlin, A.J.; Rounsley, S.D.; Young, S. Plantagora: Modeling whole genome sequencing and assembly of plant genomes. PLoS One 2011, 6, e28436.
35. Wang, L.; Li, P.; Brutnell, T.P. Exploring plant transcriptomes using ultra high-throughput sequencing. Brief. Funct. Genomics 2010, 9, 118–128.
36. Vandepoele, K.; Quimbaya, M.; Casneuf, T.; de Veylder, L.; van de Peer, Y. Unraveling transcriptional control in Arabidopsis using cis-regulatory elements and coexpression networks. Plant Physiol. 2009, 150, 535–546.
37. He, F.; Zhou, Y.; Zhang, Z. Deciphering the Arabidopsis floral transition process by integrating a protein-protein interaction network and gene expression data. Plant Physiol. 2010, 153, 1492–1505.
38. Alvarez, J.M.; Vidal, E.A.; Gutierrez, R.A. Integration of local and systemic signaling pathways for plant N responses. Curr. Opin. Plant Biol. 2012, 15, 185–191.
39. Proost, S.; van Bel, M.; Sterck, L.; Billiau, K.; van Parys, T.; van de Peer, Y.; Vandepoele, K. PLAZA: A comparative genomics resource to study gene and genome evolution in plants. Plant Cell 2009, 21, 3718–3731.
40. Wegrzyn, J.L.; Lee, J.M.; Tearse, B.R.; Neale, D.B. TreeGenes: A forest tree genome database. Int. J. Plant Genomics 2008, 2008, 412875.
41. Fernandez-Pozo, N.; Canales, J.; Guerrero-Fernandez, D.; Villalobos, D.P.; Diaz-Moreno, S.M.; Bautista, R.; Flores-Monterroso, A.; Guevara, M.A.; Perdiguero, P.; Collada, C.; et al. EuroPineDB: A high-coverage web database for maritime pine transcriptome. BMC Genomics 2011, 12, 366.
42. Rengel, D.; San Clemente, H.; Servant, F.; Ladouce, N.; Paux, E.; Wincker, P.; Couloux, A.; Sivadon, P.; Grima-Pettenati, J. A new genomic resource dedicated to wood formation in Eucalyptus. BMC Plant Biol. 2009, 9, 36.
43. Gonzalez-Ibeas, D.; Blanca, J.; Roig, C.; Gonzalez-To, M.; Pico, B.; Truniger, V.; Gomez, P.; Deleu, W.; Cano-Delgado, A.; Arus, P.; et al. MELOGEN: An EST database for melon functional genomics. BMC Genomics 2007, 8, 306.
44. Goff, S.A.; Vaughn, M.; McKay, S.; Lyons, E.; Stapleton, A.E.; Gessler, D.; Matasci, N.; Wang, L.; Hanlon, M.; Lenards, A.; et al. The iPlant collaborative: Cyberinfrastructure for plant biology. Front. Plant Sci. 2011, 2, 34.31–34.16.
45. Katari, M.S.; Nowicki, S.D.; Aceituno, F.F.; Nero, D.; Kelfer, J.; Thompson, L.P.; Cabello, J.M.; Davidson, R.S.; Goldberg, A.P.; Shasha, D.E.; et al. VirtualPlant: A software platform to support systems biology research. Plant Physiol. 2010, 152, 500–515.
46. Lapitan, N.L.V. Organization and evolution of higher plant nuclear genome. Genome 1992, 35, 171–181.
47. Janicki, M.; Rooke, R.; Yang, G. Bioinformatics and genomic analysis of transposable elements in eukaryotic genomes. Chromosome Res. 2011, 19, 787–808.
48. Wicker, T.; Sabot, F.; Hua-Van, A.; Bennetzen, J.L.; Capy, P.; Chalhoub, B.; Flavell, A.; Leroy, P.; Morgante, M.; Panaud, O.; et al. A unified classification system for eukaryotic transposable elements. Nat. Rev. Genet. 2007, 8, 973–982.
49. Bousios, A.; Darzentas, N.; Tsaftaris, A.; Pearce, S.R. Highly conserved motifs in non-coding regions of Sirevirus retrotransposons: The key for their pattern of distribution within and across plants? BMC Genomics 2010, 11, 89.

50. Treangen, T.J.; Salzberg, S.L. Repetitive DNA and next-generation sequencing: Computational challenges and solutions. Nat. Rev. Genet. 2012, 13, 36–46.
51. Schatz, M.C.; Delcher, A.L.; Salzberg, S.L. Assembly of large genomes using second-generation sequencing. Genome Res. 2010, 20, 1165–1173.
52. Hochholdinger, F.; Hoecker, N. Towards the molecular basis of heterosis. Trends Plant Sci. 2007, 12, 427–432.
53. Tuskan, G.A.; Difazio, S.; Jansson, S.; Bohlmann, J.; Grigoriev, I.; Hellsten, U.; Putnam, N.; Ralph, S.; Rombauts, S.; Salamov, A.; et al. The genome of black cottonwood, Populus trichocarpa (Torr. & Gray). Science 2006, 313, 1596–1604.
54. Jaillon, O.; Aury, J.M.; Noel, B.; Policriti, A.; Clepet, C.; Casagrande, A.; Choisne, N.; Aubourg, S.; Vitulo, N.; Jubin, C.; et al. The grapevine genome sequence suggests ancestral hexaploidization in major angiosperm phyla. Nature 2007, 449, 463–467.
55. Kelley, D.R.; Salzberg, S.L. Detection and correction of false segmental duplications caused by genome mis-assembly. Genome Biol. 2010, 11, R28.
56. Comai, L. The advantages and disadvantages of being polyploid. Nat. Rev. Genet. 2005, 6, 836–846.
57. Meyers, L.A.; Levin, D.A. On the abundance of polyploids in flowering plants. Evolution 2006, 60, 1198–1206.
58. Bento, M.; Gustafson, J.P.; Viegas, W.; Silva, M. Size matters in Triticeae polyploids: Larger genomes have higher remodeling. Genome 2011, 54, 175–183.
59. Tang, H.; Bowers, J.E.; Wang, X.; Ming, R.; Alam, M.; Paterson, A.H. Synteny and collinearity in plant genomes. Science 2008, 320, 486–488.
60. Potato Genome Sequencing, C.; Xu, X.; Pan, S.; Cheng, S.; Zhang, B.; Mu, D.; Ni, P.; Zhang, G.; Yang, S.; Li, R.; et al. Genome sequence and analysis of the tuber crop potato. Nature 2011, 475, 189–195.
61. Shulaev, V.; Sargent, D.J.; Crowhurst, R.N.; Mockler, T.C.; Folkerts, O.; Delcher, A.L.; Jaiswal, P.; Mockaitis, K.; Liston, A.; Mane, S.P.; et al. The genome of woodland strawberry (Fragaria vesca). Nat. Genet. 2011, 43, 109–116.
62. Heslop-Harrison, J.S. Comparative genome organization in plants: From sequence and markers to chromatin and chromosomes. Plant Cell 2000, 12, 617–636.
63. Giussani, L.M.; Cota-Sanchez, J.H.; Zuloaga, F.O.; Kellogg, E.A. A molecular phylogeny of the grass subfamily Panicoideae (Poaceae) shows multiple origins of C4 photosynthesis. Am. J. Bot. 2001, 88, 1993–2012.
64. Sappl, P.G.; Heazlewood, J.L.; Millar, A.H. Untangling multi-gene families in plants by integrating proteomics into functional genomics. Phytochemistry 2004, 65, 1517–1530.
65. Duarte, J.M.; Cui, L.; Wall, P.K.; Zhang, Q.; Zhang, X.; Leebens-Mack, J.; Ma, H.; Altman, N.; dePamphilis, C.W. Expression pattern shifts following duplication indicative of subfunctionalization and neofunctionalization in regulatory genes of Arabidopsis. Mol. Biol. Evol. 2006, 23, 469–478.
66. Fernández-Pozo, N.; Guerrero-Fernández, D.; Bautista, R.; Claros, M.G. Full-LengtherNext: A tool for fine-tuning de novo assembled transcriptomes of non-model organisms. Departamento de Biología Molecular y Bioquímica, Facultad de Ciencias, Universidad de Málaga, 29071 Málaga, Spain, and Plataforma Andaluza de Bioinformática, Centro de Supercomputación y Bioinformática, Edificio de Bioinnovación, Universidad de Málaga, 29590 Málaga, Spain. Unpublished work, to be submitted for publication, 2012.

67. Phillippy, A.M.; Schatz, M.C.; Pop, M. Genome assembly forensics: Finding the elusive mis-assembly. Genome Biol. 2008, 9, R55.
68. Lai, J.; Li, Y.; Messing, J.; Dooner, H.K. Gene movement by Helitron transposons contributes to the haplotype variability of maize. Proc. Natl. Acad. Sci. USA 2005, 102, 9068–9073.
69. Freeling, M.; Lyons, E.; Pedersen, B.; Alam, M.; Ming, R.; Lisch, D. Many or most genes in Arabidopsis transposed after the origin of the order Brassicales. Genome Res. 2008, 18, 1924–1937.
70. Lindbo, J.A.; Silva-Rosales, L.; Proebsting, W.M.; Dougherty, W.G. Induction of a highly specific antiviral state in transgenic plants: Implications for regulation of gene expression and virus resistance. Plant Cell 1993, 5, 1749–1759.
71. Huang, R.; Jaritz, M.; Guenzl, P.; Vlatkovic, I.; Sommer, A.; Tamir, I.M.; Marks, H.; Klampfl, T.; Kralovics, R.; Stunnenberg, H.G.; et al. An RNA-Seq strategy to detect the complete coding and non-coding transcriptome including full-length imprinted macro ncRNAs. PLoS One 2011, 6, e27288.
72. Carninci, P.; Kasukawa, T.; Katayama, S.; Gough, J.; Frith, M.C.; Maeda, N.; Oyama, R.; Ravasi, T.; Lenhard, B.; Wells, C.; et al. The transcriptional landscape of the mammalian genome. Science 2005, 309, 1559–1563.
73. Gore, M.A.; Chia, J.M.; Elshire, R.J.; Sun, Q.; Ersoz, E.S.; Hurwitz, B.L.; Peiffer, J.A.; McMullen, M.D.; Grills, G.S.; Ross-Ibarra, J.; et al. A first-generation haplotype map of maize. Science 2009, 326, 1115–1117.
74. Wang, X.; Tang, H.; Bowers, J.E.; Paterson, A.H. Comparative inference of illegitimate recombination between rice and sorghum duplicated genes produced by polyploidization. Genome Res. 2009, 19, 1026–1032.
75. Pruitt, R.E.; Meyerowitz, E.M. Characterization of the genome of Arabidopsis thaliana. J. Mol. Biol. 1986, 187, 169–183.
76. Murata, M.; Ogura, Y.; Motoyoshi, F. Centromeric repetitive sequences in Arabidopsis thaliana. Jpn. J. Genet. 1994, 69, 361–371.
77. Horáková, M.; Fajkus, J. TAS4—A dispersed repetitive sequence isolated from subtelomeric regions of Nicotiana tomentosiformis chromosomes. Genome 2000, 43, 273–284.
78. Kilian, A.; Stiff, C.; Kleinhofs, A. Barley telomeres shorten during differentiation but grow in callus culture. Proc. Natl. Acad. Sci. USA 1995, 92, 9555–9559.
79. Schatz, M.C.; Witkowski, J.; McCombie, W.R. Current challenges in de novo plant genome sequencing and assembly. Genome Biol. 2012, 13, 243.
80. Tomato Genome, C. The tomato genome sequence provides insights into fleshy fruit evolution. Nature 2012, 485, 635–641.
81. Garcia-Mas, J.; Benjak, A.; Sanseverino, W.; Bourgeois, M.; Mir, G.; González, V.M.; Hénaff, E.; Cámara, F.; Cozzuto, L.; Lowy, E.; et al. The genome of melon (Cucumis melo L.). Proc. Natl. Acad. Sci. USA 2012, in press.
82. SeqTrimNext. Available online: http://www.scbi.uma.es/seqtrimnext (accessed on 14 September 2012).
83. Falgueras, J.; Lara, A.J.; Fernandez-Pozo, N.; Canton, F.R.; Perez-Trabado, G.; Claros, M.G. SeqTrim: A high-throughput pipeline for pre-processing any type of sequence read. BMC Bioinformatics 2010, 11, 38.

84. Guerrero-Fernaández, D.; Falgueras, J.; Claros, M.G. SCBI_MAPREDUCE: A task-farm, practical solution in Ruby for distribution of new and legacy bioinformatics software. IEEE Trans. Parallel. Distr. Syst. 2012, submitted for publication.
85. Paszkiewicz, K.; Studholme, D.J. De novo assembly of short sequence reads. Brief. Bioinform. 2010, 11, 457–472.
86. Nakamura, K.; Oshima, T.; Morimoto, T.; Ikeda, S.; Yoshikawa, H.; Shiwa, Y.; Ishikawa, S.; Linak, M.C.; Hirai, A.; Takahashi, H.; et al. Sequence-specific error profile of Illumina sequencers. Nucleic Acids Res. 2011, 39, e90.
87. Minoche, A.E.; Dohm, J.C.; Himmelbauer, H. Evaluation of genomic high-throughput sequencing data generated on Illumina HiSeq and genome analyzer systems. Genome Biol. 2011, 12, R112.
88. Hoffmann, S.; Otto, C.; Kurtz, S.; Sharma, C.M.; Khaitovich, P.; Vogel, J.; Stadler, P.F.; Hackermuller, J. Fast mapping of short sequences with mismatches, insertions and deletions using index structures. PLoS Comput. Biol. 2009, 5, e1000502.
89. Gilles, A.; Meglecz, E.; Pech, N.; Ferreira, S.; Malausa, T.; Martin, J.F. Accuracy and quality assessment of 454 GS-FLX Titanium pyrosequencing. BMC Genomics 2011, 12, 245.
90. Rasko, D.A.; Webster, D.R.; Sahl, J.W.; Bashir, A.; Boisen, N.; Scheutz, F.; Paxinos, E.E.; Sebra, R.; Chin, C.S.; Iliopoulos, D.; et al. Origins of the E. coli strain causing an outbreak of hemolytic-uremic syndrome in Germany. N. Engl. J. Med. 2011, 365, 709–717.
91. Balzer, S.; Malde, K.; Jonassen, I. Systematic exploration of error sources in pyrosequencing flowgram data. Bioinformatics 2011, 27, i304–309.
92. Miller, J.R.; Koren, S.; Sutton, G. Assembly algorithms for next-generation sequencing data. Genomics 2010, 95, 315–327.
93. Medvedev, P.; Pham, S.; Chaisson, M.; Tesler, G.; Pevzner, P. Paired de bruijn graphs: A novel approach for incorporating mate pair information into genome assemblers. J. Comput. Biol. 2011, 18, 1625–1634.
94. Compeau, P.E.; Pevzner, P.A.; Tesler, G. How to apply de Bruijn graphs to genome assembly. Nat. Biotechnol. 2011, 29, 987–991.
95. Earl, D.; Bradnam, K.; St. John, J.; Darling, A.; Lin, D.; Fass, J.; Yu, H.O.; Buffalo, V.; Zerbino, D.R.; Diekhans, M.; et al. Assemblathon 1: A competitive assessment of de novo short read assembly methods. Genome Res. 2011, 21, 2224–2241.
96. Huang, X.; Madan, A. CAP3: A DNA sequence assembly program. Genome Res. 1999, 9, 868–877.
97. Benzekri, H.; Bautista, R.; Guerrero-Fernández, D.; Claros, M.G. Departamento de Biología Molecular y Bioquímica, Facultad de Ciencias, Universidad de Málaga, 29071 Málaga, Spain, and Plataforma Andaluza de Bioinformática, Centro de Supercomputación y Bioinformática, Edificio de Bioinnovación, Universidad de Málaga, 29590 Málaga, Spain. Unpublished work, 2012.
98. Lander, E.S.; Waterman, M.S. Genomic mapping by fingerprinting random clones: A mathematical analysis. Genomics 1988, 2, 231–239.
99. Aird, D.; Ross, M.G.; Chen, W.S.; Danielsson, M.; Fennell, T.; Russ, C.; Jaffe, D.B.; Nusbaum, C.; Gnirke, A. Analyzing and minimizing PCR amplification bias in Illumina sequencing libraries. Genome Biol. 2011, 12, R18.

100. Li, Z.; Chen, Y.; Mu, D.; Yuan, J.; Shi, Y.; Zhang, H.; Gan, J.; Li, N.; Hu, X.; Liu, B.; et al. Comparison of the two major classes of assembly algorithms: Overlap-layout-consensus and de Bruijn-graph. Brief. Funct. Genomics 2012, 11, 25–37.
101. FullLengtherNext. Available online: http://www.scbi.uma.es/fulllengthernext (accessed on 14 September 2012).
102. Loblolly Pine Genome Project. Available online: http://dendrome.ucdavis.edu/NealeLab/lpgp/ (accessed on 14 September 2012).
103. Díaz-Sala, C.; Cervera, M. Promoting a functional and comparative understanding of the conifer genome-implementing applied aspects for more productive and adapted forests (ProCoGen). BCM Proceedings 2011, 5, P158.
104. Kumar, S.; Blaxter, M.L. Comparing de novo assemblers for 454 transcriptome data. BMC Genomics 2010, 11, 571.
105. Sommer, D.D.; Delcher, A.L.; Salzberg, S.L.; Pop, M. Minimus: A fast, lightweight genome assembler. BMC Bioinformatics 2007, 8, 64.
106. Zheng, Y.; Zhao, L.; Gao, J.; Fei, Z. iAssembler: A package for de novo assembly of Roche-454/Sanger transcriptome sequences. BMC Bioinformatics 2011, 12, 453.
107. Iorizzo, M.; Senalik, D.A.; Grzebelus, D.; Bowman, M.; Cavagnaro, P.F.; Matvienko, M.; Ashrafi, H.; van Deynze, A.; Simon, P.W. De novo assembly and characterization of the carrot transcriptome reveals novel genes, new markers, and genetic diversity. BMC Genomics 2011, 12, 389.
108. Martin, J.; Bruno, V.M.; Fang, Z.; Meng, X.; Blow, M.; Zhang, T.; Sherlock, G.; Snyder, M.; Wang, Z. Rnnotator: An automated de novo transcriptome assembly pipeline from stranded RNA-Seq reads. BMC Genomics 2010, 11, 663.
109. Gnerre, S.; Maccallum, I.; Przybylski, D.; Ribeiro, F.J.; Burton, J.N.; Walker, B.J.; Sharpe, T.; Hall, G.; Shea, T.P.; Sykes, S.; et al. High-quality draft assemblies of mammalian genomes from massively parallel sequence data. Proc. Natl. Acad. Sci. USA 2011, 108, 1513–1518.
110. Simpson, J.T.; Wong, K.; Jackman, S.D.; Schein, J.E.; Jones, S.J.; Birol, I. ABySS: A parallel assembler for short read sequence data. Genome Res. 2009, 19, 1117–1123.

CHAPTER 3

GENOME WALKING BY NEXT GENERATION SEQUENCING APPROACHES

MARIATERESA VOLPICELLA, CLAUDIA LEONI, ALESSANDRA COSTANZA, IMMACOLATA FANIZZA, ANTONIO PLACIDO, and LUIGI R. CECI

3.1 INTRODUCTION

The identification of unknown nucleotide sequences starting from a previously identified DNA region can be directly obtained by a number of Genome Walking (GW) methods all having in common a final PCR amplification in which an oligonucleotide specific for the known sequence is coupled with an oligonucleotide derived from the adopted GW strategy. The numerous GW methods can be classified into three main categories, according to the first step of the whole strategy: (1) Restriction-based GW methods, requiring a restriction digestion of genomic DNA and ligation of restriction fragments to DNA-cassettes; (2) Primer-based GW methods, in which PCR amplifications are directly carried out using a variously designed combinatorial (random or degenerate primer) coupled to a sequence specific primer; (3) Extension-based GW methods, in which the extension of a sequence specific primer and subsequent 3'-tailing of the resulting single strand DNA (ssDNA) provide the substrate for the

This chapter was originally published under the Creative Commons Attribution License. Volpicella M, Leoni C, Costanza A, Fanizza I, Placido A, and Ceci LR. Genome Walking by Next Generation Sequencing Approaches. Biology. ***1** (2012), 495–507. doi:10.3390/biology1030495.*

final PCR amplification. Critical overviews of the conceived GW strategies, and their possible applications to both eukaryotic and prokaryotic genomes, have been recently reported [1–3].

GW is a highly flexible approach, allowing both the identification of specific, unique sequences (as for the analysis of single gene flanking sequences) and the analysis of large libraries (such as those obtained by insertional mutagenesis of retroviruses and transposable elements). In the latter case, GW has the potential to be powered by the enormous capacity of Next Generation Sequencing (NGS) approaches. In this review we focus on the application of NGS to GW. Among NGS technologies [4,5], currently only pyrosequencing and the sequencing by synthesis (SBS) methods have been reported as successful sequencing approaches for GW. A first report on the application of pyrosequencing (using the Roche 454 platform) to GW is that of Wang et al. [6], while the first application of SBS methods (using the Illumina technology) to GW is more recent [7]. Nowadays both technologies have been applied to GW by several research groups and, recently, specific methodological papers have been published [8–10]. In principle, however, also other NGS technologies in which parallel sequencing of amplicons is performed (such as the SOLiD "Sequencing by Oligo Ligation Detection" approach and the more recent "sequencing by synthesis" approach, known as Ion Torrent [11], by Life Technologies), can potentially be used for GW.

The first part of this review illustrates applications of pyrosequencing to GW, describing protocols adopted for the analysis of genomes of man, mouse, yeasts and plants, then the application of the SBS-Illumina technology to GW is reported. A final paragraph gives some general considerations about the possibility to use other, as yet unexplored, combinations of NGS and GW methods.

3.2 GW BY 454-PYROSEQUENCING

Wang et al. [6] were the first to associate NGS and GW technologies to analyze the insertion sites of the immunodeficiency virus (HIV) into the human genome. In this first approach, authors used a classical restriction-based GW approach, consisting essentially in the digestion of genomic

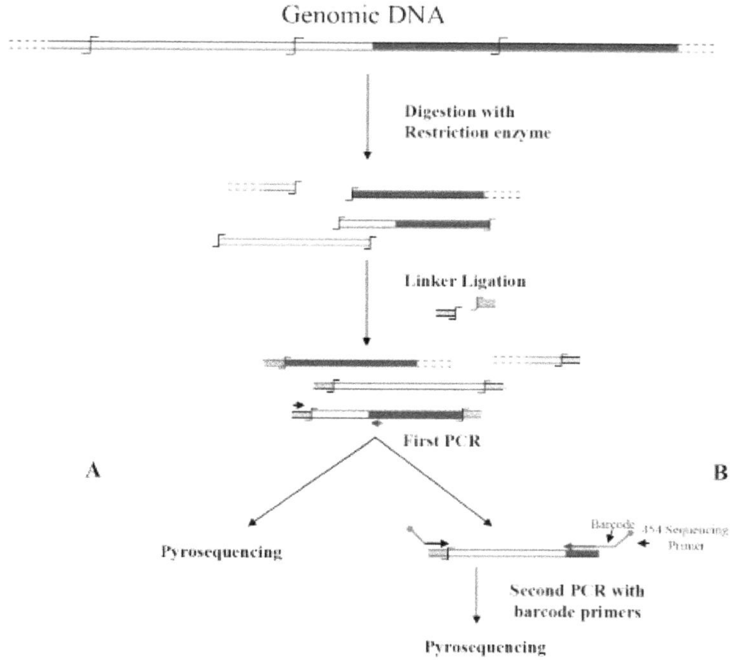

FIGURE 1: Main steps of cassette-PCR genome walking (GW) and pyrosequencing association strategies. The original strategy reported by Wang et al. [6] (A) and its improvement by the adoption of barcoded primers [13] (B) are illustrated. Black and white bars correspond to known and unknown sequences, respectively. A black horizontal arrow indicates adapter specific primers. The symbol indicates a restriction site. Barcodes and 454 sequencing traits of longer amplification primers are indicated in the figure.

DNA molecules with a restriction enzyme and subsequent ligation of restriction fragments to DNA cassettes. The library of PCR fragments obtained by amplifying the ligation products was directly used for pyrosequencing (Figure 1A). A similar approach, with the improvement of the introduction of biotinylated primers, but also with some complicated and unuseful steps, such as the ligation of a double set of adapters and a primer extension reaction to create an additional restriction site, was used to analyze the transposon flanking sequences in 1,000 petunia dTph1 insertional mutants [12]. Interestingly, however, in this paper the use of a nucleotide barcode to distinguish specific samples was firstly adopted in NGS-GW experiments. At the same time, an improved version of their first NGS-GW approach, with the inclusion of a DNA barcoding step (Figure 1B) was also published by Wang et al. [13]. One hundred and sixty thousand integration sites for lentiviral and gamma-retroviral vectors in twenty-eight tissue samples from eight different mice could be identified by this approach.

Liu et al. [14,15] combined another restriction-based GW approach, known as Digestion Ligation Amplification (DLA) with pyrosequencing. DLA-GW is characterized by the use of a single-stranded oligonucleotide adaptor to ligate to restricted genomic DNA fragments. In this case the introduction of a DNA-barcode was also employed for sequencing in a single experiment multiple independent insertional mutants of the high copy number Mu transposon from different maize Mu-stocks. The assay allowed the observation of the expected Mu/genomic junctions in approximately 94% of the 965,000 reads, demonstrating the specificity of the strategy. 324 gene hotspots for Mu insertions were detected.

The first application of the pyrosequencing-GW approach to yeast is by Guo and Levin [16] who studied the integration of the Tfl retrotransposon in the genome of Schizosaccharomyces pombe. About 600,000 sequences were analyzed, allowing the identification of more than 73,000 independent integration sites.

The above reported GW-pyrosequencing strategies adopted two restriction-based GW methods. Subsequently pyrosequencing was also applied to the Extension-based GW method known as nonrestrictive Linear Amplification-Mediated PCR (nrLAM-PCR) [9,10]. Extension-based GW methods are based on the extension of an oligonucleotide specific of the known DNA region and directed toward the unknown sequence. After the synthesis of the single-strand DNA, several strategies can be adopted to make it a suitable substrate for PCR reactions (reviewed in [1]). The basic method (known as Ligation Mediated PCR, LM PCR) was firstly introduced in 1989 by Muller and Wold [17,18] and several modifications, including the LAM-PCR, have been added since then. In the original LAM-

PCR procedure [19] a 5'-biotynylated primer is extended from the known region of the genomic DNA and then a complementary strand is synthesized on the purified extension product by random exanucleotide priming. The obtained DNA molecule is then digested with a restriction enzyme recognizing a four nucleotide site, and ligated to a DNA cassette for providing the substrate for a final PCR amplification. In the "nonrestrictive" version of the method, the product of the extension reaction is ligated to a specific single-strand linker by means of RNA ligase. The strategy was designed in order to study the critical step of vector DNA integration during gene therapy, as a possible origin of the interruption of important genes and/or activation of proto-oncogenes by vector-introduced promoter and enhancer sequences. Figure 2 shows the combined nrLAM-pyrosequencing strategy, including barcoding of amplicons to facilitate clonal identification of the insertional events.

3.3 GW BY ILLUMINA-SBS

The first association of the SBS-Illumina NGS approach to GW was reported by Gawronski et al. [7] in their study for the identification of *Haemophilus influenzae* virulence genes required to delay bacterial clearance in the lungs of mice. The assay was carried out by sequencing insertion sites of a Himar1-mariner transposon insertional library of H. influenzae after infection in lungs of 5 mice (negative selection). Transposon/bacterial DNA junctions were identified in the resulting libraries by the so-called "high-throughput insertion tracking by deep sequencing" (HITS) method, practically consisting in a massive GW analysis by the Illumina technology (Figure 3). In this strategy the fragmentation of genomic DNA by restriction enzymes (as shown in Figure 1) has been replaced by "shearing". The additional step of repairing DNA ends is however necessary before ligation of adapters. After repairing of the sheared DNA ends from transposon mutant libraries and ligation of Illumina oligonucleotide adapters, transposon-containing fragments are enriched via PCR using a biotinylated transposon primer and affinity purification on streptavidin-coated paramagnetic beads. Purified fragments are then used for sequencing according to Illumina protocols. Using

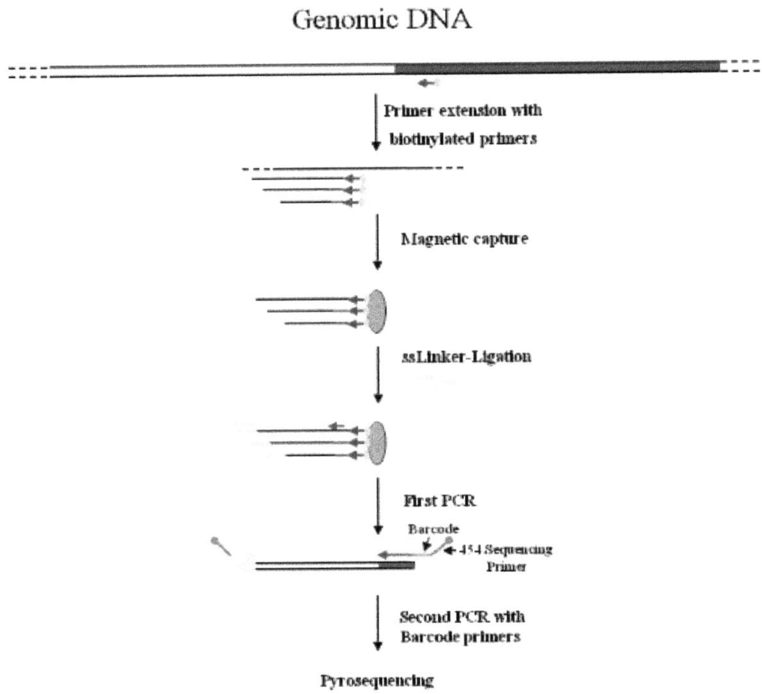

FIGURE 2: Main steps of nrLAM-PCR GW and pyrosequencing association strategy. Black and white bars correspond to known and unknown sequences, respectively. An arrow with a small oval corresponds to biotinylated primers in the known nucleotide region. A larger oval indicates a streptavidin magnetic device. Arrows represent primers designed in correspondence of the linker and known genomic region. Barcodes and 454 sequencing traits of longer amplification primers are indicated in the figure.

this approach, a large library of approximately 75,000 insertional mutants was analyzed, providing a rapid genome-wide analysis of bacterial genes required for growth/survival during infection of host organisms.

In an almost contemporary paper, a similar approach was used to investigate a large Tn5-derived bacterial transposon insertional library produced in *Salmonella enterica* for the identification of interrupted genes [20]. Sequencing was directly carried out on amplicons obtained with one primer corresponding to the transposon, and a second primer corresponding to the Illumina adaptor. Even if not mentioned by the authors, this approach stands as a classical NGS-GW approach. A Tn5-derived transposon was used to generate an estimated pool of 1.1 million transposon mutants and 370,000 unique transposon insertion sites were identified. Authors performed also a negative selection in order to identify *S. enterica* genes involved in the bacterium resistance to high concentrations of bile. 169 genes involved in bile tolerance were identified by this approach.

The same approach was successful for the analysis of an eukaryotic transposon insertional library. Li et al. [21] analyzed an inducible piggyBac (PB) transposon-based mutagenesis library in the yeast *S. pombe*.

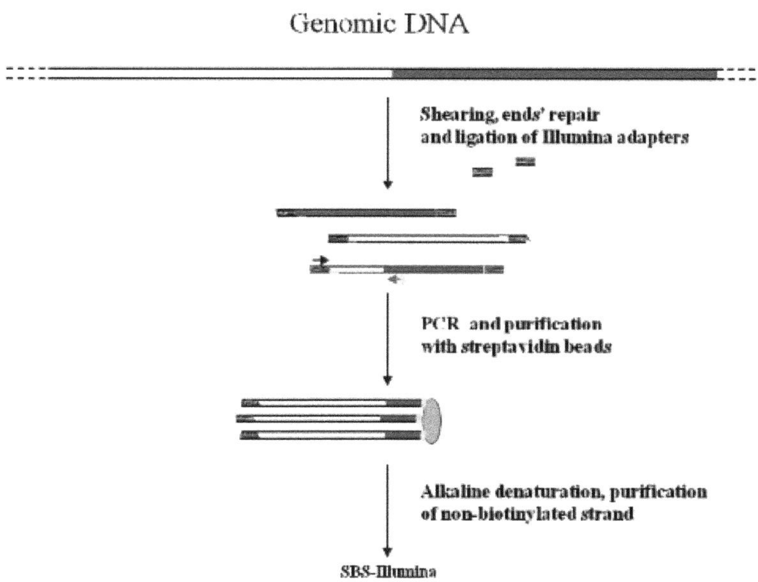

FIGURE 3: Main steps of cassette-PCR GW and sequencing by synthesis (SBS)-Illumina association strategy. Bars correspond to known and unknown sequences, respectively. A arrow with a small oval corresponds to biotinylated primers in the known nucleotide region. A black horizontal arrow indicates adapter specific primers. An oval indicates a streptavidin magnetic device.

From a mutant pool of 400,000 Arg+ Ura+ colonies, PB insertions were detected in 54% of the ORFs containing the typical TTAA PB insertion site.

Illumina-GW proved also effective in the identification of Mu transposon insertions in the genome of maize photosynthesis mutants [22]. Distinct DNA samples were pooled and analyzed thanks to the introduction of barcodes. The problem of unambiguous assignment of transposon flanking regions, which may be encountered in the analysis of complex eukaryotic genomes, was addressed by authors by using 60-mer long biotinylated primer and low-cycle PCR amplifications. The approach allowed the identification of four genes whose interruption blocks chloroplast biogenesis. In a similar approach Urbanski et al. [23] analyzed a pool of Lotus mutant plants containing the retrotransposon LORE1. In this assay, the blunt-ended sheared DNA fragments were first subjected to 3' adenylation by Taq polymerase, and then ligated to splinkerette-adaptors provided of single T-overhangs. Thanks to this strategy 3,744 plants were examined and 8,935 new LORE1 insertion sites were identified.

A different strategy for sequencing transposon/genomic DNA junctions was ideated by van Opijnen et al. [8,24] to identify the insertion sites of the mariner Himar1 mini-transposon into the genome of the Gram-positive bacterium *Streptococcus pneumoniae*. The strategy is highly innovative and must not be confused with a restriction-based/NGS approach. Indeed in this case, differing from all the restriction-based GW methods, the enzyme recognition site is not on the genomic DNA of the mutated organism, but in the transposon used for mutagenesis. The method relies on the presence of a MmeI recognition site at four bp from the left and right ends of the inverted repeats of the transposon. MmeI cuts 20 bp downstream of the recognition site and generates a random two-nucleotide 3'-overhang. After restriction of the library DNA, the ligation of an adapter with a random two-nucleotide overhang allows the amplification of unique-size fragments using transposon and adapter specific primers. Amplicons of 160 bp will contain 16 bp of flanking genomic DNA. The introduction of a DNA barcode was also employed for sequencing different samples in a single flow cell lane using the Illumina approach (Figure 4). The *S. pneumoniae* library was analyzed to categorize genes according to their relevance in bacterial fitness, allowing identification of genes essen-

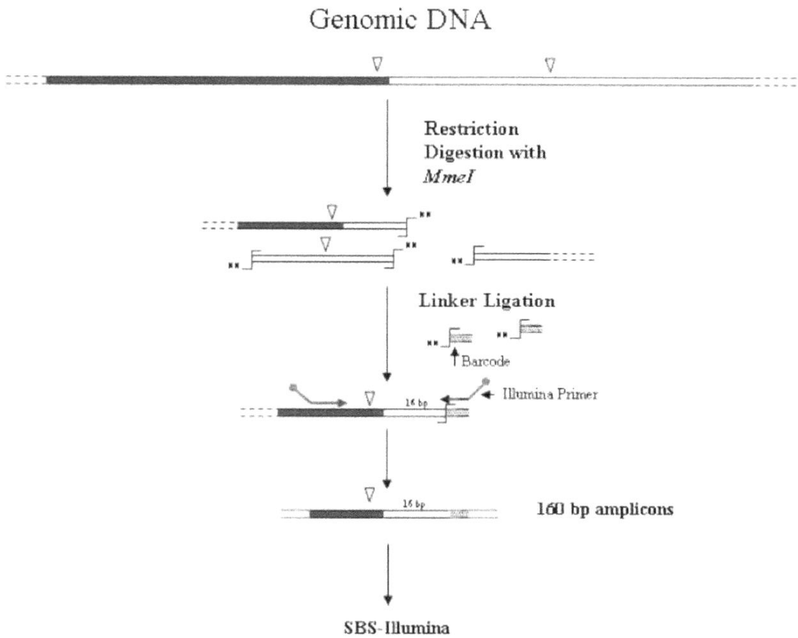

FIGURE 4: Main steps of MmeI-assisted cassette-PCR GW and SBS-Illumina association strategy. Blue and white bars correspond to known and unknown sequences, respectively. An open vertical arrow indicates the MmeI recognition site. The symbol indicates the MmeI digestion site. A black horizontal arrow indicates adapter specific primers; a blue arrow represents primers designed in the known region. Barcodes and Illumina sequencing traits of longer amplification primers are indicated in the figure.

tial for basal growth. A similar approach was also used by Goodman et al. [25] for the identification of fitness genes of the gut symbiont *Bacteroides thetaiotaomicron.*

Brett et al. [26] designed an Illumina-GW procedure for the analysis of Sleeping Beauty (SB) induced tumors in mice. The method is based on the ligation of adapters to genomic DNA fragments obtained by digestion with restriction enzymes (AluI or NlaIII). After a first PCR with transposon and adapter specific primers, a nested PCR is performed with primers provided

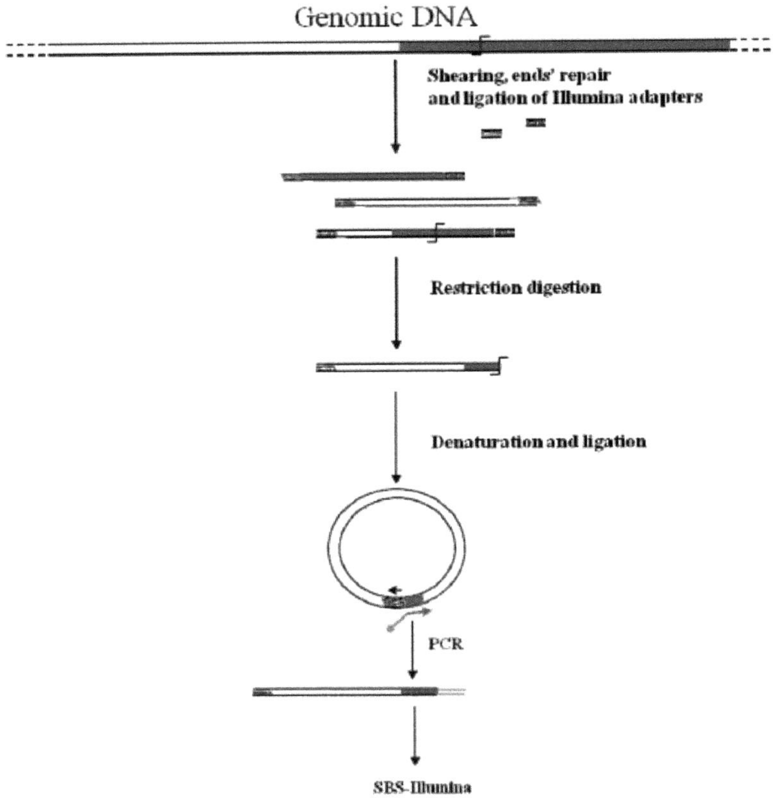

FIGURE 5: Main steps of Inverted-PCR GW and SBS-Illumina association strategy. Blue and white bars correspond to known and unknown sequences, respectively. An open vertical arrow indicates the MmeI recognition site. The symbol indicates a restriction site. A black horizontal arrow indicates adapter specific primers; a blue arrow represents primers designed in the known region. The Illumina sequencing trait of longer amplification primers is indicated in the figure.

of barcodes and tags for sequencing on Illumina platforms. Regardless of the NGS technology, the approach is similar to that described by Wang et al. [13] and illustrated in Figure 1B.

An innovative NGS-GW approach is that reported by Gallagher et al. [27] to track large numbers of transposon mutants of Pseudomonas aeruginosa. In this case authors associated an inverted-PCR GW approach (firstly described by Triglia et al. [28]) to the Illumina NGS technology (Figure 5). Genomic DNA from a mutant pool is sheared, end-repaired and ligated to one Illumina adapter. DNA fragments are then digested by a proper restriction enzyme cutting the transposon near the junction with the unknown DNA. After size-selection, restriction fragments are denaturated and single-strand DNA self-ligated. A second PCR step with divergent primers corresponding to the adaptor and to the transposon allows the identification of transposon/genome junctions. Illumina sequencing can be performed by introducing into the transposon primer the required Illumina adaptor sequence. The procedure was tested by screening a pool of 100,000 mutants for the identification of tobramycin resistance genes. A total of 117 resistance genes were identified, including previously unknown ones.

3.4 OTHER POSSIBLE NGS-GW PAIRING AND GENERAL CONCLUSIONS

The possibility to combine other NGS and GW strategies, in addition to those already reported, is examined in this paragraph. This analysis, besides constituting a preliminary base for future applications, also gives the opportunity to draw some general conclusions about the NGS and GW strategies to combine for optimal results.

Taking into account that all the GW methods have a final PCR amplification step, NGS methods relying on the synthesis of large arrays of amplicons as substrate for sequencing can in principle be combined. Even if currently there are no articles reporting the application of the SOLiD and Ion Torrent NGS technologies to GW methods, the possibility to construct large barcoded libraries has been recently illustrated for both methods, making them eligible for large, multiplex, GW applications. For example, the construction of barcoded libraries for SOLiD sequencer was reported by Farias-Hesson et al. [29], who prepared 32 libraries by ligation of end-repaired sheared DNA molecules to SOLiD adaptors, one of which provided with 6-bp barcode. This procedure should be also adaptable to multiple sequence analysis of the same targeted region (i.e., a transposon) amongst several biological samples (see also the Applied Biosystem Application note "SOLiD System Barcoding" [30]). Additional examples of the use of barcoding to analyze the same targeted region in several biological samples by SOLiD technology can be found in some recent papers [31–34]. The possibility to apply the Ion Torrent sequencing technology to GW can be deduced from the genotyping-by-sequencing approach exemplified for the complex genomes of barley and wheat and available at the Invitrogen website [35]. The lower cost of the Personal Genome Machines by Ion Torrent compared to SOLiD platforms, together with their reduced times for amplicon sequencing, can make them probably preferred for most NGS applications, including GW.

From the analysis of the GW-NGS studies reported in the previous paragraphs, it comes out that pyrosequencing has been exclusively used for eukaryotic genomes, while the SBS-Illumina approach has been adopted for both prokaryotic and eukaryotic genomes. However, a definitive indication about the most appropriate NGS technology to use in GW insertional mutation analysis is not really relevant at this time, since all the NGS approaches reported above have larger sequencing capacity than that needed for identification of insertion sites. This is true even for high-copy transposons in eukaryotic genomes, as the Mu transposon in maize, which has been tracked by both GW-pyrosequencing [14,15] and GW-Illumina approaches [22]. Accordingly, in all the cases the introduction of DNA barcodes allowed the analysis of multiple samples in single sequencing experiments. In the case of SB-induced tumors, however, Brett et al. [26] reported about the higher capacity of insertion sites detection by SBS-Illumina compared to 454-pyrosequencing. Even if some experimental details were different, comparison of the two sequencing strategies in the analysis of ligation-mediated PCR libraries showed an increased sensitivity of the Illumina approach respect to pyrosequencing of about 50 fold, allowing the detection of rare transposon insertion events. In the case of bacteria with high GC-content, Ion Torrent approach has been found more reliable than Illumina, with a more stable quality of sequencing data (reviewed in Liu et al. [36]).

Also for the GW method to combine with NGS technologies, data available indicate that both restriction-based and extension-based approaches are suitable for adaptation to NGS technologies. The first case is exemplified by the cassette-ligation (Figure 1), its "shearing-adapted" version (Figure 3), and by the self-ligation (Figure 5) methods. The latter approach is shown in Figure 2. Association of primer-based GW methods to NGS approaches appears more difficult. Primer-based GW strategies start with a preliminary PCR carried out with a primer containing degenerate sequences (possibly annealing in the unknown region of the genome) and a primer corresponding to the known sequence (i.e., a gene region or an insertion sequence). Subsequently PCR products are further selected by hemi-nested PCR amplifications using a second primer specific for the known region. Even if in this last step nested primer containing DNA barcodes and NGS sequencing regions can be designed, making the association to NGS approaches theoretically possible, currently there are no reports about the combined use of the two methods.

Another critical parameter to be considered in the choice of the GW approach is its sensitivity. Some specific comparative studies have been carried out and discussed in a recent review article [1]. Generally speaking, however, the use of solid-phase purification of biotinylated DNA fragments has highly contributed to increasing the sensitivity of GW methods. For example in studies on the integration of foreign DNA into salmon genome, it has been found that classical anchored PCR amplifications, combined with solid-phase purification, allow identification of about two copies of the target sequence in 25 ng of DNA [37]. Several of the above reported GW-NGS approaches include streptavidin purification of biotinylated DNA molecules [9,10,12,22].

Finally, the need of proper bioinformatic resources for the analysis of GW-NGS data cannot be disregarded. A typical analysis pipeline includes: sorting of sequences according to the incorporated barcode for the identification of sequences belonging to individual amplicons; removal of foreign genome sequences (linkers, transposons, vectors); clustering and counting of identical sequences and mapping of insertion sites. Good examples of bioinformatic analysis with description of the employed tools and filtering parameters are given in several articles on GW-NGS applications [7,9,12,14,15,20–23,26,27].

In conclusion, GW-NGS approaches have the potentiality to make insertional mutagenesis a high throughput screening approach of wide use in functional genomics, replacing more classical methods such as microarray hybridization. In addition, the combined methods showed interesting applications in medicine for the study of the integration mechanisms of retrovirus and DNA vector for gene therapy.

REFERENCES

1. Leoni, C.; Volpicella, M.; de Leo, F.; Gallerani, R.; Ceci, L.R. Genome walking in eukaryotes. FEBS J. 2011, 278, 3953–3977.
2. Tonooka, Y.; Fujishima, M. Comparison and critical evaluation of pcr-mediated methods to walk along the sequence of genomic DNA. Appl. Microbiol. Biotechnol. 2009, 85, 37–43.
3. Kotik, M. Novel genes retrieved from environmental DNA by polymerase chain reaction: Current genome-walking techniques for future metagenome applications. J. Biotechnol. 2009, 144, 75–82.
4. Shendure, J.; Ji, H. Next-generation DNA sequencing. Nat. Biotechnol. 2008, 26, 1135–1145.
5. Metzker, M.L. Sequencing technologies—The next generation. Nat. Rev. Genet. 2010, 11, 31–46.
6. Wang, G.P.; Ciuffi, A.; Leipzig, J.; Berry, C.C.; Bushman, F.D. HIV integration site selection: Analysis by massively parallel pyrosequencing reveals association with epigenetic modifications. Genome Res. 2007, 17, 1186–1194.
7. Gawronski, J.D.; Wong, S.M.; Giannoukos, G.; Ward, D.V.; Akerley, B.J. Tracking insertion mutants within libraries by deep sequencing and a genome-wide screen for haemophilus genes required in the lung. Proc. Natl. Acad. Sci. USA 2009, 106, 16422–16427.
8. Van Opijnen, T.; Camilli, A. Genome-wide fitness and genetic interactions determined by tn-seq, a high-throughput massively parallel sequencing method for microorganisms. Curr. Protoc. Microbiol. 2010, doi:10.1002/9780471729259.mc01e03s19.
9. Paruzynski, A.; Arens, A.; Gabriel, R.; Bartholomae, C.C.; Scholz, S.; Wang, W.; Wolf, S.; Glimm, H.; Schmidt, M.; von Kalle, C. Genome-wide high-throughput integrome analyses by nrlam-pcr and next-generation sequencing. Nat. Protoc. 2010, 5, 1379–1395.

10. Bartholomae, C.C.; Glimm, H.; von Kalle, C.; Schmidt, M. Insertion site pattern: Global approach by linear amplification-mediated PCR and mass sequencing. Methods Mol. Biol. 2012, 859, 255–265.
11. Rothberg, J.M.; Hinz, W.; Rearick, T.M.; Schultz, J.; Mileski, W.; Davey, M.; Leamon, J.H.; Johnson, K.; Milgrew, M.J.; Edwards, M.; et al. An integrated semiconductor device enabling non-optical genome sequencing. Nature 2011, 475, 348–352.
12. Vandenbussche, M.; Janssen, A.; Zethof, J.; van Orsouw, N.; Peters, J.; van Eijk, M.J.; Rijpkema, A.S.; Schneiders, H.; Santhanam, P.; de Been, M.; et al. Generation of a 3D indexed petunia insertion database for reverse genetics. Plant J. 2008, 54, 1105–1114.
13. Wang, G.P.; Garrigue, A.; Ciuffi, A.; Ronen, K.; Leipzig, J.; Berry, C.; Lagresle-Peyrou, C.; Benjelloun, F.; Hacein-Bey-Abina, S.; Fischer, A.; et al. DNA bar coding and pyrosequencing to analyze adverse events in therapeutic gene transfer. Nucleic Acids Res. 2008, 36, e49.
14. Liu, S.; Dietrich, C.R.; Schnable, P.S. Dla-based strategies for cloning insertion mutants: Cloning the gl4 locus of maize using mu transposon tagged alleles. Genetics 2009, 183, 1215–1225.
15. Liu, S.; Yeh, C.T.; Ji, T.; Ying, K.; Wu, H.; Tang, H.M.; Fu, Y.; Nettleton, D.; Schnable, P.S. Mu transposon insertion sites and meiotic recombination events co-localize with epigenetic marks for open chromatin across the maize genome. PLoS Genet. 2009, 5, e1000733.
16. Guo, Y.; Levin, H.L. High-throughput sequencing of retrotransposon integration provides a saturated profile of target activity in schizosaccharomyces pombe. Genome Res. 2010, 20, 239–248.
17. Mueller, P.R.; Wold, B. In vivo footprinting of a muscle specific enhancer by ligation mediated pcr. Science 1989, 246, 780–786.
18. Pfeifer, G.P.; Steigerwald, S.D.; Mueller, P.R.; Wold, B.; Riggs, A.D. Genomic sequencing and methylation analysis by ligation mediated pcr. Science 1989, 246, 810–813.
19. Schmidt, M.; Zickler, P.; Hoffmann, G.; Haas, S.; Wissler, M.; Muessig, A.; Tisdale, J.F.; Kuramoto, K.; Andrews, R.G.; Wu, T.; et al. Polyclonal long-term repopulating stem cell clones in a primate model. Blood 2002, 100, 2737–2743.
20. Langridge, G.C.; Phan, M.D.; Turner, D.J.; Perkins, T.T.; Parts, L.; Haase, J.; Charles, I.; Maskell, D.J.; Peters, S.E.; Dougan, G.; et al. Simultaneous assay of every salmonella typhi gene using one million transposon mutants. Genome Res. 2009, 19, 2308–2316.
21. Li, J.; Zhang, J.M.; Li, X.; Suo, F.; Zhang, M.J.; Hou, W.; Han, J.; Du, L.L. A piggybac transposon-based mutagenesis system for the fission yeast schizosaccharomyces pombe. Nucleic Acids Res. 2011, 39, e40.
22. Williams-Carrier, R.; Stiffler, N.; Belcher, S.; Kroeger, T.; Stern, D.B.; Monde, R.A.; Coalter, R.; Barkan, A. Use of illumina sequencing to identify transposon insertions underlying mutant phenotypes in high-copy mutator lines of maize. Plant J. 2010, 63, 167–177.

23. Urbanski, D.F.; Malolepszy, A.; Stougaard, J.; Andersen, S.U. Genome-wide lore1 retrotransposon mutagenesis and high-throughput insertion detection in lotus japonicus. Plant J. 2012, 69, 731–741.
24. Van Opijnen, T.; Bodi, K.L.; Camilli, A. Tn-seq: High-throughput parallel sequencing for fitness and genetic interaction studies in microorganisms. Nat. Methods 2009, 6, 767–772.
25. Goodman, A.L.; McNulty, N.P.; Zhao, Y.; Leip, D.; Mitra, R.D.; Lozupone, C.A.; Knight, R.; Gordon, J.I. Identifying genetic determinants needed to establish a human gut symbiont in its habitat. Cell Host Microbe 2009, 6, 279–289.
26. Brett, B.T.; Berquam-Vrieze, K.E.; Nannapaneni, K.; Huang, J.; Scheetz, T.E.; Dupuy, A.J. Novel molecular and computational methods improve the accuracy of insertion site analysis in sleeping beauty-induced tumors. PLoS One 2011, 6, e24668.
27. Gallagher, L.A.; Shendure, J.; Manoil, C. Genome-scale identification of resistance functions in pseudomonas aeruginosa using tn-seq. MBio 2011, 2, e00315-10.
28. Triglia, T.; Peterson, M.G.; Kemp, D.J. A procedure for in vitro amplification of DNA segments that lie outside the boundaries of known sequences. Nucleic Acids Res. 1988, 16, 8186.
29. Farias-Hesson, E.; Erikson, J.; Atkins, A.; Shen, P.; Davis, R.W.; Scharfe, C.; Pourmand, N. Semi-automated library preparation for high-throughput DNA sequencing platforms. J. Biomed. Biotechnol. 2010, 2010, 617469.
30. SOLiDTM System Barcoding. Available online: http://www3.appliedbiosystems.com/cms/groups/mcb_marketing/documents/generaldocuments/cms_057554.pdf (accessed on 1 August 2012).
31. Harakalova, M.; Mokry, M.; Hrdlickova, B.; Renkens, I.; Duran, K.; van Roekel, H.; Lansu, N.; van Roosmalen, M.; de Bruijn, E.; Nijman, I.J.; et al. Multiplexed array-based and in-solution genomic enrichment for flexible and cost-effective targeted next-generation sequencing. Nat. Protoc. 2011, 6, 1870–1886.
32. Lotta, L.A.; Wang, M.; Yu, J.; Martinelli, I.; Yu, F.; Passamonti, S.M.; Consonni, D.; Pappalardo, E.; Menegatti, M.; Scherer, S.E.; et al. Identification of genetic risk variants for deep vein thrombosis by multiplexed next-generation sequencing of 186 hemostatic/pro-inflammatory genes. BMC Med. Genomics 2012, 5, 7.
33. Harakalova, M.; Nijman, I.J.; Medic, J.; Mokry, M.; Renkens, I.; Blankensteijn, J.D.; Kloosterman, W.; Baas, A.F.; Cuppen, E. Genomic DNA pooling strategy for next-generation sequencing-based rare variant discovery in abdominal aortic aneurysm regions of interest-challenges and limitations. J. Cardiovasc. Transl. Res. 2011, 4, 271–280.
34. Tu, J.; Ge, Q.; Wang, S.; Wang, L.; Sun, B.; Yang, Q.; Bai, Y.; Lu, Z. Pair-barcode high-throughput sequencing for large-scale multiplexed sample analysis. BMC Genomics 2012, 13, 43.
35. Genotyping by Sequencing (GBS) on Barley Using Ion PGM™ Sequencer: A Feasibility Study. Available online: http://www.invitrogen.com/etc/medialib/images/agricultural-biotechnology/pdf. Par.0661.File.tmp/Genotyping-by-sequencing-on-barley-Ion-PGM.pdf. (accessed on 1 August 2012).

36. Liu, L.; Li, Y.; Li, S.; Hu, N.; He, Y.; Pong, R.; Lin, D.; Lu, L.; Law, M. Comparison of next-generation sequencing systems. J. Biomed. Biotechnol. 2012, 2012, 251364.
37. Nielsen, C.R.; Berdal, K.G.; Holst-Jensen, A. Anchored pcr for possible detection and characterisation of foreign integrated DNA at near single molecule level. Eur. Food Res. Technol. 2008, 226, 949–956..

CHAPTER 4

MOLECULAR TOOLS FOR EXPLORING POLYPLOID GENOMES IN PLANTS

RICCARDO AVERSANO, MARIA RAFFAELLA ERCOLANO, IMMACOLATA CARUSO, CARLO FASANO, DANIELE ROSELLINI, and DOMENICO CARPUTO

4.1 POLYPLOIDY IN THE PLANT KINGDOM: OCCURRENCE AND SIGNIFICANCE

The genome sizes of eukaryotes can differ 10,000-fold and part of these differences may be attributed to changes in the ploidy level. Polyploids are organisms having more than two complete sets of chromosomes in their cells. They are common in angiosperms, where at least 70% of the species experienced one or more events of genome doubling during their evolutionary history [1,2]. Many crop species are polyploids (Table 1), and it was stated that "life on earth is predominantly a polyploid phenomenon and civilization depends mainly on use of polyploid tissues—noteworthy is the endosperm of cereals" [3]. Polyploidization is considered a major evolutionary force in plants. It is a definitive cause of sympatric speciation due to the immediate reproductive isolation between newly formed polyploids and their parents [4]. Polyploidization also makes it possible to

This chapter was originally published under the Creative Commons Attribution License. Aversano R, Ercolano MR, Caruso I, Fasano C, Rosellini D, and Carputo D. Molecular Tools for Exploring Polyploid Genomes in Plants. International Journal of Molecular Sciences ***13** (2012), 10316–10335. doi:10.3390/ijms130810316.*

overcome hybrid sterility and produce viable offspring following interspecific hybridization. However, there are still several open questions related to polyploidy and polyploidization. For example, Soltis et al. [5] reported that polyploidy frequency in angiosperms is high, even if the number of lineages that really derived from genome-wide duplication (WGD) events is still largely unknown. Similarly, it is not clear if polyploidy causes a change in the interaction with herbivores and pollinators.

TABLE 1: Examples of polyploid crops. The somatic chromosome number is reported in brackets.

Crop	Species
Cereals	*Triticum aestivum* (6× = 42); *T. durum* (4× = 28); *Avena sativa* (6× = 42); *A. nuda* (6× = 42)
Forage grasses	*Dactylis glomerata* (4× = 28); *Festuca arundinacea* (4× = 28); *Agropyrum repens* (4× = 28); *Paspalum dilatatum* (4× = 40)
Legumes	*Medicago sativa* (4× = 32); *Lupinus alba* (4× = 40); *Trifolium repens* (4× = 32); *Arachis hypogaea* (4× = 40); *Lotus corniculatus* (4× = 32); *Glycine max* (4× = 40)
Industrial plants	*Nicotiana tabacum* (4× = 48); *Coffea spp.* (4× = 44 fino a 8×); *Brassica napus* (4× = 38); *Saccharum officinalis* (8× = 80); *Gossypium hirsutum* (4× = 52)
Tuber plants	*Solanum tuberosum* (4× = 48); *Ipomoea batatas* (6× = 96); *Dioscorea sativa* (6× = 60)
Fruit trees	*Prunus domestica* (6× = 48); *Musa spp.* (3× = 33; 4× = 44); *Citrus aurantifolia* (3× = 27); *Actinidia deliciosa* (4× = 116); *P. cerasus* (4× = 32)

Based on their origin, polyploids are classified into two major groups, autopolyploids and allopolyploids [6]. The former are the result of doubling homologous genomes (e.g., autotetraploid AAAA) from a single, or closely related, species. The latter are the result of hybridization between different species. Therefore, they combine two or more different genomes (e.g., allotetraploid AABB). As a consequence, autopolyploids may form multivalents at meiosis, and have polysomic inheritance. By contrast allopolyploids show bivalent pairing and have disomic inheritance [7]. There are several mechanisms that may lead to an increase in ploidy level in plants. However, there is strong circumstantial evidence that sexual poly-

FIGURE 1: Phenotypic variation between diploids and tetraploids in *Solanum commersonii* (a, b) and in *Medicago sativa* (c, d). A diploid (2n = 2× = 24) clone of *S. commersonii* was subjected to oryzaline treatment, an antimitotic drug commonly employed to induce chromosome doubling in plants. The autotetraploid (2n = 4× = 48) genotype displayed larger size at both whole plant (a) and leaf (b) level. (c) Diploid M. sativa subsp. coerulea (2n = 2× = 16, left) and its cultivated tetraploid counterpart, *M. sativa subsp. sativa* (2n = 4× = 32, right) differ clearly for flower size. (d) Leaves of diploid (upper row) and tetraploid (bottom row) plants obtained from crossing two diploid *M. sativa* plants producing both n and 2n gametes. Leaves are the best qualitative component of forage: tetraploid *M. sativa* has larger leaves and is cultivated.

ploidization through gametes with unreduced chromosome number (2n gametes) represents the main route for polyploidization [8,9]. 2n gametes generally result from the expression of mutations affecting micro- and megasporogenesis. Such mutations have been extensively studied in a number of genera, including *Solanum, Medicago, Manihot, Malus, Arachis, Lolium,* and *Agropyrum*, and have been generally attributed to the action of single recessive genes [10]. The first gene (AtPS1) involved in 2n gamete production has been identified in *A. thaliana* [11]. AtPS1 mutants display an anomalous (parallel, fused, tripolar) orientation of spindles at metaphase II of male meiosis, leading to the production of 2n pollen. Similarly, 2n gametes have been observed in the jason (jas) [12], switch1(swi1)/dyad [13,14], osd1 and tam (CYCAS1;2) mutants [15,16].

In a given species and habitat, the acquisition of the polyploid condition may bring about several advantages. Following polyploidization, often novel phenotypes appear or variation exceeds the range observed in the diploid parental species. Fawcett et al. [17] suggested that polyploids had a better chance to survive the Cretaceous-Tertiary extinction event. Phenotypic advantages may include, among the others, changes in morphology, physiology and secondary metabolism that confer an increased fitness. Some of these traits, such as increased drought tolerance, pathogen resistance, longer flowering time, larger vegetative and reproductive organs (Figure 1) may represent important plant breeding targets and, therefore, increase the potential use of polyploids in agriculture. From a genetic point of view, the most significant advantages associated with polyploidy are probably heterosis and gene redundancy [18]. Heterosis is due to non-additive inheritance of traits in a newly formed polyploid compared to its parents. Notably, it can be present also at the gametophytic level. The main factors that affect non-additive inheritance are likely novel regulatory interactions and allelic dosage [19]. Gene redundancy promotes neofunctionalization of duplicated genes, in the long term, but also immediately protects against deleterious recessive alleles. In a recent treatment, Mayrose et al. [20] shed more light on the evolutionary dynamics and consequences of polyploidy. The authors, computing the net diversification rates of polyploid lineages for ferns, lycophytes, gymnosperms and angiosperms, hypothesized that polyploidy is often an evolutionary dead end. However, the possible longer-term evolutionary success of those polyploids that survive, needs to be tested. Several studies provided evidence that extensive and reproducible genetic and epigenetic changes are possible following polyploidization [9]. They include DNA and histone methylation, DNA elimination, gene losses, gene neo- and subfunctionalization, translocations, amplification and reduction of repetitive sequences and alteration of gene expressions through different mechanisms (methylation-mediated silencing, transposon activation, intergenomic interactions, etc.). Given the recent advances in the field of plant molecular biology and biotechnology, through this review we provide an overview of the most suitable technologies and strategies that can allow, at the molecular level, efficient studies on polyploid plants, possibly promoting research in areas that have been ignored or underestimated so far. Reference to some recent significant findings will also be made.

4.2 METHODS FOR GENOME ANALYSIS

A combination of genetic mapping, molecular cytogenetics, sequence and comparative analysis has shed new light and opened perspectives on the nature of ploidy evolution at all timescales, from the base of the plant kingdom, to intra- and interspecific hybridization events associated with plant domestication and breeding. Strong evidences on the mechanisms of genomic modification have come from the use of physical analysis of chromosomes by in situ hybridization techniques and from genome-wide molecular marker analyses.

4.2.1 IN SITU HYBRIDIZATION

In situ hybridization represents the bridge between the chromosomal and molecular level of genome investigation. In recent years it has received a renewed interest for detecting chromosome rearrangements. It is very powerful for reliable identification of chromosomes, allowing the positioning of unique sequences and repetitive DNAs along the chromosome(s). Fluorescent in situ hybridization (FISH) is based on fluorescent labels linked to DNA probes and visualized under a fluorescence microscope. Genomic in situ hybridization (GISH) involves the use of total genomic DNA of species as a probe on chromosomes, thus leading to whole genome discrimination rather than the localization of specific sequences. There are several examples on the use of these techniques. Studies on the distribution of four tandem repeats in allotetraploids *Tragopogon mirus, Tragopogon miscellus* and their diploid parents provided evidence that chromosomal rearrangements did not occur following polyploidization, as suggested by the additive patterns of polyploids [21]. By contrast, in newly synthesized allotetraploid genotypes of *B. napus,* Szadkowski et al. [22] demonstrated extensive genome remodeling due to homeologous pairing between the chromosomes of the A and C genomes. Based on high-resolution cytogenetic maps, Wang et al. [23] demonstrated that genome size difference between the A and D sub-genomes in allotetraploid cotton was mainly associated with uneven expansion or contraction between different regions of homoeologous chromosomes. Recently, Chester and co-workers [24] combined GISH and FISH analysis to demonstrate that in natural populations of *T. miscellus* extensive chromosomal variation (mainly due to chromosome substitutions and homeologous rearrangements) was present up to the 40th generation following polyploidization.

4.2.2 MOLECULAR MARKER-BASED GENETIC MAPPING

Genetic mapping in polyploids presents unique problems with respect to diploid species, since segregations and statistical methods are much more complicated, and large segregating populations are needed to obtain reliable genetic distance estimates [25]. Important results were obtained using classical Restriction Fragment Length Polymorphism markers (RFLP) in *B. napus* [22,26–28], *Draba norvegica* and *T. miscellus* [29], and wheat [30] or employing Amplified Fragment Length Polymorphism (AFLP) in *Arabidopsis* [31], *Brassica* [32], and *Spartina* [33].

Autopolyploids are particularly intractable because segregation depends on chromosome pairing behavior (preferential vs. random pairing) and double reduction [34]. A simplification that has been generally adopted for mapping dominant molecular markers is to only utilize the so-called single dose markers from each parent, i.e., those segregating 1:1 in the mapping population (for example, a population obtained from the cross Mmmm × mmmm in a tetraploid species) [35,36]. Statistical models for QTL mapping have been developed for polyploids with bivalent pairing, taking into account preferential vs. random chromosome pairing behavior by incorporating a preferential pairing factor [37,38]. Molecular marker segregation allows estimating chromosome pairing behavior. For instance, significant preferential pairing was evidenced in a population obtained by crossing *M. sativa* subsp. falcata × *M. sativa* subsp. sativa autotetraploid genotypes [39], whereas the data of Julier et al. [40] did not support preferential pairing in a mapping study involving cultivated *M. sativa* genotypes. In Salix, an allotetraploid origin was inferred based on molecular marker segregation supporting preferential pairing [41]. The recourse to diploid species in search of synteny can facilitate genetic mapping of agronomic traits in complex polyploids [42].

Recently, improvements have been proposed in polysomic inheritance analysis through the determination of the number of allele copies in microsatellite loci [43,44]. In particular, a method named Microsatellite Allele Dose and Configuration Establishment (MADCE) was proposed to establish the exact dose and configuration of microsatellite alleles in allo-octoploid cultivated strawberry (Fragaria × ananassa) [44]. This technique enhances the utility of SSR markers in polyploids by overcoming the limitation of single dose markers. A recently introduced type of molecular marker, Diversity Arrays Technology (DArT), was applied to mapping in a tetraploid recombinant inbred line population obtained by crossing wild *Avena magna* ($2n = 4\times = 28$) with cultivated *A. sativa* ($2n = 6\times = 42$) [45].

4.2.3 METHYLATION SENSITIVE MOLECULAR MARKERS

The use of an AFLP-like method using restriction enzymes sharing the same recognition site but having differential sensitivity to DNA methylation (isoschizomeres) was proved to be efficient and reliable for the determination of genome-wide DNA methylation patterns [46]. This technique, termed Methylation-Sensitive Amplified Polymorphism (MSAP), is based on the use of the isoschizomers HpaII and MspI both recognizing the 5′CCGG sequence, but affected by the methylation state of the outer or inner cytosine residues. Using this method, noteworthy results were obtained in newly synthesized polyploids. In *Arabidopsis*, Madlung et al. [47] demonstrated that frequent changes occurred in F4 allotetraploids when compared with the parents. Changes involved both increases and decreases in methylation, but no overall hyper- or hypomethylation. Similarly, alterations in cytosine methylation in wheat occurred in about 13% of the loci, either in F1 hybrids or in allopolyploids. Notably, alterations in methylation patterns affected both repetitive DNA sequences and low-copy DNA in approximately equal proportions [25]. On the other hand, lack of rapid DNA methylation changes at symmetric CCGG sites was hypothesized in allopolyploid cotton [48]. A similar behavior was observed also in *Brassica* [49] and sugarcane [50].

4.2.4 COMPARATIVE GENOME ANALYSIS

Comparative genomics research has gained importance as a powerful tool for addressing both fundamental and applied questions in genome evolution [51–53]. The implementation of these methodologies, however, requires consideration of the variable rates at which different aspects of genome evolution occur [52]. Comparing different wheat species, genomic rearrangements originating by illegitimate DNA recombination were identified as a major evolutionary mechanism [54,55]. Innes et al. [56], comparing homologous regions in several related legume species, demonstrated that retroelements were the largest contributor to duplicated regions. Comparative analysis of *Brassica oleracea* triplicated segments showed that 35% of the genes were lost. Retained genes were dosage-sensitive and not randomly located. Duplicates of transcription factors and members of signal transduction pathways were significantly over-retained following WGD, whereas these same functional gene categories exhibited lower retention rates following smaller scale duplications [57]. For instance, in four independent polyploid wheat lineages, recurrent deletions of Puroindoline (Pin) gene at the grain Hardness (Ha) locus were identified [58].

Phylogenetic and taxonomic studies have been conducted in order to pinpoint the exact placement of the ancient polyploidy events within lineages and to determine when novel genes resulting from polyploidy have enabled adaptive processes. Recent genomic investigations not only indicated that polyploidy is ubiquitous among angiosperms, but also suggested several ancient WGD events [59–62], even in basal angiosperm lineages. Phylogenetic reconstruction with completely sequenced genomes suggested that genome doubling led to a dramatic increase in species richness in several angiosperms, including *Poaceae, Solanaceae* and *Brassicaceae*, thus contributing to the dominance of seed plants and angiosperms [63].

To date, only a few reports investigated the fate of the genome after polyploid formation [64]. The probability of fixation and maintenance of duplicated genes depends on many variables. Transposable elements may play a key role in fuelling genome reorganization and functional changes following allopolyploidization. A pivotal example of using comparative approaches to investigate the role of retroelements in polyploids is provided by Wawrzynski et al. [65]. The authors, investigating the nonautonomous retrotransposon replication in soybean estimated a much greater impact of such transposable elements on genome size than previously appreciated. More recently, in wheat Kraitshtein et al. [66] reported a retrotransposition bursts in subsequent generations. By contrast, no evidence for a transposition burst was found in different allopolyploid species [31,67,68]. Comparative approaches in which genetic events are considered both in a phylogenetic and genetics framework should be conceptualized and modeled.

4.2.5 HIGH-THROUGHPUT DNA SEQUENCING AND HIGH RESOLUTION MELTING (HRM) ANALYSIS

High-throughput DNA sequencing associated with computational analysis provides general solutions for the genetic analysis of polyploids [69]. However, ploidy is a substantial challenge in sequencing and assembly of plant genomes. A number of biological factors influence the feasibility of discrimination, including the degree of gene family complexity, and the reproductive system. Of course, the level of knowledge concerning the progenitor diploid species is also very important. To date, all attempts to sequence polyploids have relied on either a reduction in ploidy or a physical separation of chromosomes. Attempts to sequence a heterozygous diploid potato genome (RH89-039-16) were challenging due to the high degree of heterozygosity [70]. In order to bypass the difficulties of sequencing the polyploid genome of cultivated strawberry (*Fragaria x ananassa*), woodland strawberry (*Fragaria vesca*) was sequenced [71]. In *B. napus*, the polyploidy issue was addressed by sequencing leaf transcriptome across a mapping population and representative ancestors of the parents of the population [72]. The Wheat Genome Initiative (http://www.wheatgenome.org/) has focused on flow cytometry separation of individual or groups of homeologous chromosomes [73]. To better understand the nature and extent of variation in functionally relevant regions of a polyploid genome, a sequence capture assay to compare exonic sequences of allotetraploid wheat accessions was developed [74]. In cultivated wheat gene duplications were predominant, while in wild wheat mainly deletions were identified. Exon capture proved to be a powerful approach for variant discovery in polyploids. This technique has the potential to identify variation that can play a critical role in the origin of new adaptations and important agronomic traits.

A wealth of SNP detection approaches has been applied to study polyploidy in plants. Akhunov et al. [75], using the Illumina GoldenGate assay, identified a high number of SNPs in tetraploid and hexaploid wheat. More recently, Allen et al. [76] from Illumina GAIIx data identified more than 14,000 putative SNPs in 6225 distinct hexaploid bread wheat reference sequences. In elite inbred maize lines, more than 1 million SNPs have been identified an Illumina sequencing platform [77]. In the heterozygous polyploid sugarcane, a targeted SNP discovery approach based on 454 sequencing technology was developed by Bundock et al. [78]. Using a 454 and Illumina expressed sequence tag sequencing of the parental diploid species of the allotetraploid *Tragopogon miscellus*, Buggs et al. [79] identified more than 7,700 SNPs differing between the two progenitor genomes the allotetraploid derived from. The Sequenom MassARRAY iPlex platform [80] was used by to validate 92 SNP markers at the genomic level that were diagnostic for the two parental genomes. SNP discovery was also pursed through 454 technology coupled to High Resolution Melting (HRM) curve analysis in tetraploid alfalfa (*Medicago sativa*) [81]. HRM is a technique that can identify mismatches, even for single bases, in amplicons containing heteroduplex molecules [82], and is emerging as a powerful tool for polyploid genetics [83]. It was demonstrated that the 454 system is a cost-effective approach for SNP discovery targeted to genes of interest in polyploid genomes, and that HRM can identify different alleles in polyploids [66,68]. Salmon et al. [84] detected homoeologous SNPs in *G. arboreum* (A genome), *G. raimondii* (D genome), and *G. hirsutum* (AD genomes). The authors estimated that the proportion of genome in *G. hirsutum* that has experienced non reciprocal homoeologous exchanges since the origin of polyploid cotton 1–2 Mya was between 1.8% and 1.9%. SNPs have also been discovered in transcriptome sequences of polyploidy *B. napus* [85]. Next-generation sequencing has been used to mine SNPs in elite wheat germplasm [76].

4.3 METHODS FOR GENE EXPRESSION AND REGULATION ANALYSIS

Several methods have been developed for quantifying gene transcription and regulation patterns in polyploids. Although studying gene expression changes in allopolyploids is more complicated than in autopolyploids, most studies on ploidy-related gene expression changes were carried out on synthetic allopolyploids. Indeed, genome merger and doubling can determine widespread transcriptome modifications, generating cascades of novel expression patterns, regulatory interactions, and new phenotypic variations that subsequent natural selection may act upon.

4.3.1 NORTHERN HYBRIDIZATION AND CDNA-AFLP

When sequence information was still scanty, comprehensive transcript-profiling for quantitatively measuring gene expression variation was carried out by Northern blot analysis. This technique involves the use of electrophoresis to separate RNA samples and detection with a labeled probe complementary to a specific RNA target sequence. There are only a few examples on the use of this type of analysis in polyploids. It was used by Guo and co-workers [86] to investigate the dosage effects of 18 genes in an autopolyploid maize series (1×, 2×, 3×, and 4×). Expression levels of genes were dependent on chromosomes dosage, although some varied their expression in response to the "odd" or "even" ploidy. By contrast, several examples are available on the use of cDNA-AFLP. It is a PCR-based technique, which relies on digestion of cDNA by two restriction enzymes and ligation of specific adapters. A set of specific primers designed for these adapters allow simultaneous amplification of fragments under stringent conditions. In synthetic allotetraploids between *A. thaliana* and

A. arenosa, Comai et al. [87] found 20 suppressed genes out of 700 examined. Similarly, Lee and Chen [88], by extending the analysis also to *A. suecica* (a natural allopolyploid likely formed through pollination of *A. arenosa* with 2n gametes from *A. thaliana*) were able to identify a set of 10 different genes differentially expressed in *A. suecica* and its progenitors. In synthetic *Triticum aestivum* allohexaploids [89], about 8% of transcripts displayed altered expression, and >95% of them were reduced or absent. Similar gene expression changes have been found in cotton allopolyploids [90]. Another example on the use of this technique is offered by the work carried out by Tate et al. [91]. In newly synthesized *Tragopogon* allotetraploids, preferential expression of parental homeologues displayed a correlation with a loss of parental genomic fragments. Notably, such changes were not observed in newly developed *Tragopogon* F1 hybrids, implying that they arose following genome duplication.

4.3.2 SINGLE-STRAND CONFORMATIONAL POLYMORPHISM (SSCP) ANALYSIS

This technique detects sequence variations (single-point mutations and other small-scale DNA changes) through electrophoretic mobility differences. DNA that contains a sequence mutation (even a single base pair change) displays a different measurable mobility compared to reference DNA when electrophoresed in non-denaturing, or partially denaturing conditions. Due to these features, SSCP was employed to distinguish between homoeologous cDNA molecules. This approach has been applied to *A. suecica* [88], cotton [90] and wheat [92] leading to the finding that, basically, genes duplicated by polyploidy are rarely expressed at similar levels, and that there is a biased expression or silencing of some homeologous gene pairs.

4.3.3 MICROARRAYS

Technologies to monitor gene expression achieved a breakthrough through the introduction of microarrays [93]. Genome and transcriptome sequencing have speed up probe development, which consequently resulted in the commercial availability of whole-genome microarrays for many model and crop species. This also gave the possibility to design custom arrays at affordable costs. Approaches of comparative expression profiling have mainly focused on synthetic allotetraploids revealing both additive and non-additive gene expression. The former occurs when gene expression level in the tetraploid is either the sum of the parental values or equal to the mid-parent value (MPV). For instance, tanscriptional profiling of re-synthesized *A. suecica* lines from newly created autotetraploid *A. thaliana* and the natural tetraploid *A. arenosa* revealed that, albeit most of genes were additively expressed (from 65% to 95%), more than 1,400 genes diverged from the MPV. The combination of diverged parental genomes in a common nucleus during allopolyploidization implies the reunion of previously diverged regulatory hierarchies, which likely entails non-additive gene expression. This hypothesis has been validated by genome-wide expression analyses also in synthetic polyploids of wheat, cotton, *Senecio*, *Brassica*, and *Spartina* [94–99]. These studies demonstrated that allopolyploid plants exhibit considerable transcriptome alterations as compared with their diploid progenitors. Transcriptome analyses of autopolyploids suggested that there are less dramatic alterations of gene expression compared to allopolyploids [100]. Expression profiling analyses of autotetraploid *A. thaliana* of two different accessions revealed that transcriptome alterations caused by autopolyploidy depend on genome or genetic composition [101]. Microarray analysis provided evidence that ~10% of the ~9,000 potato genes tested displayed expression changes (within the twofold level) among a potato autopolyploid series (1×, 2×, and 4×) [102]. A similar twofold level change was detected in a corn ploidy series (1×–4×) [103].

DNA microarray technology has been also used to profile expression of noncoding RNA molecules naturally occurring in the plant genomes, such as micro RNA (miRNA). They are a class of 20–24 nucleotide small RNAs that repress their target genes by mRNA degradation or translational repression. Therefore, identification and quantification of miRNAs is deemed essential to understanding an organism's or tissue's gene regulatory network [104]. MicroRNA expression profiling was performed with custom designed chips in both natural *A. suecica* and resynthesized

Arabidopsis genotypes. It indicated that many miRNA and trans-acting siRNA (tasiRNA), i.e., endogenous siRNAs that direct the cleavage of non-identical transcripts, are non-additively expressed [105]. Among the differentially expressed miRNAs, miR163 is severely repressed in leaves and flowers of *A. arenosa* and allotetraploids, but is highly expressed in *A. thaliana*. Analysis by Ng et al. [106] demonstrated that miR163 expression differences results from cis-acting effects, as well as from trans-acting repressor(s) that are present in *A. arenosa* and allotetraploids but absent in *A. thaliana*.

4.3.4 HIGH THROUGHPUT RNA SEQUENCING

Next-generation sequencing (NGS) technologies are changing the ways in which gene expression is studied. The principle behind these applications of high throughput sequencing technologies, which have been termed RNA-seq, is simple: complex RNA samples are directly sequenced to determine their content. Therefore, unlike hybridization-based data requiring the estimation of RNA amount by image analysis, RNA-seq data consists of absolute numbers of reads from each gene. These data are highly suitable for the analysis of gene expression since, not relying on probes, they are less error-prone than previous methods and allow to determine absolute expression levels [107]. Sequencing-based methods also permit the genome-wide study of small RNA expression. In *T. miscellus* allopolyploids, Buggs et al. [79] profiled almost 3000 SNP markers using an Illumina RNA-seq approach to study differential expression of duplicate homologous genes derived from the parental genomes. The authors found expression biases among tissues in the diploid parents (*T. dubius* and *T. pratensis*) in comparison to the natural allopolyploids, as well as uniform expression in F1 and first-generation synthetic allopolyploids. To explain the observed "transcriptomic shock", they hypothesized a loosening of gene expression regulation, which may set the stage for gradual evolution of novel patterns of expression in the early generations of polyploidy. Croate and Doyle [108] used quantitative reverse transcriptase-polymerase chain reaction and RNAseq in allopolyploid *Glycine dolichocarpa* and its diploid progenitors. They inferred dosage responses for several thousand genes and showed that most of them had partial dosage compensation. In *G. max*, RNA-seq allowed the identification of the gene family likely contributing to differences in photosynthetic rate between the allotetraploid and its progenitors [109]. The authors also provided evidences that the tetraploid appeared to use the "redundant" gene copies in novel ways. In *A. thaliana* allopolyploids, transcriptome profiling was carried out by Ha and colleagues [105] through high-throughput cDNA pyrosequencing. For the first time, these authors gained insight into small RNA expression diversity and evolution in closely related species as well as in interspecific hybrids. The data suggested a role for small RNAs in buffering against genomic shock in *Arabidopsis* interspecific hybrids and allopolyploids. In particular, they seem to have a central role in maintaining genome and chromatin stability as well as in modulating non additive gene expression. In addition, Ha et al. [105] found that repeat- and transposon-associated siRNAs (rasiRNA and TE-siRNA, respectively) were highly divergent between *A. thaliana* and *A. arenosa* and their non additive gene expression in allopolyploids were not correlated. By contrast, miRNA and tasiRNA sequences were conserved between species, but their expression patterns were highly variable between the allotetraploids and their progenitors.

4.4 METHODS FOR PROTEIN ANALYSIS

Compared with genomic and gene expression variations, changes in proteins and gene products in polyploids and their progenitors were rarely examined. An early study on genome-wide protein profiling was performed in maize lines of different ploidies by sodium dodecyl sulfate–polyacrylamide gel electrophoresis (SDS–PAGE) [110]. This is a technique for the separation of proteins according to their molecular weight, in the presence of a reducing agent (2-mercaptoethanol). Data obtained showed that expression per genome for most maize proteins did not change with ploidy, even though ploidy-modulated expression changes were detected for a few proteins. A more powerful method to analyze complex protein mixtures is the protein two-dimensional electrophoresis (2-DE) analysis, in which proteins are separated according to their isoelectric point and mass. In diploid, tetraploid and hexaploid wheat 2-DE experiments showed that the expression of homeologous proteins in hexaploid wheat depended on interactions among the parental A, B and D genomes [111]. Other recent studies using 2-DE indicated numerous and unbiased variations of proteins in newly synthesized *B. napus* [112,113] and wheat hybrids [114]. Using protein 2-DE coupled with mass spectrometry (MS) assays, a recent study in maize showed a positive correlation of differentially expressed proteins with ploidy levels [115]. The highest correlations were found in diploid–hexaploid and tetraploid–hexaploid comparisons. Recently, Ng and co-workers [116] were able to study quantitative changes in the proteome of *Arabidopsis* autopolyploids and allotetraploids and their progenitors using the isobaric tags for relative and absolute quantification (iTRAQ) technique, coupled with mass spectrometry. The levels of protein divergence ranged from ~18% between *A. thaliana* and *A. arenosa* to ~7% between an *A. thaliana* diploid and autotetraploid. In F1- and F8-resynthesized allotetraploids the proteomic divergence relative to MPV was intermediate (~8%). These data suggest that, during polyploidization, rapid changes occurring in post-transcriptional regulation and translational modifications of proteins can lead to high protein discrepancy between species.

4.5 CONCLUSIONS AND PERSPECTIVES

With the speed of technology improvement and the application of genomic tools, polyploidy research is undergoing a renaissance. It can be expected that comprehensive studies using multidisciplinary approaches will push the boundaries of current methodologies to translate the knowledge gained into practical applications. Particularly significant will be the high-throughput genome-wide approaches to unraveling the genetic and epigenetic consequences of polyploidization and the availability of phenotyping platforms. They are all reaching an unprecedented level of resolution at relatively affordable costs to the point that genotyping-by-sequencing [117] and targeted sequence capture [118] are now feasible also for high diversity, large genome species. NGS not only will extend the possibilities of gene and marker discovery, but will enable genome-wide quantification of gene expression. It will also allow direct genome-scale investigation of chromatin and DNA methylation cross-talk, by ChIP-Seq, bisulfite sequencing, etc. Characterizing transcripts through sequencing is advantageous to circumvent problems posed by highly redundant and extremely large genomes. It should be pointed out that the rapid pace at which new sequencing technologies are emerging is generating a growing disparity between the rate of data generation and its full and biologically meaningful analysis. However, there are outstanding examples addressing successful strategies for dealing with these challenges [109,118–120].

Genetic mapping can exploit robust statistical models, and will be crucial for identifying the genes underlying the polyploidization process in the bulk of the fast growing genome sequence information. Merging results from genetic, genomics and proteomics investigations will help to understand to what extent polyploid genome flexibility is associated with amplified responses to selection. We have recently hypothesized that defense response plasticity of potato could be correlated to gene number and category and cluster organization [121]. Understanding polyploid evolution requires knowledge to be integrated at the population level, and will have not only to rely on suitable experimental designs, but also on surveys of variation at multiple levels. Recent and forthcoming sequencing technologies are providing a wealth of genomic data to be released soon, also for wild species that can be employed for evolutionary studies. Until recently, sequencing complex genomes was considered very challenging due, for example, to the difficulties in discriminating among paralogous, hortologous, and homoeologous sequences. However, the availability of the genome sequence of the ancestors, the reduction in the ploidy level or the physical separation of chromosomes offered the possibility to circumvent these challenges and examples of polyploid genomes fully sequenced have become available [122,123]. The accumulation of knowledge on polyploid formation, maintenance, and divergence at the whole-genome and subgenome levels will not only help plant biologists to understand how plants have evolved and diversified, but also assist plant breeders in designing new strategies for crop improvement.

REFERENCES

1. Masterson, J. Stomatal size in fossil plants: Evidence for polyploidy in majority of angiosperms. Science 1994, 264, 421–424.

2. Wendel, J.F. Genome evolution in polyploids. Plant. Mol. Biol. 2000, 42, 225–249.
3. Bennett, M.D. Perspectives on polyploidy in plants—Ancient and neo. Biol. J. Linn. Soc. 2004, 82, 411–423.
4. Hendry, A.P.; Bolnick, D.I.; Berner, D.; Peichel, C.L. Along the speciation continuum in sticklebacks. J. Fish. Biol. 2009, 75, 2000–2036.
5. Soltis, D.E.; Buggs, R.J.A.; Doyle, J.J.; Soltis, P.S. What we still don't know about polyploidy. Taxon 2010, 59, 1387–1403.
6. Ramsey, J.; Schemske, D.W. Pathways, mechanisms, and rates of polyploid formation in flowering plants. Annu. Rev. Ecol. Syst. 1998, 29, 467–501.
7. Ramsey, J.; Schemske, D.W. Neopolyploidy in flowering plants. Annu. Rev. Ecol. Syst. 2002, 33, 589–639.
8. Carputo, D.; Frusciante, L.; Peloquin, S.J. The role of 2n gametes and endosperm balance number in the origin and evolution of polyploids in the tuber-bearing Solanums. Genetics 2003, 163, 287–294.
9. Chen, Z.J. Genetic and epigenetic mechanisms for gene expression and phenotypic variation in plant polyploids. Annu. Rev. Plant. Biol. 2007, 58, 377–406.
10. Bretagnolle, F.; Thompson, J.D. Tansley review no. 78. Gametes with the stomatic chromosome number: Mechanisms of their formation and role in the evolution of autopolypoid plants. New Phytol. 1995, 129, 1–22.
11. D'Erfurth, I.; Jolivet, S.; Froger, N.; Catrice, O.; Novatchkova, M.; Simon, M.; Jenczewski, E.; Mercier, R. Mutations in ATPS1 (*Arabidopsis* Thaliana Parallel Spindle 1) lead to the production of diploid pollen grains. PLoS Genet. 2008, 4, e1000274.
12. Erilova, A.; Brownfield, L.; Exner, V.; Rosa, M.; Twell, D.; Scheid, O.M.; Hennig, L.; Köhler, C. Imprinting of the polycomb group gene MEDEA serves as a ploidy sensor in *Arabidopsis*. PLoS Genet. 2009, 5, e1000663.
13. Mercier, R.; Vezon, D.; Bullier, E.; Motamayor, J.C.; Sellier, A.; Lefèvre, F.; Pelletier, G.; Horlow, C. SWITCH1 (SWI1): A novel protein required for the establishment of sister chromatid cohesion and for bivalent formation at meiosis. Genes Dev. 2001, 15, 1859–1871.
14. Agashe, B.; Prasad, C.K.; Siddiqi, I. Identification and analysis of DYAD: A gene required for meiotic chromosome organisation and female meiotic progression in *Arabidopsis*. Development 2002, 129, 3935–3943.
15. D'Erfurth, I.; Jolivet, S.; Froger, N.; Catrice, O.; Novatchkova, M.; Mercier, R. Turning meiosis into mitosis. PLoS Biol. 2009, 7, e1000124.
16. D'Erfurth, I.; Cromer, L.; Jolivet, S.; Girard, C.; Horlow, C.; Sun, Y.; To, J.P.C.; Berchowitz, L.E.; Copenhaver, G.P.; Mercier, R. The CYCLIN-A CYCA1;2/TAM is required for the meiosis I to meiosis II transition and cooperates with OSD1 for the prophase to first meiotic division transition. PLoS Genet. 2010, 6, e1000989.
17. Fawcett, J.A.; Maere, S.; van de Peer, Y. Plants with double genomes might have had a better chance to survive the cretaceous-tertiary extinction event. Proc. Natl. Acad. Sci. USA 2009, 106, 5737–5742.
18. Comai, L. The advantages and disadvantages of being polyploid. Nat. Rev. Genet. 2005, 6, 836–846.
19. Osborn, T.C.; Pires, J.C.; Birchler, J.A.; Auger, D.L.; Chen, Z.J.; Lee, H.-S.; Comai, L.; Madlung, A.; Doerge, R.W.; Colot, V.; et al. Understanding mechanisms of novel gene expression in polyploids. Trends Genet. 2003, 19, 141–147.

20. Mayrose, I.; Zhan, S.H.; Rothfels, C.J.; Magnuson-Ford, K.; Barker, M.S.; Rieseberg, L.H.; Otto, S.P. Recently formed polyploid plants diversify at lower rates. Science 2011, 333, doi:10.1126/science.1207205.
21. Pires, J.C.; Lim, K.Y.; Kovarík, A.; Matyásek, R.; Boyd, A.; Leitch, A.R.; Leitch, I.J.; Bennett, M.D.; Soltis, P.S.; Soltis, D.E. Molecular cytogenetic analysis of recently evolved Tragopogon (Asteraceae) allopolyploids reveal a karyotype that is additive of the diploid progenitors. Am. J. Bot. 2004, 91, 1022–1035.
22. Szadkowski, E.; Eber, F.; Huteau, V.; Lode, M.; Huneau, C.; Belcram, H.; Coriton, O.; Manzanares-Dauleux, M.J.; Delourme, R.; King, G.J.; et al. The first meiosis of resynthesized Brassica napus, a genome blender. New Phytol. 2010, 186, 102–112.
23. Wang, K.; Guo, W.; Yang, Z.; Hu, Y.; Zhang, W.; Zhou, B.; Stelly, D.; Chen, Z.; Zhang, T. Structure and size variations between 12A and 12D homoeologous chromosomes based on high-resolution cytogenetic map in allotetraploid cotton. Chromosoma 2010, 119, 255–266.
24. Chester, M.; Gallagher, J.P.; Symonds, V.V.; Cruz da Silva, A.V.; Mavrodiev, E.V.; Leitch, A.R.; Soltis, P.S.; Soltis, D.E. Extensive chromosomal variation in a recently formed natural allopolyploid species, Tragopogon miscellus (Asteraceae). Proc. Natl. Acad. Sci. USA 2012, 109, 1176–1181.
25. Porceddu, A.; Albertini, E.; Barcaccia, G.; Falistocco, E.; Falcinelli, M. Linkage mapping in apomictic and sexual Kentucky bluegrass (Poa pratensis L.) genotypes using a two way pseudo-testcross strategy based on AFLP and SAMPl markers. Theor. Appl. Genet. 2002, 104, 273–280.
26. Gaeta, R.T.; Pires, J.C.; Iniguez-Luy, F.; Leon, E.; Osborn, T.C. Genomic changes in resynthesized Brassica napus and their effect on gene expression and phenotype. Plant Cell 2007, 19, 3403–3417.
27. Nicolas, S.D.; Leflon, M.; Monod, H.; Eber, F.; Coriton, O.; Huteau, V.; Chèvre, A.-M.; Jenczewski, E. Genetic regulation of meiotic cross-overs between related genomes in Brassica napus haploids and hybrids. Plant Cell 2009, 21, 373–385.
28. Cifuentes, M.; Eber, F.; Lucas, M.-O.; Lode, M.; Chèvre, A.-M.; Jenczewski, E. Repeated polyploidy drove different levels of crossover suppression between homoeologous chromosomes in Brassica napus allohaploids. Plant Cell 2010, 22, 2265–2276.
29. Leitch, I.J.; Bennett, M.D. Polyploidy in angiosperms. Trends Plant Sci. 1997, 2, 470–476.
30. Liu, B.; Vega, J.M.; Feldman, M. Rapid genomic changes in newly synthesized amphiploids of Triticum and Aegilops. II. Changes in low-copy coding DNA sequences. Genome 1998, 41, 535–542.
31. Madlung, A.; Tyagi, A.P.; Watson, B.; Jiang, H.; Kagochi, T.; Doerge, R.W.; Martienssen, R.; Comai, L. Genomic changes in synthetic *Arabidopsis* polyploids. Plant J. 2005, 41, 221–230.
32. Song, K.; Lu, P.; Tang, K.; Osborn, T.C. Rapid genome change in synthetic polyploids of Brassica and its implications for polyploid evolution. Proc. Natl. Acad. Sci. USA 1995, 92, 7719–7723.
33. Salmon, A.; Ainouche, M.L.; Wendel, J.F. Genetic and epigenetic consequences of recent hybridization and polyploidy in Spartina (Poaceae). Mol. Ecol. 2005, 14, 1163–1175.

34. Wu, K.K.; Burnquist, W.; Sorrells, M.E.; Tew, T.L.; Moore, P.H.; Tanksley, S.D. The detection and estimation of linkage in polyploids using single-dose restriction fragments. Theor. Appl. Genet. 1992, 83, 294–300.
35. Hackett, C.A.; Bradshaw, J.E.; Meyer, R.C.; McNicol, J.W.; Milbourne, D.; Waugh, R. Linkage analysis in tetraploid species: A simulation study. Genet. Res. 1998, 71, 143–154.
36. Ripol, M.I.; Churchill, G.A.; da Silva, J.A.G.; Sorrells, M. Statistical aspects of genetic mapping in autopolyploids. Gene 1999, 235, 31–41.
37. Wu, R.; Ma, C.-X.; Casella, G. A bivalent polyploid model for mapping quantitative trait loci in outcrossing tetraploids. Genetics 2004, 166, 581–595.
38. Li, J.; Das, K.; Liu, J.; Fu, G.; Li, Y.; Tobias, C.; Wu, R. Statistical models for genetic mapping in polyploids: Challenges and opportunities. Methods Mol. Biol. 2012, 871, 245–261.
39. Ma, J.F.; Shen, R.; Zhao, Z.; Wissuwa, M.; Takeuchi, Y.; Ebitani, T.; Yano, M. Response of rice to Al stress and identification of quantitative trait loci for Al tolerance. Plant Cell Physiol. 2002, 43, 652–659.
40. Julier, B.; Flajoulot, S.; Barre, P.; Cardinet, G.; Santoni, S.; Huguet, T.; Huyghe, C. Construction of two genetic linkage maps in cultivated tetraploid alfalfa (Medicago sativa) using microsatellite and AFLP markers. BMC Plant Biol. 2003, 3, doi:10.1186/1471-2229-3-9.
41. Barcaccia, G.; Meneghetti, S.; Albertini, E.; Triest, L.; Lucchin, M. Linkage mapping in tetraploid willows: Segregation of molecular markers and estimation of linkage phases support an allotetraploid structure for Salix alba × Salix fragilis interspecific hybrids. Heredity 2003, 90, 169–180.
42. Le Cunff, L.; Garsmeur, O.; Raboin, L.M.; Pauquet, J.; Telismart, H.; Selvi, A.; Grivet, L.; Philippe, R.; Begum, D.; Deu, M.; et al. Diploid/polyploid syntenic shuttle mapping and haplotype-specific chromosome walking toward a rust resistance gene (Bru1) in highly polyploid sugarcane (2n~12x~115). Genetics 2008, 180, 649–660.
43. Esselink, G.D.; Nybom, H.; Vosman, B. Assignment of allelic configuration in polyploids using the MAC-PR (microsatellite DNA allele counting—peak ratios) method. Theor. Appl. Genet. 2004, 109, 402–408.
44. Van Dijk, T.; Noordijk, Y.; Dubos, T.; Bink, M.; Meulenbroek, B.; Visser, R.; van de Weg, E. Microsatellite allele dose and configuration establishment (MADCE): An integrated approach for genetic studies in allopolyploids. BMC Plant Biol. 2012, 12, doi:10.1186/1471 2229-12-25.
45. Oliver, R.; Jellen, E.; Ladizinsky, G.; Korol, A.; Kilian, A.; Beard, J.; Dumlupinar, Z.; Wisniewski-Morehead, N.; Svedin, E.; Coon, M.; et al. New Diversity Arrays Technology (DArT) markers for tetraploid oat (Avena magna Murphy et Terrell) provide the first complete oat linkage map and markers linked to domestication genes from hexaploid A. sativa L. Theor. Appl. Genet. 2011, 123, 1159–1171.
46. Reyna-López, G.E.; Simpson, J.; Ruiz-Herrera, J. Differences in DNA methylation patterns are detectable during the dimorphic transition of fungi by amplification of restriction polymorphisms. Mol. Gen. Genet. 1997, 253, 703–710.
47. Madlung, A.; Masuelli, R.W.; Watson, B.; Reynolds, S.H.; Davison, J.; Comai, L. Remodeling of DNA methylation and phenotypic and transcriptional changes in synthetic *Arabidopsis* allotetraploids. Plant Physiol. 2002, 129, 733–746.

48. Liu, B.; Brubaker, C.L.; Mergeai, G.; Cronn, R.C.; Wendel, J.F. Polyploid formation in cotton is not accompanied by rapid genomic changes. Genome 2001, 44, 321–330.
49. Axelsson, T.; Bowman, C.M.; Sharpe, A.G.; Lydiate, D.J.; Lagercrantz, U. Amphidiploid Brassica juncea contains conserved progenitor genomes. Genome 2000, 43, 679–688.
50. Jannoo, N.; Grivet, L.; Chantret, N.; Garsmeur, O.; Glaszmann, J.C.; Arruda, P.; D'Hont, A. Orthologous comparison in a gene-rich region among grasses reveals stability in the sugarcane polyploid genome. Plant J. 2007, 50, 574–585.
51. Paterson, A.H. Leafing through the genomes of our major crop plants: Strategies for capturing unique information. Nat. Rev. Genet. 2006, 7, 174–184.
52. Schranz, M.E.; Song, B.-H.; Windsor, A.J.; Mitchell-Olds, T. Comparative genomics in the Brassicaceae: A family-wide perspective. Curr. Opin. Plant Biol. 2007, 10, 168–175.
53. Margulies, E.H.; Birney, E. Approaches to comparative sequence analysis: Towards a functional view of vertebrate genomes. Nat. Rev. Genet. 2008, 9, 303–313.
54. Chantret, N.; Salse, J.; Sabot, F.; Rahman, S.; Bellec, A.; Laubin, B.; Dubois, I.; Dossat, C.; Sourdille, P.; Joudrier, P.; et al. Molecular basis of evolutionary events that shaped the hardness locus in diploid and polyploid wheat species (Triticum and Aegilops). Plant Cell 2005, 17, 1033–1045.
55. Gao, S.; Gu, Y.; Wu, J.; Coleman-Derr, D.; Huo, N.; Crossman, C.; Jia, J.; Zuo, Q.; Ren, Z.; Anderson, O.; et al. Rapid evolution and complex structural organization in genomic regions harboring multiple prolamin genes in the polyploid wheat genome. Plant Mol. Biol. 2007, 65, 189–203.
56. Innes, R.W.; Ameline-Torregrosa, C.; Ashfield, T.; Cannon, E.; Cannon, S.B.; Chacko, B.; Chen, N.W.G.; Couloux, A.; Dalwani, A.; Denny, R.; et al. Differential accumulation of retroelements and diversification of NB-LRR disease resistance genes in duplicated regions following polyploidy in the ancestor of soybean. Plant Physiol. 2008, 148, 1740–1759.
57. Town, C.D.; Cheung, F.; Maiti, R.; Crabtree, J.; Haas, B.J.; Wortman, J.R.; Hine, E.E.; Althoff, R.; Arbogast, T.S.; Tallon, L.J.; et al. Comparative genomics of Brassica oleracea and *Arabidopsis* thaliana reveal gene loss, fragmentation, and dispersal after polyploidy. Plant Cell 2006, 18, 1348–1359.
58. Li, W.; Huang, L.; Gill, B.S. Recurrent deletions of puroindoline genes at the grain hardness locus in four independent lineages of polyploid wheat. Plant Physiol. 2008, 146, 200–212.
59. De Bodt, S.; Maere, S.; van de Peer, Y. Genome duplication and the origin of angiosperms. Trends Ecol. Evol. 2005, 20, 591–597.
60. Soltis, D.E.; Bell, C.D.; Kim, S.; Soltis, P.S. Origin and early evolution of angiosperms. Ann. N. Y. Acad. Sci. 2008, 1133, 3–25.
61. Velasco, R.; Zharkikh, A.; Affourtit, J.; Dhingra, A.; Cestaro, A.; Kalyanaraman, A.; Fontana, P.; Bhatnagar, S.K.; Troggio, M.; Pruss, D.; et al. The genome of the domesticated apple (Malus domestica Borkh.). Nat. Genet. 2010, 42, 833–839.
62. The Tomato Genome Consortium. The tomato genome sequence provides insights into fleshy fruit evolution. Nature 2012, 485, 635–641.

63. Jiao, Y.; Wickett, N.J.; Ayyampalayam, S.; Chanderbali, A.S.; Landherr, L.; Ralph, P.E.; Tomsho, L.P.; Hu, Y.; Liang, H.; Soltis, P.S.; et al. Ancestral polyploidy in seed plants and angiosperms. Nature 2011, 473, 97–100.
64. Levasseur, A.; Pontarotti, P. The role of duplications in the evolution of genomes highlights the need for evolutionary-based approaches in comparative genomics. Biol. Direct 2011, 6, doi:10.1186/1745-6150-6-11.
65. Wawrzynski, A.; Ashfield, T.; Chen, N.W.G.; Mammadov, J.; Nguyen, A.; Podicheti, R.; Cannon, S.B.; Thareau, V.; Ameline-Torregrosa, C.; Cannon, E.; et al. Replication of nonautonomous retroelements in soybean appears to be both recent and common. Plant Physiol. 2008, 148, 1760–1771.
66. Kraitshtein, Z.; Yaakov, B.; Khasdan, V.; Kashkush, K. Genetic and epigenetic dynamics of a retrotransposon after allopolyploidization of wheat. Genetics 2010, 186, 801–812.
67. Ainouche, M.; Fortune, P.; Salmon, A.; Parisod, C.; Grandbastien, M.A.; Fukunaga, K.; Ricou, M.; Misset, M.T. Hybridization, polyploidy and invasion: Lessons from Spartina (Poaceae). Biol. Invasions 2009, 11, 1159–1173.
68. Beaulieu, J.; Jean, M.; Belzile, F. The allotetraploid *Arabidopsis* thaliana–*Arabidopsis* lyrata subsp. petraea as an alternative model system for the study of polyploidy in plants. Mol. Genet. Genomics 2009, 281, 421–435.
69. Kaur, S.; Francki, M.G.; Forster, J.W. Identification, characterization and interpretation of single-nucleotide sequence variation in allopolyploid crop species. Plant Biotechnol. J. 2012, 10, 125–138.
70. Hamilton, J.P.; Buell, R.C. Advances in plant genome sequencing. Plant J. 2012, 70, 177–190.
71. Shulaev, V.; Sargent, D.J.; Crowhurst, R.N.; Mockler, T.C.; Folkerts, O.; Delcher, A.L.; Jaiswal, P.; Mockaitis, K.; Liston, A.; Mane, S.P.; et al. The genome of woodland strawberry (Fragaria vesca). Nat. Genet. 2011, 43, 109–116.
72. Bancroft, I.; Morgan, C.; Fraser, F.; Higgins, J.; Wells, R.; Clissold, L.; Baker, D.; Long, Y.; Meng, J.; Wang, X.; et al. Dissecting the genome of the polyploid crop oilseed rape by transcriptome sequencing. Nat. Biotech. 2011, 29, 762–766.
73. Paux, E.; Sourdille, P.; Salse, J.; Saintenac, C.; Choulet, F.; Leroy, P.; Korol, A.; Michalak, M.; Kianian, S.; Spielmeyer, W.; et al. A physical map of the 1-gigabase bread wheat chromosome 3B. Science 2008, 322, 101–104.
74. Saintenac, C.; Jiang, D.; Akhunov, E. Targeted analysis of nucleotide and copy number variation by exon capture in allotetraploid wheat genome. Genome Biol. 2011, 12, doi:10.1186/gb-2011- 12-9-r88.
75. Akhunov, E.; Nicolet, C.; Dvorak, J. Single nucleotide polymorphism genotyping in polyploid wheat with the Illumina GoldenGate assay. Theor. Appl. Genet. 2009, 119, 507–517.
76. Allen, A.; Islamovic, E.; Kaur, J.; Gold, S.; Shah, D.; Smith, T.J. Transgenic maize plants expressing the Totivirus antifungal protein, KP4, are highly resistant to corn smut. Plant Biotechnol. J. 2011, 9, 857–864.
77. Lai, J.; Li, R.; Xu, X.; Jin, W.; Xu, M.; Zhao, H.; Xiang, Z.; Song, W.; Ying, K.; Zhang, M.; et al. Genome-wide patterns of genetic variation among elite maize inbred lines. Nat. Genet. 2010, 42, 1027–1030.

78. Bundock, P.C.; Eliott, F.G.; Ablett, G.; Benson, A.D.; Casu, R.E.; Aitken, K.S.; Henry, R.J. Targeted single nucleotide polymorphism (SNP) discovery in a highly polyploid plant species using 454 sequencing. Plant Biotechnol. J. 2009, 7, 347–354.
79. 79. Buggs, R.J.; Chamala, S.; Wu, W.; Gao, L.; May, G.D.; Schnable, P.S.; Soltis, D.E.; Soltis, P.S.; Barbazuk, W.B. Characterization of duplicate gene evolution in the recent natural allopolyploid Tragopogon miscellus by next-generation sequencing and Sequenom iPLEX MassARRAY genotyping. Mol. Ecol. 2010, 19, S132–S146.
80. Gabriel, S.; Ziaugra, L.; Tabbaa, D. SNP Genotyping Using the Sequenom iPLEX MassARRAY Platform. In Current Protocols in Human Genetics; John Wiley & Sons, Inc.: Hoboken, NJ, USA, 2009; Chapter 2.12, pp. 1–16.
81. Han, Y.; Kang, Y.; Torres-Jerez, I.; Cheung, F.; Town, C.; Zhao, P.; Udvardi, M.; Monteros, M. Genome-wide SNP discovery in tetraploid alfalfa using 454 sequencing and high resolution melting analysis. BMC Genomics 2011, 12, doi:10.1186/1471-2164-12-350.
82. Cho, M.H.; Ciulla, D.; Klanderman, B.J.; Raby, B.A.; Silverman, E.K. High-resolution melting curve analysis of genomic and whole-genome amplified DNA. Clin. Chem. 2008, 54, 2055–2058.
83. Han, Y.; Khu, D.M.; Monteros, M.J. High-resolution melting analysis for SNP genotyping and mapping in tetraploid alfalfa (Medicago sativa L.). Mol. Breed. 2012, 29, 489–501.
84. Salmon, A.; Flagel, L.; Ying, B.; Udall, J.A.; Wendel, J.F. Homoeologous nonreciprocal recombination in polyploid cotton. New Phytol. 2010, 186, 123–134.
85. Trick, M.; Long, Y.; Meng, J.; Bancroft, I. Single nucleotide polymorphism (SNP) discovery in the polyploid Brassica napus using Solexa transcriptome sequencing. Plant Biotechnol. J. 2009, 7, 334–346.
86. Guo, M.; Davis, D.; Birchler, J.A. Dosage effects on gene expression in a maize ploidy series. Genetics 1996, 142, 1349–1355.
87. Comai, L.; Tyagi, A.P.; Winter, K.; Holmes-Davis, R.; Reynolds, S.H.; Stevens, Y.; Byers, B. Phenotypic instability and rapid gene silencing in newly formed *Arabidopsis* allotetraploids. Plant Cell 2000, 12, 1551–1567.
88. Lee, H.-S.; Chen, Z.J. Protein-coding genes are epigenetically regulated in *Arabidopsis* polyploids. Proc. Natl. Acad. Sci. USA 2001, 98, 6753–6758.
89. He, P.; Friebe, B.R.; Gill, B.S.; Zhou, J.-M. Allopolyploidy alters gene expression in the highly stable hexaploid wheat. Plant Mol. Biol. 2003, 52, 401–414.
90. Adams, K.L.; Cronn, R.; Percifield, R.; Wendel, J.F. Genes duplicated by polyploidy show unequal contributions to the transcriptome and organ-specific reciprocal silencing. Proc. Natl. Acad. Sci. USA 2003, 100, 4649–4654.
91. Tate, J.A.; Ni, Z.; Scheen, A.-C.; Koh, J.; Gilbert, C.A.; Lefkowitz, D.; Chen, Z.J.; Soltis, P.S.; Soltis, D.E. Evolution and expression of homeologous loci in Tragopogon miscellus (Asteraceae), a recent and reciprocally formed allopolyploid. Genetics 2006, 173, 1599–1611.
92. Bottley, A.; Xia, G.M.; Koebner, R.M.D. Homoeologous gene silencing in hexaploid wheat. Plant J. 2006, 47, 897–906.
93. Schena, M.; Shalon, D.; Davis, R.W.; Brown, P.O. Quantitative monitoring of gene expression patterns with a complementary DNA microarray. Science 1995, 270, 467–470.

94. Chagué, V.; Just, J.; Mestiri, I.; Balzergue, S.; Tanguy, A.-M.; Huneau, C.; Huteau, V.; Belcram, H.; Coriton, O.; Jahier, J.; et al. Genome-wide gene expression changes in genetically stable synthetic and natural wheat allohexaploids. New Phytol. 2010, 187, 1181–1194.
95. Pumphrey, M.; Bai, J.; Laudencia-Chingcuanco, D.; Anderson, O.; Gill, B.S. Non-additive expression of homoeologous genes is established upon polyploidization in hexaploid wheat. Genetics 2009, 181, 1147–1157.
96. Chaudhary, B.; Hovav, R.; Flagel, L.; Mittler, R.; Wendel, J. Parallel expression evolution of oxidative stress-related genes in fiber from wild and domesticated diploid and polyploid cotton (Gossypium). BMC Genomics 2009, 10, doi:10.1186/1471-2164-10-378.
97. Marmagne, A.; Brabant, P.; Thiellement, H.; Alix, K. Analysis of gene expression in resynthesized Brassica napus allotetraploids: Transcriptional changes do not explain differential protein regulation. New Phytol. 2010, 186, 216–227.
98. Chelaifa, H.; Monnier, A.; Ainouche, M. Transcriptomic changes following recent natural hybridization and allopolyploidy in the salt marsh species Spartina × townsendii and Spartina anglica (Poaceae). New Phytol. 2010, 186, 161–174.
99. Buggs, R.J.A.; Doust, A.N.; Tate, J.A.; Koh, J.; Soltis, K.; Feltus, F.A.; Paterson, A.H.; Soltis, P.S.; Soltis, D.E. Gene loss and silencing in Tragopogon miscellus (Asteraceae): Comparison of natural and synthetic allotetraploids. Heredity 2009, 103, 73–81.
100. Doyle, J.J.; Flagel, L.E.; Paterson, A.H.; Rapp, R.A.; Soltis, D.E.; Soltis, P.S.; Wendel, J.F. Evolutionary genetics of genome merger and doubling in plants. Annu. Rev. Genet. 2008, 42, 443–461.
101. Yu, Z.; Haberer, G.; Matthes, M.; Rattei, T.; Mayer, K.F.X.; Gierl, A.; Torres-Ruiz, R.A. Impact of natural genetic variation on the transcriptome of autotetraploid *Arabidopsis* thaliana. Proc. Natl. Acad. Sci. USA 2010, 107, 17809–17814.
102. Stupar, R.M.; Bhaskar, P.B.; Yandell, B.S.; Rensink, W.A.; Hart, A.L.; Ouyang, S.; Veilleux, R.E.; Busse, J.S.; Erhardt, R.J.; Buell, C.R.; et al. Phenotypic and transcriptomic changes associated with potato autopolyploidization. Genetics 2007, 176, 2055–2067.
103. Riddle, N.; Jiang, H.; An, L.; Doerge, R.; Birchler, J. Gene expression analysis at the intersection of ploidy and hybridity in maize. Theor. Appl. Genet. 2010, 120, 341–353.
104. Havecker, E.R. Detection of small RNAs and microRNAs using deep sequencing technology. Methods Mol. Biol. 2011, 732, 55–68.
105. Ha, M.; Lu, J.; Tian, L.; Ramachandran, V.; Kasschau, K.D.; Chapman, E.J.; Carrington, J.C.; Chen, X.; Wang, X.-J.; Chen, Z.J. Small RNAs serve as a genetic buffer against genomic shock in *Arabidopsis* interspecific hybrids and allopolyploids. Proc. Natl. Acad. Sci. USA 2009, 106, 17835–17840.
106. Ng, D.W.-K.; Zhang, C.; Miller, M.; Palmer, G.; Whiteley, M.; Tholl, D.; Chen, Z.J. Cis- and trans-regulation of miR163 and target genes confers natural variation of secondary metabolites in two *Arabidopsis* species and their allopolyploids. Plant Cell 2011, 23, 1729–1740.
107. Wang, Z.; Gerstein, M.; Snyder, M. RNA-Seq: A revolutionary tool for transcriptomics. Nat. Rev. Genet. 2009, 10, 57–63.

108. Croate, E.C.; Doyle, J.J. Quantifying whole transcriptome size, a prerequisite for understanding transcriptome evolution across species: An example from a plant allopolyploid. Genome Biol. Evol. 2010, 2, 534–546.
109. Ilut, D.C.; Coate, J.E.; Luciano, A.K.; Owens, T.G.; May, G.D.; Farmer, A.; Doyle, J.J. A comparative transcriptomic study of an allotetraploid and its diploid progenitors illustrates the unique advantages and challenges of RNA-seq in plant species. Am. J. Bot. 2012, 99, 383–396.
110. Birchler, J.A.; Newton, K.J. Modulation of protein levels in chromosomal dosage series of maize: The biochemical basis of aneuploid syndromes. Genetics 1981, 99, 247–266.
111. Islam, N.; Tsujimoto, H.; Hirano, H. Proteome analysis of diploid, tetraploid and hexaploid wheat: Towards understanding genome interaction in protein expression. Proteomics 2003, 3, 549–557.
112. Albertin, W.; Balliau, T.; Brabant, P.; Chèvre, A.-M.; Eber, F.; Malosse, C.; Thiellement, H. Numerous and rapid nonstochastic modifications of gene products in newly synthesized Brassica napus allotetraploids. Genetics 2006, 173, 1101–1113.
113. Albertin, W.; Langella, O.; Joets, J.; Négroni, L.; Zivy, M.; Damerval, C.; Thiellement, H. Comparative proteomics of leaf, stem, and root tissues of synthetic Brassica napus. Proteomics 2009, 9, 793–799.
114. Song, X.; Ni, Z.; Yao, Y.; Xie, C.; Li, Z.; Wu, H.; Zhang, Y.; Sun, Q. Wheat (Triticum aestivum L.) root proteome and differentially expressed root proteins between hybrid and parents. Proteomics 2007, 7, 3538–3557.
115. Yao, H.; Kato, A.; Mooney, B.; Birchler, J. Phenotypic and gene expression analyses of a ploidy series of maize inbred Oh43. Plant Mol. Biol. 2011, 75, 237–251.
116. 116. Ng, D.W.K.; Zhang, C.; Miller, M.; Shen, Z.; Briggs, S.P.; Chen, Z.J. Proteomic divergence in *Arabidopsis* autopolyploids and allopolyploids and their progenitors. Heredity 2012, 108, 419–430.
117. Elshire, R.J.; Glaubitz, J.C.; Sun, Q.; Poland, J.A.; Kawamoto, K.; Buckler, E.S.; Mitchell, S.E. A robust, simple genotyping-by-sequencing (GBS) approach for high diversity species. PLoS One 2011, 6, e19379.
118. Grover, C.E.; Salmon, A.; Wendel, J.F. Targeted sequence capture as a powerful tool for evolutionary analysis. Am. J. Bot. 2012, 99, 312–319.
119. Cronn, R.; Knaus, B.J.; Liston, A.; Maughan, P.J.; Parks, M.; Syring, J.V.; Udall, J. Targeted enrichment strategies for next-generation plant biology. Am. J. Bot. 2012, 99, 291–311.
120. Kvam, V.M.; Liu, P.; Si, Y. A comparison of statistical methods for detecting differentially expressed genes from RNA-seq data. Am. J. Bot. 2012, 99, 248–256.
121. Ercolano, M.R. University of Naples Federico II, Portici (NA), Italy. Unpublished work, 2012.
122. Schmutz, J.; Cannon, S.B.; Schlueter, J.; Ma, J.; Mitros, T.; Nelson, W.; Hyten, D.L.; Song, Q.; Thelen, J.J.; Cheng, J.; et al. Genome sequence of the palaeopolyploid soybean. Nature 2010, 463, 178–183.
123. Potato Genome Sequencing Consortium. Genome sequence and analysis of the tuber crop potato. Nature 2012, 475, 189–195.

CHAPTER 5

WHEAT GENOMICS: PRESENT STATUS AND FUTURE PROSPECTS

P. K. GUPTA, R. R. MIR, A. MOHAN, and J. KUMAR

5.1 INTRODUCTION

Wheat is one of the most important staple food crops of the world, occupying 17% (one sixth) of crop acreage worldwide, feeding about 40% (nearly half) of the world population and providing 20% (one fifth) of total food calories and protein in human nutrition. Although wheat production during the last four decades had a steady significant increase, a fatigue has been witnessed during the last few years, leading to the lowest current global wheat stocks ever since 1948/49. Consequently, wheat prices have also been soaring, reaching the highest level of US $ 10 a bushel as against US $ 4.50 a year ago (http://www.planetark.com/dailynewsstory.cfm/newsid/44968/story.htm). As against this, it is projected that, in order to meet growing human needs, wheat grain production must increase at an annual rate of 2%, without any additional land to become available for this crop [1]. In order to meet this challenge, new level of understanding of the structure and function of the wheat genome is required.

This chapter was originally published under the Creative Commons Attribution License. Gupta PK, Mir RR, Mohan A, and Kumar J. Wheat Genomics: Present Status and Future Prospects. International Journal of Plant Genomics. ***2008*** *(2008), 36 pages. doi:10.1155/2012/251364.*

Wheat is adapted to temperate regions of the world and was one of the first crops to be domesticated some 10000 years ago. At the cytogenetics level, common wheat is known to have three subgenomes (each subgenome has 7 chromosomes, making n = 21) that are organized in seven homoeologous groups, each homoeologous group has three closely related chromosomes, one from each of the three related subgenomes. The diploid progenitors of the A, B, and D subgenomes have been identified, although there has always been a debate regarding the progenitor of B genome (reviewed in [1]). It has also been found that common wheat behaves much like a diploid organism during meiosis, but its genome can tolerate aneuploidy because of the presence of triplicate genes. These features along with the availability of a large number of aneuploids [particularly including a complete set of monosomics, a set of 42 compensating nullisomic-tetrasomics and a complete set of 42 ditelocentrics developed by Sears [2]] and more than 400 segmental deletion lines [developed later by Endo and Gill [3]] facilitated greatly the wheat genomics research.

Molecular tools have recently been used in a big way for cytogenetic studies in wheat, so that all recent cytogenetic studies in wheat now have a molecular component, thus paving the path for wheat genomics research. However, these studies in the area of molecular cytogenetics have been relatively difficult in bread wheat due to its three closely related subgenomes and a large genome (1C = >16 billion base pairs) with high proportion (>80%) of repetitive DNA. Despite this, significant progress in the area of molecular cytogenetics and cytogenomics of wheat has been made during the last two decades, thus making it amenable to genomics research. For instance, molecular maps in bread wheat, emmer wheat, and einkorn wheat utilizing a variety of molecular markers are now available, where gene rich regions (GRRs) and recombination hotspots have also been identified (for a review, see [4, 5]).

In recent years, a number of initiatives have been taken to develop new tools for wheat genomics research. These include construction of large insert libraries and development of massive EST collections, genetic and physical molecular maps, and gene targeting systems. For instance, the number of wheat ESTs has increased from a mere ~5 in 1999 [6] to a massive >1 240 000 in January 2008 (http://www.ncbi.nlm.nih.gov/), thus forming the largest EST collection in any crop as a resource for genome analysis. These ESTs are being used for a variety of activities including development of functional molecular markers, preparation of transcript maps, and construction of cDNA arrays. A variety of molecular markers that were developed either from ESTs or from genomic DNA also helped to discover relationships between genomes [7] and to compare marker-trait associations in different crops. Comparative genomics, involving major crop grasses including wheat, has also been used not only to study evolutionary relationships, but also to design crop improvement programs [8]. Functional genomics research in wheat, which though lagged far behind relative to that in other major food crops like maize and rice, has also recently witnessed significant progress. For instance, RNA interference, TILLING, and "expression genetics" leading to mapping of eQTLs have been used to identify functions of individual genes [9]. This allowed development of sets of candidate genes for individual traits, which can be used for understanding the biology of these traits and for development of perfect diagnostic marker(s) to be used not only for map-based cloning of genes, but also for MAS [9, 10]. In order to sequence the GRRs of wheat genome, a multinational collaborative program named International Genome Research on Wheat (IGROW) was earlier launched, which later took the shape of International Wheat Genome Sequencing Consortium (IWGSC) [11]. This will accelerate the progress on genome sequencing and will allow analysis of structure and function of the wheat genome. Keeping the above background in mind, Somers [12] identified the following five thrust areas of research for wheat improvement: (i) genetic mapping, (ii) QTL analysis, (iii) molecular breeding, (iv) association mapping, and (v) software development. In this communication, we briefly review the recent advances in all these areas of wheat genomics and discuss their impact on wheat improvement programs.

5.2 MOLECULAR MAPS OF WHEAT GENOME

5.2.1 MOLECULAR GENETIC MAPS

Although some efforts toward mapping of molecular markers on wheat genome were initially made during late 1980s [13], a systematic construction of molecular maps in wheat started only in 1990, with the organization of International Triticeae Mapping Initiative (ITMI), which coordinated the construction of molecular maps of wheat genome. Individual groups (headed by R Appels, PJ Sharp, ME Sorrells, J Dvorak, BS Gill, GE Hart, and MD Gale) prepared the maps for chromosomes belonging to each of the seven different homoeologous groups. A detailed account on mapping of chromosomes of individual homoeologous groups and that of the whole wheat genome is available elsewhere [14]; an updated version is available at GrainGenes (http://wheat.pw.usda.gov/), and summarized in Table 1. Integrated or composite maps involving more than one type of molecular markers have also been prepared in wheat (particularly the SSR, AFLP, SNP, and DArT markers (see Table 1)). Consensus maps, where map information from multiple genomes or multiple maps was merged into a single comprehensive map, were also prepared in wheat [15, 16]. On these maps, classical and newly identified genes of economic importance are being placed to facilitate marker-assisted selection (MAS). Many genes controlling a variety of traits (both qualitative and quantitative) have already been tagged/mapped using a variety of molecular markers (for references, see [14, 17]). The density of wheat genetic maps was improved with the development of microsatellite (SSR) markers leading to construction of SSR maps of wheat [18–20]. Later, Somers et al. [16] added more SSR markers to these earlier maps and prepared a high-density SSR consensus map. At present, >2500 mapped genomic SSR (gSSR) markers are available in wheat, which will greatly facilitate the preparation of high-density genetic maps, so that we will be able to identify key recombination events in breeding populations and fine-map genes. In addition to gSSRs, more then 300 EST-SSR could also be placed on the genetic map of wheat genome

TABLE 1: A list of some important molecular maps developed in wheat.

Map type/class of wheat	Population used for mapping	No. of loci mapped	Genetic map length (cM)	Reference
RFLP maps				
Diploid wheat (D-genome)	[*T. tauschii* (TA1691 var. meyeri x TA1704 var. typica)]	152	1554	[25]
Diploid wheat (D-genome)	[*Aegilops tauschii* var. meyeri(TA1691) *Ae. tauschii* var. typica (TA 1704)]	546	—	[26]
SSR maps				
Bread wheat	ITMI RILs (W7984 x Opata85)	279	—	[18]
Bread wheat	RILs (Synthetic x Opata)	1235	2569	[16]
Bread wheat	RILs (W7984 x Opata85)	1406	2654	[27]
Bread wheat	DHs (Kitamoe x Munstertaler)	464	3441	[28]
Bread wheat*	RILs (Chuan-Mai18 x Vigour18)	244	3150	[29]
AFLP maps				
Bread wheat*	RILs (Wangshuibai x Alondra's)	250	2430	[30]
Composite maps				
Einkorn wheat	(*T. monococcumssp* . monococcum DV92 *T. monococcum* ssp. aegilopoides C3116) (marker loci-mainly RFLPs)	3335	714	[31]
Einkorn wheat	RILs (*Triticum boeoticum T. monococcum*) marker loci-RFLPs, SSR	177	1262	[5]
Durum wheat	RILs (*T. durum* var. Messapia *T. turgidium* var. MG4343) (marker loci-RFLP, Glu3B, others)	213	1352	[32]
Durum wheat	RILs (*T. durum* var. Messapia *T. turgidium* var. MG4343) (marker loci-AFLPs, RFLPs)	88	2063	[33]
Durum wheat	RILs (Jennah Khetifa x Cham10 (marker loci-RFLPs, SSRs, AFLPs)	206	3598	[34]

TABLE 1: *Cont.*

Map type/class of wheat	Population used for mapping	No. of loci mapped	Genetic map length (cM)	Reference
Durum wheat*	RILs (Omrabi 5 T dicoccoides 600545) (marker loci-SSRs, AFLPs)	312	2289	[35]
Bread wheat	RILs (*T. aestivum* L. var. Forno *T. spelta* L. var. Oberkulmer) (marker loci-RFLPs, SSRs)	230	2469	[36]
Bread wheat*	DHs (CM-82036 x Remus) (marker loci-RFLPs, AFLPs, SSRs, etc.)	384	1860	[37]
Bread wheat*	DHs (Savannah x Senat) (marker loci-SSRs, AFLPs)	345 (17)	2300	[38]
Bread wheat*	RILs (Renan x Récital) (marker loci-SSRs, RFPLs, AFLPs)	265 (17)	2722	[39, 40]
Bread wheat	F5s(Arina x Forno) (marker loci-RFLPs, SSRs)	396	3086	[41]
Bread wheat	DHs (Courtot x Chinese Spring) (marker loci-RFLPs, SSRs, AFLPs)	659	3685	[42]
Bread wheat*	DHs (Frontana x Remus) (marker loci-SSRs, STSs, AFLPs, etc.)	535	2840	[43]
Bread wheat	RILs (Grandin x BR34) (marker loci-TRAPs, SSRs)	352	3045	[44]
Bread wheat*	DHs (Spring x SQ1) (marker loci-AFLPs, SSRs)	567	3521	[45]
Bread wheat*	RILs (Dream x Lynx) (marker loci-SSRs, STSs, AFLPs)	283 (17)	1734	[46]
Bread wheat*	DHs (AC Karma x 87E03-S2B1) (marker loci-STSs, SSRs, etc.)	167 (15)	2403	[47]
Bread wheat*	DHs (Trident x Molineux) (marker loci-SSRs, STSs, RFLPs, etc.)	251	3061	[48]
Bread wheat*	DH (Arina x Riband) (marker loci-AFLPs, SSRs)	279	1199	[49]
Bread wheat*	DHs (RL4452 x AC Domain) (marker loic-SSRs, genes, etc.)	369	2793	[50]
Bread wheat*	RILs (Chuan 35050 x Shannong 483) (marker loci-SSRs, EST-SSRs, ISSRs, SRAPs,TRAPs, Glu loci)	381	3636	[51]
Bread wheat*	DHs (Shamrock x Shango) (marker loci-SSRs, DArTs)	263	1337	[52]
Bread wheat*	DHs Cranbrook x Halberd (Marker loci-SSRs, RFLPs, AFLPs, DArTs, STSs)	749	2937	[53]

**These are framework linkage map prepared for QTL analyses.*

[21–23]. However, more markers are still needed, particularly for preparation of high-density physical maps for gene cloning [24]. Availability of a number of molecular markers associated each with individual traits will also facilitate marker-assisted selection (MAS) during plant breeding.

In addition to random DNA markers (RDM), gene targeted markers (GTMs) and functional markers (FMs) are also being used in wheat to facilitate identification of genes responsible for individual traits and to improve possibilities of using MAS in wheat breeding. As a corollary, functional markers (FMs) are also being developed from the available gene sequences [10]. These markers were also used to construct transcript and molecular functional maps. Recently, microarray-based high-throughput diversity array technology (DArT) markers were also developed and used for preparing genetic maps in wheat [53, 54]. Large-scale genotyping for dozens to thousands of SNPs is also being undertaken using several high-density platforms including Illumina's GoldenGate and ABI's SNaPshot platforms (http://wheat.pw.usda.gov/SNP/new/index.shtml). The genotyping activity may be extended further through the use of Solexa's high throughput and low-cost resequencing technology.

5.2.2 MOLECULAR MARKER-BASED PHYSICAL MAPS

Molecular markers in bread wheat have also been used for the preparation of physical maps, which were then compared with the available genetic maps involving same markers. These maps allowed comparisons between genetic and physical distances to give information about variations in recombination frequencies and cryptic structural changes (if any) in different regions of individual chromosomes. Several methods have been employed for the construction of physical maps.

5.2.2.1 DELETION MAPPING

In wheat, physical mapping of genes to individual chromosomes began with the development of aneuploids [55], which led to mapping of genes to individual chromosomes. Later, deletion lines of wheat chromosomes developed by Endo and Gill [3] were extensively used as a tool for physical mapping of molecular markers. Using these deletion stocks, genes for morphological characters were also mapped to physical segments of wheat chromosomes directly in case of unique and genome specific markers or indirectly in case of duplicate or triplicate loci through the use of intergenomic polymorphism between the A, B, and D subgenomes (see Table 2 for details of available physical maps). In addition to physical mapping of genomic SSRs, ESTs and EST-SSRs were also subjected to physical mapping (see Table 2). As a part of this effort, a major project (funded by National Science Foundation, USA) on mapping of ESTs in wheat was successfully completed by a consortium of 13 laboratories in USA leading to physical mapping of ~16000 EST loci (http://wheat.pw.usda.gov/NSF/progress_mapping.html; [56] (see Table 2)).

5.2.2.2 *IN SILICO PHYSICAL MAPPING*

As many as 16000 wheat EST loci assigned to deletion bins, as mentioned above, constitute a useful source for in silico mapping, so that markers with known sequences can be mapped to wheat chromosomes through sequence similarity with mapped EST loci available at GrainGene database (http://wheat.pw.usda.gov/GG2/blast.shtml). Using the above approach, Parida et al. [80] were able to map 157 SSR containing wheat unique sequences (out of 429 class I unigene-derived microsatellites (UGMS) markers developed in wheat) to chromosome bins. These bin-mapped UGMS markers provide valuable information for a targeted mapping of genes for useful traits, for comparative genomics, and for sequencing of gene-rich regions of the wheat genome. Another set of 672 loci belonging to 275 EST-SSRs of wheat and rye was assigned to individual bins through in silico and wet-lab approaches by Mohan et al. [79]. A few cDNA clones associated with QTL for FHB resistance in wheat were also successfully mapped using in silico approach [81].

5.2.2.3 *RADIATION-HYBRID MAPPING*

Radiation hybrid (RH) mapping was first described by Goss and Harris [82] and was initially used by Cox et al. [83] for physical mapping in

TABLE 2: Deletion-based physical maps of common wheat.

Homoeologous group/ chromosome/arm	Marker loci mapped	No. of deletion stocks used	Reference
1	19 RFLPs	18	[57]
1	50 RFLPs	56	[58]
2	30 RFLPs	21	[59]
2	43 SSRs	25	[60]
3	29 RFLPs	25	[61]
4	40 RFLPs	39	[62]
5	155 RFLPs	65	[63]
5	245 RFLPs, 3 SSRs	36	[64]
5S	100 RFLPs	17	[65]
5A	22 RFLPs	19	[66]
6	24 RFLPs	26	[67]
6	210 RFLPs	45	[68]
6S	82 RFLPs	14	[69]
7	16 RFLPs	41	[70]
7	91 RFLPs, 6 RAPDs	54	[71]
6B, 2D, and 7D	16 SSRs	13	[72]
1BS	24 AFLPs	8	[73]
4DL	61 AFLPs, 2 SSRs, 2 RFLPs	8	[74]
1BS	22 ESTs	2	[75]
Whole genome	725 SSRs	118	[76]
Whole genome	260 BARC	117	[27]
Whole genome	313 SSRs	162	[77]
Whole genome	16000 ESTs	101	http://wheat.pw.usda.gov/NSF/progressmapping.html
Whole genome	266 eSSRs	105	[78]
Whole genome	672 EST-SSRs	101	[79]

animals/humans. In wheat, the approach has been used at North Dakota State University (NDSU) utilizaing addition and substituition of individual D-genome chromosomes into tetraploid durum wheat. For RH mapping of 1D, durum wheat alien substitution line for chromosome 1D (DWRH-1D), harboring nuclear-cytoplasmic compatibility gene scsae was used. These RH lines initially allowed detection of 88 radiation-induced breaks involving 39 1D specific markers. Later, this 1D RH map was further expanded to a resolution of one break every 199 kb of DNA, utilizing 378 markers [84]. Using the same approach, construction of radiation hybrid map for chromosome 3B is currently in progress (S. Kianian personal communication).

5.2.3 BAC-BASED PHYSICAL MAPS

BAC-based physical map of wheat D genome is being constructed using the diploid species, *Aegilops tauschii*, with the aim to identify and map genes and later sequence the gene-rich regions (GRRs). For this purpose, a large number of BACs were first fingerprinted and assembled into contigs. Fingerprint contigs (FPCs) and the data related to physical mapping of the D genome are available in the database (http://wheat.pw.usda.gov/PhysicalMapping/index.html). BACs belonging to chromosome 3B are also being fingerprinted (with few BACs already anchored to wheat bins), and a whole genome BAC-based physical map of hexaploid wheat is proposed to be constructed under the aegis of IWGSC in its pilot studies (see later).

5.3 IN SITU HYBRIDIZATION STUDIES IN WHEAT

In bread wheat, in situ hybridization (ISH) involving radioactively labeled probes was initially used to localize repetitive DNA sequences, rRNA and alien DNA segments [104–106]. Later, fluorescence in situ hybridization (FISH), multicolor FISH (McFISH, simultaneous detection of more than one probe), and genome in situ hybridization (GISH, total genomic DNA as probe) were used in several studies. FISH with some repeated sequences as probes was used for identification of individual chromosomes [107–

110]. FISH was also utilized to physically map rRNA multigene family [111, 112], RFLP markers [110, 113], and unique sequences [114–116] and also for detecting and locating alien chromatin introgressed into wheat [117–119].

A novel high-resolution FISH strategy using super-stretched flow-sorted chromosomes was also used (extended DNA fibre-FISH; [120–122]) to fine map DNA sequences [123, 124] and to confirm integration of transgenes into the wheat genome [125].

Recently, BACs were also utilized as probes for the so called BAC-FISH which helped not only in the discrimination between the three subgenomes, but also in the identification of intergenomic translocations, molecular cytogenetic markers, and individual chromosomes [126]. BAC-FISH also helped in localization of genes (BACs carrying genes) and in studying genome evolution and organization among wheat and its relatives [110, 127, 128].

5.4 MAP-BASED CLONING IN WHEAT

In wheat, a number of genes for some important traits including disease resistance, vernalization response, grain protein content, free threshing habit, and tolerance to abiotic stresses have been recently cloned/likely to be cloned via map-based cloning (see Table 3). The first genes to be isolated from wheat by map-based cloning included three resistance genes, against fungal diseases, including leaf rust (Lr21; [88, 129, 130] and Lr10; [87]) and powdery mildew (Pm3b ; [94]). A candidate gene for the Q locus conferring free threshing character to domesticated wheat was also cloned [92]. This gene influences many other domestication-related traits like glume shape and tenacity, rachis fragility, plant height, spike length, and ear-emergence time. Another important QTL, Gpc-B1, associated with increased grain protein, zinc, and iron content has been cloned, which will contribute in breeding enhanced nutritional value wheat in future [96]. Cloning of three genes for vernalization response (*VRN1, VRN2, VRN3*) helped in postulating a hypothetical model summarizing interactions among these three genes [89–91, 131].

TABLE 3: Genes already cloned or likely to be cloned through map-based cloning in wheat.

Gene/QTL	Trait	Reference
Lr1	Leaf rust resistance	[85, 86]
Lr10	Leaf rust resistance	[87]
Lr21	Leaf rust resistance	[88]
VRN1	Vernalization response	[89]
VRN2	Vernalization response	[90]
VRN3	Vernalization response	[91]
Q	Free threshing character	[92, 93]
Pm3b	Powdery mildew resistance	[94, 95]
GPC-B1	High grain protein content	[96, 97]
Qfhs.Ndsu-3bs	Fusarium head blight resistance	[98]
Yr5	Resistance to stripe rust	[99]
B	Boron tolerance	[100]
Fr2	Frost resistance	http://www.agronomy.ucdavis.edu/Dubcovsky
EPS-1	Flowering time	http://www.agronomy.ucdavis.edu/Dubcovsky
Tsn1	Host-selective toxin Ptr ToxA	[101]
Ph1	Chromosome pairing locus	[102]
Sr2	Stem rust resistance	[103]

5.5 EST Databases and their Uses During the last 8–10 years, more than 1240455 wheat ESTs have become available in the public domain as in January 2008 (http://www.ncbi.nlm.nih.gov/). A number of cDNA libraries have been used for this purpose. These ESTs proved to be an enormous resource for a variety of studies including development of functional molecular markers (particularly SSRs and SNPs), construction of a DNA chip, gene expression, genome organization, and comparative genomics research.

5.5.1 EST-DERIVED SSRS

Wheat ESTs have been extensively used for SSR mining (1SSR/10.6kb; [80]), so that in our own laboratory and elsewhere detected by author, a large number of SSRs have already been developed from EST sequences [22, 78, 80, 132–134]. These EST-SSRs served as a valuable source for a variety of studies including gene mapping, marker-aided selection (MAS), and eventually positional cloning of genes. The ESTs and EST-derived SSRs were also subjected to genetic and physical mapping (see above).

Since EST-SSRs are derived from the expressed portion of the genome, which is relatively more conserved, these markers show high level of transferability among species and genera [133, 135]. However, the transferability of wheat EST-SSRs to closely related triticeae species (*Triticum* and *Aegilops* species) is higher as compared to more distant relatives such as barley, maize, rice, sorghum, oats, and rye. The EST-SSRs thus also prove useful in comparative mapping, transfer of markers to orphanage wild species, and for genetic diversity estimates [79, 132, 134, 136–139].

5.5.2 EST-DERIVED SNPS AND THE INTERNATIONAL SNP CONSORTIUM

In recent years, single nucleotide polymorphisms (SNPs) have become the markers of choice. Therefore, with the aim to discover and map SNPs in tetraploid and hexaploid wheats, an International Wheat SNP Consortium was constituted, and comprehensive wheat SNP database was developed (http://wheat.pw.usda.gov/SNP/new/index.shtml). Approximately 6000 EST unigenes from the database of mapped ESTs and other EST databases were distributed to consortium members for locating SNPs, for designing conserved primers for these SNPs and for validation of these SNP. Considerable progress has been made in this direction in different laboratories; the project data are accessible through http://wheat.pw.usda.gov/SNP/snpdb.html. In May 2006, the database contained 17174 primers (forward and reverse), 1102 wheat polymorphic loci, and 2224 polymorphic sequence tagged sites in diploid ancestors of polyploid wheat. Zhang et al. [140] also reported 246 gene loci with SNPs and/or small insertions/deletions from wheat homoeologous group 5. Another set of 101 SNPs (1SNP/212 bp) was discovered from genomic sequence analysis in 26-bread wheat lines and one synthetic line (http://urgi.versailles.inra.fr/GnpSNP/, [141]).

TABLE 4: BAC libraries available in wheat.

Species (accession)	Coverage	Restriction site	No. of clones (clone size in kb)	Curator
T. monococcum (DV92)	5.6 X	Hind III	276000 (115)	J. Dubcovsky
T. dicoccoides (Langdon)	5.0 X	Hind III	516000 (130)	J. Dubcovsky
T. urartu (G1812)	4.9 X	BamH I	163200 (110)	J. Dvorak
Ae. tauschii (AL8/78)	2.2 X	EcoR I	54000 (167)	H.B. Zhang
Ae. tauschii (AL8/78)	2.2 X	Hind III	59000 (189)	H.B. Zhang
Ae. tauschii (AL8/78)	3.2 X	Hind III	52000 (190)	H.B. Zhang
Ae. tauschii (AL8/78)	2.8 X	BamH I	59000 (149)	H.B. Zhang
Ae. tauschii (AL8/78)	2.4 X	BamH I	76000 (174)	H.B. Zhang
Ae. tauschii (Aus 18913)	4.2 X	Hind III	144000 (120)	E. Lagudah
Ae. tauschii (AS75)	4.1 X	BamH I	181248 (115)	J. Dvorak
Ae. speltoides (2-12-4-8-1-1-1)	5.4 X	BamH I	237312 (115)	J. Dvorak
T. aestivum (Glenlea)	3.1 X	BamH I & Hind III	656640 (80)	S. Cloutier
T. aestivum (Renan)	3.2 X	Hind III	478840 (150)	B. Chalhoub
T. aestivum (Renan)	2.2 X	EcoR I	285312 (132)	B. Chalhoub
T. aestivum (Renan)	1.5 X	BamH I	236160 (122)	B. Chalhoub
T. aestivum (Chinese Spring)		Hind III	950000 (54)	Y. Ogihara
T. aestivum (Chinese Spring)	< 4%	Mlu I	>12000 (45)	K. Willars
Not I	>1000			
T. aestivum (Chinese Spring) 3B	6.2 X	Hind III	67968 (103)	J. Dolezel & B. Chalhoub
T. aestivum, (Chinese Spring) 1D, 4D & 6D	3.4 X	Hind III	87168 (85)	J. Dolezel & B. Chalhoub
T. aestivum (Pavon) 1BS	14.5 X	Hind III	65280 (82)	J. Dolezel & B. Chalhoub
T. aestivum (AVS-Yr5)	3.6 X	Hind III	422400 (140)	X.M. Chen
T. aestivum (Norstar)	5.5 X	Hind III	1200000 (75)	R. Chibbar

5.6 BAC/BIBAC RESOURCES

BAC/BIBAC libraries have been produced in diploid, tetraploid, and hexaploid wheats (see Table 4). Chromosome-specific BAC libraries were also prepared in hexaploid wheat [142–144]. These BAC resources proved useful for a variety of studies including map-based cloning (see Table 3), organization of wheat genome into gene-rich and gene-poor regions that are loaded with retroelements [8, 145–147], and for physical mapping and sequencing of wheat genome (http://wheatdb.ucdavis.edu:8080/wheatdb/, [11]).

5.7 GENE DISTRIBUTION IN WHEAT: GENE-RICH AND GENE-POOR REGIONS

Genetic and physical maps of the wheat genome, discussed above, have been utilized for a study of gene distribution within the genome [58, 63, 148]. In order to identify and demarcate the gene-containing regions, 3025 loci including 252 phenotypically characterized genes and 17 quantitative trait loci (QTL) were physically mapped with the help of deletion stocks [149, 150]. It was shown that within the genome, genes are not distributed randomly and that there are gene-rich regions (GRRs) and gene-poor regions (GPRs), not only within the wheat genome, but perhaps in all eukaryotes (for reviews, see [4, 151]).

In wheat genome, 48 GRRs containing 94% of gene markers were identified with an average of ~7 such GRRs (range 5–8) per homoeologous group. It was also shown that different wheat chromosomes differed for number and location of GRRs, with 21 GRRs on the short arms containing 35% of the wheat genes, and the remaining 27 GRRs on the long arms containing about 59% of the genes. The GRRs also vary in their size and in gene-density with a general trend of increased gene-density toward the distal parts of individual chromosome arms. This is evident from the fact that more than 80% of the total marker loci were mapped in the distal half of the chromosomes and ~58% mapped in the distal 20%.

Among 48 GRRs, there were 18 GRRs (major GRRs), which contained nearly 60% of the wheat genes, covering only 11% of the genome, suggesting a very high density of genes in these GRRs, although the number and density of genes in these 18 GRRs was also variable [149, 150]. It has also been shown that the size of GRRs decreases and the number of GRRs increases, as the genome size increases from rice to wheat [4]. For instance, the average size of gene clusters in rice is ~300 kb as compared to less than 50 kb in wheat and barley. However, no correlation was observed between the chromosome size and the proportion of genes or the size of the GRRs. For instance, group 3 has the longest chromosomes among the wheat homoeologous groups but contained only 13% of the genes compared to group 5 chromosomes that contained 20% of genes [150].

For the chromosomes of homoeologous group 1, the distribution of genes and recombination rates have been studied in a relatively greater detail. Each chromosome of this group (1A, 1B, 1D) has eight GRRs (ranging in size from 3 Mb to 35 Mb), occupying ~119 Mb of the 800-Mb-long chromosome. Using this homoeologous group, it was confirmed that the GRRs differ in the number of genes and gene-density even within a chromosome or its arms. For instance, the "1S0.8 region" is the smallest of all GRRs, but has the highest gene-density, which is ~12 times that in the "1L1.0 region."

The distribution of GRRs has also been compared with the distribution of chromosome breaks involved in the generation of deletion stocks that are currently available and have been used for physical mapping of wheat genome. It was found that the breakpoints are nonrandom, and occur more frequently around the GRRs (one break every 7 Mb; [58, 67]); they seem to occur around GRRs twice as frequently as one would expect on random basis (one break every 16 Mb; [149]). Consequently, GRRs interspersed by <7-Mb-long GPRs will not be resolved and better resolution would be needed to partition the currently known GRRs into mini-GRRs and GPRs.

It has also been inferred that perhaps in eukaryotic genomes, the "gene-poor" regions preferentially enlarged during evolution, as is obvious in wheat, where large, essentially, "gene-empty" blocks of up to 192 Mb are common. Taking polyploidy into account, 30% gene-rich part of the genome is still ~4 times larger than the entire rice genome [149]. Therefore, gene distribution within the currently defined GRRs of wheat would probably be similar to that in the rice genome, except that the gene-clusters would be smaller and the interspersing "gene-empty" regions would be larger, similar to barley as described above. It has also been shown that the "gene-empty" regions of the higher eukaryotic genomes are mainly comprised of retrotransposons and pseudogenes [152, 153]. The proportion of retrotransposons is significantly higher than pseudogenes, especially in the larger genomes, like those of maize and bread wheat.

5.8 VARIABLE RECOMBINATION RATES

The recombination rate has also been recently shown to vary in different regions of the wheat genome. This was demonstrated through a comparison of consensus physical and genetic maps involving 428 common markers [149, 150]. Recombination in the distal regions was generally found to be much higher than that in the proximal half of individual chromosomes, and a strong suppression of recombination was observed in the centromeric regions. Recombination rate among GRRs present in the distal half of the chromosome was highly variable with higher recombination in some proximal GRRs than in the distal GRRs [149, 150]. The gene poor-regions accounted for only ~5% of recombination.

It has also been reported that the distribution of recombination rates along individual chromosomes is uneven in all eukaryotes studied so far (for more references, see [154, 155]). Among cereals, the average frequency of recombination in rice (with the smallest genome) is translated into a genetic distance of about 0.003 cM per kb with a range of 0 to 0.06 cM per kb (http://rgp.dna.affrc.go.jp/Publicdata.html) and that of wheat (the largest genome) is 0.0003 cM per kb with a range from 0 to 0.007 cM per kb. Non-recombinogenic regions were observed in yeast as well as in rice, but the highest recombination rate for a region appears to be ~35-fold lower in rice and ~140-fold lower in bread wheat (relative to yeast). It may be due to differences in the resolution of recombination rates, which is 400 kb in rice (in wheat the resolution is much lower than in rice), whereas the resolution in recombination hotspots in yeast may be as high as only <1 kb in length. Due to averaging over larger regions, recombination in hotspots in rice and wheat may appear to be low relative to that in yeast [4, 150, 151, 156].

5.9 FLOW CYTOGENETICS AND MICRODISSECTION OF CHROMOSOMES IN WHEAT

Flow cytogenetics and microdissection facilitated physical dissection of the large wheat genome into smaller and defined segments for the purpose of gene discovery and genome sequencing. Flow karyotypes of wheat chromosomes were also prepared [157–159]. DNA obtained from the flow-sorted chromosomes has been used for the construction of chromosome-specific large-insert DNA libraries, as has been done for chromosome 4A [157, 159]. Later, all individual 42 chromosome arms involving 21 wheat chromosomes were also sorted out using flow cytometry [160]. In another study, it was also possible to microdissect 5BL isochromosomes from meiotic cells and to use their DNA with degenerate oligonucleotide primer PCR (DOP-PCR) to amplify chromosome arm-specific DNA sequences. These amplified PCR sequences were then used as probes for exclusive painting of 5BL [161].

Flow sorting in wheat has also been used for efficient construction of bacterial artificial chromosome (BAC) libraries for individual chromosomes [143, 162]. The use of these chromosome- and chromosome arm-specific BAC libraries is expected to have major impact on wheat genomics research [1]. For instance, the availability of 3B-specific BAC library facilitated map-based cloning of agronomically important genes such as major QTL for Fusarium head blight resistance [98]. Flow cytometry can also be used to detect numerical and structural changes in chromosomes and for the detection of alien chromosomes or segments thereof (reviewed in detail by [163]). For instance, a 1BL.1RS translocation could be detected by a characteristic change in the flow karyotype [164]. In addition, DNA from flow-sorted chromosomes can be used for hybridization on DNA arrays and chips, with the aim of assigning DNA sequences to specific chromosome arms. This technique will be extensively used now with the availability of Affymetrix wheat GeneChip [165].

5.10 WHEAT GENE SPACE SEQUENCING

International Triticeae Mapping Initiative (ITMI), at its meeting held at Winnipeg, Canada during June 1–4, 2003, took the first initiative toward whole genome sequencing (WGS) in wheat and decided to launch a project that was described as International Genome Research of Wheat (IGROW) by B. S. Gill. A workshop on wheat genome sequencing was later organized in Washington, DC during November 11–13, 2003, which was followed by another meeting of IGROW during the National Wheat Workers Workshop organized at Kansas, USA, during Feb 22–25, 2004 [166]. Consequently, IGROW developed into an International Wheat Genome Sequencing Consortium (IWGSC). Chinese Spring (common wheat) was selected for WGS, since it already had ample genetic and molecular resources [1].

Three phases were proposed for sequencing the wheat genome: pilot, assessment, and scale up. The first phase was recommended for 5 years and is mainly focused on the short-term goal of IWGSC, involving physical and genetic mapping along with sample sequencing of the wheat genome aimed at better understanding of the wheat genome structure. The assessment phase will involve determining which method(s) can be used in a cost-effective manner to generate the sequence of the wheat genome. After a full assessment, the scale-up phase will involve the deployment of optimal methods on the whole genome, obtaining the genome sequence and annotation, which is the long-term goal of IWGSC. With the availability of new sequencing technologies provided by 454/Roche and those provided by Illumina/Solexa and ABI SOLiD [167]; sequencing of gene space of the wheat genome, which was once thought to be almost impossible, should become possible within the foreseeable future.

First pilot project for sequencing of gene space of wheat genome, led by INRA in France, was initiated in 2004 using the largest wheat chromosome, 3B (1GB = 2x the rice genome) of hexaploid wheat as a model. As many as 68000 BAC clones from a 3B chromosome specific BAC library [143] were fingerprinted and assembled into contigs, which were then anchored to wheat bins, covering ~80% of chromosome 3B. Currently, one or more of these contigs are being sequenced [11], which will demonstrate the feasibility of large-scale sequencing of complete gene space of wheat genome.

5.11 FUNCTIONAL GENOMICS

The determination of the functions of all the genes in a plant genome is the most challenging task in the postgenomic era of plant biology. However, several techniques or platforms, like serial analysis of gene expression

(SAGE), massively parallel signature sequencing (MPSS), and micro- and macroarrays, are now available in several crops for the estimation of mRNA abundance for large number of genes simultaneously. The microarrays have also been successfully used in wheat for understanding alterations in the transcriptome of hexaploid wheat during grain development, germination and plant development under abiotic stresses [168, 169]. Recently, a comparison was made between Affymetrix GeneChip Wheat Genome Array (an in-house custom-spotted complementary DNA array) and quantitative reverse transcription-polymerase chain reaction (RT-PCR) for the study of gene expression in hexaploid wheat [170]. Also, functional genomics approach in combination with "expression genetics" or "genetical genomics" provides a set of candidate genes that can be used for understanding the biology of a trait and for the development of perfect or diagnostic marker(s) to be used in map-based cloning of genes and MAS [9]. A similar example was provided by Jordan et al. [9], when they identified regions of wheat genome controlling seed development by mapping 542 eQTLs, using a DH mapping popultion that was earlier used for mapping of SSRs and QTL analysis of agronomic and seed quality traits [171]. Expression analysis using mRNA from developing seeds from the same mapping population was also conducted using Affymetrix GeneChip Wheat Genome Array [172].

5.11.1 RNA INTERFERENCE FOR WHEAT FUNCTIONAL GENOMICS

RNA interference (RNAi), which was the subject of the 2006 Nobel Prize in Physiology or Medicine, is also being extensively utilized for improvement of crop plants [173]. This technique does not involve introduction of foreign genes and thus provides an alternative to the most controversial elements of genetic modification. Plans in Australia are underway, where the knowledge gained from RNAi approach will be used for developing similar wheats by conventional method of plant breeding, as suggested by CSIRO scientists for developing high-fibre wheat [174]. In bread wheat, in particular, the technology provides an additional advantage of silencing all genes of a multigene family including homoeoloci for individual genes, which are often simultaneously expressed, leading to a high degree of functional gene redundancy [175]. It has been shown that delivery of specific dsRNA into single epidermal cells in wheat transiently interfered with gene function [176, 177]. Yan et al. [90] and Loukoianov et al. [178] used RNAi for stable transformation and to demonstrate that RNAi-mediated reduction of *VRN2* and *VRN1* transcript levels, respectively, accelerated and delayed flowering initiation in winter wheat. Similarly, Regina et al. [179] used RNAi to generate high-amylose wheat. However, none of the above studies reported long-term phenotypic stability of RNAi-mediated gene silencing over several generations, neither did they report any molecular details on silencing of homoeologous genes. However, Travella et al. [180] showed RNAi results in stably inherited phenotypes suggesting that RNAi can be used as an efficient tool for functional genomic studies in polyploid wheat. They introduced dsRNA-expressing constructs containing fragments of genes encoding Phytoene Desaturase (PDS) or the signal transducer of ethylene, Ethylene Insensitive 2 (EIN2) and showed stably inherited phenotypes of transformed wheat plants that were similar to mutant phenotypes of the two genes in diploid model plants. Synthetic microRNA constructs can also be used as an alternative to large RNA fragments for gene silencing, as has been demonstrated for the first time in wheat by Yao et al. [181] by discovering and predicting targets for 58 miRNAs, belonging to 43 miRNA families (20 of these are conserved and 23 are novel to wheat); more importantly four of these miRNAs are monocot specific. This study will serve as a foundation for the future functional genomic studies. The subject of the use of RNAi for functional genomics in wheat has recently been reviewed [173].

5.11.2 TILLING IN WHEAT

Recently, Targeting Induced Local Lesions IN Genomes (TILLING) was developed as a reverse genetic approach to take advantage of DNA sequence information and to investigate functions of specific genes [182]. TILLING was initially developed for model plant Arabidopsis thaliana [183] having fully sequenced diploid genome and now has also been successfully used in complex allohexaploid genome of wheat, which was once considered most challenging candidate for reverse genetics [184].

To demonstrate the utility of TILLING for complex genome of bread wheat, Slade et al. [185] created TILLING library in both bread and durum wheat and targeted waxy locus, a well characterized gene in wheat encoding granule bound starch synthase I (GBSSI). Loss of all copies of this gene results in the production of waxy starch (lacking amylose). Production of waxy wheat by traditional breeding was difficult due to lack of genetic variation at one of the waxy loci. However, targeting waxy loci by TILLING [185], using locus specific PCR primers led to identification of 246 alleles (196 alleles in hexaploid and 50 alleles in tetraploid) using 1920 cultivars of wheat (1152 hexaploid and 768 tetraploid). This made available novel genetic diversity at waxy loci and provided a way for allele mining in important germplasm of wheat. The approach also allowed evaluation of a triple homozygous mutant line containing mutations in two waxy loci (in addition to a naturally occurring deletion of the third locus) and exhibiting a near waxy phenotype.

Another example of on-going research using TILLING in wheat is the development of EMS mutagenised populations of *T. aestivum* (cv. Cadenza, 4200 lines, cv. Paragon, 6000 lines), *T. durum* (cv. Cham1, 4,200 lines), and *T. monococcum* (Accession DV92, 3000 lines) under the Wheat Genetic Improvement Network (WGIN; funded by Defra and BBSRC in the UK and by the EU Optiwheat programme). The aim of this program is to search noval variant alleles for Rht-b1c,RAR-1, SGT-1, and NPR-1 genes (personal communication: andy.phillips@bbsrc.ac.uk and Simon.Orford@bbsrc.ac.uk).

The above examples provide proof-of-concept for TILLING other genes, whose mutations may be desired in wheat or other crops. However, homoeolog-specific primers are required in order to identify new alleles via TILLING in wheat. In case of waxy, the sequences of the three homoeologous sequences were already known, which facilitated primer designing, but TILLING of other genes may require cloning and sequencing of these specific genes in order to develop homoeolog-specific target primers.

5.12 COMPARATIVE GENOMICS

In cereals, a consensus map of 12 grass genomes including wheat is now available, representing chromosome segments of each genome relative to those in rice on the basis of mapping of anchor DNA markers [186]. Some of the immediate applications of comparative genomics in wheat include a study of evolution [187] and isolation/characterization of genes using the model genome of rice. The genes, which have been examined using comperative genomics approach include the pairing gene, Ph1 [102, 188], gene(s) controlling preharvest sprouting (PHS; [189]), receptor-like kinase loci [190], gene for grain hardness [191], genes for glume coloration and pubescence (Bg, Rg; [192]), and the Pm3 gene, responsible for resistance against powdery mildew [187].

Conservation of Colinearity and Synteny

Among cereals, using molecular markers, colinearity was first reported among A, B, and D subgenomes of wheat [13, 193], and later in the high-gene density regions of wheat and barley. At the Lrk10 locus in wheat and its orthologous region in barley, a gene density of one gene per 4-5 kb was observed, which was similar to that found in *A. thaliana* [6]. Conservation of colinearity between homoeologous A genomes of diploid einkorn wheat and the hexaploid was also exploited for chromosome walking leading to cloning of candidate gene for the leaf rust resistance locus Lr10 in bread wheat [194]. Lr10 locus along with LMW/HMW loci of diploid wheat, when compared with their orthologs from tetraploid and hexaploid wheats, was found to be largely conserved except some changes that took place in intergenic regions [195–197]. On the basis of divergence of intergenic DNA (mostly transposable elements), tetraploid and hexaploid wheats were shown to have diverged about 800000 years ago [197]. Similarly, the divergence of diploid from the tetraploid/hexaploid lineage was estimated to have occurred about 2.6–3 million years ago [195, 196].

Notwithstanding the above initial demonstration of colinearity using molecular markers, later studies based on genome sequences suggested disruption of microcolinearity in many regions thus complicating the use of rice as a model for cross-species transfer of information in these genomic regions. For instance, Guyot et al. [198] conducted an in silico study and reported a mosaic conservation of genes within a novel colinear region in wheat chromosome 1AS and rice chromosome 5S. Similarly, Sorrells et al. [199] while comparing 4485 physically mapped wheat ESTs to rice genome sequence data belonging to 2251 BAC/PAC clones, resolved numerous chromosomal rearrangements. The above findings also received support from sequence analysis of the long arm of rice chromosome 11 for rice-wheat synteny [200].

More recently, the grass genus *Brachypodium* is emerging as a better model system for wheat belonging to the genus *Triticum*, because of a more recent divergence of these two genera (35–40 million years) relative to wheat-rice divergence [201–203]. Also, sequence of Brachypodium, which is likely to become available in the near future, may help further detailed analyses of colinearity and synteny among grass genomes. This has already been demonstrated through a comparison of 371 kb sequence of *B. sylvaticum* with orthologous regions from rice and wheat [204]. In this region, *Brachypodium* and wheat showed perfect macrocolinearity, but rice was shown to contain ~220 kb inversion relative to *Brachypodium* sequence. Also, in Ph1 region, more orthologous genes were identified between the related species *B. sylvaticum* and wheat than between wheat and rice, thus once again demonstrating relative utility of *Brachypodium* genome as a better model than rice genome for wheat comparative genomics [102, 188].

5.13 EPIGENETICS IN WHEAT

Epigenetics refers to a heritable change that is not a result of a change in DNA sequence, but, instead, results due to a chemical modification of nucleotides in the DNA or its associated histone proteins in the chromatin. Several studies have recently been intiated to study the epigenetic modifications in the wheat genome. For instance, methylation-sensitive amplified polymorphism (MSAP) has been used to analyze the levels of DNA methylation at four different stages (2d, 4d, 8d, and 30d after pollination) of seed development in bread wheat [205]. It was found that 36–38% of CCGG sites were either fully methylated at the internal C's and/

or hemimethylated at the external C's at the four corresponding stages. Similarly, Shitsukawa et al. [206] also studied genetic and epigenetic alterations among three homoeologs in the two class E-type wheat genes for flower development, namely, wheat SEPALLATA (WSEP) and wheat LEAFY HULL STERILE1 (WLHS1). Analyses of gene structure, expression patterns, and protein functions showed that no alterations were present in the WSEP homoeologs. By contrast, the three WLHS1 homoeologs showed genetic and epigenetic alterations. It was shown that WLHS1-B was predominantly silenced by cytosine methylation, suggesting that the expression of three homoeologous genes is differentially regulated by genetic or epigenetic mechanisms. Similar results were reported for several other genes like TaHd1 involved in photoperiodic flowering pathway, Ha for grain hardness, and TaBx for benzoxazinone biosynthesis [207–209].

A prebreeding program in wheat (along with barley and canola) based on epigenetically modified genes has also been initiated in Australia at CSIRO, under the leadership of Dr. Liz Dennis and Dr. Jim Peacock, with the support from Dr. Ben Trevaskis (http://www.grdc.com.au/director/events/groundcover?item_id=A5B55D1DED8B9C20860C0CDE8C6EE077&article_id=A97C28B1F1614E34835D6BDB8CBDC75C). This pioneering work will involve vernalization, the mechanism that allows winter crops to avoid flowering until spring, when long days and mild conditions favor seed setting and grain filling. They plan to breed varieties with a wider range of heading dates and improved frost tolerance during flowering. In wheat (as also in other cereals), the epigenetic component is also built around *VRN1* gene, which plays a role analogous to that of Flowering Locus C (FLC) in *Arabidopsis* and canola. *VRN1* is one of the most important determinants of heading dates in winter cereals including wheat and also accounts for difference between winter and spring wheat varieties. It has been shown that during vegetative growth, *VRN1* is repressed epigenetically; this repression is lifted in spring, allowing the protein encoded by *VRN1* to activate other genes involved in reproduction. As many as ~3000 wheat varieties are being looked at for variation in their *VRN1* gene so as to breed better combinations of heading date and frost tolerance (http://www.grdc.com.au/director/events/groundcover?item_id=A5B55D1DED8B9C20860C0CDE8C6EE077&article_id=A97C28B1F1614E34835D6BDB8CBDC75C).

Wheat Allopolyploidy and Epigenetics

Polyploidization induces genetic and epigenetic modifications in the genomes of higher plants including wheat (reviewed in [210, 211]). Elimination of noncoding and low-copy DNA sequences has been reported in synthetic allopolyploids of *Triticum* and *Aegilops* species [212–214]. In two other studies, patterns of cytosine methylation were also examined throughout the genome in two synthetic allotetraploids, using methylation-sensitive amplification polymorphism (MSAP; [215, 216]). This analysis indicated that the parental patterns of methylation were altered in the allotetraploid in 13% of the genomic DNA analyzed. Gene silencing and activation were also observed when 3072 transcribed loci were analyzed, using cDNA-AFLP [217, 218]. This study demonstrated new, nonadditive patterns of gene expression in allotetraploid, as indicated by the fact that 48 transcripts disappeared and 12 transcripts that were absent in the diploid parents, appeared in the allotetraploid. These results were found reproducible in two independent synthetic allotetraploids. The disappearance of transcripts could be related to gene silencing rather than gene loss and was partly associated with cytosine methylation. In another similar study involving artificially synthesized hexaploid wheats and their parents, down-regulation of some genes and activation of some other genes, selected in a nonrandom manner, was observed [219]. The genome-wide genetic and epigenetic alterations triggered by allopolyploidy thus suggested plasticity of wheat genome. The reproducibility of genetic and epigenetic events indicated a programmed rather than a chaotic response and suggests that allopolyploidy is sensed in a specific way that triggers specific response rather than a random mutator response [218].

5.14 QUANTITATIVE TRAIT LOCI (QTL) AND PROTEIN QUANTITATIVE LOCI (PQLS) IN WHEAT

A large number of QTL studies for various traits have been conducted in bread wheat, leading to mapping of QTL for these traits on different chromosomes. In most of these studies, either single marker regression approach or QTL interval mapping has been utilized. Although most of these studies involved mapping of QTL with main effects only, there are also reports of QTL, which have no main effects but have significant digenic epistatic interactions and QTL x environment interactions [220–222]. A detailed account of studies involving gene tagging and QTL analyses for various traits conducted in wheat is available elsewhere [14, 223]. More up-to-date accounts on QTL studies (summarized in Table 5) are also available for disease resistance [224], for resistance against abiotic stresses [225], grain size, and grain number [226], and for several other traits including yield and yield contributing characters, plant type, and flowering time [222, 227]. Advanced backcross QTL (AB-QTL) analysis, proposed by Tanksley and Nelson [228], has also been utilized in wheat to identify QTL for a number of traits including yield and yield components, plant height, and ear emergence [129, 229]. More recently AB-QTL analysis was practiced for the identification of QTL for baking quality traits in two BC_2F_3 populations of winter wheat [230].

Quantitative variation in protein spots was also used for detection of protein quantitative loci (PQL) in wheat. For instance, in a study, 170-amphiphilic protein spots that were specific to either of the two parents of ITMIpop were used for genotyping 101 inbred lines; 72 out of these 170 proteins spots were assigned to 15 different chromosomes, with highest number of spots mapped to Group-1 chromosomes. QTL mapping approaches were also used to map PQL; 96 spots out of the 170 specific ones showed at least one PQL. These PQL were distributed throughout the genome. With the help of MALDI-TOF spectrometry and database search, functions were also assigned to 93 specific and 41 common protein spots. It was shown in the above study that majority of these proteins are associated with membranes and/or play a role in plant defense against external invasions [231].

5.15 RECENT INSIGHTS INTO THE ORIGIN/EVOLUTION OF WHEAT GENOMES

In the genomics era, the subject of origin and evolution of bread wheat has also been revisited. This gave new insights into the identity of progenitors of the three subgenomes (A, B, D) of bread wheat, and into the genome alterations, which presumably accompanied the course of its evolution and domestication (see Figure 1). These aspects of evolution of bread wheat will be discussed briefly in this section.

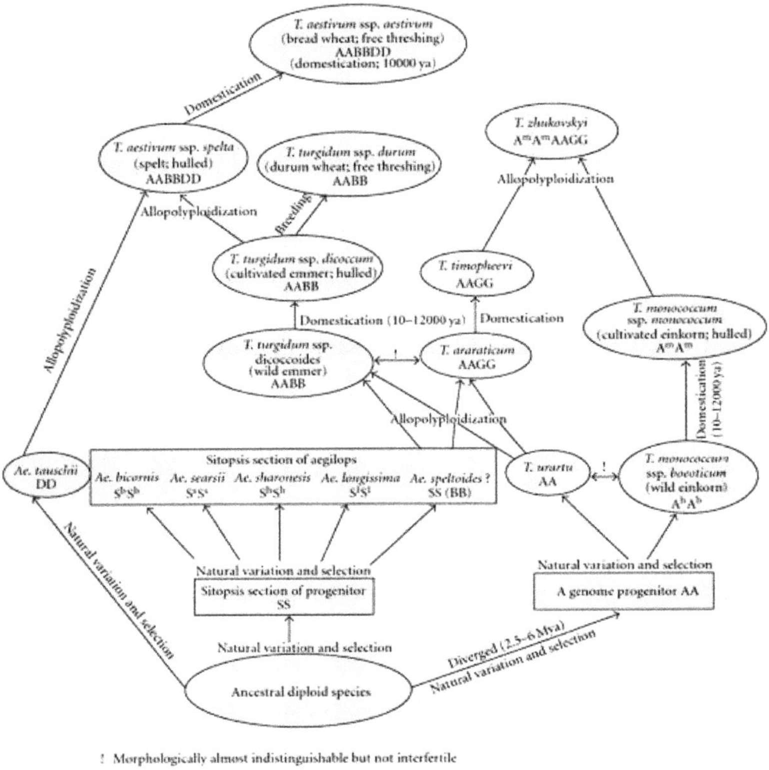

FIGURE 1: Schematic representation of the evolutionary history of wheat species (Triticum and Aegilops).

5.15.1 ORIGIN OF A, B, AND D SUBGENOMES

As mentioned earlier, bread wheat is a segmental allohexaploid having three closely related subgenomes A, B, and D. Initial analysis of the three subgenomes of bread wheat was mainly based on studies involving chromosome pairing in interspecific hybrids, and karyotype analysis in bread wheat as well as in the probable donors of the subgenomes (for reviews, see [232–236]). However, more recently, molecular markers and DNA sequence data have been used for the analysis of these subgenomes (see [237–239]). As a result, we have known with some degree of certainty that T. urartu (2 n = 14) is the donor of subgenome A and *Ae. tauschii* (synonyms, *T. tauschii, Ae. squarrosa*) is the donor of subgenome D; this has recently been confirmed through analysis of DNA sequences of two genes, namely, Acc-1 (plastid acetyl-CoA carboxylase) and Pgk-1 (plastid 3-phosphoglycerate kinase) [240]. In contrast to this, although *Ae. speltoides* was once considered as the probable donor of the B subgenome ([241], for a review, see [237]), studies carried out later showed that *Ae. speltoides* more closely resembles the subgenome G of *T. timopheevii* rather than to the subgenome B of bread wheat. DNA sequences of the above genes, Acc-1 and Pgk-1 also proved to be of no help in identification of the progenitor of the subgenome B. There is, thus still no unanimity on the progenitor of the subgenome B of bread wheat (for more details, see [242]), and there are speculations that the donor of the subgenome B might have lost its identity during evolution and may never be discovered.

DNA sequences of genes other than the above two genes have also been used for the study of origin and evolution of the component subgenomes of bread wheat. For instance, in one such study, sequences from 14 loci (2 sequences from each of the 7 chromosomes) belonging to the subgenome B of bread wheat, when compared with those from five diploid species (from section Sitopsis) closely related to the B subgenome of bread wheat, indicated that the B subgenome of bread wheat and the genomes of the above five diploid species diverged greatly after the origin of tetraploid wheat [243]. The above study also received support from the recent evidence of independent origins of wheat B and G subgenomes [244]. In this study, ~70 AFLP loci were used to sample diversity among 480 wheat lines collected from their natural habitats, which encompassed the entire range of habitats for all S genome *Aegilops* species. Also, a comparison of 59 *Aegilops* representatives of S genome diversity with 2x, 4x chromosome number, and 11 nulli-tetrasomic wheat lines at 375 AFLP loci suggested that B genome chromosomes of 6x wheat were derived from chromosomes of *Ae. speltoides*, and no other species. Further, an analysis of the haplotypes at nuclear and chloroplast loci ACC1, G6PDH, GPT, PGK1, Q, *VRN1*, and ndhF for 70 *Aegilops* and *Triticum* lines (0.73 Mb sequenced) revealed that both B and G genomes of polyploid wheats are unique samples of A. speltoides haplotype diversity. However, it is likely that due to the outbreeding nature of A. speltoides, no modern A. speltoides lines have preserved the B donor genotype in its ancestral state. The above findings can be incorporated into a broader scheme of wheat genome evolution (see Figure 1) with resolved positions of the B genome relative to S progenitors and G sisters. Similar analysis of the D subgenome and its progenitor showed that the D subgenome had more than one allele for a single locus derived from a progenitor, suggesting that hexaploid wheat perhaps originated from tetraploid wheat more than once utilizing different sources of Ae. tauschii [245]. Also, it was realized that major part of the large genome (16000 Mb) of bread wheat is composed of transposable elements (TEs). Therefore, the role of TEs in the evolution of bread-wheat and allied genomes has also been examined [246, 247]. In these studies, some specific sequences from A and B genomes of diploid species were located, respectively, in B- and A-subgenomes of bread wheat, suggesting the role of TEs in transfer of sequences between A and B subgenomes. A bioinformatics approach was also used on a large genomic region (microgenomic approach) sequenced from *T. monococcum* (AA) and *Ae. tauschii* (DD). This approach allowed a comparison of variation within coding regions with that in the noncoding regions of the subgenomes.

5.15.2 ALTERATIONS THAT ACCOMPANIED DOMESTICATION

Domestication of most crop plants including wheat involved transition from short day, small-seeded plants with natural seed dispersal to photoperiod insensitive, large-seeded nonshattering plants. A study of genetic loci underlying domestication-related traits in *T. dicoccoides* was also conduced [430], where seven domestication syndrome factors (DSFs) were proposed, each affecting 5–11 traits. Following conclusions were made with respect to the domestication-related QTL. (i) Some of these QTL had strong effect and were clustered. (ii) Strong QTL were mainly associated with GRRs, where recombination rates are high. (iii) These QTL predominantly occurred in the A genome, suggesting that A genome has played a more important role than the B genome in evolution during domestication;

this is understandable, because einkorn diploid wheat (*T. monococcum*) carrying the A genome was the first wheat to be domesticated, so that most of the domestication related traits in different wheats must have been selected within the A genome. Similar studies involving study of evolution during domestication were also conducted in hexaploid wheats for seed size, free threshing habit, rachis stiffness, photoperiod insensitivity, and so forth (for a review, see [431]). In wheat, a primary component of domestication syndrome was the loss of spike shattering, controlled by Br (brittle rachis) loci on chromosome 3A and 3B [414]. Other traits of wheat domestication syndrome shared by all domesticated wheats are the soft glumes, increased seed size, reduced number of tillers, more erect growth, and reduced dormancy [432]. A gene GPC-B1, which is an early regulator of senescence with pleiotropic effects on grain nutrient content, has also been found to affect seed size [96]. However, in some genotypes and environments, the accelerated grain maturity conferred by functional GPC-B1 allele has been found associated with smaller seeds [433], suggesting that indirect selection for large seeds may explain the fixation of the nonfunctional GPC-B1 allele in both durum and bread wheats [96]. Among many genes relevant to wheat domestication syndrome, only Q and GPC-B1 have been successfully isolated so far, suggesting a need for systematic effort to clone other genes, since it is possible that genetic variation at these loci might have played an important role in the success of wheat as a modern crop.

5.16 APPLICATION OF GENOMICS TO MOLECULAR BREEDING OF WHEAT

5.16.1 ASSOCIATION MAPPING IN WHEAT

Association mapping is a high-resolution method for mapping QTL based on linkage disequilibrium (LD) and holds great promise for genetic dissection of complex traits. It offers several advantages, which have been widely discussed [434, 435]. In wheat, some parts of the genome relative to other parts are more amenable to LD/association mapping for QTL detection and fine mapping, since the level of LD is variable across the length of a chromosome. As we know, LD decay over longer distances will facilitate initial association of trait data with the haplotypes in a chromosome region and LD decay over short distances will facilitate fine mapping of QTL [12].

Several studies involving association mapping in wheat have been conducted in the recent past. For instance, association mapping has been conducted for kernel morphology and milling quality [436] and for the quantity of a high-molecular-weight glutenin [141, 437]. In another study, 242 diversity array technology (DArT) markers were utilized for association mapping of genes/QTL controlling resistance against stem rust (SR), leaf rust (LR), yellow rust (YR), powdery mildew (PM), and those controlling grain yield (GY). Phenotypic data from five historical CIMMYT elite spring wheat yield trials (ESWYT) conducted in a large number of international environments were utilized for this purpose and two linear mixed models were applied to assess marker-trait associations after a study of population structure and additive genetic covariance between relatives [438]. A total of 122, 213, 87, 63, and 61 DArT markers were found to be significantly associated with YR, GY, LR, SR, and PM, respectively. Association analysis was also conducted between markers in the region of a major QTL responsible for resistance to *Stagonospora nodorum* (causing glume blotch); it was concluded that association mapping had a marker resolution, which was 390-fold more powerful than QTL analysis conducted using an RIL mapping population [439]. Such high-resolution mapping of traits and/or QTL to the level of individual genes, using improved statistical methods, will provide new possibilities for studying molecular and biochemical basis of quantitative trait variation and will help to identify specific targets for crop improvement.

5.16.2 MARKER-ASSISTED SELECTION IN WHEAT

A large number of marker-trait associations determined during the last decades facilitated the use of molecular markers for marker-assisted selection (MAS) in bread wheat, which is gaining momentum in several countries.

In particular, major programs involving MAS in wheat are currently underway in USA, Australia, and at CIMMYT in Mexico. In USA, a wheat MAS consortium comprosing more than 20 wheat-breeding programs was constituted at the end of 2001. The objective of this consortium was to apply and to integrate MAS in public wheat breeding programs [440]. Under these programs, MAS has been utilized for transfer of as many as 27 different insect and pest resistance genes and 20 alleles with beneficial effects on bread making and pasta quality into ~180 lines adapted to the primary US production regions. These programs led to release of germplasm consisting of 45 MAS-derived lines [441]. Similarly, the program in Australia involved improvement of 20 different traits (including resistance to some abiotic stresses) and has already led to release of some improved cultivars ([442], Peter Langridge personal communication). Among these traits, MAS has become a method of choice for those agronomically important traits, where conventional bioassays were expensive and unconvincing, as was the case in selection for cereal cyst nematodes resistance carried out by Agriculture Victoria [443]. In addition to this, MAS has been incorporated in backcross breeding in order to introgress QTL for improvement of transpiration efficiency and for negative selection for undesirable traits such as yellow flour color [444]. Australian scientists also conducted a computer simulation in order to design a genetically effective and economically efficient marker-assisted wheat-breeding strategy for a specific outcome. This investigation involved an integration of both restricted backcrossing and doubled haploid (DH) technology. Use of MAS at the BC_1F_1 followed by MAS in haploids derived from pollen of B (prior to chromosome doubling) led to reduction of cost of marker-assisted breeding up to 40% [445]. Later, this MAS strategy was validated practically in a marker-assisted wheat-breeding program in order to improve quality and resistance against rust disease (for review, see [446]). At CIMMYT, markers associated with 25 different genes governing insect pest resistance, protein quality, homoeologous pairing, and other agronomic characters are currently being utilized in wheat breeding programs in order to develop improved wheat cultivars [447]. Some of the markers used in these programs are perfect markers that have been developed from available nucleotide sequences of these genes. In future, large-scale sequencing of GRRs (gene-rich regions), to be undertaken by IWGSC, will also facilitate isolation of important genes for production of improved transgenic crops, and for development of "perfect markers" for agronomically important traits to be used in MAS [448, 449].

5.17 ORGANELLAR GENOMES AND THEIR ORGANIZATION

The genomes of wheat chloroplast and mitochondrion have also been subjected to a detailed study during the last decade. The results of these studies will be briefly discussed in this section.

5.17.1 CHLOROPLAST GENOME

In bread wheat, 130–155 chloroplasts, each containing 125–170 circular DNA molecules (135 kb), are present in each mesophyll cell, thus making 16000–26000 copies of cpDNA within a cell. This makes 5–7% of the cellular DNA in the leaf and 10–14% of the DNA in a mesophyll cell. In the related diploid species, there are 4900–6600 copies and in tetraploid species, there are 9600–12400 copies of cpDNA per mesophyll cell.

The wheat chloroplast genome, like all other plant chloroplast genomes, has two inverted repeat regions, each copy (21-kb-long) separated from the other by two single copy regions (12.8 kb, 80.2 kb). The gene content of wheat chloroplast is the same as those of rice and maize plastomes, however some structural divergence was reported in the gene coding regions, due to illegitimate recombination between two short direct repeats and/or replication slippage; this included the presence of some hotspot regions for length mutations. The study of deletion patterns of open reading frames (ORFs) in the inverted-repeat regions and in the borders between the inverted repeats and the small single-copy regions supports the view that wheat and rice are related more closely to each other than to maize (see [450, 451]). Deletions, insertions, and inversions have also been detected during RFLP analysis of cpDNA, which gave eleven different cpDNA types, in the genus *Triticum*, the bread wheat sharing entirely the cpDNA type with durum wheats, but not with that of any of the diploid species. The cpDNA of *Ae. speltoides* showed maximum similarity to those of *T. aestivum, T. timopheevii,* and *T. zhukovskyi,* suggesting that *Ae. speltoides* should be the donor of the B subgenome of common wheat [452].

5.17.2 MITOCHONDRIAL GENOME

Wheat mtDNA is larger (430 kb) than cpDNA (135 kb) with a minimum of 10 repeats but encodes only 30–50% polypeptides relative to cpDNA. Thus, large amount of mtDNA is noncoding, there being about 50 genes involved in RNA synthesis [453]. Mitochondrial genome of Chinese Spring has been sequenced using 25 cosmid clones of mitochondrial DNA, selected on the basis of their gene content. This led to the identification of 55 (71) genes including the following: 18 genes (20) for electron transport system, 4 genes for mitochondrial biogenesis, 11 genes for ribosomal proteins, 2 genes for splicing and other function, 3 genes (10) for rRNAs, and 17 genes (24) for tRNAs (the numerals in parentheses represent number of genes, taking multiple copies of a gene as separate genes). When mitochondrial gene maps were compared among wheat, rice, and maize, no major synteny was found between them other than a block of two to five genes. Therefore, mitochondrial genes seem to have thoroughly reshuffled during speciation of cereals. In contrast, chloroplast genes show perfect synteny among wheat, rice, and maize [451].

5.18 CONCLUSIONS

Significant progress during the last two decades has been made in different areas of wheat genomics research. These include development of thousands of molecular markers (including RFLPs, SSRs, AFLPs, SNPs, and DArT markers), construction of molecular genetic and physical maps (including radiation hybrid maps for some chromosomes) with reasonably high density of markers, development of more than 1 million ESTs and their use for developing functional markers, and the development of BAC/BIBAC resources for individual chromosomes and entire subgenomes to facilitate genome sequencing. Functional genomics approaches like TILL-

ING, RNAi, and epigenetics have also been utilized successfully, and a number of genes/QTL have been cloned to be used in future wheat improvement programs. Organellar genomes including chloroplast and mitochondrial genomes have been fully sequenced, and we are at the threshold of initiating a major program of sequencing the gene space of the whole nuclear genome in this major cereal. The available molecular tools also facilitated a revisit of the wheat community to the problem of origin and evolution of the wheat genome and helped QTL analysis (including studies involving LD and association mapping) for identification of markers associated with all major economic traits leading to the development of major marker-aided selection (MAS) programs for wheat improvement in several countries.

REFERENCES

1. B. S. Gill, R. Appels, A.-M. Botha-Oberholster, et al., "A workshop report on wheat genome sequencing: international genome research on wheat consortium," Genetics, vol. 168, no. 2, pp. 1087–1096, 2004.
2. E. R. Sears, "Nullisomic-tetrasomic combinations in hexaploid wheat," in Chromosome Manipulation and Plant Genetics, R. Riley and K. R. Lewis, Eds., pp. 29–45, Oliver and Boyd, Edinburgh, UK, 1966.
3. T. R. Endo and B. S. Gill, "The deletion stocks of common wheat," Journal of Heredity, vol. 87, no. 4, pp. 295–307, 1996.
4. K. S. Gill, "Gene distribution in cereal genomes," in Cereal Genomics, P. K. Gupta and R. K. Varshney, Eds., pp. 361–385, Kluwer Academic Publishers, Dordrecht, The Netherlands, 2004.
5. K. Singh, M. Ghai, M. Garg, et al., "An integrated molecular linkage map of diploid wheat based on a Triticum boeoticum×T. monococcum RIL population," Theoretical and Applied Genetics, vol. 115, no. 3, pp. 301–312, 2007.
6. C. Feuillet and B. Keller, "Comparative genomics in the grass family: molecular characterization of grass genome structure and evolution," Annals of Botany, vol. 89, no. 1, pp. 3–10, 2002.
7. M. D. Gale and K. M. Devos, "Plant comparative genetics after 10 years," Science, vol. 282, no. 5389, pp. 656–659, 1998.
8. K. M. Devos, "Updating the 'crop circle'," Current Opinion in Plant Biology, vol. 8, no. 2, pp. 155–162, 2005.
9. M. C. Jordan, D. J. Somers, and T. W. Banks, "Identifying regions of the wheat genome controlling seed development by mapping expression quantitative trait loci," Plant Biotechnology Journal, vol. 5, no. 3, pp. 442–453, 2007.
10. M. Bagge, X. Xia, and T. Lübberstedt, "Functional markers in wheat," Current Opinion in Plant Biology, vol. 10, no. 2, pp. 211–216, 2007.

11. P. Moolhuijzen, D. S. Dunn, M. Bellgard, et al., "Wheat genome structure and function: genome sequence data and the international wheat genome sequencing consortium," Australian Journal of Agricultural Research, vol. 58, no. 6, pp. 470–475, 2007.
12. D. J. Somers, "Molecular breeding and assembly of complex genotypes in wheat," in Frontiers of Wheat Bioscience. The 100 Memorial Issue of Wheat Information Service, K. Tsunewaki, Ed., pp. 235–246, Kihara Memorial Yokohama Foundation for the Advancement of Life Sciences, Yokohama, Japan, 2005.
13. S. Chao, P. J. Sharp, A. J. Worland, E. J. Warham, R. M. D. Koebner, and M. D. Gale, "RFLP-based genetic maps of wheat homoeologous group 7 chromosomes," Theoretical and Applied Genetics, vol. 78, no. 4, pp. 495–504, 1989.
14. P. K. Gupta, R. K. Varshney, P. C. Sharma, and B. Ramesh, "Molecular markers and their applications in wheat breeding," Plant Breeding, vol. 118, no. 5, pp. 369–390, 1999.
15. R. Appels, "A consensus molecular genetic map of wheat-a cooperative international effort," in Proceedings of the 10th International Wheat Genetics Symposium, N. E. Pogna, Ed., pp. 211–214, Paestum, Italy, September 2003.
16. D. J. Somers, P. Isaac, and K. Edwards, "A high-density microsatellite consensus map for bread wheat (Triticum aestivum L.)," Theoretical and Applied Genetics, vol. 109, no. 6, pp. 1105–1114, 2004.
17. R. A. McIntosh, K. M. Devos, J. Dubcovsky, C. F. Morris, and W. J. Rogers, "Catalogue of gene symbols for wheat," 2003, http://wheat.pw.usda.gov/ggpages/wgc/2003upd.html.
18. M. S. Röder, V. Korzun, K. Wendehake, et al., "A microsatellite map of wheat," Genetics, vol. 149, no. 4, pp. 2007–2023, 1998.
19. E. Pestsova, M. W. Ganal, and M. S. Röder, "Isolation and mapping of microsatellite markers specific for the D genome of bread wheat," Genome, vol. 43, no. 4, pp. 689–697, 2000.
20. P. K. Gupta, H. S. Balyan, K. J. Edwards, et al., "Genetic mapping of 66 new microsatellite (SSR) loci in bread wheat," Theoretical and Applied Genetics, vol. 105, no. 2-3, pp. 413–422, 2002.
21. L. F. Gao, R. L. Jing, N. X. Huo, et al., "One hundred and one new microsatellite loci derived from ESTs (EST-SSRs) in bread wheat," Theoretical and Applied Genetics, vol. 108, no. 7, pp. 1392–1400, 2004.
22. J.-K. Yu, T. M. Dake, S. Singh, et al., "Development and mapping of EST-derived simple sequence repeat markers for hexaploid wheat," Genome, vol. 47, no. 5, pp. 805–818, 2004.
23. N. Nicot, V. Chiquet, B. Gandon, et al., "Study of simple sequence repeat (SSR) markers from wheat expressed sequence tags (ESTs)," Theoretical and Applied Genetics, vol. 109, no. 4, pp. 800–805, 2004.
24. J. W. Snape and G. Moore, "Reflections and opportunities: gene discovery in the complex wheat genome," in Wheat Production in Stressed Environments, H. T. Buck, Ed., pp. 677–684, Springer, Dordrecht, The Netherlands, 2007.
25. K. S. Gill, E. L. Lubbers, B. S. Gill, W. J. Raupp, and T. S. Cox, "A genetic linkage map of Triticum tauschii (DD) and its relationship to the D genome of bread wheat (AABBDD)," Genome, vol. 34, no. 3, pp. 362–374, 1991.

26. E. V. Boyko, B. S. Gill, L. Mickelson-Young, et al., "A high-density genetic linkage map of Aegilops tauschii, the D-genome progenitor of bread wheat," Theoretical and Applied Genetics, vol. 99, no. 1-2, pp. 16–26, 1999.
27. Q. J. Song, J. R. Shi, S. Singh, et al., "Development and mapping of microsatellite (SSR) markers in wheat," Theoretical and Applied Genetics, vol. 110, no. 3, pp. 550–560, 2005.
28. A. Torada, M. Koike, K. Mochida, and Y. Ogihara, "SSR-based linkage map with new markers using an intraspecific population of common wheat," Theoretical and Applied Genetics, vol. 112, no. 6, pp. 1042–1051, 2006.
29. W. Spielmeyer, J. Hyles, P. Joaquim, et al., "A QTL on chromosome 6A in bread wheat (Triticum aestivum) is associated with longer coleoptiles, greater seedling vigour and final plant height," Theoretical and Applied Genetics, vol. 115, no. 1, pp. 59–66, 2007.
30. X. Zhang, M. Zhou, L. Ren, et al., "Molecular characterization of Fusarium head blight resistance from wheat variety Wangshuibai," Euphytica, vol. 139, no. 1, pp. 59–64, 2004.
31. J. Dubcovsky, M.-C. Luo, G.-Y. Zhong, et al., "Genetic map of diploid wheat, Triticum monococcum L., and its comparison with maps of Hordeum vulgare L.," Genetics, vol. 143, no. 2, pp. 983–999, 1996.
32. A. Blanco, M. P. Bellomo, A. Cenci, et al., "A genetic linkage map of durum wheat," Theoretical and Applied Genetics, vol. 97, no. 5-6, pp. 721–728, 1998.
33. C. Lotti, S. Salvi, A. Pasqualone, R. Tuberosa, and A. Blanco, "Integration of AFLP markers into an RFLP-based map of durum wheat," Plant Breeding, vol. 119, no. 5, pp. 393–401, 2000.
34. M. M. Nachit, I. Elouafi, M. A. Pagnotta, et al., "Molecular linkage map for an intraspecific recombinant inbred population of durum wheat (Triticum turgidum L. var. durum)," Theoretical and Applied Genetics, vol. 102, no. 2-3, pp. 177–186, 2001.
35. I. Elouafi and M. M. Nachit, "A genetic linkage map of the Durum×Triticum dicoccoides backcross population based on SSRs and AFLP markers, and QTL analysis for milling traits," Theoretical and Applied Genetics, vol. 108, no. 3, pp. 401–413, 2004.
36. M. M. Messmer, M. Keller, S. Zanetti, and B. Keller, "Genetic linkage map of a wheat×spelt cross," Theoretical and Applied Genetics, vol. 98, no. 6-7, pp. 1163–1170, 1999.
37. H. Buerstmayr, M. Lemmens, L. Hartl, et al., "Molecular mapping of QTLs for Fusarium head blight resistance in spring wheat. I. Resistance to fungal spread (type II resistance)," Theoretical and Applied Genetics, vol. 104, no. 1, pp. 84–91, 2002.
38. L. Eriksen, F. Borum, and A. Jahoor, "Inheritance and localisation of resistance to Mycosphaerella graminicola causing septoria tritici blotch and plant height in the wheat (Triticum aestivum L.) genome with DNA markers," Theoretical and Applied Genetics, vol. 107, no. 3, pp. 515–527, 2003.
39. C. Groos, N. Robert, E. Bervas, and G. Charmet, "Genetic analysis of grain protein-content, grain yield and thousand-kernel weight in bread wheat," Theoretical and Applied Genetics, vol. 106, no. 6, pp. 1032–1040, 2003.

40. C. Groos, E. Bervas, E. Chanliaud, and G. Charmet, "Genetic analysis of bread-making quality scores in bread wheat using a recombinant inbred line population," Theoretical and Applied Genetics, vol. 115, no. 3, pp. 313–323, 2007.
41. S. Paillard, T. Schnurbusch, M. Winzeler, et al., "An integrative genetic linkage map of winter wheat (Triticum aestivum L.)," Theoretical and Applied Genetics, vol. 107, no. 7, pp. 1235–1242, 2003.
42. P. Sourdille, T. Cadalen, H. Guyomarc'h, et al., "An update of the Courtot × Chinese Spring intervarietal molecular marker linkage map for the QTL detection of agronomic traits in wheat," Theoretical and Applied Genetics, vol. 106, no. 3, pp. 530–538, 2003.
43. B. Steiner, M. Lemmens, M. Griesser, U. Scholz, J. Schondelmaier, and H. Buerstmayr, "Molecular mapping of resistance to Fusarium head blight in the spring wheat cultivar Frontana," Theoretical and Applied Genetics, vol. 109, no. 1, pp. 215–224, 2004.
44. Z. H. Liu, J. A. Anderson, J. Hu, T. L. Friesen, J. B. Rasmussen, and J. D. Faris, "A wheat intervarietal genetic linkage map based on microsatellite and target region amplified polymorphism markers and its utility for detecting quantitative trait loci," Theoretical and Applied Genetics, vol. 111, no. 4, pp. 782–794, 2005.
45. S. A. Quarrie, A. Steed, C. Calestani, et al., "A high-density genetic map of hexaploid wheat (Triticum aestivum L.) from the cross Chinese Spring × SQ1 and its use to compare QTLs for grain yield across a range of environments," Theoretical and Applied Genetics, vol. 110, no. 5, pp. 865–880, 2005.
46. M. Schmolke, G. Zimmermann, H. Buerstmayr, et al., "Molecular mapping of Fusarium head blight resistance in the winter wheat population Dream/Lynx," Theoretical and Applied Genetics, vol. 111, no. 4, pp. 747–756, 2005.
47. X. Q. Huang, S. Cloutier, L. Lycar, et al., "Molecular detection of QTLs for agronomic and quality traits in a doubled haploid population derived from two Canadian wheats (Triticum aestivum L.)," Theoretical and Applied Genetics, vol. 113, no. 4, pp. 753–766, 2006.
48. K. J. Williams, K. L. Willsmore, S. Olson, M. Matic, and H. Kuchel, "Mapping of a novel QTL for resistance to cereal cyst nematode in wheat," Theoretical and Applied Genetics, vol. 112, no. 8, pp. 1480–1486, 2006.
49. R. Draeger, N. Gosman, A. Steed, et al., "Identification of QTLs for resistance to Fusarium head blight, DON accumulation and associated traits in the winter wheat variety Arina," Theoretical and Applied Genetics, vol. 115, no. 5, pp. 617–625, 2007.
50. C. A. McCartney, D. J. Somers, B. D. McCallum, et al., "Microsatellite tagging of the leaf rust resistance gene Lr16 on wheat chromosome 2BSc," Molecular Breeding, vol. 15, no. 4, pp. 329–337, 2005.
51. S. Li, J. Jia, X. Wei, et al., "A intervarietal genetic map and QTL analysis for yield traits in wheat," Molecular Breeding, vol. 20, no. 2, pp. 167–178, 2007.
52. J. R. Simmonds, L. J. Fish, M. A. Leverington-Waite, Y. Wang, P. Howell, and J. W. Snape, "Mapping of a gene (Vir) for a non-glaucous, viridescent phenotype in bread wheat derived from Triticum dicoccoides, and its association with yield variation," Euphytica, vol. 159, no. 3, pp. 333–341, 2008.

53. M. Akbari, P. Wenzl, V. Caig, et al., "Diversity arrays technology (DArT) for high-throughput profiling of the hexaploid wheat genome," Theoretical and Applied Genetics, vol. 113, no. 8, pp. 1409–1420, 2006.
54. K. Semagn, Å. Bjørnstad, H. Skinnes, A. G. Marøy, Y. Tarkegne, and M. William, "Distribution of DArT, AFLP, and SSR markers in a genetic linkage map of a doubled-haploid hexaploid wheat population," Genome, vol. 49, no. 5, pp. 545–555, 2006.
55. E. R. Sears, "The aneuploids of common wheat," University of Missouri Agriculture Experiment Station, Bulleten, vol. 572, pp. 1–58, 1954.
56. L. L. Qi, B. Echalier, S. Chao, et al., "A chromosome bin map of 16,000 expressed sequence tag loci and distribution of genes among the three genomes of polyploid wheat," Genetics, vol. 168, no. 2, pp. 701–712, 2004.
57. R. S. Kota, K. S. Gill, B. S. Gill, and T. R. Endo, "A cytogenetically based physical map of chromosome 1B in common wheat," Genome, vol. 36, no. 3, pp. 548–554, 1993.
58. K. S. Gill, B. S. Gill, T. R. Endo, and T. Taylor, "Identification and high-density mapping of gene-rich regions in chromosome group 1 of wheat," Genetics, vol. 144, no. 4, pp. 1883–1891, 1996.
59. D. E. Delaney, S. Nasuda, T. R. Endo, B. S. Gill, and S. H. Hulbert, "Cytologically based physical maps of the group-2 chromosomes of wheat," Theoretical and Applied Genetics, vol. 91, no. 4, pp. 568–573, 1995.
60. M. S. Röder, V. Korzun, B. S. Gill, and M. W. Ganal, "The physical mapping of microsatellite markers in wheat," Genome, vol. 41, no. 2, pp. 278–283, 1998.
61. D. E. Delaney, S. Nasuda, T. R. Endo, B. S. Gill, and S. H. Hulbert, "Cytologically based physical maps of the group 3 chromosomes of wheat," Theoretical and Applied Genetics, vol. 91, no. 5, pp. 780–782, 1995.
62. L. Mickelson-Young, T. R. Endo, and B. S. Gill, "A cytogenetic ladder-map of the wheat homoeologous group-4 chromosomes," Theoretical and Applied Genetics, vol. 90, no. 7-8, pp. 1007–1011, 1995.
63. K. S. Gill, B. S. Gill, T. R. Endo, and E. V. Boyko, "Identification and high-density mapping of gene-rich regions in chromosome group 5 of wheat," Genetics, vol. 143, no. 2, pp. 1001–1012, 1996.
64. J. D. Faris, K. M. Haen, and B. S. Gill, "Saturation mapping of a gene-rich recombination hot spot region in wheat," Genetics, vol. 154, no. 2, pp. 823–835, 2000.
65. L. L. Qi and B. S. Gill, "High-density physical maps reveal that the dominant male-sterile gene Ms3 is located in a genomic region of low recombination in wheat and is not amenable to map-based cloning," Theoretical and Applied Genetics, vol. 103, no. 6-7, pp. 998–1006, 2001.
66. Y. Ogihara, K. Hasegawa, and H. Tsujimoto, "High-resolution cytological mapping of the long arm of chromosome 5A in common wheat using a series of deletion lines induced by gametocidal (Gc) genes of Aegilops speltoides," Molecular and General Genetics, vol. 244, no. 3, pp. 253–259, 1994.
67. K. S. Gill, B. S. Gill, and T. R. Endo, "A chromosome region-specific mapping strategy reveals gene-rich telomeric ends in wheat," Chromosoma, vol. 102, no. 6, pp. 374–381, 1993.

68. Y. Weng, N. A. Tuleen, and G. E. Hart, "Extended physical maps and a consensus physical map of the homoeologous group-6 chromosomes of wheat (Triticum aestivum L. em Thell.)," Theoretical and Applied Genetics, vol. 100, no. 3-4, pp. 519–527, 2000.
69. Y. Weng and M. D. Lazar, "Comparison of homoeologous group-6 short arm physical maps of wheat and barley reveals a similar distribution of recombinogenic and gene-rich regions," Theoretical and Applied Genetics, vol. 104, no. 6-7, pp. 1078–1085, 2002.
70. J. E. Werner, T. R. Endo, and B. S. Gill, "Towards a cytogenetically based physical map of the wheat genome," Proceedings of the National Academy of Sciences of the United States of America, vol. 89, pp. 11307–11311, 1992.
71. U. Hohmann, T. R. Endo, K. S. Gill, and B. S. Gill, "Comparison of genetic and physical maps of group 7 chromosomes from Triticum aestivum L," Molecular and General Genetics, vol. 245, no. 5, pp. 644–653, 1994.
72. R. K. Varshney, M. Prasad, J. K. Roy, M. S. Röder, H. S. Balyan, and P. K. Gupta, "Integrated physical maps of 2DL, 6BS and 7DL carrying loci for grain protein content and pre-harvest sprouting tolerance in bread wheat," Cereal Research Communications, vol. 29, no. 1-2, pp. 33–40, 2001.
73. H. Zhang, S. Nasuda, and T. R. Endo, "Identification of AFLP markers on the satellite region of chromosome 1BS in wheat," Genome, vol. 43, no. 5, pp. 729–735, 2000.
74. M. A. Rodriguez Milla and J. P. Gustafson, "Genetic and physical characterization of chromosome 4DL in wheat," Genome, vol. 44, no. 5, pp. 883–892, 2001.
75. D. Sandhu, D. Sidhu, and K. S. Gill, "Identification of expressed sequence markers for a major gene-rich region of wheat chromosome group 1 using RNA fingerprinting-differential display," Crop Science, vol. 42, no. 4, pp. 1285–1290, 2002.
76. P. Sourdille, S. Singh, T. Cadalen, et al., "Microsatellite-based deletion bin system for the establishment of genetic-physical map relationships in wheat (Triticum aestivum L.)," Functional and Integrative Genomics, vol. 4, no. 1, pp. 12–25, 2004.
77. A. Goyal, R. Bandopadhyay, P. Sourdille, T. R. Endo, H. S. Balyan, and P. K. Gupta, "Physical molecular maps of wheat chromosomes," Functional & Integrative Genomics, vol. 5, no. 4, pp. 260–263, 2005.
78. J. H. Peng and N. L. V. Lapitan, "Characterization of EST-derived microsatellites in the wheat genome and development of eSSR markers," Functional and Integrative Genomics, vol. 5, no. 2, pp. 80–96, 2005.
79. A. Mohan, A. Goyal, R. Singh, H. S. Balyan, and P. K. Gupta, "Physical mapping of wheat and rye expressed sequence tag-simple sequence repeats on wheat chromosomes," Crop Science, vol. 47, supplement 1, pp. S3–S13, 2007.
80. S. K. Parida, K. A. Raj Kumar, V. Dalal, N. K. Singh, and T. Mohapatra, "Unigene derived microsatellite markers for the cereal genomes," Theoretical and Applied Genetics, vol. 112, no. 5, pp. 808–817, 2006.
81. K. Hill-Ambroz, C. A. Webb, A. R. Matthews, W. Li, B. S. Gill, and J. P. Fellers, "Expression analysis and physical mapping of a cDNA library of Fusarium head blight infected wheat spikes," Crop Science, vol. 46, supplement 1, pp. S15–S26, 2006.

82. S. J. Goss and H. Harris, "New method for mapping genes in human chromosomes," Nature, vol. 255, no. 5511, pp. 680–684, 1975.
83. D. R. Cox, M. Burmeister, E. R. Price, S. Kim, and R. M. Myers, "Radiation hybrid mapping: a somatic cell genetic method for constructing high-resolution maps of mammalian chromosomes," Science, vol. 250, no. 4978, pp. 245–250, 1990.
84. V. Kalavacharla, K. Hossain, Y. Gu, et al., "High-resolution radiation hybrid map of wheat chromosome 1D," Genetics, vol. 173, no. 2, pp. 1089–1099, 2006.
85. H.-Q. Ling, Y. Zhu, and B. Keller, "High-resolution mapping of the leaf rust disease resistance gene Lr1 in wheat and characterization of BAC clones from the Lr1 locus," Theoretical and Applied Genetics, vol. 106, no. 5, pp. 875–882, 2003.
86. S. Cloutier, B. D. McCallum, C. Loutre, et al., "Leaf rust resistance gene Lr1, isolated from bread wheat (Triticum aestivum L.) is a member of the large psr567 gene family," Plant Molecular Biology, vol. 65, no. 1-2, pp. 93–106, 2007.
87. C. Feuillet, S. Travella, N. Stein, L. Albar, A. Nublat, and B. Keller, "Map-based isolation of the leaf rust disease resistance gene Lr10 from the hexaploid wheat (Triticum aestivum L.) genome," Proceedings of the National Academy of Sciences of the United States of America, vol. 100, no. 25, pp. 15253–15258, 2003.
88. L. Huang, S. A. Brooks, W. Li, J. P. Fellers, H. N. Trick, and B. S. Gill, "Map-based cloning of leaf rust resistance gene Lr21 from the large and polyploid genome of bread wheat," Genetics, vol. 164, no. 2, pp. 655–664, 2003.
89. L. Yan, A. Loukoianov, G. Tranquilli, M. Helguera, T. Fahima, and J. Dubcovsky, "Positional cloning of the wheat vernalization gene *VRN1*," Proceedings of the National Academy of Sciences of the United States of America, vol. 100, no. 10, pp. 6263–6268, 2003.
90. L. Yan, A. Loukoianov, A. Blechl, et al., "The wheat VRN2 gene is a flowering repressor down-regulated by vernalization," Science, vol. 303, no. 5664, pp. 1640–1644, 2004.
91. L. Yan, D. Fu, C. Li, et al., "The wheat and barley vernalization gene VRN3 is an orthologue of FT," Proceedings of the National Academy of Sciences of the United States of America, vol. 103, no. 51, pp. 19581–19586, 2006.
92. K. J. Simons, J. P. Fellers, H. N. Trick, et al., "Molecular characterization of the major wheat domestication gene Q," Genetics, vol. 172, no. 1, pp. 547–555, 2006.
93. J. D. Faris, J. P. Fellers, S. A. Brooks, and B. S. Gill, "A bacterial artificial chromosome contig spanning the major domestication locus Q in wheat and identification of a candidate gene," Genetics, vol. 164, no. 1, pp. 311–321, 2003.
94. N. Yahiaoui, P. Srichumpa, R. Dudler, and B. Keller, "Genome analysis at different ploidy levels allows cloning of the powdery mildew resistance gene Pm3b from hexaploid wheat," Plant Journal, vol. 37, no. 4, pp. 528–538, 2004.
95. S. Brunner, P. Srichumpa, N. Yahiaoui, and B. Keller, "Positional cloning and evolution of powdery mildew resistance gene at Pm3 locus of hexaploid wheat," in Proceedings of the Plant & Animal Genome XIII Conference, p. 73, Town & Country Convention Center, San Diego, Calif, USA, January 2005.
96. C. Uauy, A. Distelfeld, T. Fahima, A. Blechl, and J. Dubcovsky, "A NAC gene regulating senescence improves grain protein, zinc, and iron content in wheat," Science, vol. 314, no. 5803, pp. 1298–1301, 2006.

97. A. Distelfeld, C. Uauy, S. Olmos, A. R. Schlatter, J. Dubcovsky, and T. Fahima, "Microcolinearity between a 2-cM region encompassing the grain protein content locus Gpc-6B1 on wheat chromosome 6B and a 350-kb region on rice chromosome 2," Functional & Integrative Genomics, vol. 4, no. 1, pp. 59–66, 2004.
98. S. Liu, M. O. Pumphery, X. Zhang, et al., "Towards positional cloning of Qfhs.ndsu-3BS, a major QTL for Fusarium head blight resistance in wheat," in Proceedings of the Plant & Animal Genome XIII Conference, p. 71, Town & Country Convention Center, San Diego, Calif, USA, January 2005.
99. P. Ling, X. zChen, D. Q. Le, and K. G. Campbell, "Towards cloning of the Yr5 gene for resistance to wheat stripe rust resistance," in Proceedings of the Plant & Animal Genomes XIII Conference, Town & Country Convention Center, San Diego, Calif, USA, January 2005.
100. T. Schnurbusch, N. C. Collins, R. F. Eastwood, T. Sutton, S. P. Jefferies, and P. Langridge, "Fine mapping and targeted SNP survey using rice-wheat gene colinearity in the region of the Bo1 boron toxicity tolerance locus of bread wheat," Theoretical and Applied Genetics, vol. 115, no. 4, pp. 451–461, 2007.
101. H.-J. Lu, J. P. Fellers, T. L. Friesen, S. W. Meinhardt, and J. D. Faris, "Genomic analysis and marker development for the Tsn1 locus in wheat using bin-mapped ESTs and flanking BAC contigs," Theoretical and Applied Genetics, vol. 112, no. 6, pp. 1132–1142, 2006.
102. S. Griffiths, R. Sharp, T. N. Foote, et al., "Molecular characterization of Ph1 as a major chromosome pairing locus in polyploid wheat," Nature, vol. 439, no. 7077, pp. 749–752, 2006.
103. R. S. Kota, W. Spielmeyer, R. A. McIntosh, and E. S. Lagudah, "Fine genetic mapping fails to dissociate durable stem rust resistance gene Sr2 from pseudo-black chaff in common wheat (Triticum aestivum L.)," Theoretical and Applied Genetics, vol. 112, no. 3, pp. 492–499, 2006.
104. R. B. Flavell and D. B. Smith, "The role of homoeologous group 1 chromosomes in the control of rRNA genes in wheat," Biochemical Genetics, vol. 12, no. 4, pp. 271–279, 1974.
105. W. L. Gerlach and W. J. Peacock, "Chromosomal locations of highly repeated DNA sequences in wheat," Heredity, vol. 44, no. 2, pp. 269–276, 1980.
106. W. L. Gerlach, E. S. Dennis, and W. J. Peacock, "Molecular cytogenetics of wheat," in Cytogenetics of Crop Plant, M. S. Swaminathan, P. K. Gupta, and U. Sinha, Eds., pp. 191–212, MacMillan, Bomby, India, 1983.
107. Y. Mukai, Y. Nakahara, and M. Yamamoto, "Simultaneous discrimination of the three genomes in hexaploid wheat by multicolor fluorescence in situ hybridization using total genomic and highly repeated DNA probes," Genome, vol. 36, no. 3, pp. 489–494, 1993.
108. C. Pedersen and P. Langridge, "Identification of the entire chromosome complement of bread wheat by two-colour FISH," Genome, vol. 40, no. 5, pp. 589–593, 1997.
109. E. D. Badaeva, A. V. Amosova, O. V. Muravenko, et al., "Genome differentiation in Aegilops. 3. Evolution of the D-genome cluster," Plant Systematics and Evolution, vol. 231, no. 1–4, pp. 163–190, 2002.
110. P. Zhang, W. Li, B. Friebe, and B. S. Gill, "Simultaneous painting of three genomes in hexaploid wheat by BAC-FISH," Genome, vol. 47, no. 5, pp. 979–987, 2004.

111. Y. Mukai, T. R. Endo, and B. S. Gill, "Physical mapping of the 5S rRNA multigene family in common wheat," Journal of Heredity, vol. 81, no. 4, pp. 290–295, 1990.
112. Y. Mukai, T. R. Endo, and B. S. Gill, "Pysical mapping of the 18S.26S rRNA multigene family in common wheat: identification of a new locus," Chromosoma, vol. 100, no. 2, pp. 71–78, 1991.
113. X.-F. Ma, K. Ross, and J. P. Gustafson, "Physical mapping of restriction fragment length polymorphism (RFLP) markers in homoeologous groups 1 and 3 chromosomes of wheat by in situ hybridization," Genome, vol. 44, no. 3, pp. 401–412, 2001.
114. S. Rahman, A. Regina, Z. Li, et al., "Comparison of starch-branching enzyme genes reveals evolutionary relationships among isoforms. Characterization of a gene for starch-branching enzyme IIa from the wheat D genome donor Aegilops tauschii," Plant Physiology, vol. 125, no. 3, pp. 1314–1324, 2001.
115. Z. Li, F. Sun, S. Xu, et al., "The structural organisation of the genes encoding class II starch synthase of wheat and barley and the evolution of the genes encoding starch syuthases in plants," Functional & Integrative Genomics, vol. 3, no. 1-2, pp. 76–85, 2003.
116. K.-M. Turnbull, M. Turner, Y. Mukai, et al., "The organization of genes tightly linked to the Ha locus in Aegilops tauschii, the D-genome donor to wheat," Genome, vol. 46, no. 2, pp. 330–338, 2003.
117. Y. Mukai and B. S. Gill, "Detection of barley chromatin added to wheat by genomic in situ hybridization," Genome, vol. 34, no. 3, pp. 448–452, 1991.
118. T. Schwarzacher, K. Anamthawat-Jónsson, G. E. Harrison, et al., "Genomic in situ hybridization to identify alien chromosomes and chromosome segments in wheat," Theoretical and Applied Genetics, vol. 84, no. 7-8, pp. 778–786, 1992.
119. M. Biagetti, F. Vitellozzi, and C. Ceoloni, "Physical mapping of wheat-Aegilops longissima breakpoints in mildew-resistant recombinant lines using FISH with highly repeated and low-copy DNA probes," Genome, vol. 42, no. 5, pp. 1013–1019, 1999.
120. M. Yamamoto and Y. Mukai, "High-resolution mapping in wheat and rye by FISH on extended DNA fibres," in Proceedings of the 9th International Wheat Genetics Symposium, A. E. Slinkard, Ed., vol. 1, pp. 12–16, Saskatoon, Canada, August 1998.
121. M. Yamamoto and Y. Mukai, "High-resolution physical mapping of the secalin-1 locus of rye on extended DNA fibers," Cytogenetic and Genome Research, vol. 109, no. 1–3, pp. 79–82, 2005.
122. U. C. Lavania, M. Yamamoto, and Y. Mukai, "Extended chromatin and DNA fibers from active plant nuclei for high-resolution FISH," Journal of Histochemistry & Cytochemistry, vol. 51, no. 10, pp. 1249–1253, 2003.
123. K.-N. Fukui, G. Suzuki, E. S. Lagudah, et al., "Physical arrangement of retrotransposon-related repeats in centromeric regions of wheat," Plant & Cell Physiology, vol. 42, no. 2, pp. 189–196, 2001.
124. M. Valárik, J. Bartoš, P. Kovářová, M. Kubaláková, J. H. de Jong, and J. Doležel, "High-resolution FISH on super-stretched flow-sorted plant chromosomes," Plant Journal, vol. 37, no. 6, pp. 940–950, 2004.
125. S. A. Jackson, P. Zhang, W. P. Chen, et al., "High-resolution structural analysis of biolistic transgene integration into the genome of wheat," Theoretical and Applied Genetics, vol. 103, no. 1, pp. 56–62, 2001.

126. P. Zhang, B. Friebe, and B. Gill, "Potential and limitations of BAC-FISH mapping in wheat," in Proceedings of the Plant, Animal & Microbe Genomes X Conference, p. 272, Town & Country Convention Center, San Diego, Calif, USA, January 2002.
127. D. Papa, C. A. Miller, G. R. Anderson, et al., "FISH physical mapping of DNA sequences associated with RWA resistance in wheat and barley," in Proceedings of the Plant & Animal Genome VIII Conference, p. 36, Town & Country Hotel, San Diego, Calif, USA, January 2000.
128. P. Zhang, W. Li, J. Fellers, B. Friebe, and B. S. Gill, "BAC-FISH in wheat identifies chromosome landmarks consisting of different types of transposable elements," Chromosoma, vol. 112, no. 6, pp. 288–299, 2004.
129. X. Q. Huang, H. Cöster, M. W. Ganal, and M. S. Röder, "Advanced backcross QTL analysis for the identification of quantitative trait loci alleles from wild relatives of wheat (Triticum aestivum L.)," Theoretical and Applied Genetics, vol. 106, no. 8, pp. 1379–1389, 2003.
130. X. Q. Huang, L. X. Wang, M. X. Xu, and M. S. Röder, "Microsatellite mapping of the powdery mildew resistance gene Pm5e in common wheat (Triticum aestivum L.)," Theoretical and Applied Genetics, vol. 106, no. 5, pp. 858–865, 2003.
131. G. P. Yan, X. M. Chen, R. F. Line, and C. R. Wellings, "Resistance gene-analog polymorphism markers co-segregating with the Yr5 gene for resistance to wheat stripe rust," Theoretical and Applied Genetics, vol. 106, no. 4, pp. 636–643, 2003.
132. P. K. Gupta, S. Rustgi, S. Sharma, R. Singh, N. Kumar, and H. S. Balyan, "Transferable EST-SSR markers for the study of polymorphism and genetic diversity in bread wheat," Molecular Genetics and Genomics, vol. 270, no. 4, pp. 315–323, 2003.
133. L. F. Gao, J. Tang, H. Li, and J. Jia, "Analysis of microsatellites in major crops assessed by computational and experimental approaches," Molecular Breeding, vol. 12, no. 3, pp. 245–261, 2003.
134. R. Bandopadhyay, S. Sharma, S. Rustgi, et al., "DNA polymorphism among 18 species of Triticum-Aegilops complex using wheat EST-SSRs," Plant Science, vol. 166, no. 2, pp. 349–356, 2004.
135. R. K. Varshney, R. Sigmund, A. Börner, et al., "Interspecific transferability and comparative mapping of barley EST-SSR markers in wheat, rye and rice," Plant Science, vol. 168, no. 1, pp. 195–202, 2005.
136. J.-K. Yu, M. La Rota, R. V. Kantety, and M. E. Sorrells, "EST derived SSR markers for comparative mapping in wheat and rice," Molecular Genetics and Genomics, vol. 271, no. 6, pp. 742–751, 2004.
137. L. Y. Zhang, M. Bernard, P. Leroy, C. Feuillet, and P. Sourdille, "High transferability of bread wheat EST-derived SSRs to other cereals," Theoretical and Applied Genetics, vol. 111, no. 4, pp. 677–687, 2005.
138. J. Tang, L. Gao, Y. Cao, and J. Jia, "Homologous analysis of SSR-ESTs and transferability of wheat SSR-EST markers across barley, rice and maize," Euphytica, vol. 151, no. 1, pp. 87–93, 2006.
139. K. Chabane, O. Abdalla, H. Sayed, and J. Valkoun, "Assessment of EST-microsatellites markers for discrimination and genetic diversity in bread and durum wheat landraces from Afghanistan," Genetic Resources and Crop Evolution, vol. 54, no. 5, pp. 1073–1080, 2007.

140. W. Zhang, S. Chao, E. D. Akhunov, et al., "Discovery of SNPs for wheat homoeologous group 5 and polymorphism among US adapted wheat germplasm," in Proceedings of the Plant & Animal Genome XI Conference, p. 184, San Diego, Calif, USA, January 2007.
141. C. Ravel, S. Praud, A. Murigneux, et al., "Single-nucleotide polymorphism frequency in a set of selected lines of bread wheat (Triticum aestivum L.)," Genome, vol. 49, no. 9, pp. 1131–1139, 2006.
142. J. Janda, J. Bartoš, J. Šafář, et al., "Construction of a subgenomic BAC library specific for chromosomes 1D, 4D and 6D of hexaploid wheat," Theoretical and Applied Genetics, vol. 109, no. 7, pp. 1337–1345, 2004.
143. J. Šafář, J. Bartoš, J. Janda, et al., "Dissecting large and complex genomes: flow sorting and BAC cloning of individual chromosomes from bread wheat," The Plant Journal, vol. 39, no. 6, pp. 960–968, 2004.
144. J. Janda, J. Šafář, M. Kubaláková, et al., "Advanced resources for plant genomics: a BAC library specific for the short arm of wheat chromosome 1B," The Plant Journal, vol. 47, no. 6, pp. 977–986, 2006.
145. T. Wicker, N. Stein, L. Albar, C. Feuillet, E. Schlagenhauf, and B. Keller, "Analysis of a contiguous 211 kb sequence in diploid wheat (Triticum monococcum L.) reveals multiple mechanisms of genome evolution," The Plant Journal, vol. 26, no. 3, pp. 307–316, 2001.
146. S. A. Brooks, L. Huang, B. S. Gill, and J. P. Fellers, "Analysis of 106 kb of contiguous DNA sequence from the D genome of wheat reveals high gene density and a complex arrangement of genes related to disease resistance," Genome, vol. 45, no. 5, pp. 963–972, 2002.
147. E. Paux, D. Roger, E. Badaeva, et al., "Characterizing the composition and evolution of homoeologous genomes in hexaploid wheat through BAC-end sequencing on chromosome 3B," The Plant Journal, vol. 48, no. 3, pp. 463–474, 2006.
148. D. Sandhu, J. A. Champoux, S. N. Bondareva, and K. S. Gill, "Identification and physical localization of useful genes and markers to a major gene-rich region on wheat group 1S chromosomes," Genetics, vol. 157, no. 4, pp. 1735–1747, 2001.
149. M. Erayman, D. Sandhu, D. Sidhu, M. Dilbirligi, P. S. Baenziger, and K. S. Gill, "Demarcating the gene-rich regions of the wheat genome," Nucleic Acids Research, vol. 32, no. 12, pp. 3546–3565, 2004.
150. K. S. Gill, "Structural organization of the wheat genome," in Frontiers of Wheat Bioscience: The 100th Memorial Issue of Wheat Information Service, K. Tsunewaki, Ed., pp. 151–167, Kihara Memorial Yokohama Foundation for the Advancement of Life Sciences, Yokohama, Japan, 2005.
151. D. Sidhu and K. S. Gill, "Distribution of genes and recombination in wheat and other eukaryotes," Plant Cell, Tissue and Organ Culture, vol. 79, no. 3, pp. 257–270, 2005.
152. A. Barakat, N. Carels, and G. Bernardi, "The distribution of genes in the genomes of Gramineae," Proceedings of the National Academy of Sciences of the United States of America, vol. 94, no. 13, pp. 6857–6861, 1997.
153. C. Feuillet and B. Keller, "High gene density is conserved at syntenic loci of small and large grass genomes," Proceedings of the National Academy of Sciences of the United States of America, vol. 96, no. 14, pp. 8265–8270, 1999.

154. D. Sandhu and K. S. Gill, "Gene-containing regions of wheat and the other grass genomes," Plant Physiology, vol. 128, no. 3, pp. 803–811, 2002.
155. D. Sandhu and K. S. Gill, "Structural and functional organization of the '1S0.8 gene-rich region' in the Triticeae," Plant Molecular Biology, vol. 48, no. 5-6, pp. 791–804, 2002.
156. P. K. Gupta, P. L. Kulwal, and S. Rustgi, "Wheat cytogenetics in the genomics era and its relevance to breeding," Cytogenetic and Genome Research, vol. 109, no. 1–3, pp. 315–327, 2005.
157. M. L. Wang, A. R. Leitch, T. Schwarzacher, J. S. Heslop-Harrison, and G. Moore, "Construction of a chromosome-enriched Hpall library from flow-sorted wheat chromosomes," Nucleic Acids Research, vol. 20, no. 8, pp. 1897–1901, 1992.
158. J.-H. Lee, K. Arumuganathan, Y. Yen, S. Kaeppler, H. Kaeppler, and P. S. Baenziger, "Root tip cell cycle synchronization and metaphase-chromosome isolation suitable for flow sorting in common wheat (Triticum aestivum L.)," Genome, vol. 40, no. 5, pp. 633–638, 1997.
159. J. Vrána, M. Kubalákova, H. Simková, J. Cíhalíková, M. A. Lysák, and J. Dolezel, "Flow sorting of mitotic chromosomes in common wheat (Triticum aestivum L.)," Genetics, vol. 156, no. 4, pp. 2033–2041, 2000.
160. K. S. Gill, K. Arumuganathan, and J.-H. Lee, "Isolating individual wheat (Triticum aestivum) chromosome arms by flow cytometric analysis of ditelosomic lines," Theoretical and Applied Genetics, vol. 98, no. 8, pp. 1248–1252, 1999.
161. J. M. Vega, S. Abbo, M. Feldman, and A. A. Levy, "Chromosome painting in plants: in situ hybridization with a DNA probe from a specific microdissected chromosome arm of common wheat," Proceedings of the National Academy of Sciences of the United States of America, vol. 91, no. 25, pp. 12041–12045, 1994.
162. B. Chalhoub, H. Belcram, and M. Caboche, "Efficient cloning of plant genomes into bacterial artificial chromosome (BAC) libraries with larger and more uniform insert size," Plant Biotechnology Journal, vol. 2, no. 3, pp. 181–188, 2004.
163. J. Doležel, M. Kubaláková, J. Bartoš, and J. Macas, "Flow cytogenetics and plant genome mapping," Chromosome Research, vol. 12, no. 1, pp. 77–91, 2004.
164. M. Kubaláková, J. Vrána, J. Číhalíková, H. Šimková, and J. Doležel, "Flow karyotyping and chromosome sorting in bread wheat (Triticum aestivum L.)," Theoretical and Applied Genetics, vol. 104, no. 8, pp. 1362–1372, 2002.
165. J. Doležel, M. Kubaláková, P. Suchankova, et al., "Flow cytogenetic analysis of the wheat genome," in Frontiers of Wheat Bioscience: The 100th Memorial Issue of Wheat Information Service, K. Tsunewaki, Ed., pp. 3–15, Yokohama Publishers, Yokohama, Japan, 2005.
166. B. S. Gill, "International genome research on wheat (IGROW)," in Proceedings of the National Wheat Workers Workshop, Kansas City, Mo, USA, February 2004.
167. P. K. Gupta, "Ultrafast and low-cost DNA sequencing methods for applied genomics research," Proceedings of the National Academy of Sciences, India. In press.
168. I. D. Wilson, G. L. A. Barker, R. W. Beswick, et al., "A transcriptomics resource for wheat functional genomics," Plant Biotechnology Journal, vol. 2, no. 6, pp. 495–506, 2004.

169. I. D. Wilson, G. L. Barker, C. Lu, et al., "Alteration of the embryo transcriptome of hexaploid winter wheat (Triticum aestivum cv. Mercia) during maturation and germination," Functional and Integrative Genomics, vol. 5, no. 3, pp. 144–154, 2005.
170. R. Poole, G. Barker, I. D. Wilson, J. A. Coghill, and K. J. Edwards, "Measuring global gene expression in polyploidy; a cautionary note from allohexaploid wheat," Functional & Integrative Genomics, vol. 7, no. 3, pp. 207–219, 2007.
171. C. A. McCartney, D. J. Somers, D. G. Humphreys, et al., "Mapping quantitative trait loci controlling agronomic traits in the spring wheat cross RL4452 × 'AC Domain'," Genome, vol. 48, no. 5, pp. 870–883, 2005.
172. C. A. McCartney, D. J. Somers, O. Lukow, et al., "QTL analysis of quality traits in the spring wheat cross RL4452 × 'AC domain'," Plant Breeding, vol. 125, no. 6, pp. 565–575, 2006.
173. D. Fu, C. Uauy, A. Blechl, and J. Dubcovsky, "RNA interference for wheat functional gene analysis," Transgenic Research, vol. 16, no. 6, pp. 689–701, 2007.
174. A. Salleh, "Gene silencing yields high-fibre wheat," February 2006, ABC Science online.
175. K. Mochida, Y. Yamazaki, and Y. Ogihara, "Discrimination of homoeologous gene expression in hexaploid wheat by SNP analysis of contigs grouped from a large number of expressed sequence tags," Molecular Genetics and Genomics, vol. 270, no. 5, pp. 371–377, 2003.
176. P. Schweizer, J. Pokorny, P. Schulze-Lefert, and R. Dudler, "Double-stranded RNA interferes with gene function at the single-cell level in cereals," The Plant Journal, vol. 24, no. 6, pp. 895–903, 2000.
177. A. B. Christensen, H. Thordal-Christensen, G. Zimmermann, et al., "The Germin-like protein GLP4 exhibits superoxide dismutase activity and is an important component of quantitative resistance in wheat and barley," Molecular Plant-Microbe Interactions, vol. 17, no. 1, pp. 109–117, 2004.
178. A. Loukoianov, L. Yan, A. Blechl, A. Sanchez, and J. Dubcovsky, "Regulation of VRN-1 vernalization genes in normal and transgenic polyploid wheat," Plant Physiology, vol. 138, no. 4, pp. 2364–2373, 2005.
179. A. Regina, A. Bird, D. Topping, et al., "High-amylose wheat generated by RNA interference improves indices of large-bowel health in rats," Proceedings of the National Academy of Sciences of the United States of America, vol. 103, no. 10, pp. 3546–3551, 2006.
180. S. Travella, T. E. Klimm, and B. Keller, "RNA interference-based gene silencing as an efficient tool for functional genomics in hexaploid bread wheat," Plant Physiology, vol. 142, no. 1, pp. 6–20, 2006.
181. G. Yao, J. Zhang, L. Yang, et al., "Genetic mapping of two powdery mildew resistance genes in einkorn (Triticum monococcum L.) accessions," Theoretical and Applied Genetics, vol. 114, no. 2, pp. 351–358, 2007.
182. S. Henikoff, B. J. Till, and L. Comai, "TILLING. Traditional mutagenesis meets functional genomics," Plant Physiology, vol. 135, no. 2, pp. 630–636, 2004.
183. B. J. Till, S. H. Reynolds, E. A. Greene, et al., "Large-scale discovery of induced point mutations with high-throughput TILLING," Genome Research, vol. 13, no. 3, pp. 524–530, 2003.

184. A. J. Slade and V. C. Knauf, "TILLING moves beyond functional genomics into crop improvement," Transgenic Research, vol. 14, no. 2, pp. 109–115, 2005.
185. A. J. Slade, S. I. Fuerstenberg, D. Loeffler, M. N. Steine, and D. Facciotti, "A reverse genetic, nontransgenic approach to wheat crop improvement by TILLING," Nature Biotechnology, vol. 23, no. 1, pp. 75–81, 2005.
186. K. M. Devos and M. D. Gale, "Genome relationships: the grass model in current research," The Plant Cell, vol. 12, no. 5, pp. 637–646, 2000.
187. T. Wicker, N. Yahiaoui, and B. Keller, "Contrasting rates of evolution in Pm3 loci from three wheat species and rice," Genetics, vol. 177, no. 2, pp. 1207–1216, 2007.
188. N. Huo, Y. Q. Gu, G. R. Lazo, et al., "Construction and characterization of two BAC libraries from Brachypodium distachyon, a new model for grass genomics," Genome, vol. 49, no. 9, pp. 1099–1108, 2006.
189. M. D. Gale, J. E. Flintham, and K. M. Devos, "Cereal comparative genetics and preharvest sprouting," Euphytica, vol. 126, no. 1, pp. 21–25, 2002.
190. C. Feuillet, A. Penger, K. Gellner, A. Mast, and B. Keller, "Molecular evolution of receptor-like kinase genes in hexaploid wheat. Independent evolution of orthologs after polyploidization and mechanisms of local rearrangements at paralogous loci," Plant Physiology, vol. 125, no. 3, pp. 1304–1313, 2001.
191. N. Chantret, A. Cenci, F. Sabot, O. Anderson, and J. Dubcovsky, "Sequencing of the Triticum monococcum Hardness locus reveals good microcolinearity with rice," Molecular genetics and genomics, vol. 271, no. 4, pp. 377–386, 2004.
192. E. K. Khlestkina, T. A. Pshenichnikova, M. S. Röder, E. A. Salina, V. S. Arbuzova, and A. Börner, "Comparative mapping of genes for glume colouration and pubescence in hexaploid wheat (Triticum aestivum L.)," Theoretical and Applied Genetics, vol. 113, no. 5, pp. 801–807, 2006.
193. K. M. Devos, M. D. Atkinson, C. N. Chinoy, C. J. Liu, and M. D. Gale, "RFLP-based genetic map of the homoeologous group 3 chromosomes of wheat and rye," Theoretical and Applied Genetics, vol. 83, no. 8, pp. 931–939, 1992.
194. N. Stein, C. Feuillet, T. Wicker, E. Schlagenhauf, and B. Keller, "Subgenome chromosome walking in wheat: a 450-kb physical contig in Triticum monococcum L. spans the Lr10 resistance locus in hexaploid wheat (Triticum aestivum L.)," Proceedings of the National Academy of Sciences of the United States of America, vol. 97, no. 24, pp. 13436–13441, 2000.
195. T. Wicker, N. Yahiaoui, R. Guyot, et al., "Rapid genome divergence at orthologous low molecular weight glutenin loci of the A and Am genomes of wheat," The Plant Cell, vol. 15, no. 5, pp. 1186–1197, 2003.
196. E. Isidore, B. Scherrer, B. Chalhoub, C. Feuillet, and B. Keller, "Ancient haplotypes resulting from extensive molecular rearrangements in the wheat A genome have been maintained in species of three different ploidy levels," Genome Research, vol. 15, no. 4, pp. 526–536, 2005.
197. Y. Q. Gu, J. Salse, D. Coleman-Derr, et al., "Types and rates of sequence evolution at the high-molecular-weight glutenin locus in hexaploid wheat and its ancestral genomes," Genetics, vol. 174, no. 3, pp. 1493–1504, 2006.
198. R. Guyot, N. Yahiaoui, C. Feuillet, and B. Keller, "In silico comparative analysis reveals a mosaic conservation of genes within a novel colinear region in wheat chro-

mosome 1AS and rice chromosome 5S," Functional & Integrative Genomics, vol. 4, no. 1, pp. 47–58, 2004.
199. M. E. Sorrells, M. La Rota, C. E. Bermudez-Kandianis, et al., "Comparative DNA sequence analysis of wheat and rice genomes," Genome Research, vol. 13, no. 8, pp. 1818–1827, 2003.
200. N. K. Singh, S. Raghuvanshi, S. K. Srivastava, et al., "Sequence analysis of the long arm of rice chromosome 11 for rice-wheat synteny," Functional and Integrative Genomics, vol. 4, no. 2, pp. 102–117, 2004.
201. J. Draper, L. A. J. Mur, G. Jenkins, et al., "Brachypodium distachyon. A new model system for functional genomics in grasses," Plant Physiology, vol. 127, no. 4, pp. 1539–1555, 2001.
202. R. Hasterok, A. Marasek, I. S. Donnison, et al., "Alignment of the genomes of Brachypodium distachyon and temperate cereals and grasses using bacterial artificial chromosome landing with fluorescence in situ hybridization," Genetics, vol. 173, no. 1, pp. 349–362, 2006.
203. J. P. Vogel, Y. Q. Gu, P. Twigg, et al., "EST sequencing and phylogenetic analysis of the model grass Brachypodium distachyon," Theoretical and Applied Genetics, vol. 113, no. 2, pp. 186–195, 2006.
204. E. Bossolini, T. Wicker, P. A. Knobel, and B. Keller, "Comparison of orthologous loci from small grass genomes Brachypodium and rice: implications for wheat genomics and grass genome annotation," The Plant Journal, vol. 49, no. 4, pp. 704–717, 2007.
205. Y. Xie, Z. Ni, Y. Yao, Y. Yin, Q. Zhang, and Q. Sun, "Analysis of differential cytosine methylation during seed development in wheat," in Proceedings of the Plant Genomics in China VIII, p. 60, Shanghai, China, August 2007.
206. N. Shitsukawa, C. Tahira, K.-I. Kassai, et al., "Genetic and epigenetic alteration among three homoeologous genes of a class E MADS box gene in hexaploid wheat," Plant Cell, vol. 19, no. 6, pp. 1723–1737, 2007.
207. Y. Nemoto, M. Kisaka, T. Fuse, M. Yano, and Y. Ogihara, "Characterization and functional analysis of three wheat genes with homology to the CONSTANS flowering time gene in transgenic rice," The Plant Journal, vol. 36, no. 1, pp. 82–93, 2003.
208. N. Chantret, J. Salse, F. Sabot, et al., "Molecular basis of evolutionary events that shaped the hardness locus in diploid and polyploid wheat species (Triticum and Aegilops)," The Plant Cell, vol. 17, no. 4, pp. 1033–1045, 2005.
209. T. Nomura, A. Ishihara, R. C. Yanagita, T. R. Endo, and H. Iwamura, "Three genomes differentially contribute to the biosynthesis of benzoxazinones in hexaploid wheat," Proceedings of the National Academy of Sciences of the United States of America, vol. 102, no. 45, pp. 16490–16495, 2005.
210. L. Comai, "Genetic and epigenetic interactions in allopolyploid plants," Plant Molecular Biology, vol. 43, no. 2-3, pp. 387–399, 2000.
211. Z. J. Chen and Z. Ni, "Mechanisms of genomic rearrangements and gene expression changes in plant polyploids," BioEssays, vol. 28, no. 3, pp. 240–252, 2006.
212. M. Feldman, B. Liu, G. Segal, S. Abbo, A. A. Levy, and J. M. Vega, "Rapid elimination of low-copy DNA sequences in polyploid wheat: a possible mechanism for differentiation of homoeologous chromosomes," Genetics, vol. 147, no. 3, pp. 1381–1387, 1997.

213. B. Liu, J. M. Vega, G. Segal, S. Abbo, M. Rodova, and M. Feldman, "Rapid genomic changes in newly synthesized amphiploids of Triticum and Aegilops—I: changes in low-copy noncoding DNA sequences," Genome, vol. 41, no. 2, pp. 272–277, 1998.
214. B. Liu, J. M. Vega, and M. Feldman, "Rapid genomic changes in newly synthesized amphiploids of Triticum and Aegilops—II: changes in low-copy coding DNA sequences," Genome, vol. 41, no. 4, pp. 535–542, 1998.
215. L. Z. Xiong, C. G. Xu, M. A. S. Maroof, and Q. Zhang, "Patterns of cytosine methylation in an elite rice hybrid and its parental lines, detected by a methylation-sensitive amplification polymorphism technique," Molecular and General Genetics, vol. 261, no. 3, pp. 439–446, 1999.
216. H. Shaked, K. Kashkush, H. Özkan, M. Feldman, and A. A. Levy, "Sequence elimination and cytosine methylation are rapid and reproducible responses of the genome to wide hybridization and allopolyploidy in wheat," Plant Cell, vol. 13, no. 8, pp. 1749–1759, 2001.
217. K. Kashkush, M. Feldman, and A. A. Levy, "Gene loss, silencing and activation in a newly synthesized wheat allotetraploid," Genetics, vol. 160, no. 4, pp. 1651–1659, 2002.
218. A. A. Levy and M. Feldman, "Genetic and epigenetic reprogramming of the wheat genome upon allopolyploidization," Biological Journal of the Linnean Society, vol. 82, no. 4, pp. 607–613, 2004.
219. P. He, B. R. Friebe, B. S. Gill, and J.-M. Zhou, "Allopolyploidy alters gene expression in the highly stable hexaploid wheat," Plant Molecular Biology, vol. 52, no. 2, pp. 401–414, 2003.
220. P. L. Kulwal, R. Singh, H. S. Balyan, and P. K. Gupta, "Genetic basis of pre-harvest sprouting tolerance using single-locus and two-locus QTL analyses in bread wheat," Functional & Integrative Genomics, vol. 4, no. 2, pp. 94–101, 2004.
221. P. L. Kulwal, N. Kumar, A. Kumar, R. K. Gupta, H. S. Balyan, and P. K. Gupta, "Gene networks in hexaploid wheat: interacting quantitative trait loci for grain protein content," Functional & Integrative Genomics, vol. 5, no. 4, pp. 254–259, 2005.
222. N. Kumar, P. L. Kulwal, H. S. Balyan, and P. K. Gupta, "QTL mapping for yield and yield contributing traits in two mapping populations of bread wheat," Molecular Breeding, vol. 19, no. 2, pp. 163–177, 2007.
223. P. Langridge, E. S. Lagudah, T. A. Holton, R. Appels, P. J. Sharp, and K. J. Chalmers, "Trends in genetic and genome analyses in wheat: a review," Australian Journal of Agricultural Research, vol. 52, no. 11-12, pp. 1043–1077, 2001.
224. A. Jahoor, L. Eriksen, and G. Backes, "QTLs and genes for disease resistance in barley and wheat," in Cereal Genomics, P. K. Gupta and R. K. Varshney, Eds., pp. 199–251, Kluwer Academic Publishers, Dordrecht, The Netherlands, 2004.
225. R. Tuberosa and S. Salvi, "QTLs and genes for tolerance to abiotic stresses in cereals," in Cereal Genomics, P. K. Gupta and R. K. Varshney, Eds., pp. 253–315, Kluwer Academic Publishers, Dordrecht, The Netherlands, 2004.
226. P. K. Gupta, S. Rustgi, and N. Kumar, "Genetic and molecular basis of grain size and grain number and its relevance to grain productivity in higher plants," Genome, vol. 49, no. 6, pp. 565–571, 2006.
227. W. Li and B. S. Gill, "Genomics for cereal improvement," in Cereal Genomics, P. K. Gupta and R. K. Varshney, Eds., pp. 585–634, Kluwer Academic Publishers, Dordrecht, The Netherlands, 2004.

228. S. D. Tanksley and J. C. Nelson, "Advanced backcross QTL analysis: a method for the simultaneous discovery and transfer of valuable QTLs from unadapted germplasm into elite breeding lines," Theoretical and Applied Genetics, vol. 92, no. 2, pp. 191–203, 1996.
229. X. Q. Huang, H. Kempf, M. W. Canal, and M. S. Röder, "Advanced backcross QTL analysis in progenies derived from a cross between a German elite winter wheat variety and a synthetic wheat (Triticum aestivum L.)," Theoretical and Applied Genetics, vol. 109, no. 5, pp. 933–943, 2004.
230. A. Kunert, A. A. Naz, O. Dedeck, K. Pillen, and J. Léon, "AB-QTL analysis in winter wheat—I: synthetic hexaploid wheat (T. turgidum ssp. dicoccoides×T. tauschii) as a source of favourable alleles for milling and baking quality traits," Theoretical and Applied Genetics, vol. 115, no. 5, pp. 683–695, 2007.
231. N. Amiour, M. Merlino, P. Leroy, and G. Branlard, "Chromosome mapping and identification of amphiphilic proteins of hexaploid wheat kernels," Theoretical and Applied Genetics, vol. 108, no. 1, pp. 62–72, 2003.
232. R. B. Flavell, M. D. Bennett, A. G. Seal, and J. Hutchinson, "Chromosome structure and organisation," in Wheat Breeding, Its Scientific Basis, F. G. H. Lupton, Ed., pp. 211–268, Chapman & Hall, London, UK, 1987.
233. G Kimber, "The B genome of wheat: the present status," in Cytogenetics of Crop Plants, M. S. Swaminathan, P. K. Gupta, and U. Sinha, Eds., pp. 213–224, Macmillan, Delhi, India, 1983.
234. G. Kimber and E. R. Sears, "Evolution in the genus Triticum and the origin of cultivated wheat," in Wheat and Wheat Improvement, E. G. Heyne, Ed., pp. 154–164, American Society of Agronomy, Madison, Wis, USA, 1987.
235. M. Feldman, F. G. H. Lupton, and T. E. Miller, "Wheats," in Evolution of Crops, J. Smartt and N. W. Simmonds, Eds., pp. 184–192, Longman Scientific, London, UK, 2nd edition, 1995.
236. B. S. Gill and B. Friebe, "Cytogenetics, phylogeny and evolution of cultivated wheats," in Bread Wheat, Improvement and Production, B. C. Curtis, S. Rajaram, and H. G. Macpherson, Eds., Plant Production and Protection Series 30, FAO, Rome, Italy, 2002.
237. Y. Yen, P. S. Baenziger, and R. Morris, "Genomic constitution of bread wheat: current status," in Methods of Genome Analysis in Plants, P. P. Jauhar, Ed., pp. 359–373, CRC Press, Boca Raton, Fla, USA, 1996.
238. A. A. Levy and M. Feldman, "The impact of polyploidy on grass genome evolution," Plant Physiology, vol. 130, no. 4, pp. 1587–1593, 2002.
239. K. S. Caldwell, J. Dvorak, E. S. Lagudah, et al., "Sequence polymorphism in polyploid wheat and their D-genome diploid ancestor," Genetics, vol. 167, no. 2, pp. 941–947, 2004.
240. S. Huang, A. Sirikhachornkit, X. Su, et al., "Genes encoding plastid acetyl-CoA carboxylase and 3-phosphoglycerate kinase of the Triticum/Aegilops complex and the evolutionary history of polyploid wheat," Proceedings of the National Academy of Sciences of the United States of America, vol. 99, no. 12, pp. 8133–8138, 2002.
241. B. Maestra and T. Naranjo, "Homoeologous relationships of Aegilops speltoides chromosomes to bread wheat," Theoretical and Applied Genetics, vol. 97, no. 1-2, pp. 181–186, 1998.

242. E. Nevo, A. B. Korol, A. Beiles, and T. Fahima, Evolution of Wild Emmer and Wheat Improvement, Springer, Berlin, Germany, 2002.
243. N. K. Blake, B. R. Lehfeldt, M. Lavin, and L. E. Talbert, "Phylogenetic reconstruction based on low copy DNA sequence data in an allopolyploid: the B genome of wheat," Genome, vol. 42, no. 2, pp. 351–360, 1999.
244. B. Kilian, H. Özkan, O. Deusch, et al., "Independent wheat B and G genome origins in outcrossing Aegilops progenitor haplotypes," Molecular Biology and Evolution, vol. 24, no. 1, pp. 217–227, 2007.
245. L. E. Talbert and N. K. Blake, "Comparative DNA sequence analysis and the origin of wheat," in Proceedings of the Plant & Animal Genomes VIII Conference, Town & Country Convention Center, San Diego, Calif, USA, January 2000.
246. F. Sabot, B. Laubin, L. Amilhat, P. Leroy, P. Sourdille, and M. Bernard, "Evolution history of the Triticum sp. through the study of transposable elements," in Proceedings of the Plant & Animal Genome XII Conference, p. 421, Town & Country Convention Center, San Diego, Calif, USA, January 2004.
247. A. H. Schulman, P. K. Gupta, and R. K. Varshney, "Organization of retrotransposons and microsatellites in cereal genomes," in Cereal Genomics, P. K. Gupta and R. K. Varshney, Eds., pp. 83–118, Kluwer Academic Publishers, Dordrecht, The Netherlands, 2004.
248. G. Schachermayr, H. Siedler, M. D. Gale, H. Winzeler, M. Winzeler, and B. Keller, "Identification and localization of molecular markers linked to the Lr9 leaf rust resistance gene of wheat," Theoretical and Applied Genetics, vol. 88, no. 1, pp. 110–115, 1994.
249. C. Feuillet, M. Messmer, G. Schachermayr, and B. Keller, "Genetic and physical characterization of the LR1 leaf rust resistance locus in wheat (Triticum aestivum L.)," Molecular and General Genetics, vol. 248, no. 5, pp. 553–562, 1995.
250. G. M. Schachermayr, M. M. Messmer, C. Feuillet, H. Winzeler, M. Winzeler, and B. Keller, "Identification of molecular markers linked to the Agropyron elongatum-derived leaf rust resistance gene Lr24 in wheat," Theoretical and Applied Genetics, vol. 90, no. 7-8, pp. 982–990, 1995.
251. G. Schachermayr, C. Feuillet, and B. Keller, "Molecular markers for the detection of the wheat leaf rust resistance gene Lr10 in diverse genetic backgrounds," Molecular Breeding, vol. 3, no. 1, pp. 65–74, 1997.
252. S. Naik, K. S. Gill, V. S. Prakasa Rao, et al., "Identification of a STS marker linked to the Aegilops speltoides-derived leaf rust resistance gene Lr28 in wheat," Theoretical and Applied Genetics, vol. 97, no. 4, pp. 535–540, 1998.
253. F. Sacco, E. Y. Suárez, and T. Naranjo, "Mapping of the leaf rust resistance gene Lr3 on chromosome 6B of Sinvalocho MA wheat," Genome, vol. 41, no. 5, pp. 686–690, 1998.
254. R. Seyfarth, C. Feuillet, G. Schachermayr, M. Winzeler, and B. Keller, "Development of a molecular marker for the adult plant leaf rust resistance gene Lr35 in wheat," Theoretical and Applied Genetics, vol. 99, no. 3-4, pp. 554–560, 1999.
255. M. Helguera, I. A. Khan, and J. Dubcovsky, "Development of PCR markers for the wheat leaf rust resistance gene Lr47," Theoretical and Applied Genetics, vol. 100, no. 7, pp. 1137–1143, 2000.

256. M. Aghaee-Sarbarzeh, H. Singh, and H. S. Dhaliwal, "A microsatellite marker linked to leaf rust resistance transferred from Aegilops triuncialis into hexaploid wheat," Plant Breeding, vol. 120, no. 3, pp. 259–261, 2001.
257. R. Prins, J. Z. Groenewald, G. F. Marais, J. W. Snape, and R. M. D. Koebner, "AFLP and STS tagging of Lr19, a gene conferring resistance to leaf rust in wheat," Theoretical and Applied Genetics, vol. 103, no. 4, pp. 618–624, 2001.
258. W. J. Raupp, S. Singh, G.L. Brown-Guedira, and B. S. Gill, "Cytogenetic and molecular mapping of the leaf rust resistance gene Lr39 in wheat," Theoretical and Applied Genetics, vol. 102, no. 2-3, pp. 347–352, 2001.
259. S. Seah, H. Bariana, J. Jahier, K. Sivasithamparam, and E. S. Lagudah, "The introgressed segment carrying rust resistance genes Yr17, Lr37 and Sr38 in wheat can be assayed by a cloned disease resistance gene-like sequence," Theoretical and Applied Genetics, vol. 102, no. 4, pp. 600–605, 2001.
260. C. Neu, N. Stein, and B. Keller, "Genetic mapping of the Lr20-Pm1 resistance locus reveals suppressed recombination on chromosome arm 7AL in hexaploid wheat," Genome, vol. 45, no. 4, pp. 737–744, 2002.
261. D. P. Cherukuri, S. K. Gupta, A. Charpe, et al., "Identification of a molecular marker linked to an Agropyron elongatum-derived gene Lr19 for leaf rust resistance in wheat," Plant Breeding, vol. 122, no. 3, pp. 204–208, 2003.
262. H.-Q. Ling, J. Qiu, R. P. Singh, and B. Keller, "Identification and genetic characterization of an Aegilops tauschii ortholog of the wheat leaf rust disease resistance gene Lr1," Theoretical and Applied Genetics, vol. 109, no. 6, pp. 1133–1138, 2004.
263. D. P. Cherukuri, S. K. Gupta, A. Charpe, et al., "Molecular mapping of Aegilops speltoides derived leaf rust resistance gene Lr28 in wheat," Euphytica, vol. 143, no. 1-2, pp. 19–26, 2005.
264. W. Spielmeyer, R. A. McIntosh, J. Kolmer, and E. S. Lagudah, "Powdery mildew resistance and Lr34/Yr18 genes for durable resistance to leaf and stripe rust cosegregate at a locus on the short arm of chromosome 7D of wheat," Theoretical and Applied Genetics, vol. 111, no. 4, pp. 731–735, 2005.
265. C. W. Hiebert, J. B. Thomas, and B. D. McCallum, "Locating the broad-spectrum wheat leaf rust resistance gene Lr52(LrW) to chromosome 5B by a new cytogenetic method," Theoretical and Applied Genetics, vol. 110, no. 8, pp. 1453–1457, 2005.
266. S. K. Gupta, A. Charpe, K. V. Prabhu, and Q. M. R. Haque, "Identification and validation of molecular markers linked to the leaf rust resistance gene Lr19 in wheat," Theoretical and Applied Genetics, vol. 113, no. 6, pp. 1027–1036, 2006.
267. E. Bossolini, S. G. Krattinger, and B. Keller, "Development of simple sequence repeat markers specific for the Lr34 resistance region of wheat using sequence information from rice and Aegilops tauschii," Theoretical and Applied Genetics, vol. 113, no. 6, pp. 1049–1062, 2006.
268. C. W. Hiebert, J. B. Thomas, D. J. Somers, B. D. McCallum, and S. L. Fox, "Microsatellite mapping of adult-plant leaf rust resistance gene Lr22a in wheat," Theoretical and Applied Genetics, vol. 115, no. 6, pp. 877–884, 2007.
269. J.-W. Qiu, A. C. Schürch, N. Yahiaoui, et al., "Physical mapping and identification of a candidate for the leaf rust resistance gene Lr1 of wheat," Theoretical and Applied Genetics, vol. 115, no. 2, pp. 159–168, 2007.

270. D. E. Obert, A. K. Fritz, J. L. Moran, S. Singh, J. C. Rudd, and M. A. Menz, "Identification and molecular tagging of a gene from PI 289824 conferring resistance to leaf rust (Puccinia triticina) in wheat," Theoretical and Applied Genetics, vol. 110, no. 8, pp. 1439–1444, 2005.
271. T. Schnurbusch, S. Paillard, A. Schori, et al., "Dissection of quantitative and durable leaf rust resistance in Swiss winter wheat reveals a major resistance QTL in the Lr34 chromosomal region," Theoretical and Applied Genetics, vol. 108, no. 3, pp. 477–484, 2004.
272. I. N. Leonova, L. I. Laikova, O. M. Popova, O. Unger, A. Börner, and M. S. Röder, "Detection of quantitative trait loci for leaf rust resistance in wheat—T. timopheevii/T. tauschii introgression lines," Euphytica, vol. 155, no. 1-2, pp. 79–86, 2007.
273. R. P. Singh, J. C. Nelson, and M. E. Sorrells, "Mapping Yr28 and other genes for resistance to stripe rust in wheat," Crop Science, vol. 40, no. 4, pp. 1148–1155, 2000.
274. G. L. Sun, T. Fahima, A. B. Korol, et al., "Identification of molecular markers linked to the Yr15 stripe rust resistance gene of wheat originated in wild emmer wheat, Triticum dicoccoides," Theoretical and Applied Genetics, vol. 95, no. 4, pp. 622–628, 1997.
275. J. H. Peng, T. Fahima, M. S. Röder, et al., "Microsatellite tagging of the stripe-rust resistance gene YrH52 derived from wild emmer wheat, Triticum dicoccoides, and suggestive negative crossover interference on chromosome 1B," Theoretical and Applied Genetics, vol. 98, no. 6-7, pp. 862–872, 1999.
276. A. Börner, M. S. Röder, O. Unger, and A. Meinel, "The detection and molecular mapping of a major gene for non-specific adult-plant disease resistance against stripe rust (Puccinia striiformis) in wheat," Theoretical and Applied Genetics, vol. 100, no. 7, pp. 1095–1099, 2000.
277. J. H. Peng, T. Fahima, M. S. Röder, et al., "High-density molecular map of chromosome region harboring stripe-rust resistance genes YrH52 and Yr15 derived from wild emmer wheat, Triticum dicoccoides," Genetica, vol. 109, no. 3, pp. 199–210, 2001.
278. Z. X. Shi, X. M. Chen, R. F. Line, H. Leung, and C. R. Wellings, "Development of resistance gene analog polymorphism markers for the Yr9 gene resistance to wheat stripe rust," Genome, vol. 44, no. 4, pp. 509–516, 2001.
279. J. Ma, R. Zhou, Y. Dong, L. Wang, X. Wang, and J. Jia, "Molecular mapping and detection of the yellow rust resistance gene Yr26 in wheat transferred from Triticum turgidum L. using microsatellite markers," Euphytica, vol. 120, no. 2, pp. 219–226, 2001.
280. L. Wang, J. Ma, R. Zhou, X. Wang, and J. Jia, "Molecular tagging of the yellow rust resistance gene Yr10 in common wheat, P.I.178383 (Triticum aestivum L.)," Euphytica, vol. 124, no. 1, pp. 71–73, 2002.
281. C. Uauy, J. C. Brevis, X. Chen, et al., "High-temperature adult-plant (HTAP) stripe rust resistance gene Yr36 from Triticum turgidum ssp. dicoccoides is closely linked to the grain protein content locus Gpc-B1," Theoretical and Applied Genetics, vol. 112, no. 1, pp. 97–105, 2005.

282. G. Q. Li, Z. F. Li, W. Y. Yang, et al., "Molecular mapping of stripe rust resistance gene YrCH42 in Chinese wheat cultivar Chuanmai 42 and its allelism with Yr24 and Yr26," Theoretical and applied genetics, vol. 112, no. 8, pp. 1434–1440, 2006.
283. Z. F. Li, T. C. Zheng, Z. H. He, et al., "Molecular tagging of stripe rust resistance gene YrZH84 in Chinese wheat line Zhou 8425B," Theoretical and Applied Genetics, vol. 112, no. 6, pp. 1098–1103, 2006.
284. H. S. Bariana, N. Parry, I. R. Barclay, et al., "Identification and characterization of stripe rust resistance gene Yr34 in common wheat," Theoretical and Applied Genetics, vol. 112, no. 6, pp. 1143–1148, 2006.
285. C. Wang, Y. Zhang, D. Han, et al., "SSR and STS markers for wheat stripe rust resistance gene Yr26," Euphytica, vol. 159, no. 3, pp. 359–366, 2008.
286. S. Mallard, D. Gaudet, A. Aldeia, et al., "Genetic analysis of durable resistance to yellow rust in bread wheat," Theoretical and Applied Genetics, vol. 110, no. 8, pp. 1401–1409, 2005.
287. M. J. Christiansen, B. Feenstra, I. M. Skovgaard, and S. B. Andersen, "Genetic analysis of resistance to yellow rust in hexaploid wheat using a mixture model for multiple crosses," Theoretical and Applied Genetics, vol. 112, no. 4, pp. 581–591, 2006.
288. J. G. Paull, M. A. Pallotta, P. Langridge, and T. T. The, "RFLP markers associated with Sr22 and recombination between chromosome 7A of bread wheat and the diploid species Triticum boeoticum," Theoretical and Applied Genetics, vol. 89, no. 7-8, pp. 1039–1045, 1994.
289. W. Spielmeyer, P. J. Sharp, and E. S. Lagudah, "Identification and validation of markers linked to broad-spectrum stem rust resistance gene Sr2 in wheat (Triticum aestivum L.)," Crop Science, vol. 43, no. 1, pp. 333–336, 2003.
290. P. A. Cuthbert, D. J. Somers, and A. Brulé-Babel, "Mapping of Fhb2 on chromosome 6BS: a gene controlling Fusarium head blight field resistance in bread wheat (Triticum aestivum L.)," Theoretical and Applied Genetics, vol. 114, no. 3, pp. 429–437, 2007.
291. W. Bourdoncle and H. W. Ohm, "Quantitative trait loci for resistance to Fusarium head blight in recombinant inbred wheat lines from the cross huapei 57-2/Patterson," Euphytica, vol. 131, no. 1, pp. 131–136, 2003.
292. I. A. del Blanco, R. C. Frohberg, R. W. Stack, W. A. Berzonsky, and S. F. Kianian, "Detection of QTL linked to Fusarium head blight resistance in Sumai 3-derived North Dakota bread wheat lines," Theoretical and Applied Genetics, vol. 106, no. 6, pp. 1027–1031, 2003.
293. F. Lin, Z. X. Kong, H. L. Zhu, et al., "Mapping QTL associated with resistance to Fusarium head blight in the Nanda2419 × Wangshuibai population—I: type II resistance," Theoretical and Applied Genetics, vol. 109, no. 7, pp. 1504–1511, 2004.
294. F. Lin, S. L. Xue, Z. Z. Zhang, et al., "Mapping QTL associated with resistance to Fusarium head blight in the Nanda2419 × Wangshuibai population—II: type I resistance," Theoretical and Applied Genetics, vol. 112, no. 3, pp. 528–535, 2006.
295. S. Paillard, T. Schnurbusch, R. Tiwari, et al., "QTL analysis of resistance to Fusarium head blight in Swiss winter wheat (Triticum aestivum L.)," Theoretical and Applied Genetics, vol. 109, no. 2, pp. 323–332, 2004.

296. J. Gilsinger, L. Kong, X. Shen, and H. Ohm, "DNA markers associated with low Fusarium head blight incidence and narrow flower opening in wheat," Theoretical and Applied Genetics, vol. 110, no. 7, pp. 1218–1225, 2005.
297. G. Jia, P. Chen, G. Qin, et al., "QTLs for Fusarium head blight response in a wheat DH population of Wangshuibai/Alondra's'," Euphytica, vol. 146, no. 3, pp. 183–191, 2005.
298. X. Chen, J. D. Faris, J. Hu, et al., "Saturation and comparative mapping of a major Fusarium head blight resistance QTL in tetraploid wheat," Molecular Breeding, vol. 19, no. 2, pp. 113–124, 2007.
299. G.-L. Jiang, Y. Dong, J. Shi, and R. W. Ward, "QTL analysis of resistance to Fusarium head blight in the novel wheat germplasm CJ 9306—II: resistance to deoxynivalenol accumulation and grain yield loss," Theoretical and Applied Genetics, vol. 115, no. 8, pp. 1043–1052, 2007.
300. A. Klahr, G. Zimmermann, G. Wenzel, and V. Mohler, "Effects of environment, disease progress, plant height and heading date on the detection of QTLs for resistance to Fusarium head blight in an European winter wheat cross," Euphytica, vol. 154, no. 1-2, pp. 17–28, 2007.
301. X. Shen and H. Ohm, "Molecular mapping of Thinopyrum-derived Fusarium head blight resistance in common wheat," Molecular Breeding, vol. 20, no. 2, pp. 131–140, 2007.
302. W. Zhou, F. L. Kolb, G. Bai, G. Shaner, and L. L. Domier, "Genetic analysis of scab resistance QTL in wheat with microsatellite and AFLP markers," Genome, vol. 45, no. 4, pp. 719–727, 2002.
303. W.-C. Zhou, F. L. Kolb, G.-H. Bai, L. L. Domier, L. K. Boze, and N. J. Smith, "Validation of a major QTL for scab resistance with SSR markers and use of marker-assisted selection in wheat," Plant Breeding, vol. 122, no. 1, pp. 40–46, 2003.
304. L. Hartl, H. Weiss, U. Stephan, F. J. Zeller, and A. Jahoor, "Molecular identification of powdery mildew resistance genes in common wheat (Triticum aestivum L.)," Theoretical and Applied Genetics, vol. 90, no. 5, pp. 601–606, 1995.
305. J. Jia, K. M. Devos, S. Chao, T. E. Miller, S. M. Reader, and M. D. Gale, "RFLP-based maps of the homoeologous group-6 chromosomes of wheat and their application in the tagging of Pm12, a powdery mildew resistance gene transferred from Aegilops speltoides to wheat," Theoretical and Applied Genetics, vol. 92, no. 5, pp. 559–565, 1996.
306. Z. Liu, Q. Sun, Z. Ni, and T. Yang, "Development of SCAR markers linked to the Pm21 gene conferring resistance to powdery mildew in common wheat," Plant Breeding, vol. 118, no. 3, pp. 215–219, 1999.
307. P. Sourdille, P. Robe, M.-H. Tixier, G. Doussinault, M.-T. Pavoinc, and M. Bernard, "Location of Pm3g, a powdery mildew resistance allele in wheat, by using a monosomic analysis and by identifying associated molecular markers," Euphytica, vol. 110, no. 3, pp. 193–198, 1999.
308. X. Q. Huang, S. L. K. Hsam, F. J. Zeller, G. Wenzel, and V. Mohler, "Molecular mapping of the wheat powdery mildew resistance gene Pm24 and marker validation for molecular breeding," Theoretical and Applied Genetics, vol. 101, no. 3, pp. 407–414, 2000.

309. J. K. Rong, E. Millet, J. Manisterski, and M. Feldman, "A new powdery mildew resistance gene: introgression from wild emmer into common wheat and RFLP-based mapping," Euphytica, vol. 115, no. 2, pp. 121–126, 2000.
310. W. J. Tao, D. Liu, J. Y. Liu, Y. Feng, and P. Chen, "Genetic mapping of the powdery mildew resistance gene Pm6 in wheat by RFLP analysis," Theoretical and Applied Genetics, vol. 100, no. 3-4, pp. 564–568, 2000.
311. K. Järve, H. O. Peusha, J. Tsymbalova, S. Tamm, K. M. Devos, and T. M. Enno, "Chromosomal location of a Triticum timopheevii-derived powdery mildew resistance gene transferred to common wheat," Genome, vol. 43, no. 2, pp. 377–381, 2000.
312. V. Mohler, S. L. K. Hsam, F. J. Zeller, and G. Wenzel, "An STS marker distinguishing the rye-derived powdery mildew resistance alleles at the Pm8/Pm17 locus of common wheat," Plant Breeding, vol. 120, no. 5, pp. 448–450, 2001.
313. Y. Bougot, J. Lemoine, M. T. Pavoine, D. Barloy, and G. Doussinault, "Identification of a microsatellite marker associated with Pm3 resistance alleles to powdery mildew in wheat," Plant Breeding, vol. 121, no. 4, pp. 325–329, 2002.
314. F. J. Zeller, L. Kong, L. Hartl, V. Mohler, and S. L. K. Hsam, "Chromosomal location of genes for resistance to powdery mildew in common wheat (Triticum aestivum L. em Thell.) 7. Gene Pm29 in line Pova," Euphytica, vol. 123, no. 2, pp. 187–194, 2002.
315. Z. Liu, Q. Sun, Z. Ni, E. Nevo, and T. Yang, "Molecular characterization of a novel powdery mildew resistance gene Pm30 in wheat originating from wild emmer," Euphytica, vol. 123, no. 1, pp. 21–29, 2002.
316. C. Alberto, D. Renato, T. O. Antonio, C. Carla, P. Marina, and P. Enrico, "Genetic analysis of the Aegilops longissima 3S chromosome carrying the Pm13 resistance gene," Euphytica, vol. 130, no. 2, pp. 177–183, 2003.
317. Z.-Q. Ma, J.-B. Wei, and S.-H. Cheng, "PCR-based markers for the powdery mildew resistance gene Pm4a in wheat," Theoretical and Applied Genetics, vol. 109, no. 1, pp. 140–145, 2004.
318. Y. C. Qiu, R. H. Zhou, X. Y. Kong, S. S. Zhang, and J. Z. Jia, "Microsatellite mapping of a Triticum urartu Tum. derived powdery mildew resistance gene transferred to common wheat (Triticum aestivum L.)," Theoretical and Applied Genetics, vol. 111, no. 8, pp. 1524–1531, 2005.
319. L. M. Miranda, J. P. Murphy, D. Marshall, and S. Leath, "Pm34: a new powdery mildew resistance gene transferred from Aegilops tauschii Coss. to common wheat (Triticum aestivum L.)," Theoretical and Applied Genetics, vol. 113, no. 8, pp. 1497–1504, 2006.
320. Z. Zhu, R. Zhou, X. Kong, Y. Dong, and J. Jia, "Microsatellite marker identification of a Triticum aestivum—Aegilops umbellulata substitution line with powdery mildew resistance," Euphytica, vol. 150, no. 1-2, pp. 149–153, 2006.
321. L. M. Miranda, J. P. Murphy, D. Marshall, C. Cowger, and S. Leath, "Chromosomal location of Pm35, a novel Aegilops tauschii derived powdery mildew resistance gene introgressed into common wheat (Triticum aestivum L.)," Theoretical and Applied Genetics, vol. 114, no. 8, pp. 1451–1456, 2007.

322. G. Nematollahi, V. Mohler, G. Wenzel, F. J. Zeller, and S. L. K. Hsam, "Microsatellite mapping of powdery mildew resistance allele Pm5d from common wheat line IGV1-455," Euphytica, vol. 159, no. 3, pp. 307–313, 2008.
323. W. Song, H. Xie, Q. Liu, et al., "Molecular identification of Pm12-carrying introgression lines in wheat using genomic and EST-SSR markers," Euphytica, vol. 158, no. 1-2, pp. 95–102, 2007.
324. N. Chantret, P. Sourdille, M. Röder, M. Tavaud, M. Bernard, and G. Doussinault, "Location and mapping of the powdery mildew resistance gene MlRE and detection of a resistance QTL by bulked segregant analysis (BSA) with microsatellites in wheat," Theoretical and Applied Genetics, vol. 100, no. 8, pp. 1217–1224, 2000.
325. C. Xie, Q. Sun, Z. Ni, T. Yang, E. Nevo, and T. Fahima, "Chromosomal location of a Triticum dicoccoides-derived powdery mildew resistance gene in common wheat by using microsatellite markers," Theoretical and Applied Genetics, vol. 106, no. 2, pp. 341–345, 2003.
326. Ch. Singrün, S. L. K. Hsam, F. J. Zeller, G. Wenzel, and V. Mohler, "Localization of a novel recessive powdery mildew resistance gene from common wheat line RD30 in the terminal region of chromosome 7AL," Theoretical and Applied Genetics, vol. 109, no. 1, pp. 210–214, 2004.
327. M. Keller, B. Keller, G. Schachermayr, et al., "Quantitative trait loci for resistance against powdery mildew in a segregating wheat × spelt population," Theoretical and Applied Genetics, vol. 98, no. 6-7, pp. 903–912, 1999.
328. S. Liu, C. A. Griffey, and M. A. Saghai Maroof, "Identification of molecular markers associated with adult plant resistance to powdery mildew in common wheat cultivar Massey," Crop Science, vol. 41, no. 4, pp. 1268–1275, 2001.
329. D. Mingeot, N. Chantret, P. V. Baret, et al., "Mapping QTL involved in adult plant resistance to powdery mildew in the winter wheat line RE714 in two susceptible genetic backgrounds," Plant Breeding, vol. 121, no. 2, pp. 133–140, 2002.
330. Y. Bougot, J. Lemoine, M. T. Pavoine, et al., "A major QTL effect controlling resistance to powdery mildew in winter wheat at the adult plant stage," Plant Breeding, vol. 125, no. 6, pp. 550–556, 2006.
331. D. M. Tucker, C. A. Griffey, S. Liu, G. Brown-Guedira, D. S. Marshall, and M. A. S. Maroof, "Confirmation of three quantitative trait loci conferring adult plant resistance to powdery mildew in two winter wheat populations," Euphytica, vol. 155, no. 1-2, pp. 1–13, 2007.
332. A. Laroche, T. Demeke, D. A. Gaudet, B. Puchalski, M. Frick, and R. McKenzie, "Development of a PCR marker for rapid identification of the Bt-10 gene for common bunt resistance in wheat," Genome, vol. 43, no. 2, pp. 217–223, 2000.
333. B. Fofana, D. G. Humphreys, S. Cloutier, C. A. McCartney, and D. J. Somers, "Mapping quantitative trait loci controlling common bunt resistance in a doubled haploid population derived from the spring wheat cross RL4452×AC Domain," Molecular Breeding, vol. 21, no. 3, pp. 317–325, 2008.
334. J. D. Faris, J. A. Anderson, L. J. Francl, and J. G. Jordahl, "RFLP mapping of resistance to chlorosis induction by Pyrenophora tritici-repentis in wheat," Theoretical and Applied Genetics, vol. 94, no. 1, pp. 98–103, 1997.

335. P. K. Singh, M. Mergoum, T. B. Adhikari, S. F. Kianian, and E. M. Elias, "Chromosomal location of genes for seedling resistance to tan spot and Stagonospora nodorum blotch in tetraploid wheat," Euphytica, vol. 155, no. 1-2, pp. 27–34, 2007.
336. W. Tadesse, M. Schmolke, S. L. K. Hsam, V. Mohler, G. Wenzel, and F. J. Zeller, "Molecular mapping of resistance genes to tan spot [Pyrenophora tritici-repentis race 1] in synthetic wheat lines," Theoretical and Applied Genetics, vol. 114, no. 5, pp. 855–862, 2007.
337. L. S. Arraiano, A. J. Worland, C. Ellerbrook, and J. K. M. Brown, "Chromosomal location of a gene for resistance to septoria tritici blotch (Mycosphaerella graminicola) in the hexaploid wheat 'Synthetic 6x'," Theoretical and Applied Genetics, vol. 103, no. 5, pp. 758–764, 2001.
338. M. R. Simón, F. M. Ayala, C. A. Cordo, M. S. Röder, and A. Börner, "Molecular mapping of quantitative trait loci determining resistance to septoria tritici blotch caused by Mycosphaerella graminicola in wheat," Euphytica, vol. 138, no. 1, pp. 41–48, 2004.
339. L. Ayala, M. Henry, M. van Ginkel, R. Singh, B. Keller, and M. Khairallah, "Identification of QTLs for BYDV tolerance in bread wheat," Euphytica, vol. 128, no. 2, pp. 249–259, 2002.
340. V. Aguilar, P. Stamp, M. Winzeler, et al., "Inheritance of field resistance to Stagonospora nodorum leaf and glume blotch and correlations with other morphological traits in hexaploid wheat (Triticum aestivum L.)," Theoretical and Applied Genetics, vol. 111, no. 2, pp. 325–336, 2005.
341. L. E. Talbert, P. L. Bruckner, L. Y. Smith, R. Sears, and T. J. Martin, "Development of PCR markers linked to resistance to wheat streak mosaic virus in wheat," Theoretical and Applied Genetics, vol. 93, no. 3, pp. 463–467, 1996.
342. A. A. Khan, G. C. Bergstrom, J. C. Nelson, and M. E. Sorrells, "Identification of RFLP markers for resistance to wheat spindle streak mosaic bymovirus (WSSMV) disease," Genome, vol. 43, no. 3, pp. 477–482, 2000.
343. W. Liu, H. Nie, S. Wang, et al., "Mapping a resistance gene in wheat cultivar Yangfu 9311 to yellow mosaic virus, using microsatellite markers," Theoretical and Applied Genetics, vol. 111, no. 4, pp. 651–657, 2005.
344. R. C. de la Peña, T. D. Murray, and S. S. Jones, "Identification of an RFLP interval containing Pch2 on chromosome 7AL in wheat," Genome, vol. 40, no. 2, pp. 249–252, 1997.
345. V. Huguet-Robert, F. Dedryver, M. S. Röder, et al., "Isolation of a chromosomally engineered durum wheat line carrying the Aegilops ventricosaPch1 gene for resistance to eyespot," Genome, vol. 44, no. 3, pp. 345–349, 2001.
346. J. Z. Groenewald, A. S. Marais, and G. F. Marais, "Amplified fragment length polymorphism-derived microsatellite sequence linked to the Pch1 and Ep-D1 loci in common wheat," Plant Breeding, vol. 122, no. 1, pp. 83–85, 2003.
347. Y. Weng and M. D. Lazar, "Amplified fragment length polymorphism- and simple sequence repeat-based molecular tagging and mapping of greenbug resistance gene Gb3 in wheat," Plant Breeding, vol. 121, no. 3, pp. 218–223, 2002.
348. E. Boyko, S. Starkey, and M. Smith, "Molecular genetic mapping of Gby, a new greenbug resistance gene in bread wheat," Theoretical and Applied Genetics, vol. 109, no. 6, pp. 1230–1236, 2004.

349. Y. Weng, W. Li, R. N. Devkota, and J. C. Rudd, "Microsatellite markers associated with two Aegilops tauschii-derived greenbug resistance loci in wheat," Theoretical and Applied Genetics, vol. 110, no. 3, pp. 462–469, 2005.
350. L. C. Zhu, C. M. Smith, A. Fritz, E. Boyko, P. Voothuluru, and B. S. Gill, "Inheritance and molecular mapping of new greenbug resistance genes in wheat germplasms derived from Aegilops tauschii," Theoretical and Applied Genetics, vol. 111, no. 5, pp. 831–837, 2005.
351. Z.-Q. Ma, B. S. Gill, M. E. Sorrells, and S. D. Tanksley, "RELP markers linked to two Hessian fly-resistance genes in wheat (Triticum aestivum L.) from Triticum tauschii (coss.) Schmal," Theoretical and Applied Genetics, vol. 85, no. 6-7, pp. 750–754, 1993.
352. I. Dweikat, H. W. Ohm, S. Mackenzie, F. Patterson, S. Cambron, and R. Ratcliffe, "Association of a DNA marker with Hessian fly resistance gene H9 in wheat," Theoretical and Applied Genetics, vol. 89, no. 7-8, pp. 964–968, 1994.
353. I. Dweikat, H. W. Ohm, F. Patterson, and S. Cambron, "Identification of RAPD markers for 11 Hessian fly resistance genes in wheat," Theoretical and Applied Genetics, vol. 94, no. 3-4, pp. 419–423, 1997.
354. Y. W. Seo, J. W. Johnson, and R. L. Jarret, "A molecular marker associated with the H21 Hessian fly resistance gene in wheat," Molecular Breeding, vol. 3, no. 3, pp. 177–181, 1997.
355. I. Dweikat, W. Zhang, and H. W. Ohm, "Development of STS markers linked to Hessian fly resistance gene H6 in wheat," Theoretical and Applied Genetics, vol. 105, no. 5, pp. 766–770, 2002.
356. X. M. Liu, B. S. Gill, and M.-S. Chen, "Hessian fly resistance gene H13 is mapped to a distal cluster of resistance genes in chromosome 6DS of wheat," Theoretical and Applied Genetics, vol. 111, no. 2, pp. 243–249, 2005.
357. T. Wang, S. S. Xu, M. O. Harris, J. Hu, L. Liu, and X. Cai, "Genetic characterization and molecular mapping of Hessian fly resistance genes derived from Aegilops tauschii in synthetic wheat," Theoretical and Applied Genetics, vol. 113, no. 4, pp. 611–618, 2006.
358. H. X. Zhao, X. M. Liu, and M.-S. Chen, "H22, a major resistance gene to the Hessian fly (Mayetiola destructor), is mapped to the distal region of wheat chromosome 1DS," Theoretical and Applied Genetics, vol. 113, no. 8, pp. 1491–1496, 2006.
359. L. Kong, S. E. Cambron, and H. W. Ohm, "Hessian fly resistance genes H16 and H17 are mapped to a resistance gene cluster in the distal region of chromosome 1AS in wheat," Molecular Breeding, vol. 21, no. 2, pp. 183–194, 2008.
360. X. M. Liu, C. M. Smith, B. S. Gill, and V. Tolmay, "Microsatellite markers linked to six Russian wheat aphid resistance genes in wheat," Theoretical and Applied Genetics, vol. 102, no. 4, pp. 504–510, 2001.
361. C. A. Miller, A. Altinkut, and N. L. V. Lapitan, "A microsatellite marker for tagging Dn2, a wheat gene conferring resistance to the Russian wheat aphid," Crop Science, vol. 41, no. 5, pp. 1584–1589, 2001.
362. X. M. Liu, C. M. Smith, and B. S. Gill, "Identification of microsatellite markers linked to Russian wheat aphid resistance genes Dn4 and Dn6," Theoretical and Applied Genetics, vol. 104, no. 6-7, pp. 1042–1048, 2002.

363. K. J. Williams, J. M. Fisher, and P. Langridge, "Identification of RFLP markers linked to the cereal cyst nematode resistance gene (Cre) in wheat," Theoretical and Applied Genetics, vol. 89, no. 7-8, pp. 927–930, 1994.
364. J. Jahier, P. Abelard, A. M. Tanguy, et al., "The Aegilops ventricosa segment on chromosome 2AS of the wheat cultivar 'VPM1' carries the cereal cyst nematode resistance gene Cre5," Plant Breeding, vol. 120, no. 2, pp. 125–128, 2001.
365. F. C. Ogbonnaya, S. Seah, A. Delibes, et al., "Molecular-genetic characterisation of a new nematode resistance gene in wheat," Theoretical and Applied Genetics, vol. 102, no. 4, pp. 623–629, 2001.
366. D. Barloy, J. Lemoine, F. Dredryver, and J. Jahier, "Molecular markers linked to the Aegilops variabilis-derived root-knot nematode resistance gene Rkn-mn1 in wheat," Plant Breeding, vol. 119, no. 2, pp. 169–172, 2000.
367. K. J. Williams, S. P. Taylor, P. Bogacki, M. Pallotta, H. S. Bariana, and H. Wallwork, "Mapping of the root lesion nematode (Pratylenchus neglectus) resistance gene Rlnn1 in wheat," Theoretical and Applied Genetics, vol. 104, no. 5, pp. 874–879, 2002.
368. K. Kato, W. Nakamura, T. Tabiki, H. Miura, and S. Sawada, "Detection of loci controlling seed dormancy on group 4 chromosomes of wheat and comparative mapping with rice and barley genomes," Theoretical and Applied Genetics, vol. 102, no. 6-7, pp. 980–985, 2001.
369. J. Flintham, R. Adlam, M. Bassoi, M. Holdsworth, and M. D. Gale, "Mapping genes for resistance to sprouting damage in wheat," Euphytica, vol. 126, no. 1, pp. 39–45, 2002.
370. H. Miura, N. Sato, K. Kato, and Y. Amano, "Detection of chromosomes carrying genes for seed dormancy of wheat using the backcross reciprocal monosomic method," Plant Breeding, vol. 121, no. 5, pp. 394–399, 2002.
371. M. Osa, K. Kato, M. Mori, C. Shindo, A. Torada, and H. Miura, "Mapping QTLs for seed dormancy and the Vp1 homologue on chromosome 3A in wheat," Theoretical and Applied Genetics, vol. 106, no. 8, pp. 1491–1496, 2003.
372. P. L. Kulwal, N. Kumar, A. Gaur, et al., "Mapping of a major QTL for pre-harvest sprouting tolerance on chromosome 3A in bread wheat," Theoretical and Applied Genetics, vol. 111, no. 6, pp. 1052–1059, 2005.
373. D. Mares, K. Mrva, J. Cheong, et al., "A QTL located on chromosome 4A associated with dormancy in white- and red-grained wheats of diverse origin," Theoretical and Applied Genetics, vol. 111, no. 7, pp. 1357–1364, 2005.
374. L. R. Joppa, C. Du, G. E. Hart, and G. A. Hareland, "Mapping gene(s) for grain protein in tetraploid wheat (Triticum turgidum L.) using a population of recombinant inbred chromosome lines," Crop Science, vol. 37, no. 5, pp. 1586–1589, 1997.
375. M. Prasad, N. Kumar, P. L. Kulwal, et al., "QTL analysis for grain protein content using SSR markers and validation studies using NILs in bread wheat," Theoretical and Applied Genetics, vol. 106, no. 4, pp. 659–667, 2003.
376. A. Blanco, R. Simeone, and A. Gadaleta, "Detection of QTLs for grain protein content in durum wheat," Theoretical and Applied Genetics, vol. 112, no. 7, pp. 1195–1204, 2006.

377. G. D. Parker, K. J. Chalmers, A. J. Rathjen, and P. Langridge, "Mapping loci associated with flour colour in wheat (Triticum aestivum L.)," Theoretical and Applied Genetics, vol. 97, no. 1-2, pp. 238–245, 1998.
378. G. D. Parker, K. J. Chalmers, A. J. Rathjen, and P. Langridge, "Mapping loci associated with milling yield in wheat (Triticum aestivum L.)," Molecular Breeding, vol. 5, no. 6, pp. 561–568, 1999.
379. M. R. Perretant, T. Cadalen, G. Charmet, et al., "QTL analysis of bread-making quality in wheat using a doubled haploid population," Theoretical and Applied Genetics, vol. 100, no. 8, pp. 1167–1175, 2000.
380. G. Charmet, N. Robert, G. Branlard, L. Linossier, P. Martre, and E. Triboï, "Genetic analysis of dry matter and nitrogen accumulation and protein composition in wheat kernels," Theoretical and Applied Genetics, vol. 111, no. 3, pp. 540–550, 2005.
381. W. Ma, R. Appels, F. Bekes, O. Larroque, M. K. Morell, and K. R. Gale, "Genetic characterisation of dough rheological properties in a wheat doubled haploid population: additive genetic effects and epistatic interactions," Theoretical and Applied Genetics, vol. 111, no. 3, pp. 410–422, 2005.
382. M. Arbelbide and R. Bernardo, "Mixed-model QTL mapping for kernel hardness and dough strength in bread wheat," Theoretical and Applied Genetics, vol. 112, no. 5, pp. 885–890, 2006.
383. O. Dobrovolskaya, V. S. Arbuzova, U. Lohwasser, M. S. Röder, and A. Börner, "Microsatellite mapping of complementary genes for purple grain colour in bread wheat (Triticum aestivum) L.," Euphytica, vol. 150, no. 3, pp. 355–364, 2006.
384. J. C. Nelson, C. Andreescu, F. Breseghello, et al., "Quantitative trait locus analysis of wheat quality traits," Euphytica, vol. 149, no. 1-2, pp. 145–159, 2006.
385. F. Chen, Z. Luo, Z. Zhang, G. Xia, and H. Min, "Variation and potential value in wheat breeding of low-molecular-weight glutenin subunit genes cloned by genomic and RT-PCR in a derivative of somatic introgression between common wheat and Agropyron elongatum," Molecular Breeding, vol. 20, no. 2, pp. 141–152, 2007.
386. C. J. Pozniak, R. E. Knox, F. R. Clarke, and J. M. Clarke, "Identification of QTL and association of a phytoene synthase gene with endosperm colour in durum wheat," Theoretical and Applied Genetics, vol. 114, no. 3, pp. 525–537, 2007.
387. T. Cadalen, P. Sourdille, G. Charmet, et al., "Molecular markers linked to genes affecting plant height in wheat using a doubled-haploid population," Theoretical and Applied Genetics, vol. 96, no. 6-7, pp. 933–940, 1998.
388. V. Korzun, M. S. Röder, M. W. Ganal, A. J. Worland, and C. N. Law, "Genetic analysis of the dwarfing gene (Rht8) in wheat—I: molecular mapping of Rht8 on the short arm of chromosome 2D of bread wheat (Triticum aestivum L.)," Theoretical and Applied Genetics, vol. 96, no. 8, pp. 1104–1109, 1998.
389. M. H. Ellis, D. G. Bonnett, and G. J. Rebetzke, "A 192bp allele at the Xgwm261 locus is not always associated with the Rht8 dwarfing gene in wheat (Triticum aestivum L.)," Euphytica, vol. 157, no. 1-2, pp. 209–214, 2007.
390. V. Kuraparthy, S. Sood, H. S. Dhaliwal, P. Chhuneja, and B. S. Gill, "Identification and mapping of a tiller inhibition gene (tin3) in wheat," Theoretical and Applied Genetics, vol. 114, no. 2, pp. 285–294, 2007.

391. K. Kosuge, N. Watanabe, T. Kuboyama, et al., "Cytological and microsatellite mapping of mutant genes for spherical grain and compact spikes in durum wheat," Euphytica, vol. 159, no. 3, pp. 289–296, 2008.
392. K. Kato, H. Miura, and S. Sawada, "QTL mapping of genes controlling ear emergence time and plant height on chromosome 5A of wheat," Theoretical and Applied Genetics, vol. 98, no. 3-4, pp. 472–477, 1999.
393. P. Sourdille, J. W. Snape, T. Cadalen, et al., "Detection of QTLs for heading time- and photoperiod response in wheat using a doubled-haploid population," Genome, vol. 43, no. 3, pp. 487–494, 2000.
394. E. Hanocq, M. Niarquin, E. Heumez, M. Rousset, and J. Le Gouis, "Detection and mapping of QTL for earliness components in a bread wheat recombinant inbred lines population," Theoretical and Applied Genetics, vol. 110, no. 1, pp. 106–115, 2004.
395. X. Xu, G. Bai, B. F. Carver, and G. E. Shaner, "A QTL for early heading in wheat cultivar Suwon 92," Euphytica, vol. 146, no. 3, pp. 233–237, 2005.
396. E. Hanocq, A. Laperche, O. Jaminon, A.-L. Lainé, and J. Le Gouis, "Most significant genome regions involved in the control of earliness traits in bread wheat, as revealed by QTL meta-analysis," Theoretical and Applied Genetics, vol. 114, no. 3, pp. 569–584, 2007.
397. K. Kato, H. Miura, and S. Sawada, "Mapping QTLs controlling grain yield and its components on chromosome 5A of wheat," Theoretical and Applied Genetics, vol. 101, no. 7, pp. 1114–1121, 2000.
398. B. Narasimhamoorthy, B. S. Gill, A. K. Fritz, J. C. Nelson, and G. L. Brown-Guedira, "Advanced backcross QTL analysis of a hard winter wheat × synthetic wheat population," Theoretical and Applied Genetics, vol. 112, no. 5, pp. 787–796, 2006.
399. F. M. Kirigwi, M. van Ginkel, G. Brown-Guedira, B. S. Gill, G. M. Paulsen, and A. K. Fritz, "Markers associated with a QTL for grain yield in wheat under drought," Molecular Breeding, vol. 20, no. 4, pp. 401–413, 2007.
400. H. Kuchel, K. J. Williams, P. Langridge, H. A. Eagles, and S. P. Jefferies, "Genetic dissection of grain yield in bread wheat—I: QTL analysis," Theoretical and Applied Genetics, vol. 115, no. 8, pp. 1029–1041, 2007.
401. Z. Ma, D. Zhao, C. Zhang, et al., "Molecular genetic analysis of five spike-related traits in wheat using RIL and immortalized F2 populations," Molecular Genetics and Genomics, vol. 277, no. 1, pp. 31–42, 2007.
402. N. Kumar, P. L. Kulwal, A. Gaur, et al., "QTL analysis for grain weight in common wheat," Euphytica, vol. 151, no. 2, pp. 135–144, 2006.
403. K. Kato, H. Miura, M. Akiyama, M. Kuroshima, and S. Sawada, "RFLP mapping of the three major genes, Vrn1, Q and B1, on the long arm of chromosome 5A of wheat," Euphytica, vol. 101, no. 1, pp. 91–95, 1998.
404. Z. S. Peng, C. Yen, and J. L. Yang, "Chromosomal location of genes for supernumerary spikelet in bread wheat," Euphytica, vol. 103, no. 1, pp. 109–114, 1998.
405. E. Salina, A. Börner, I. Leonova, et al., "Microsatellite mapping of the induced sphaerococcoid mutation genes in Triticum aestivum," Theoretical and Applied Genetics, vol. 100, no. 5, pp. 686–689, 2000.
406. L. Bullrich, M. L. Appendino, G. Tranquilli, S. Lewis, and J. Dubcovsky, "Mapping of a thermo-sensitive earliness per se gene on Triticum monococcum chromosome 1Am," Theoretical and Applied Genetics, vol. 105, no. 4, pp. 585–593, 2002.

407. E. K. Khlestkina, E. G. Pestsova, M. S. Röder, and A. Börner, "Molecular mapping, phenotypic expression and geographical distribution of genes determining anthocyanin pigmentation of coleoptiles in wheat (Triticum aestivum L.)," Theoretical and Applied Genetics, vol. 104, no. 4, pp. 632–637, 2002.
408. Q. H. Xing, Z. G. Ru, C. J. Zhou, et al., "Genetic analysis, molecular tagging and mapping of the thermo-sensitive genic male-sterile gene (wtms1) in wheat," Theoretical and Applied Genetics, vol. 107, no. 8, pp. 1500–1504, 2003.
409. C.-G. Chu, J. D. Faris, T. L. Friesen, and S. S. Xu, "Molecular mapping of hybrid necrosis genes Ne1 and Ne2 in hexaploid wheat using microsatellite markers," Theoretical and Applied Genetics, vol. 112, no. 7, pp. 1374–1381, 2006.
410. O. Dobrovolskaya, T. A. Pshenichnikova, V. S. Arbuzova, U. Lohwasser, M. S. Röder, and A. Börner, "Molecular mapping of genes determining hairy leaf character in common wheat with respect to other species of the Triticeae," Euphytica, vol. 155, no. 3, pp. 285–293, 2007.
411. S. Houshmand, R. E. Knox, F. R. Clarke, and J. M. Clarke, "Microsatellite markers flanking a stem solidness gene on chromosome 3BL in durum wheat," Molecular Breeding, vol. 20, no. 3, pp. 261–270, 2007.
412. M. Keller, Ch. Karutz, J. E. Schmid, et al., "Quantitative trait loci for lodging resistance in a segregating wheat × spelt population," Theoretical and Applied Genetics, vol. 98, no. 6-7, pp. 1171–1182, 1999.
413. L. Hai, H. Guo, S. Xiao, et al., "Quantitative trait loci (QTL) of stem strength and related traits in a doubled-haploid population of wheat (Triticum aestivum L.)," Euphytica, vol. 141, no. 1-2, pp. 1–9, 2005.
414. V. J. Nalam, M. I. Vales, C. J. W. Watson, S. F. Kianian, and O. Riera-Lizarazu, "Map-based analysis of genes affecting the brittle rachis character in tetraploid wheat (Triticum turgidum L.)," Theoretical and Applied Genetics, vol. 112, no. 2, pp. 373–381, 2006.
415. G. J. Rebetzke, M. H. Ellis, D. G. Bonnett, and R. A. Richards, "Molecular mapping of genes for coleoptile growth in bread wheat (Triticum aestivum L.)," Theoretical and Applied Genetics, vol. 114, no. 7, pp. 1173–1183, 2007.
416. G. Zhang and M. Mergoum, "Molecular mapping of kernel shattering and its association with Fusarium head blight resistance in a Sumai3 derived population," in Theoretical and Applied Genetics, vol. 115, pp. 757–766, October 2007.
417. K. Kato, S. Kidou, H. Miura, and S. Sawada, "Molecular cloning of the wheat CK2α gene and detection of its linkage with Vrn-A1 on chromosome 5A," Theoretical and Applied Genetics, vol. 104, no. 6-7, pp. 1071–1077, 2002.
418. Q. Liu, Z. Ni, H. Peng, W. Song, Z. Liu, and Q. Sun, "Molecular mapping of a dominant non-glaucousness gene from synthetic hexaploid wheat (Triticum aestivum L.): molecular mapping of non-glaucousness gene in wheat," Euphytica, vol. 155, no. 1-2, pp. 71–78, 2007.
419. A. Carrera, V. Echenique, W. Zhang, et al., "A deletion at the Lpx-B1 locus is associated with low lipoxygenase activity and improved pasta color in durum wheat (Triticum turgidum ssp. durum)," Journal of Cereal Science, vol. 45, no. 1, pp. 67–77, 2007.
420. X. Y. He, Z. H. He, L. P. Zhang, et al., "Allelic variation of polyphenol oxidase (PPO) genes located on chromosomes 2A and 2D and development of functional

markers for the PPO genes in common wheat," Theoretical and Applied Genetics, vol. 115, no. 1, pp. 47–58, 2007.
421. S. Nakamura, T. Komatsuda, and H. Miura, "Mapping diploid wheat homologues of Arabidopsis seed ABA signaling genes and QTLs for seed dormancy," Theoretical and Applied Genetics, vol. 114, no. 7, pp. 1129–1139, 2007.
422. R. Raman, H. Raman, and P. Martin, "Functional gene markers for polyphenol oxidase locus in bread wheat (Triticum aestivum L.)," Molecular Breeding, vol. 19, no. 4, pp. 315–328, 2007.
423. D.-L. Yang, R.-L. Jing, X.-P. Chang, and W. Li, "Identification of quantitative trait loci and environmental interactions for accumulation and remobilization of water-soluble carbohydrates in wheat (Triticum aestivum L.) stems," Genetics, vol. 176, no. 1, pp. 571–584, 2007.
424. V. Mohler, R. Lukman, S. Ortiz-Islas, et al., "Genetic and physical mapping of photoperiod insensitive gene Ppd-B1 in common wheat," Euphytica, vol. 138, no. 1, pp. 33–40, 2004.
425. H. Raman, R. Raman, R. Wood, and P. Martin, "Repetitive indel markers within the ALMT1 gene conditioning aluminium tolerance in wheat (Triticum aestivum L.)," Molecular Breeding, vol. 18, no. 2, pp. 171–183, 2006.
426. L.-L. Zhou, G.-H. Bai, H.-X. Ma, and B. F. Carver, "Quantitative trait loci for aluminum resistance in wheat," Molecular Breeding, vol. 19, no. 2, pp. 153–161, 2007.
427. S. P. Jefferies, M. A. Pallotta, J. G. Paull, et al., "Mapping and validation of chromosome regions conferring boron toxicity tolerance in wheat (Triticum aestivum)," Theoretical and Applied Genetics, vol. 101, no. 5-6, pp. 767–777, 2000.
428. B. Tóth, G. Galiba, E. Fehér, J. Sutka, and J. W. Snape, "Mapping genes affecting flowering time and frost resistance on chromosome 5B of wheat," Theoretical and Applied Genetics, vol. 107, no. 3, pp. 509–514, 2003.
429. L. Ma, E. Zhou, N. Huo, R. Zhou, G. Wang, and J. Jia, "Genetic analysis of salt tolerance in a recombinant inbred population of wheat (Triticum aestivum L.)," Euphytica, vol. 153, no. 1-2, pp. 109–117, 2007.
430. J. Peng, Y. Ronin, T. Fahima, et al., "Domestication quantitative trait loci in Triticum dicoccoides, the progenitor of wheat," Proceedings of the National Academy of Sciences of the United States of America, vol. 100, no. 5, pp. 2489–2494, 2003.
431. C. Pozzi, L. Rossini, A. Vecchietti, and F. Salamini, "Gene and genome changes during domestication of cereals," in Cereal Genomics, P. K. Gupta, R. K. Varshney, et al., Eds., pp. 165–198, Kluwer Academic Publishers, Dordrecht, The Netherlands, 2004.
432. J. Dubcovsky and J. Dvorak, "Genome plasticity a key factor in the success of polyploid wheat under domestication," Science, vol. 316, no. 5833, pp. 1862–1866, 2007.
433. C. Uauy, J. C. Brevis, and J. Dubcovsky, "The high grain protein content gene Gpc-B1 accelerates senescence and has pleiotropic effects on protein content in wheat," Journal of Experimental Botany, vol. 57, no. 11, pp. 2785–2794, 2006.
434. S. A. Flint-Garcia, A.-C. Thuillet, J. Yu, et al., "Maize association population: a high-resolution platform for quantitative trait locus dissection," The Plant Journal, vol. 44, no. 6, pp. 1054–1064, 2005.

435. J. Yu and E. S. Buckler, "Genetic association mapping and genome organization of maize," Current Opinion in Biotechnology, vol. 17, no. 2, pp. 155–160, 2006.
436. F. Breseghello and M. E. Sorrells, "Association mapping of kernel size and milling quality in wheat (Triticum aestivum L.) cultivars," Genetics, vol. 172, no. 2, pp. 1165–1177, 2006.
437. C. Ravel, S. Praud, and A. Murigneux, "Identification of Glu-B1-1 as a candidate gene for the quantity of high-molecular-weight glutenin in bread wheat (Triticum aestivum L.) by means of an association study," Theoretical and Applied Genetics, vol. 112, no. 4, pp. 738–743, 2006.
438. J. Crossa, J. Burgueño, S. Dreisigacker, et al., "Association analysis of historical bread wheat germplasm using additive genetic covariance of relatives and population structure," Genetics, vol. 177, no. 3, pp. 1889–1913, 2007.
439. L. Tommasini, T. Schnurbusch, D. Fossati, F. Mascher, and B. Keller, "Association mapping of Stagonospora nodorum blotch resistance in modern European winter wheat varieties," Theoretical and Applied Genetics, vol. 115, no. 5, pp. 697–708, 2007.
440. J. Dubcovsky, "Marker-assisted selection in public breeding programs: the wheat experience," Crop Science, vol. 44, no. 6, pp. 1895–1898, 2004.
441. M. E. Sorrells, "Application of new knowledge, technologies, and strategies to wheat improvement," Euphytica, vol. 157, no. 3, pp. 299–306, 2007.
442. H. A. Eagles, H. S. Bariana, F. C. Ogbonnaya, et al., "Implementation of markers in Australian wheat breeding," Australian Journal of Agricultural Research, vol. 52, no. 11-12, pp. 1349–1356, 2001.
443. F. C. Ogbonnaya, N. C. Subrahmanyam, O. Moullet, et al., "Diagnostic DNA markers for cereal cyst nematode resistance in bread wheat," Australian Journal of Agricultural Research, vol. 52, no. 11-12, pp. 1367–1374, 2001.
444. S. Landjeva, V. Korzun, and A. Börner, "Molecular markers: actual and potential contributions to wheat genome characterization and breeding," Euphytica, vol. 156, no. 3, pp. 271–296, 2007.
445. H. Kuchel, G. Ye, R. Fox, and S. Jefferies, "Genetic and economic analysis of a targeted marker-assisted wheat breeding strategy," Molecular Breeding, vol. 16, no. 1, pp. 67–78, 2005.
446. H. Kuchel, R. Fox, J. Reinheimer, et al., "The successful application of a marker-assisted wheat breeding strategy," Molecular Breeding, vol. 20, no. 4, pp. 295–308, 2007.
447. H. M. William, R. Trethowan, and E. M. Crosby-Galvan, "Wheat breeding assisted by markers: CIMMYT's experience," Euphytica, vol. 157, no. 3, pp. 307–319, 2007.
448. C. Lange and J. C. Whittaker, "On prediction of genetic values in marker-assisted selection," Genetics, vol. 159, no. 3, pp. 1375–1381, 2001.
449. N. Radovanovic and S. Cloutier, "Gene-assisted selection for high molecular weight glutenin subunits in wheat doubled haploid breeding programs," Molecular Breeding, vol. 12, no. 1, pp. 51–59, 2003.
450. C. M. Bowman, C. J. Howe, and T. A. Dyer, "Molecular mechanisms contributing to the evolution of (wheat) chloroplast genomes," in Proceedings of the 7th International Wheat Genetics Symposium, T. E. Miller and R. M. D. Koebner, Eds., pp. 69–73, Cambridge, UK, July 1988.

451. Y. Ogihara, "Genome science of polyploid wheat," in Frontiers of Wheat Bioscience. The 100th Memorial Issue of Wheat Information Service, K. Tsunewaki, Ed., pp. 169–184, Kihara Memorial Yokohama Foundation for the Advancement of Life Sciences, Yokohama, Japan, 2005.
452. K. Tsunewaki, "Plasmon differentiation in Triticum and Aegilops revealed by cytoplasmic effects on the wheat genome manifestation," in Proceedings of the US-Japan Symposium on Classical and Molecular Cytogenetic Analysis of Cereal Genomes, W. J. Raupp and B. S. Gill, Eds., pp. 38–48, Kansas Agricultural Experiment Station, Manhattan, NY, USA, 1995.
453. K. J. Newton, "Plant mitochondrial genomes: organization, expression and variation," Annual Review of Plant Physiology and Plant Molecular Biology, vol. 39, pp. 503–532, 1988.

A table has been omitted in this version of the article. To view this table, please visit the original version of the article.

CHAPTER 6

DEVELOPMENT IN RICE GENOME RESEARCH BASED ON ACCURATE GENOME SEQUENCE

TAKASHI MATSUMOTO, JIANZHONG WU,
BALTAZAR A. ANTONIO, and TAKUJI SASAKI

6.1 INTRODUCTION

Food security is a major issue as we aspire toward sustainable development. In spite of continuous increases in agricultural production due to the introduction of improved crop cultivars and the wide use of affordable technologies, more than 800 million people still do not have access to sufficient food to meet their dietary needs [1]. Cereal crops are basic source of food for humankind, with 85% of total crop production represented by maize, wheat, and rice. These three crops provide more than half of the protein and energy required for daily life. However, increase of world agricultural production in 2006 was less than 1%, which was due to decrease in cereal production [2]. On the other hand, the world's population is expected to reach 9 billion by 2050 [3]. It is therefore necessary to provide food security to this growing population in the midst of global environmental problems that deprive us of much arable land and biodiversity.

This chapter was originally published under the Creative Commons Attribution License. Matsumoto T, Wu J, Antonio BA, and Sasaki T. Development in Rice Genome Research Based on Accurate Genome Sequence. International Journal of Plant Genomics ***2008*** *(2008), 9 pages. doi:10.1155/2008/348621.*

Worldwide transformation of agriculture was first achieved with the Green Revolution, which led to significant increases in agricultural production. It began in the 1940s with the cultivation of a high-yielding dwarf wheat cultivar with resistance to pests and diseases. The Green Revolution for rice in the 1960s, based on the cultivar IR8, also dramatically increased rice production and helped food production to keep pace with population growth.

Now, the second Green Revolution, which will be based on genomics, is expected to pave the way for the leap in crop production. The availability of the rice genome sequence allowed the development of innovative approaches to increasing production. In the last 10 years, the basic syntenic relationships in gene content and gene order within the grass family have been established [4–6]. Therefore, the rice genome could be used as a reference genome for understanding the evolution of cereal crops and could provide a basis for their improvement [7, 8].

Among plants, only the *Arabidopsis* [9] and rice [10] genome sequences have been completed so far. A positionally confirmed, quality-validated genome sequence is obligatory required as a reference for the efficient use of sequence information, particularly in comparative analysis. Hence, the genome sequence derived from *Oryza sativa ssp. japonica* cv. Nipponbare has been recognized as a gold standard for understanding the genetics and biology of rice at the molecular level and in the breeding and genetic manipulation of cereal crops.

This chapter presents a past history of the rice genome sequencing efforts and a present endeavor for analysis of the genome sequence to clarify its structure and function. Approach to the "difficult" regions whose functions are the maintenance and regulation of chromosomes—notably the centromeres and telomeres—is described. Application of the new sequencing technology toward comparative studies among genus *Oryza* is also described in the context of the rice genome as a reference.

6.2 GENOME SEQUENCING THROUGH INTERNATIONAL COLLABORATION

The International Rice Genome Sequencing Project (IRGSP) was established in 1997. The 10 member countries agreed to sequence a standard rice cultivar (Nipponbare), to use common resources, and to share sequencing of the 12 rice chromosomes by using a map-based clone-by-clone strategy (http://rgp.dna.affrc.go.jp/E/IRGSP/index.html). For construction of sequence-ready physical maps, two complementary approaches were used. The Rice Genome Research Program (RGP) in Japan anchored the genomic clones using expressed sequence tags/sequence-tagged sites (EST/STS) and genetic markers from the genetic and transcript maps of rice [11, 12]. The Clemson University Genomics Institute, the Arizona Genomics Institute, and the Arizona Genomics Computational Laboratory used a high-throughput bacterial artificial chromosome (BAC) fingerprint and automatic BAC contig assembly system using FPC software [13], and anchored the assembled contigs on the rice genome by hybridization-based screening [14]. The sequence-ready physical maps generated from use of these two strategies covered more than 95% of the rice genome, and 92% to 100% of each chromosome. A total of 3453 PAC/BAC clones forming the minimum tiling path were selected for sequencing. DNA from a BAC/PAC clone was purified and fragmented by sonication. The ends of 2000 subclones of each clone were sequenced with capillary sequencers and assembled using the phred/phrap assembler [15]. The genome sequences of each PAC/BAC clone at the high-throughput genomic (HTG) phase 2 category were submitted to the DNA Data Bank of Japan (DDBJ). By December 2002, almost all the clone sequences corresponding to the minimum tiling path were sequenced to at least HTG phase 2. As a result, a high-quality draft sequence representing 366 Mb of the rice genome was released in the public database [16]. Thereafter, the IRGSP continued with the arduous task of finishing: gap-filling, improving base read quality, and resolving misassemblies (Figure 1).

In December 2004, the high-quality map-based sequence of the rice genome at HTG phase 3 category was completed and released in the public domain [10]. The sequence, ca. 370 Mb in total, covered nearly 95% of the total estimated size of the genome and about 99% of the euchromatic regions. The sequence also included three centromeres, parts of the rDNA regions, and regions for various transposable elements (corresponding up to 35% in the total genome). This comprehensive, relatively accurate sequence of the rice genome, is currently considered the gold standard.

In contrast to the hierarchical clone-by-clone strategy used by the IRGSP, a whole-genome shotgun (WGS) sequencing strategy is widely used

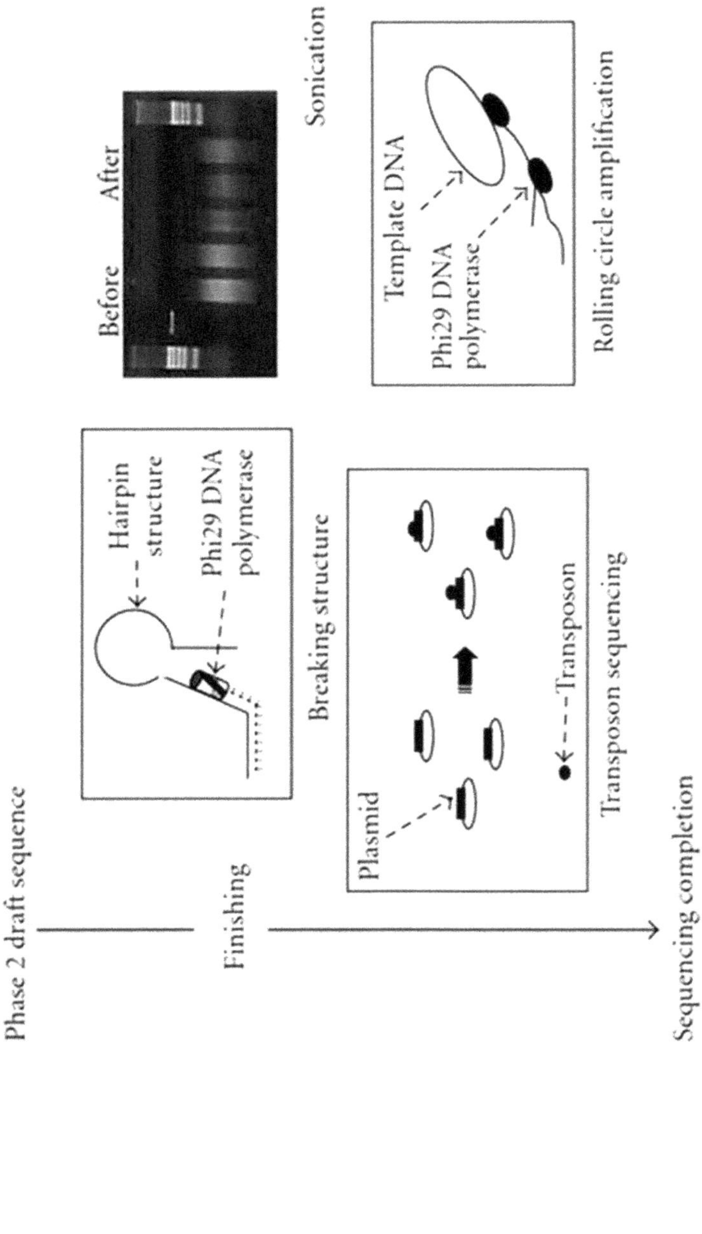

FIGURE 1: Four steps used for the finishing process to sequence completion.

in many sequencing projects [17]. In this strategy, a high-throughput computer program to reproduce the entire genome sequence assembles millions of shotgun sequences from the total genome. This method was used in sequencing the 2.9-Gb human genome [18]. Two independent groups used the WGS strategy to sequence the rice genome. The Beijing Genome Institute assembled shotgun sequences of the indica line 93-11 with 4× [19] and later 6× [20] genome coverage. A private company, Syngenta (Basel, Switzerland), also used the WGS strategy to sequence the Nipponbare [21]. This WGS sequence of Nipponbare was further improved by reassembling the shotgun sequences and combining the japonica and indica (line 99-11) sequences, resulting in 433 Mb of sequence composed of 50 233 contigs of Nipponbare [20]. Nearly 99% of the rice full-length cDNAs [22] have been localized in these latest assemblies [20] of the japonica and indica genome.

The effectiveness of the WGS sequencing strategy was compared with that of the hierarchal clone-by-clone sequencing approach [23, 24]. Although WGS assembly could readily provide an overview of the genome structure with a practical level of accuracy, misassembly could result in nonhomologous, misaligned, or duplicated coverage and some mismatches even in the genic regions. Moreover, repeat sequences could not be properly assigned to their original positions in the genome. In the case of rice, which has a lot of repeat sequences, WGS sequencing is therefore not a highly reliable strategy as it creates misassembly, particularly in duplicated regions. It is therefore important to have a highly accurate map-based sequence, which can be obtained by the hierarchical clone-by-clone strategy. Today projects aiming at obtaining entire genome sequences of gramineae plants are progressing [25–29]. All the projects, either using WGS strategy or clone-by-clone strategy, regard the completed rice genome sequence as sequence reference in reconstruction of chromosome sequences, emphasizing the importance of "gold standard."

6.3 DECIPHERING THE GENOME THROUGH ANNOTATION

Detecting the gene-coding regions within the genome sequence is one of the most efficient ways to characterize the structure and function of the genome. RGP constructed an annotation system that facilitates gene detection of the genome sequence in a timely manner. The Rice Genome Automated Annotation System, or RiceGAAS (http://ricegaas.dna.affrc.go.jp [30]), was designed as a fully automated system for annotating rice genome sequences. It retrieves rice sequences from GenBank and analyzes them with gene prediction programs such as Genscan [31] and FgeneSH (http://www.softberry.com/berry.phtml) and with BLAST [32] for similarity to proteins, rice ESTs, and rice full-length cDNAs to generate the most accurate gene models on the basis of available information (Figure 2). A similar automatic annotation pipeline was established by TIGR (http://rice.tigr.org/tdb/e2k1/osa1/data_download.shtml), and gene models are improved with rice ESTs and transcripts [33]. Both sets of gene models are published on the Web to accelerate gene analysis. With increasing data on nucleic acids and proteins in the public databases, regular re-evaluation and update of these gene models is necessary. In this respect, one of the advantages of these full-computational approaches is that whole gene sets can be relatively easily revised.

RGP has also developed a manual annotation system to facilitate curation of the gene models by human annotators (http://rgp.dna.affrc.go.jp/genomicdata/AnnSystem.html). This pipeline directly takes the output generated from RiceGAAS for in-depth analysis with in-house editing tools. Each gene model is manually edited to improve the prediction accuracy. The gene models for each BAC or PAC clone are released to the public domain through the DDBJ/EMBL/GenBank database. All data can be accessed through the central database whole genome annotation (WhoGA) on our website at http://rgp.dna.affrc.go.jp. Initially, only the six chromosomes (1, 2, 6, 7, 8, and 9) assigned to RGP were manually curated. Recently, curation of the rest was completed, so the manual annotation of the entire genome is now available. After removal of clone overlaps, a total of 57724 genes were predicted, including many hypothetical genes predicted by a single prediction program. Among them, 24056 gene models are supported by full-length cDNAs. All the gene models are ordered and organized in a genome browser.

Apart from these individual activities, the IRGSP conceived the establishment of the Rice Annotation Project (RAP), a community standard annotation project, in 2004. Genes were annotated at regular jamboree-style

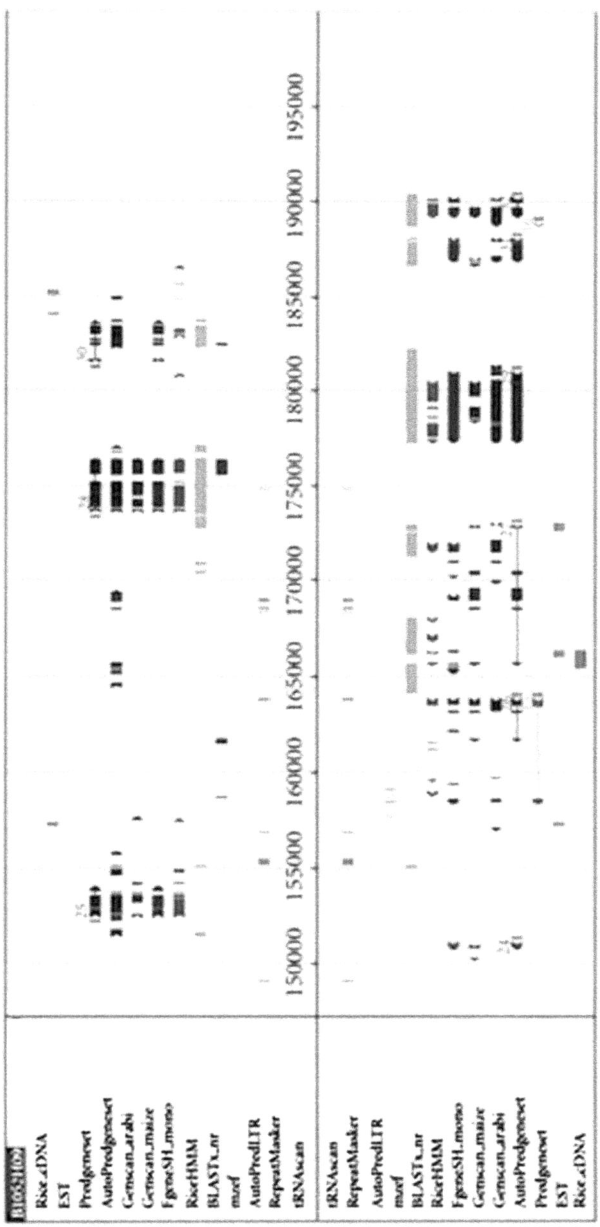

FIGURE 2: RiceGAAS annotation view, showing results from application of gene prediction software and similarity searches. Upper box: a DNA strand from left (5') to right (3'). Lower box: from right (5') to left (3').

annotation meetings to facilitate the manual curation of all gene models in rice. The National Institute of Agrobiological Sciences has been leading this project, collaborating with IRGSP members and many international and Japanese laboratories. So far, three RAP meetings have been held, at which gene models, chiefly constructed by mapping full-length cDNAs on the latest rice genome assemblies, have been manually curated. This collaboration confirmed 32000 curated genes, most of which have some degree of evidence [34]. The RAP-database (RAP-DB, http://rapdb.lab.nig.ac.jp) will be further improved with the integration of other annotation and functional genomics data.

6.4 UNCOVERED TERRITORY—EXPLORATION OF THE MISSING REGIONS

At the time of completion of the genome in 2004, IRGSP published nearly 371 Mb of high-quality DNA sequences, leaving about 5% of its estimated 389 Mb to be sequenced [10]. These unsequenced genomic regions existed as 62 gaps, including the telomeres and centromeres in all but two out of 12 chromosomes. One of the main reasons for the presence of these gaps was that no more clones with sequence extension into gap regions could be selected from any Nipponbare genomic resources, including BAC and PAC libraries (both based on partial digestion of DNA fragments) and fosmid libraries (based on physically sheared DNA fragments), containing a total of 630000 clones. For unknown reasons, specific genomic regions could not be cloned or maintained by using the above vectors in bacteria. In addition, a number of regions in the genome contain highly repeated sequences, making it difficult to construct a correct and complete physical map. However, analysis of sequences from these complicated genomic regions is not futile. Researchers have reported the importance of heterochromatic regions in silencing gene expression [35]. Cytological analysis has been used to define the distribution of such heterochromatin along each rice chromosome [36]. Through the IRGSP efforts, 2 of the 12 centromeres and 14 of the 24 telomeres have been completely or partly sequenced (http://rgp.dna.affrc.go.jp/E/IRGSP/index.html). Here, we focus on both regions because they play essential roles in chromosome maintenance or segregation.

6.4.1 COMPOSITION AND STRUCTURE OF RICE CENTROMERES

Because of the relatively small amount of centromeric satellite DNA in rice, significant progress has been made in genomic and molecular studies of the structures, functions, and evolution of rice centromeres. Two centromeres, derived from chromosomes 4 and 8, have been completely sequenced, revealing the complicated composition and structure of the first centromeres to have been sequenced among eukaryotes [37–39]. Repetitive sequences occupy ~60% of the whole region (~2 Mb) of the centromere of chromosome 8 (Cen8). The majority of copies of the 155-bp centromeric satellite repeat CentO, totaling 68.5-kb, occur in three large clusters in the center, separated by centromere-specific retrotransposon of rice (CRR) sequences. Numerous sequences of other transposable elements were also found in its surrounding region. Cen8 contains an ~750-kb core domain that binds rice CENH3, the centromere-specific H3 histone [37]. It is surprising to find transcriptionally active genes even within the core domain of Cen8. A similar result was found in Cen3, where a much bigger region (~1881 kb) has been found to have associations with CENH3 [40]. As a chromosomal site for kinetochore assembly that plays an important role in the faithful segregation of sister chromatids during cell division, the centromere has functions that are well conserved among all higher eukaryotes. Inter- and extrachromosomal analysis of the centromeres has, however, revealed the divergence of DNA components and organization patterns even among closely related species. The amount of CentO satellite DNA in the centromere of individual chromosomes varies from 60 kb to 1.9 Mb in *O. sativa* [41]. The number and organization of CentO clusters within the core region differ markedly between Cen4 and Cen8 in the Nipponbare genome. Cen8 has only three CentO tracts (clusters) with 442 copies of the 155-bp tandem repeat distributed within a 75-kb region, whereas Cen4 has up to 18 tracts but only 379 copies of the repeat within a 124-kb region [38, 39] (Figure 3). CentO repeats, on the other hand, are absent from several wild rice species, such as *Oryza brachyantha* [42]. It would be interesting to sequence and compare the compositional and structural changes in centromeres between different *Oryza* species in the future, since in-depth analysis of the Cen8 and Cen4 sequences has demonstrated segmental duplication

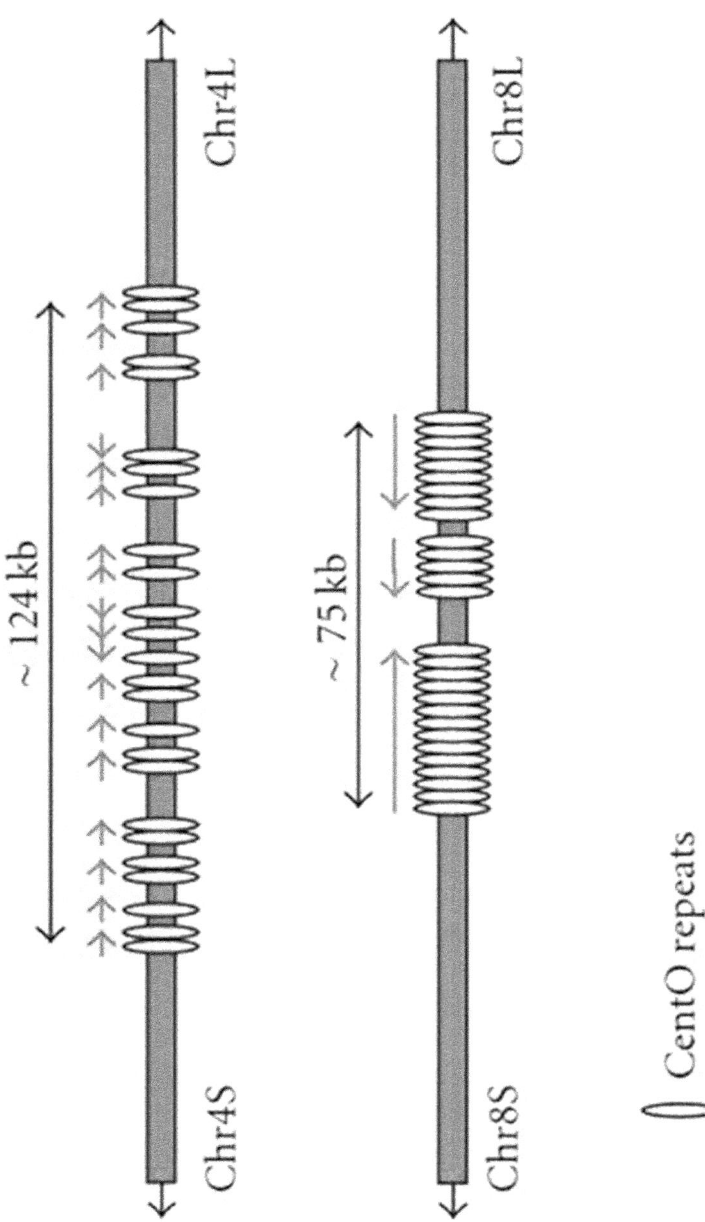

FIGURE 3: Structural comparisons of CentO domains between Nipponbare chromosomes 4 and 8. Yellow ovals and red arrows indicate the position of CentO arrays and the direction of the 155-bp tandem repeats within each array, respectively. Length of arrays ranges from 477 to 8571 bp in chromosome 4 and 7616 to 34589 bp in chromosome 8.

and inversion of centromeric DNA [43]. First glimpse of this analysis was performed in sequencing the centromere region of chromosome.8 from *O. brachyantha,* revealing positional shift of centromere [44].

Rice is now becoming a model for centromere and heterochromatin research [38, 45, 46]. Further research will lead to insights into the evolutionary dynamics, processes, and molecular mechanisms of plant centromeres.

6.4.2 COMPOSITION AND STRUCTURE OF RICE TELOMERES

Like those of centromeres, the composition and structure of telomere regions in rice have also been analyzed. Telomeres form the ends of linear eukaryotic chromosomes, serving as protective caps that prevent end-to-end fusion, recombination, and degradation of chromosomal ends [47]. The telomeres of most eukaryotes consist of an array of repeats that contain similar sequences but vary in length. For example, telomere DNA has a conserved sequence of 5′-TTAGGG-3′ in humans and 5′-TTGGGG-3′ in Tetrahymena (a ciliate protozo) [48, 49]. The first plant telomere DNA was isolated from Arabidopsis thaliana and shows tandemly repeated arrays of 5′-TTTAGGG-3′ [50]. Rice telomeres consist of the same repeat [51]. Sequencing and extensive analysis of seven rice chromosomal ends revealed several basic features that could provide a platform for analyzing and understanding the telomere structures and functions. All seven rice telomeres revealed contain highly conserved TTTAGGG sequences in tandem repeats, although deletions, insertions, and substitutions of single nucleotides or inverted copies were found within the arrayed repeats, particularly in the region of the junction between the telomere and subtelomere. Fluorescent in situ hybridization and terminal restriction fragment analyses suggest that the rice telomeres are a bit longer than those of *Arabidopsis* but much shorter than those of *Nicotiana tabacum,* ranging in a length from 5 to 20 kb, thus hinting at the genetic control of telomere length in plants [52, 53]. Interestingly, variation in telomere length is observed not only among different chromosomes, but also between different species within *Oryza*; this variation should provide useful information for future studies of telomere evolution. Gene annotation in the 7 rice subtelomere regions (each within 500 kb) demonstrated that the genomic region adjacent to the chromosome terminus is gene-rich (1 gene per 5.9 kb on average). Since nearly half of these annotated genes match rice full-length cDNAs, these rice subtelomeres could be considered to have high transcriptional activity. Recently, seven new rice telomeres were partly sequenced, and their sequences have been submitted to DDBJ (Table 1; http://rgp.dna.affrc.go.jp/E/publicdata/telomere2007/index.html). Among the above 14 chromosomal ends, the telomere and subtelomere regions on the short arm of chromosome 9 show some specific compositional and structural features. Sequencing and analysis of the fosmid clone OSJNOa063K24 revealed that the telomere repeats are co-localized with the ribosomal RNA gene (rDNA) cluster [54]. Besides the telomere-specific repeat and the long rDNA array (sized in megabases), the content of repetitive sequences such as retrotransposons within the 500-kb region proximal to the centromere is relatively high, suggesting that much of the short arm of rice chromosome 9 is heterochromatic. Rice telomere reverse transcriptase has also been isolated [55]. It will be interesting to conduct future studies using rice as a model of telomere research, as has been done for centromeres, especially to reveal how telomere length (shortening or elongation) is regulated and whether the telomere repeats and structure affect the expression of genes in the subtelomere region. The sequence resources obtained from the telomere and centromere regions of rice chromosomes should thus provide an unprecedented opportunity for future study, particularly to construct an artificial chromosome for use in both molecular and applied biology in plant science.

6.5 GENOME SEQUENCE FOR EVOLUTIONARY GENOMICS IN RICE

Rice is believed to have been domesticated from a wild relative 0.2 Mya [56] or 0.44 Mya [57]. Asian cultivated rice (*O. sativa L.*) has two subspecies, indica and japonica. Both are important as modern crops, and there are many phenotypic variations among them, conferring adaptation to many different environmental and cultural conditions. Crossing of these subspecies has produced new cultivars of agricultural importance. Knowing the differences at the molecular level would widen the capacity for rice breeding. RGP constructed a BAC library of Kasalath, an indica cultivar, generating 78427 high-quality BAC end sequences from 47194 BAC clones, and mapped these end sequences on Nipponbare chromosome sequences [58]. Mapping of 12170 clones allowed the construction of 450 Kasalath BAC contigs covering 308.5 Mb. Single-nucleotide polymorphism (SNP) frequency in the BAC end sequences and corresponding Nipponbare sequences was 0.71% on average. Sequencing of part of the Kasalath genome is in progress and could in future elucidate the precise gene dynamics in evolution and domestication. Results of Kasalath BAC physical maps are shown on RGP homepage (http://rgp.dna.affrc.go.jp/E/publicdata/kasalathendmap/index.html). Figure 4 is an example of a computer-generated Kasalath BAC physical map. BLAST searches for Kasalath BAC-end sequence screening could be performed through website (http://rgp.dna.affrc.go.jp/blast/runblast.html). Other approaches [59, 60] could identify positions of SNPs for high-density SNP markers.

TABLE 1: Mapped and sequenced rice telomeres.

Clone name	Accession no.	Copies*	Chr
OSJNOa264G09	AP008219	17	1S
OSJNOa183H18	AP006851	52	2S
OSJNOa246I10	AP008220	69	2L
OSJNOa070P15	AP009053	27	3S
OSJNOa083A04	AP009055	75	3L
OSJNOa076I12	AP009056	129	4S
OSJNOa281H13	AP009057	68	4L
OSJNOa070B13	AP009052	53	5S
OSJNOa230J22	AP006854	37	6L
OSJNOa219C16	AP008222	17	7S
OSJNOa136M23	AP008223	127	7L
OSJNOa162K02	AP008224	55	8S
OSJNOa063K24	AP009051	162	9S
OSJNOa073B23	AP009054	62	10S

Copies of telomere-specific repeats detected from the sequenced clones.

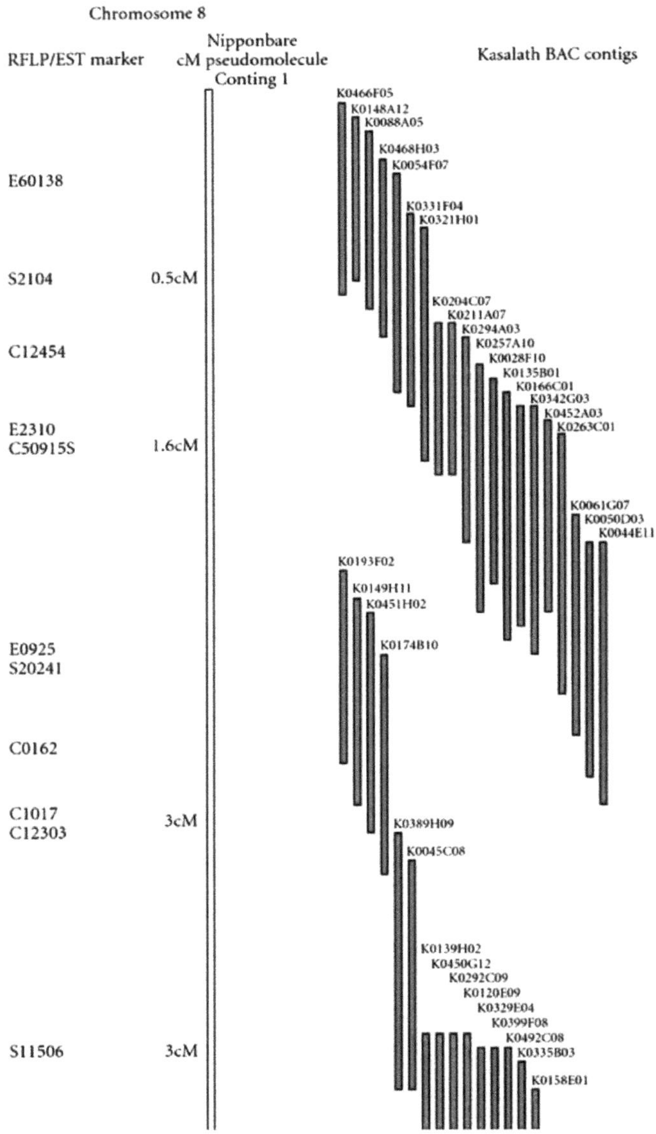

FIGURE 4: In silico physical map of Kasalath chromosome 8, based on the Nipponbare sequence. Green vertical bars indicate BAC clones (with K numbers) mapped against Nipponbare genome sequence (shown at left with landmarks).

It had long been a mystery how Asian rice originated from its wild progenitor, *Oryza rufipogon*. Recently, the origin has been clarified by comparison of retrotransposon [56], retroposon [61], chloroplast [62], and gene [63] sequences along the evolutionary lineages. These studies show evidence of multiple independent domestications of the two major subspecies. Further molecular studies of domestication will show how the crop and humans coevolved.

The genus *Oryza* has 23 species [64], but only two species (*O. sativa* in Asia and *O. glaberrima* in Africa) are domesticated and cultivated. This fact is remarkable given that rice grows under a wide variety of natural conditions. Consequently, many genetic resources might be waiting to be developed. Study of the wild relatives might reveal new genes for hybridization, improved yield, and sustainable production. The Oryza Map Alignment Project (OMAP, OMAP, http://www.omap.org/index.html) of the USA and China aims at the establishment of an experimental platform to unravel and understand the evolution, physiology, and biochemistry of the genus. The Arizona Genomics Institute has constructed 12 BAC libraries from the AA (the same as sativa species) to HHKK (remote species from sativa) species genomes. Computer-based mapping and filter hybridization screening provided high-density cross-species physical maps [65, 66].

6.6 IMPACT OF NEW SEQUENCING TECHNOLOGIES

The genome sequences of *O. sativa* and its progenitors are expected to show extensive base substitutions and rearrangements. Therefore, it would be difficult to reconstruct the genome sequences of wild rice relatives from cultivars. As resequencing with the conventional Sanger methodology can take much time and effort, a new pyrosequencing technology was developed. Massively parallel short reads from pyrosequencing analysis [67] could sequence more than 20 million bases with much less cost and less time than with Sanger analysis. In collaboration with 454 Life Sciences and Roche Diagnostics, we compared pyrosequencer and Sanger sequence data. Eight BAC clones which include OR_CBa0076I05, OR_CBa0091G05, OR_CBa0094N06, OR_CBa0004O24, OR_CBa0063M01,

OR_CBa0075G04, OR_CBa0034E23, and OR_CBa0010H05 from *O. rufipogon* IRGC105491 (AA species) were chosen from a fingerprint contig of the OMAP BAC library (OR_CBa-FPC contig 51). This contig corresponds to an 800-kb region of the short arm of Nipponbare chromosome 6 and is expected to contain two genes for rice flowering (Hd3a and RFT1). DNA of each BAC clone was purified individually and then mixed for pyrosequencing on a GS20 genome analyzer (Roche). The output from this analysis (ca. 20× coverage) contained 286639 reads. Of these, 169130 reads were mapped and 16123462 bases were aligned to the corresponding Nipponbare sequences, forming 1422 mapped contigs that cover 57.5% of the entire genomic region. The average depth was 23.39 showing deep coverage.

To compare these sequences with those from Sanger sequencing, we shotgun sequenced a BAC clone OR_CBa0004O24 and assembled it with phred/phrap software to form contigs. Each contig sequence from pyrosequencing was aligned to its corresponding Sanger sequence by BLAST alignment. Statistical results from this comparison are shown in Table 2.

Comparing only high quality (HQ sequence quality score > either 30 or 40) nucleotides gave an overall error rate of 0.0409% or 0.0359%. This means that the high-coverage reads from pyrosequencing show more than 99.95% accuracy. Researchers have pointed out that pyrosequencing is more problematic in repeats and homopolymers than Sanger technology [68, 69], but we did not observe this type of discrepancy. We also compared nucleotide sequences of Hd3a (one of the rice heading date QTL, corresponding to FT gene of *Arabidopsis*) between *O. sativa* cv. Nipponbare (by Sanger method) and *O. rufipogon* (by pyrosequencing). Only 3 SNPs and no in/del were found in exons (540 coding nt), whereas many deviations (20 SNPs, 6 indels) were found in introns; this was evolutionally reasonable. This sequence conservation might indicate that Hd3a is functionally important and under purifying selection.

These results show that emerging new resequencing technologies (not only pyrosequencing but also other methods [70]), when properly used in combination with current methods, will revolutionize the cost and performance of rice genome resequencing and will help elucidate the evolution of the *Oryza* genomes.

TABLE 2: Sequence comparison of BAC clone OR_CBa0004O24, Sanger versus Pyrosequencing.

Low-quality threshold	Low-quality Score 30	threshold Score 40
Number of alignments checked	34	34
Total length of alignments	132229	132229
Total HQ bases	131759	130639
Total LQ bases	470	1590
"In/del" type discrepancy	20	20
– Sanger insertion, total	15	15
– Pyro insertion, total	5	5
LQ insertion, total	5	6
– Sanger LQ	3	3
– Pyro LQ	2	3
HQ insertion, total	15	14
– Sanger HQ insertion	12	12
– Pyro HQ insertion	3	2
"SNP" type discrepancy	60	60
– both HQ	54	47
– LQ for Sanger	5	12
– LQ for Pyro	1	1
– both LQ	0	0
Discrepancy rate (%)	0.0409	0.0359
Accuracy rate (%)	99.9591	99.9641

6.7 CONCLUSION

The rice genome sequence has become available as a reference genome, providing a basis for understanding the wide range of diversity among cultivated and wild relatives of rice. The continuous efforts in generating a high-quality sequence have paved the way for clarifying the structures of genomic regions that are difficult to analyze, including centromeres and telomeres. Comparative genomics within the genus *Oryza* has also become a feasible strategy for understanding the evolutionary events that led to the development of cultivated rice. The syntenic relationships among cereal crops must be thoroughly exploited from now on. The rice genome sequence will be the most important tool in explaining the structure and function of other cereal genomes, and its use may open new opportunities for researchers to look deeper into the synteny between rice and other cereal crops, which has been maintained for some 60 million years of evolution [6]. From a more practical aspect, the rice genome sequence could be the key for developing rice-genomics-based research in order to improve crop production and food security for humankind.

REFERENCES

1. Director-General's message, FAO, http://www.fao.org/wfd2007/wfd_resources/dg_message.html.
2. Faostat, http://faostat.fao.org/site/339/default.aspx.
3. U.S. Census Bureau, http://www.census.gov/ipc/www/idb/worldpopinfo.html.
4. M. D. Gale and K. M. Devos, "Comparative genetics in the grasses," Proceedings of the National Academy of Sciences of the United States of America, vol. 95, no. 5, pp. 1971–1974, 1998.
5. K. M. Devos and M. D. Gale, "Genome relationships: the grass model in current research," The Plant Cell, vol. 12, no. 5, pp. 637–646, 2000.
6. K. M. Devos, "Updating the 'crop circle'," Current Opinion in Plant Biology, vol. 8, no. 2, pp. 155–162, 2005.
7. M. E. Sorrells, "Cereal genomics research in the post-genomics era," in Cereal Genomics, P. K. Gupta and R. K. Varshney, Eds., pp. 559–584, Kluwer Academic Publishers, Dordrecht, The Netherlands, 2004.
8. R. Cooke, B. Piègu, O. Panaud, et al., "From rice to other cereals: comparative genomics," in Rice Functional Genomics, N. M. Upadhyaya, Ed., pp. 429–464, Springer, New York, NY, USA, 2007.
9. The Arabidopsis Genome Initiative, "Analysis of the genome sequence of the flowering plant Arabidopsis thaliana," Nature, vol. 408, no. 6814, pp. 796–815, 2000.
10. International Rice Genome Sequencing Project, "The map-based sequence of the rice genome," Nature, vol. 436, no. 7052, pp. 793–800, 2005.
11. J. Wu, T. Maehara, T. Shimokawa, et al., "A comprehensive rice transcript map containing 6591 expressed sequence tag sites," The Plant Cell, vol. 14, no. 3, pp. 525–535, 2002.
12. Y. Harushima, M. Yano, A. Shomura, et al., "A high-density rice genetic linkage map with 2275 markers using a single F2 population," Genetics, vol. 148, no. 1, pp. 479–494, 1998.

13. C. Soderlund, I. Longden, and R. Mott, "FPC: a system for building contigs from restriction fingerprinted clones," Computer Applications in the Biosciences, vol. 13, no. 5, pp. 523–535, 1997.
14. M. Chen, G. Presting, W. B. Barbazuk, et al., "An integrated physical and genetic map of the rice genome," The Plant Cell, vol. 14, no. 3, pp. 537–545, 2002.
15. B. Ewing and P. Green, "Base-calling of automated sequencer traces using phred. II. Error probabilities," Genome Research, vol. 8, no. 3, pp. 186–194, 1998.
16. IRGSP, 2002, http://rgp.dna.affrc.go.jp/E/IRGSP/Dec18_NEWS.html.
17. J. C. Venter, H. O. Smith, and L. Hood, "A new strategy for genome sequencing," Nature, vol. 381, no. 6581, pp. 364–366, 1996.
18. J. C. Venter, M. D. Adams, E. W. Myers, et al., "The sequence of the human genome," Science, vol. 291, no. 5507, pp. 1304–1351, 2001.
19. J. Yu, S. Hu, J. Wang, et al., "A draft sequence of the rice genome (Oryza sativa L. ssp. indica)," Science, vol. 296, no. 5565, pp. 79–92, 2002.
20. J. Yu, J. Wang, W. Lin, et al., "The genomes of Oryza sativa: a history of duplications," PLoS Biology, vol. 3, no. 2, p. e38, 2005.
21. S. A. Goff, D. Ricke, T.-H. Lan, et al., "A draft sequence of the rice genome (Oryza sativa L. ssp. japonica)," Science, vol. 296, no. 5565, pp. 92–100, 2002.
22. The Rice Full-Length cDNA Consortium, "Collection, mapping, and annotation of over 28,000 cDNA clones from japonica rice," Science, vol. 301, no. 5631, pp. 376–379, 2003.
23. T. Matsumoto, R. A. Wing, B. Han, and T. Sasaki, "Rice genome sequence: the foundation for understanding the genetic systems," in Rice Functional Genomics, Challenges, Progress and Prospects, N. M. Upadhyaya, Ed., pp. 5–20, Springer, Berlin, Germany, 2007.
24. J. Yu, P. Ni, and G. K.-S. Wong, "Comparing the whole-genome-shotgun and map-based sequences of the rice genome," Trends in Plant Science, vol. 11, no. 8, pp. 387–391, 2006.
25. E. Pennisi, "Corn genomics pops wide open," Science, vol. 319, no. 5868, p. 1333, 2008.
26. phtyozome, Sorghum bicolor, http://www.phytozome.net/sorghum.
27. BrachyBase, Brachypodium distachyon, http://www.brachybase.org.
28. International Barley Sequencing Consortium: Hordeum vulgare, http://www.public.iastate.edu/~imagefpc/IBSC%20Webpage/IBSC%20Template-home.html.
29. International Wheat Genome Sequencing Consortium: Triticum. Aestivum, http://www.wheatgenome.org/index.php.
30. K. Sakata, Y. Nagamura, H. Numa, et al., "RiceGAAS: an automated annotation system and database for rice genome sequence," Nucleic Acids Research, vol. 30, no. 1, pp. 98–102, 2002.
31. C. Burge and S. Karlin, "Prediction of complete gene structures in human genomic DNA," Journal of Molecular Biology, vol. 268, no. 1, pp. 78–94, 1997.
32. S. F. Altschul, W. Gish, W. Miller, E. W. Myers, and D. J. Lipman, "Basic local alignment search tool," Journal of Molecular Biology, vol. 215, no. 3, pp. 403–410, 1990.

33. B. J. Haas, A. L. Delcher, S. M. Mount, et al., "Improving the Arabidopsis genome annotation using maximal transcript alignment assemblies," Nucleic Acids Research, vol. 31, no. 19, pp. 5654–5666, 2003.
34. Rice Annotation Project, "Curated genome annotation of Oryza sativa ssp. japonica and comparative genome analysis with Arabidopsis thaliana," Genome Research, vol. 17, no. 2, pp. 175–183, 2007.
35. P. Dimitri, N. Corradini, F. Rossi, and F. Vernì, "The paradox of functional heterochromatin," BioEssays, vol. 27, no. 1, pp. 29–41, 2004.
36. Z. Cheng, C. R. Buell, R. A. Wing, M. Gu, and J. Jiang, "Toward a cytological characterization of the rice genome," Genome Research, vol. 11, no. 12, pp. 2133–2141, 2001.
37. K. Nagaki, Z. Cheng, S. Ouyang, et al., "Sequencing of a rice centromere uncovers active genes," Nature Genetics, vol. 36, no. 2, pp. 138–145, 2004.
38. J. Wu, H. Yamagata, M. Hayashi-Tsugane, et al., "Composition and structure of the centromeric region of rice chromosome 8," The Plant Cell, vol. 16, no. 4, pp. 967–976, 2004.
39. Y. Zhang, Y. Huang, L. Zhang, et al., "Structural features of the rice chromosome 4 centromere," Nucleic Acids Research, vol. 32, no. 6, pp. 2023–2030, 2004.
40. H. Yan, H. Ito, K. Nobuta, et al., "Genomic and genetic characterization of rice Cen3 reveals extensive transcription and evolutionary implications of a complex centromere," The Plant Cell, vol. 18, no. 9, pp. 2123–2133, 2006.
41. Z. Cheng, F. Dong, T. Langdon, et al., "Functional rice centromeres are marked by a satellite repeat and a centromere-specific retrotransposon," The Plant Cell, vol. 14, no. 8, pp. 1691–1704, 2002.
42. H.-R. Lee, W. Zhang, T. Langdon, et al., "Chromatin immunoprecipitation cloning reveals rapid evolutionary patterns of centromeric DNA in Oryza species," Proceedings of the National Academy of Sciences of the United States of America, vol. 102, no. 33, pp. 11793–11798, 2005.
43. H. Yan and J. Jiang, "Rice as a model for centromere and heterochromatin research," Chromosome Research, vol. 15, no. 1, pp. 77–84, 2007.
44. J. Ma, R. A. Wing, J. L. Bennetzen, and S. A. Jackson, "Evolutionary history and positional shift of a rice centromere," Genetics, vol. 177, no. 2, pp. 1217–1220, 2007.
45. A. Sharma and G. G. Presting, "Centromeric retrotransposon lineages predate the maize/rice divergence and differ in abundance and activity," Molecular Genetics and Genomics, vol. 279, no. 2, pp. 133–147, 2008.
46. H. Mizuno, K. Ito, J. Wu, et al., "Identification and mapping of expressed genes, simple sequence repeats and transposable elements in centromeric regions of rice chromosomes," DNA Research, vol. 13, no. 6, pp. 267–274, 2006.
47. J. Lingner and T. R. Cech, "Telomerase and chromosome end maintenance," Current Opinion in Genetics & Development, vol. 8, no. 2, pp. 226–232, 1998.
48. R. K. Moyzis, J. M. Buckingham, L. S. Cram, et al., "A highly conserved repetitive DNA sequence, (TTAGGG)n, present at the telomeres of human chromosomes," Proceedings of the National Academy of Sciences of the United States of America, vol. 85, no. 18, pp. 6622–6626, 1988.
49. C. W. Greider and E. H. Blackburn, "Identification of a specific telomere terminal transferase activity in tetrahymena extracts," Cell, vol. 43, no. 2, part 1, pp. 405–413, 1985.

50. E. J. Richards and F. M. Ausubel, "Isolation of a higher eukaryotic telomere from Arabidopsis thaliana," Cell, vol. 53, no. 1, pp. 127–136, 1988.
51. H. Mizuno, J. Wu, H. Kanamori, et al., "Sequencing and characterization of telomere and subtelomere regions on rice chromosomes 1S, 2S, 2L, 6L, 7S, 7L and 8S," The Plant Journal, vol. 46, no. 2, pp. 206–217, 2006.
52. J. Fajkus, A. Kovařík, R. Královics, and M. Bezděk, "Organization of telomeric and subtelomeric chromatin in the higher plant Nicotiana tabacum," Molecular and General Genetics, vol. 247, no. 5, pp. 633–638, 1995.
53. H. Kotani, T. Hosouchi, and H. Tsuruoka, "Structural analysis and complete physical map of Arabidopsis thaliana chromosome 5 including centromeric and telomeric regions," DNA Research, vol. 6, no. 6, pp. 381–386, 1999.
54. M. Fujisawa, H. Yamagata, K. Kamiya, et al., "Sequence comparison of distal and proximal ribosomal DNA arrays in rice (Oryza sativa L.) chromosome 9S and analysis of their flanking regions," Theoretical and Applied Genetics, vol. 113, no. 3, pp. 419–428, 2006.
55. K. Heller-Uszynska, W. Schnippenkoetter, and A. Kilian, "Cloning and characterization of rice (Oryza sativa L) telomerase reverse transcriptase, which reveals complex splicing patterns," The Plant Journal, vol. 31, no. 1, pp. 75–86, 2002.
56. C. Vitte, T. Ishii, F. Lamy, D. Brar, and O. Panaud, "Genomic paleontology provides evidence for two distinct origins of Asian rice (Oryza sativa L.)," Molecular Genetics and Genomics, vol. 272, no. 5, pp. 504–511, 2004.
57. J. Ma and J. L. Bennetzen, "Rapid recent growth and divergence of rice nuclear genomes," Proceedings of the National Academy of Sciences of the United States of America, vol. 101, no. 34, pp. 12404–12410, 2004.
58. S. Katagiri, J. Wu, Y. Ito, et al., "End sequencing and chromosomal in silico mapping of BAC clones derived from an indica rice cultivar, Kasalath," Breeding Science, vol. 54, no. 3, pp. 273–279, 2004.
59. C. Li, Y. Zhang, K. Ying, X. Liang, and B. Han, "Sequence variations of simple sequence repeats on chromosome-4 in two subspecies of the Asian cultivated rice," Theoretical and Applied Genetics, vol. 108, no. 3, pp. 392–400, 2004.
60. F. A. Feltus, J. Wan, S. R. Schulze, J. C. Estill, N. Jiang, and A. H. Paterson, "An SNP resource for rice genetics and breeding based on subspecies indica and japonica genome alignments," Genome Research, vol. 14, no. 9, pp. 1812–1819, 2004.
61. C. Cheng, R. Motohashi, S. Tsuchimoto, Y. Fukuta, H. Ohtsubo, and E. Ohtsubo, "Polyphyletic origin of cultivated rice: based on the interspersion pattern of SINEs," Molecular Biology and Evolution, vol. 20, no. 1, pp. 67–75, 2003.
62. S. Kawakami, K. Ebana, T. Nishikawa, Y. Sato, D. A. Vaughan, and K. Kadowaki, "Genetic variation in the chloroplast genome suggests multiple domestication of cultivated Asian rice (Oryza sativa L.)," Genome, vol. 50, no. 2, pp. 180–187, 2007.
63. J. P. Londo, Y.-C. Chiang, K.-H. Hung, T.-Y. Chiang, and B. A. Schaal, "Phylogeography of Asian wild rice, Oryza rufipogon, reveals multiple independent domestications of cultivated rice, Oryza sativa," Proceedings of the National Academy of Sciences of the United States of America, vol. 103, no. 25, pp. 9578–9583, 2006.
64. D. A. Vaughan, H. Morishima, and K. Kadowaki, "Diversity in the Oryza genus," Current Opinion in Plant Biology, vol. 6, no. 2, pp. 139–146, 2003.

65. J. S. S. Ammiraju, M. Luo, J. L. Goicoechea, et al., "The Oryza bacterial artificial chromosome library resource: construction and analysis of 12 deep-coverage large-insert BAC libraries that represent the 10 genome types of the genus Oryza," Genome Research, vol. 16, no. 1, pp. 140–147, 2006.
66. H. Kim, B. Hurwitz, Y. Yu, et al., "Construction, alignment and analysis of twelve framework physical maps that represent the ten genome types of the genus Oryza," Genome Biology, vol. 9, no. 2, article R45, 2008.
67. M. Margulies, M. Egholm, W. E. Altman, et al., "Genome sequencing in microfabricated high-density picolitre reactors," Nature, vol. 437, no. 7057, pp. 376–380, 2005.
68. T. Wicker, E. Schlagenhauf, A. Graner, T. J. Close, B. Keller, and N. Stein, "454 sequencing put to the test using the complex genome of barley," BMC Genomics, vol. 7, article 275, 2006.
69. M. J. Moore, A. Dhingra, P. S. Soltis, et al., "Rapid and accurate pyrosequencing of angiosperm plastid genomes," BMC Plant Biology, vol. 6, article 17, 2006.
70. S. T. Bennett, C. Barnes, A. Cox, L. Davies, and C. Brown, "Toward the $1000 human genome," Pharmacogenomics, vol. 6, no. 4, pp. 373–382, 2005.

CHAPTER 7

GENOMIC DATABASES FOR CROP IMPROVEMENT

KAITAO LAI, MICHAŁ T. LORENC, and DAVID EDWARDS

7.1 INTRODUCTION

The majority of DNA sequence and expressed gene sequence data generated today comes from the next- or second-generation sequencing (NGS/2GS) technologies. NGS technologies produce vast quantities of short data rather than Sanger sequencing at a relatively low cost and short time. Genomics is undergoing a revolution, driven by advances in DNA sequencing technology, and this data flood is having a major impact on approaches and strategies for crop improvement. NGS technologies have been applied for sequenced genomes of a number of cereal crop species including rice, Sorghum and maize. A quality sequence of rice that covers 95% of the 389 Mb genome has been produced [1]. The Sorghum bicolor (L.) Moench genome has been assembled in size of 730-megabase, placing ~98% of genes in their chromosomal context [2]. The draft nucleotide sequence of the 2.3-gigabase genome of maize has also been improved [3]. One of the challenges encountered by researchers is to translate this abundance of data into improved crops in the field. There remains a gap between genome data production and next-generation crop improvement

This chapter was originally published under the Creative Commons Attribution License. Lai K, Lorenc MT, and Edwards D. Genomic Databases for Crop Improvement. Agronomy **2** *(2012), 62–73. doi:10.3390/agronomy2010062.*

strategies, but this is being rapidly closed by far sighted companies and individuals with the ability to combine the ability to mine the genomic data with practical crop-improvement skills. Bioinformatics can be defined as the structuring of biological information to enable logical interrogation, and databases are a key part of the bioinformatics toolbox. Numerous databases have been developed for genomic data, on a range of platforms and to suite a variety of different purposes (see Table 1 for examples). These range from generic DNA sequence or molecular marker databases, to those hosting a variety of data for specific species.

TABLE 1: Examples of genomic databases related to crop improvement.

Database Name	Web Link	References
autoSNPdb	http://autosnpdb.appliedbioinformatics.com.au/	[4,5]
Brachypodium database	http://www.brachypodium.org/	[6]
Brassica genome gateway	http://www.brassicagenome.net	[7]
Brassica rapa genome database	http://brassicadb.org/	[8]
DNA Data Bank of Japan (DDBJ)	http://ddbj.sakura.ne.jp/	[9]
European bioinformatics institute EnsEMBL plants	http://plants.ensembl.org/	[10,11]
European Molecular Biology Laboratory (EMBL) nucleotide sequence database	http://www.ebi.ac.uk/embl/	[12,13]
GenBank	http://www.ncbi.nlm.nih.gov/genbank/	[14–16]
Graingenes	http://wheat.pw.usda.gov/	[17–19]
Gramene	http://www.gramene.org/	[20]
International Crop Information System (ICIS)	http://www.icis.cgiar.org	[21]
International Nucleotide Sequence Database Collaboration (INSDC)	http://www.insdc.org/	[9]
Legume Information System (LIS)	http://www.comparative-legumes.org/	[22,23]
MaizeGDB	http://www.maizegdb.org/	[24–26]
Maize sequence database	http://www.maizesequence.org/	[3]
Oryzabase	http://www.shigen.nig.ac.jp/rice/oryzabase/	[27]
Panzea	http://www.panzea.org/	[28]
Phytozome	http://www.phytozome.net/	[29]
PlantsDB	http://mips.helmholtzmuenchen.de/plant/genomes.jsp	[30]

TABLE 1: *Cont.*

Database Name	Web Link	References
PlantGDB	http://www.plantgdb.org/	[31,32]
The Plant Ontology	http://www.plantontology.org/	[33]
Plaza	http://bioinformatics.psb.ugent.be/plaza/	[34]
Rice Genome Annotation Project	http://rice.plantbiology.msu.edu/	[35]
SSR Primer	http://flora.acpfg.com.au/ssrprimer2/	[36]
Database Name Web Link References SSR taxonomy tree	http://appliedbioinformatics.com.au/projects/ssrtaxonomy/php/	[36]
SOL Genomics Network (SGN)	http://solgenomics.net/	[37]
SoyBase	http://soybase.org/	[38]
TAGdb	http://flora.acpfg.com.au/tagdb/	[39]
The Crop Expressed Sequence Tag database, CR-EST	http://pgrc.ipk-gatersleben.de/cr-est/	[40]
The Triticeae Repeat Sequence Database (TREP)	http://wheat.pw.usda.gov/ITMI/Repeats/	[41]
Wheat genome information	http://www.wheatgenome.info	[42]

7.1.1 GENERIC DATABASES

The largest of the DNA sequence repositories is the International Nucleotide Sequence Database Collaboration (INSDC), made up of the DNA Data Bank of Japan (DDBJ) at The National Institute of Genetics in Mishima, Japan [9], GenBank at the National Center of Biotechnology Information (NCBI) in Bethesda, USA [15,16], and the European Molecular Biology Laboratory (EMBL) Nucleotide Sequence Database, maintained at the European Bioinformatics Institute (EBI) in the UK [13]. Daily data exchange between these groups ensures coordinated international coverage [43].

Since the introduction of advanced next-generation sequencing technology, the storage and interrogation of this data is becoming an expanding challenge [44,45]. The ability to search the vast quantity of this data is made feasible by the development of custom databases such as TAGdb (http://flora.acpfg.com.au/tagdb/) [39], but it is increasingly the assembled and annotated genome data which are applied for crop-improvement applications [46].

While it is valuable to maintain all public nucleic acid sequences in one location, the size of this resource limits the ability to visualize this data. Genome viewers, which place genomic data within the context of sequenced or partially sequenced genomes, provide more context-orientated data interrogation. There are two main generic web-based tools to view plant genomes: Ensembl [10] and GBrowse [47,48]. Both are widely used and it is not uncommon to find similar genome information hosted on both systems. A key development in genome databases was the establishment and adoption of a standard file format for genome data [49], and data in the current version, GFF3 can be visualized and searched using a wide range of tools from custom GBrowse databases to stand alone bioinformatics tools such as Biomatters Geneious [50].

There are several resources which collate genome data for multiple plant species. Gramene (http://www.gramene.org/) [20] is an EnsEMBL-based genome viewer and database hosting information on a variety of crop species, but based around the rice, maize and Arabidopsis genomes [18]. A similar resource is hosted by the EBI (http://plants.ensembl.org/) [10]. PlantGDB is a resource for comparative plant genomics [31,32] and hosts sequence data for >70,000 plant species with a focus on complete sequencing of reference species, *Arabidopsis*, rice, maize and *Medicago truncatula*. Plaza (http://bioinformatics.psb.ugent.be/plaza/) [34] hosts pre-computed comparative genomics data sets for a range of species [34]. Phytozome (http://www.phytozome.net/) [29] also hosts genome data for numerous plant species and provides several genomes using the GBrowse format. With 25 complete plant genomes, phytozome is one of the most comprehensive plant genome databases currently available [18]. In addition, PlantsDB is a generic database hosting data for multiple plant species. This database is hosted by MIPS (http://mips.helmholtz-muenchen.de/plant/genomes.jsp) [30].

While genome and transcript sequence information makes up the bulk of genome data maintained within public databases, it is often the differences between individuals and varieties which are the most valuable for crop-improvement applications. A major focus of crop genetic research in recent decades has been the development of molecular genetic markers associated with important traits. Genetic markers can be assayed with a variety of techniques [51]. Early molecular genetic markers technologies such as restriction fragment length polymorphisms have been replaced by more high throughput methods, including amplified fragment length polymorphisms (AFLPs), diversity array technologies (DArT) and simple sequence repeats (SSRs) also known as microsatellites. Another important and crop-improvement-oriented database is the maize database Panzea (http://www. panzea.org/) [28], which hosts data on genomic diversity in a large germplasm collection including genetic data, trait phenotypes, allele frequencies, phenotyping environments, genetic analysis tools and so on. The Panzea database The Panzea databases comprises the genotypic and phenotypic data and genetic marker information. This database design is based on the Genomic Diversity and Phenotype Data Model (GDPDM) (http://www.maizegenetics.net/gdpdm/) [20].

An expressed sequence tag (EST) represents a short sub-sequence of a cDNA sequence. EMBL or GenBank have sub-sections for EST sequences. The crop expressed sequence tag database, CR-EST (http://pgrc. ipk-gatersleben.de/cr-est/) [40], provides access to more than 200,000 sequences derived from 41 cDNA libraries of four species: barley, wheat, pea and potato [40].

SSRs are short stretches of DNA sequence occurring as tandem repeats of mono-, di-, tri-, tetra-, penta- and hexa-nucleotides. They are highly polymorphic due to mutation affecting the number of repeat units. The value of SSRs is due to their genetic co-dominance, abundance, dispersal throughout the genome, multi-allelic variation and high reproducibility. The hypervariability of SSRs among related organisms makes them excellent markers for genotype identification, analysis of genetic diversity, phenotype mapping and marker assisted selection [52,53]. SSRs demonstrate a high degree of transferability between species, as PCR primers designed to an SSR within one species frequently amplify a corresponding locus in related species, enabling comparative genetic and genomic analysis.

With the continued advances in DNA sequencing technologies, single-nucleotide polymorphisms (SNPs) have come to dominate high throughput molecular marker analysis. SNPs are the ultimate form of molecular genetic marker, as a nucleotide base is the smallest unit of inheritance, and a SNP represents a single-nucleotide difference between two individuals at a defined location. SNPs are direct markers as the sequence information provides the exact nature of the allelic variants. Furthermore, this sequence variation can have a major impact on how the organism develops and responds to the environment. SNPs represent the most frequent type of genetic polymorphism and may therefore provide a high density of markers near a locus of interest. SNPs at any particular site could in principle involve four different nucleotide variants, but in practice they are generally biallelic. This disadvantage, when compared with multiallelic markers such as SSRs, is compensated by the relative abundance of SNPs. The high density of SNPs makes them valuable for genome mapping, and in particular they allow the generation of ultra-high density genetic maps and haplotyping systems for genes or regions of interest, and map-based positional cloning. SNPs are used routinely in crop breeding programs, for genetic diversity analysis, cultivar identification, phylogenetic analysis, characterization of genetic resources and association with agronomic traits [54,55].

SSR Primer (http://flora.acpfg.com.au/ssrprimer2/) [36] is a web-based tool that enables the real time discovery of SSRs within submitted DNA sequences, with the concomitant design of PCR primers for SSR amplification [56]. Alternatively, users may browse an SSR Taxonomy Tree (http://appliedbioinformatics.com.au/projects/ssrtaxonomy/php/) [36] to identify pre-determined SSR amplification primers for species represented within the GenBank database [36].

The SNP discovery software autoSNP [57, 58] identifies SNPs and insertion/deletion (indel) polymorphisms from bulk sequence data using two measures of confidence; redundancy, defined as the number of times a polymorphism occurs at a locus in a sequence alignment; and co-segregation of SNPs to define a haplotype. AutoSNP software has recently been extended to database format, autoSNPdb, which permits complex queries and provides detailed genomic and functional information [4,5]. Where the sequence trace files are available, the SNP discovery tool PolyPhred [59,60] can make use of the base pair quality scores to further differentiate between true SNP polymorphisms and random sequence error. The recent developments in next-generation sequence data have led to the identification of large numbers of SNPs in a range of plant genomes and these approaches are likely to dominate SNP discovery in the coming years [61].

The increased throughput for the discovery and application of molecular genetic markers has led to the requirement for databases hosting the results of molecular marker analysis. These maybe integrated within other database systems such as Gramene [20], the Legume Information System (LIS) [22,23], or Graingenes [17,18].

One of the principal uses of molecular genetic markers is the production of genetic maps and the mapping of heritable traits. While mapping data may be described as lists, graphical representations are more readily understood. The genetic map viewer CMap, developed by the GMOD consortium [62] is valuable for the validation of traits that map to the same position in different populations and also for the linkage between crop genetic maps and sequenced model genomes, enabling the identification of candidate genes for genetically mapped traits. A recent addition, CMap3D [63], enables the comparison of a larger number of maps in 3D space.

The linking of genomic data with agronomic traits remains one of the greatest challenges in the application of genome data for crop improvement [64,65]. Several databases have been developed to assist in this endeavor. The International Crop Information System (ICIS) [21] is a database system that hosts integrated management information for crop improvement, including details on diverse germplasm and traits. One challenge in developing trait databases is the establishment of functional ontologies. The Plant Ontology (http://www.plantontology.org/) [33] is a controlled vocabulary (ontology) that describes plant anatomy and morphology and stages of growth and development for all plants [33] and this database of ontologies is becoming the standard for comparative physiology and for linking genes with potential function.

7.1.2 SPECIES FOCUSED DATABASES

It would be impossible to detail all available plant genetic and genomic databases, however some of the main ones are listed below along with a brief description of their content.

GrainGenes (http://wheat.pw.usda.gov/) [18] is a genetic database for *Triticeae*, oats, and sugarcane GrainGenes (Matthews et al., 2003; Carollo et al., 2005) [18,19]. Comprehensive information includes genetic markers, map locations, alleles, key references and disease symptoms. The Triticeae Repeat Sequence Database (TREP) (http://wheat.pw.usda.gov/ITMI/Repeats/) [41] contains a collection of repetitive DNA sequences from different *Triticeae* species which can be used for the development of molecular markers.

While *Brachypodium distachyon* is not grown as a crop, this species has many qualities that make it a model for studies in temperate grasses and cereals, including a small genome (~ 300 Mbp), small physical stature, self-fertility, a short lifecycle, and simple growth requirements. The *B. distachyon* genome was sequenced in 2010 [66] and the *Brachypodium* database which includes a GBrowse based genome viewer is available at http://www.brachypodium.org/ [6].

The maize genome was sequenced in 2009 [3] and there are several databases hosting information on this important crop. These include MaizeGDB [25,26] (http://www.maizegdb.org/) [24] based on GBrowse, and maizesequence.org (http://www.maizesequence.org/) based on EnsEMBL [3].

Rice was one of the first crop genomes to be sequenced and there are now numerous resources available to mine this genomic information. Oryzabase is an integrated rice science database established in 2000 (http://www.shigen.nig.ac.jp/rice/oryzabase/) [27]. The database hosts information on genetic resources, chromosome maps, genes and rice mutants. This is complemented by a rice genome annotation project [35] which presents data using GBrowse (http://rice.plantbiology.msu.edu/) [35].

Although wheat is an extremely important crop, advances in genomics have been limited by its large and highly complex genome. Assemblies of the gene rich regions for the group 7 chromosomes have been completed [67,68], and annotated sequences, including a large number of SNP polymorphisms are available at http://www.wheatgenome.info [42].

A central portal for *Brassica* data is maintained at Brassica.info, with links to genetic marker, map and a range of diverse *Brassica* related information. The recently sequenced *Brassica rapa* genome [7] is hosted at http://brassicadb.org/ [8] in a database named BRAD [8], with a second database which contains *Brassica* repeat information at http://www.BrassicaGenome.net [7]. Both of these databases use GBrowse.

The Legume Information System (http://www.comparative-legumes.org/) [22,23] supports basic research in the legumes by relating data from multiple crop and model species, and by helping researchers traverse among various data types [22,23]. It currently hosts data for seventeen species and includes GBrowse databases for *Glycine max* (soybean), *Lotus japonicus* (birdsfoot trefoil), *Medicago truncatula* (barrel medic) and *Cajanus cajan* (pigeonpea). Lis is complemented by detailed soybean data hosted at SoyBase (http://soybase.org/) [38].

The SOL Genomics Network (SGN) (http://solgenomics.net/) [37] is a clade oriented database containing genomic, genetic, phenotypic and taxonomic information for plant genomes, with a focus on the *Euasterid* clade, which includes *Solanaceae* (e.g., tomato, potato, eggplant, pepper and petunia) and *Rubiaceae* (coffee) [37]. As well as being a resource for basic crop research, SGN maintains databases with a specific focus on giving breeders direct links to breeder-relevant tools and data.

7.2 CONCLUSIONS AND FUTURE DIRECTION

There are currently a range of databases dedicated to generic genome data or focusing on specific crops or clades. Both the type and volumes of data have increased greatly over the last few years and this trend looks to continue. Some of the early database formats are either no longer used or have limited applications [69–71], however several newer web tools are now becoming predominant. These include the GBrowse genome viewer [47,48] and associated open source bioinformatics developments as well as the EnsEMBL system [10]. As genome technology continues to advance and an increasing number of crop genomes become available, an expanding number of these databases will be developed. One of the main challenges facing crop bioinformatics researchers is to make the ever increasing volume and types of data available in a suitable format for analysis [72]. This includes new high-throughput plant phenotype data as well as the increasing volumes of genotypic diversity data. It will be the association of this diversity data with heritable phenotypes which will likely drive genome database development over the coming years [73,74]. These databases therefore will require the implementation of appropriate statistical tools for association of high-density genotype and highthroughput phenotype data.

REFERENCES AND NOTES

1. The map-based sequence of the rice genome. The map-based sequence of the rice genome. Nature 2005, 436, 793–800.
2. Paterson, A.H.; Bowers, J.E.; Bruggmann, R.; Dubchak, I.; Grimwood, J.; Gundlach, H.; Haberer, G.; Hellsten, U.; Mitros, T.; Poliakov, A.; et al. The Sorghum bicolor genome and the diversification of grasses. Nature 2009, 457, 7229, 551–556.
3. Schnable, P.S.; Ware, D.; Fulton, R.S.; Stein, J.C.; Wei, F.S.; Pasternak, S.; Liang, C.Z.; Zhang, J.W.; Fulton, L.; Graves, T.A.; et al. The B73 maize genome: Complexity, diversity, and dynamics. Science 2009, 326, 1112–1115.
4. Duran, C.; Appleby, N.; Clark, T.; Wood, D.; Imelfort, M.; Batley, J.; Edwards, D. AutoSNPdb: An annotated single nucleotide polymorphism database for crop plants. Nucl. Acid. Res. 2009, 37, D951–D953.
5. Duran, C.; Appleby, N.; Vardy, M.; Imelfort, M.; Edwards, D.; Batley, J. Single nucleotide polymorphism discovery in barley using autoSNPdb. Plant Biotechnol. J. 2009, 7, 326–333.
6. Larré, C.; Penninck, S.; Bouchet, B.; Lollier, V.; Tranquet, O.; Denery-Papini, S.; Guillon F.; Rogniaux, H. Brachypodium distachyon grain: Identification and subcellular localization of storage proteins. J. Exp. Bot. 2010, 61, 1771–1783.
7. Wang, X.; Wang, H.; Wang, J.; Sun, R.; Wu, J.; Liu, S.; Bai, Y.; Mun, J.-H.; Bancroft, I.; Cheng, F.; et al. The genome of the mesopolyploid crop species Brassica rapa. Nat. Genet. 2011, 43, 1035–1039.
8. Cheng, F.; Liu, S.; Wu, J.; Fang, L.; Sun, S.; Liu, B.; Li, P.; Hua, W.; Wang, X.; Cheng, F.; et al. BRAD, the genetics and genomics database for Brassica plants. BMC Plant Biology 2011, 11, doi:10.1186/1471-2229-11-136.
9. Sugawara, H.; Ogasawara, O.; Okubo, K.; Gojobori, T.; Tateno Y. DDBJ with new system and face. Nucl. Acids Res. 2008, 36, D22–D24.

10. Flicek, P.; Amode, M.R.; Barrell, D.; Beal, K.; Brent, S.; Chen, Y.; Clapham, P.; Coates, G.; Fairley, S.; Fitzgerald, S.; et al. Ensembl 2011. Nucl. Acid. Res. 2011, 39, D800–D806.
11. Kersey, P.; Lawson, D.; Birney, E.; Derwent, P. S.; Haimel, M.; Herrero, J.; Keenan, S.; et al. Ensembl Genomes: Extending Ensembl across the taxonomic space. Nucl. Acids Res. 2010, 38, D563–D569.
12. Kulikova, T.; Akhtar, R.; Aldebert, P.; Althorpe, N.; Andersson, M.; Baldwin, A.; et al. EMBL Nucleotide Sequence Database in 2006. Nucl. Acids Res. 2007, 35, D16–D20.
13. Sterk, P.; Kulikova, T.; Kersey, P.; Apweiler, R. The EMBL nucleotide sequence and genome reviews databases. In Methods in Molecular Biology; Edwards, D., Ed.; Humana Press: Totowa, NJ, USA, 2007; Volume 406, pp. 1–21.
14. Karsch-Mizrachi, I.; Nakamura, Y.; Cochrane, G.; The international nucleotide sequence database collaboration. Nucl. Acids Res. 2012, 40, D33–D37.
15. Benson, D.A.; Karsch-Mizrachi, I.; Lipman, D.J.; Ostell, J.; Sayers, E.W. GenBank. Nucl. Acid. Res. 2009, 37, 26–31.
16. Wheeler, D.L.; Barrett, T.; Benson, D.A.; Bryant, S.H.; Canese, K.; Chetvernin, V.; Church, D.M.; DiCuccio, M.; Edgar, R.; Federhen, S.; et al. Database resources of the national center for biotechnology information. Nucl. Acid. Res. 2008, 36, D13–D21.
17. O'Sullivan, H. GrainGenes—A genomic database for Triticeae and Avena. In Methods in Molecular Biology; Edwards, D., Ed.; Humana Press: Totowa, NJ, USA, 2007; Volume 406, pp. 301–314.
18. Carollo, V.; Matthews, D.E.; Lazo, G.R.; Blake, T.K.; Hummel, D.D.; Lui, N.; Hane, D.L.; Anderson, O.D. GrainGenes 2.0: An improved resource for the small-grains community. Plant Physiol. 2005, 139, 643–651.
19. Matthews, D.; Carollo, V.L.; Lazo, G.R.; Anderson, O.D. GrainGenes, the genome database for small-grain crops. Nucl. Acids Res. 2003, 31, 183–186.
20. Youens-Clark, K.; Buckler, E.; Casstevens, T.; Chen, C.; DeClerck, G.; Derwent, P.; Dharmawardhana, P.; Jaiswal, P.; Kersey, P.; Karthikeyan, A.S.; et al. Gramene database in 2010: Updates and extensions. Nucl. Acid. Res. 2011, 39, D1085–D1094.
21. Fox, P.N.; Skovman, B. The International Crop Information System (ICIS)—connects genebank to breeder to farmer's field. Plant adaptation and crop improvement. CAB International: Wallingford, Oxon, UK, 1996; pp. 317–326.
22. Gonzales, M.D.; Gajendran, K.; Farmer, A.D.; Archuleta, E.; Beavis, W.D. Leveraging model legume information to find candidate genes for soybean sudden death syndrome using the legume information system. In Methods in Molecular Biology; Edwards, D., Ed.; Humana Press: Totowa, NJ, USA, 2007; Volume 406, pp. 245–259.
23. Gonzales, M.D.; Archuleta, E.; Farmer, A.; Gajendran, K.; Grant, D.; Shoemaker, R.; Beavis, W.D.; Waugh, M.E. The legume information system (LIS): An integrated information resource for comparative legume biology. Nucl. Acid. Res. 2005, 33, D660–D665.

24. Schaeffer, M.L.; Harper, L.C.; Gardiner, J.M.; Andorf, C.M.; Campbell, D.A.; Cannon, E.K.; Sen, T.Z.; Lawrence, C.J. MaizeGDB: curation and outreach go hand-in-hand. Database. 2011, doi: 10.1093/database/bar022
25. Lawrence, C.J. MaizeGDB—The maize genetics and genomics database. In Methods in Molecular Biology; Edwards, D., Ed.; Humana Press: Totowa, NJ, USA, 2007; Volume 406, pp. 331–345.
26. Lawrence, C.J.; Schaeffer, M.L.; Seigfried, T.E.; Campbell, D.A.; Harper, L.C. MaizeGDB's new data types, resources and activities. Nucl. Acid. Res. 2007, 35, D895–D900.
27. Yamazaki, Y.; Sakaniwa, S.; Tsuchiya, R.; Nonomura, K.I.; Kurata, N. Oryzabase: An integrated information resource for rice science. Breed. Sci. 2010, 60, 544–548.
28. Canaran, P.; Buckler, E.S.; Glaubitz, J.C.; Stein, L.; Sun, Q.; Zhao, W.; Ware, D. Panzea: An update on new content and features. Nucl. Acids Res. 2008, 36, D1041–D1043.
29. Goodstein, D.M.; Shu, S.; Howson, R.; Neupane, R.; Hayes, R.D.; Fazo, J.; Mitros, T.; Dirks, W.; Hellsten, U.; Putnam, N.; et al. Phytozome: A comparative platform for green plant genomics. Nucl. Acid. Res. 2012, 40, D1178–D1186.
30. Mewes, H.W.; Dietmnn, S.; Frishman, D.; Gregory, R.; Mannhapt, G.; Mayer, K.F.X.; Münsterkötter, M.; Ruepp, A.; Spannagl, M.; Stümpflen, V.; Rattei, T. MIPS: analysis and annotation of genome information in 2007. Nucl. Acids Res. 2008, 36, D196–D201.
31. Brendel, V. Gene structure annotation at PlantGDB. In Methods in Molecular Biology; Edwards, D., Ed.; Humana Press: Totowa, NJ, USA, 2007; Volume 406, pp. 521–533.
32. Duvick, J.; Fu, A.; Muppirala, U.; Sabharwal, M.; Wilkerson, M.D.; Lawrence, C.J.; Lushbough, C.; Brendel, V. PlantGDB: A resource for comparative plant genomics. Nucl. Acid. Res. 2008, 36, D959–D965.
33. Avraham, S.; Tung, C.-W.; Ilic, K.; Jaiswal, P.; Kellogg, E.A.; McCouch, S.; Pujar, A.; Reiser, L.; Rhee, S.Y.; Sachs, M.M.; et al. The plant ontology database: A community resource for plant structure and developmental stages controlled vocabulary and annotations. Nucl. Acid. Res. 2008, 36, D449–D454.
34. Proost, S.; Van Bel, M.; Sterck, L.; Billiau, K.; Van Parys, T.; Van de Peer, Y.; Vandepoele, K. PLAZA: A comparative genomics resource to study gene and genome evolution in plants. Plant Cell 2009, 21, 3718–3731.
35. Ouyang, S.; Zhu, W.; Hamilton, J.; Lin, H.; Campbell, M.; Childs, K.; Thibaud-Nissen, F.; Malek, R.L.; Lee, Y.; Zheng, L.; et al. The TIGR rice genome annotation resource: Improvements and new features. Nucl. Acid. Res. 2007, 35, D883–D887.
36. Jewell, E.; Robinson, A.; Savage, D.; Erwin, T.; Love, C.G.; Lim, G.A.C.; Li, X.; Batley, J.; Spangenberg, G.C.; Edwards, D. SSRPrimer and SSR taxonomy tree: Biome SSR discovery. Nucl. Acid. Res. 2006, 34, W656–W659.
37. Bombarely, A.; Menda, N.; Tecle, I.Y.; Buels, R.M.; Strickler, S.; Fischer-York, T.; Pujar, A.; Leto, J.; Gosselin, J.; Mueller, L.A. The sol genomics network (solgenomics.net): Growing tomatoes using Perl. Nucl. Acid. Res. 2011, 39, D1149–D1155.

38. Grant, D.; Nelson, R.T.; Cannon, S.B.; Shoemaker, R.C. SoyBase, the USDA-ARS soybean genetics and genomics database. Nucl. Acids Res. 2010, 38, D843–D846.
39. Marshall, D.; Hayward, A.; Eales, D.; Imelfort, M.; Stiller, J.; Berkman, P.; Clark, T.; McKenzie, M.; Lai, K.; Duran, C.; et al. Targeted identification of genomic regions using TAGdb. Plant Methods 2010, 6, 19; doi:10.1186/1746-4811-6-19.
40. Künne, C.; Lange, M.; Funke, T.; Miehe, H.; Thiel, T.; Grosse, I.; Scholz, U. CR-EST: A resource for crop ESTs. Nucl. Acids Res. 2005, 33, D619–D621.
41. Wicker, T.; Buell, C.R. Gene and repetitive sequence annotation in the Triticeae. Plant Genet. GenomicsCrop. Model. 2009, 7, 407–425.
42. Lai, K.; Berkman, P.J.; Lorenc, M.T.; Duran, C.; Smits, L.; Manoli, S.; Stiller, J.; Edwards, D. WheatGenome.info: An integrated database and portal for wheat genome information. Plant Cell Physiol. 2011, doi: 10.1093/pcp/pcr141.
43. Edwards, D.; Hansen, D.; Stajich, J. DNA sequence databases. In Applied Bioinformatics; Edwards, D.; Stajich, J.; Hansen, D.; Eds.; Springer: New York, NY, USA, 2009; pp. 1–11.
44. Batley, J.; Edwards, D. Genome sequence data: Management, storage, and visualization. Biotechniques 2009, 46, 333–336.
45. Lee, H.; Lai, K.; Lorenc, M.T.; Imelfort, M.; Duran, C.; Edwards, D. Bioinformatics tools and databases for analysis of next generation sequence data. Brief. Funct. Genomics 2012, 11, 12–24.
46. Edwards, D.; Batley, J. Plant genome sequencing: Applications for crop improvement. Plant Biotechnol. J. 2010, 7, 1–8.
47. Arnaoudova, E.G.; Bowens, P.J.; Chui, R.G.; Dinkins, R.D.; Hesse, U.; Jaromczyk, J.W.; Martin, M.; Maynard, P.; Moore, N.; Schardl, C.L. Visualizing and sharing results in bioinformatics projects: GBrowse and GenBank exports. BMC Bioinformatics 2009, 10, A4; doi:10.1186/1471- 2105-10-S7-A4.
48. Donlin, M. Using the generic genome browser (GBrowse). Curr. Protoc. Bioinformatics 2007, doi:10.1002/0471250953.bi0909s28.
49. Reese, M.G.; Moore, B.; Batchelor, C.; Salas, F.; Cunningham, F.; Marth, G.T.; Stein, L.; Flicek, P.; Yandell, M.; Eilbeck, K. A standard variation file format for human genome sequences. Genome Biol. 2010, 11, R88; doi: 10.1186/gb-2010-11-8-r88
50. Drummond, A.J.; Ashton, B.; Buxton, S.; Cheung, M.; Cooper, A.; Duran, C.; Field, M.; Heled, J.; Kearse, M.; Markowitz, S.; et al. Geneious, Version 5.4; Biomatters Ltd.: Auckland, New Zealand. Available online: http://www.geneious.com (accessed on 17 March 2012)
51. Duran, C.; Edwards, D.; Batley, J. Molecular marker discovery and genetic map visualisation. In Applied Bioinformatics; Edwards, D., Hanson, D., Stajich, J., Eds.; Springer: New York, NY, USA, 2009.
52. Powell, W.; Machray, G.C.; Provan, J. Polymorphism revealed by simple sequence repeats. Trends Plant Sci. 1996, 1, 215–222.
53. Tautz, D. Hypervariability of simple sequences as a general source for polymorphic DNA markers. Nucl. Acid. Res. 1989, 17, 6463–6471.

54. Rafalski, A. Applications of single nucleotide polymorphisms in crop genetics. Curr. Opin. Plant Biol. 2002, 5, 94–100.
55. Batley, J.; Edwards, D. SNP applications in plants. In Association Mapping in Plants; Oraguzie, N., Rikkerink, E., Gardiner, S., De Silva, H., Eds.; Springer: New York, NY, USA, 2007; pp. 95–102.
56. Robinson, A.J.; Love, C.G.; Batley, J.; Barker, G.; Edwards, D. Simple sequence repeat marker loci discovery using SSR primer. Bioinformatics 2004, 20, 1475–1476.
57. Barker, G.; Batley, J.; O'Sullivan, H.; Edwards, K.J.; Edwards, D. Redundancy based detection of sequence polymorphisms in expressed sequence tag data using autoSNP. Bioinformatics 2003, 19, 421–422.
58. Batley, J.; Barker, G.; O'Sullivan, H.; Edwards, K.J.; Edwards, D. Mining for single nucleotide polymorphisms and insertions/deletions in maize expressed sequence tag data. Plant Physiol. 2003, 132, 84–91.
59. Bhangale, T.R.; Stephens, M.; Nickerson, D.A. Automating resequencing-based detection of insertion-deletion polymorphisms. Nat. Genet. 2006, 38, 1457–1462.
60. Stephens, M.; Sloan, J.S.; Robertson, P.D.; Scheet, P.; Nickerson, D.A. Automating sequence-based detection and genotyping of SNPs from diploid samples. Nat. Genet. 2006, 38, 375–381.
61. Imelfort, M.; Duran, C.; Batley, J.; Edwards, D. Discovering genetic polymorphisms in nextgeneration sequencing data. Plant Biotechnol. J. 2009, 7, 312–317.
62. Youens-Clark, K.; Faga, B.; Yap, I.V.; Stein, L.; Ware, D. CMap 1.01: A comparative mapping application for the Internet. Bioinformatics 2009, 25, 3040–3042.
63. Duran, C.; Boskovic, Z.; Imelfort, M.; Batley, J.; Hamilton, N.A.; Edwards, D. CMap3D: A 3D visualisation tool for comparative genetic maps. Bioinformatics 2010, 26, 273–274.
64. Edwards, D.; Batley, J. Bioinformatics: Fundamentals and applications in plant genetics, mapping and breeding. In Principles and Practices of Plant Genomics; Kole, C., Abbott, A.G., Eds.; Science Publishers, Inc.: Enfield, NH, USA, 2008; pp. 269–302.
65. Edwards, D. Bioinformatics and plant genomics for staple crops improvement. In Breeding Major Food Staples; Kang, M.S., Priyadarshan, P.M., Eds.; Blackwell: Oxford, UK, 2007; pp. 93–106.
66. The international Brachypodium initiative. Genome sequencing and analysis of the model grass Brachypodium distachyon. Nature 2010, 463, 763–768.
67. Berkman, P.J.; Skarshewski, A.; Manoli, S.; Lorenc, M.T.; Stiller, J.; Smits, L.; Lai, K.; Campbell, E.; Kubalakova, M.; et al. Sequencing wheat chromosome arm 7BS delimits the 7BS/4AL translocation and reveals homoeologous gene conservation. Theor. Appl. Genet. 2012, 124, 423–432.
68. Berkman, B.J.; Skarshewski, A.; Lorenc, M.T.; Lai, K.; Duran, C.; Ling, E.Y.S.; Stiller, J.; Smits, L.; Imelfort, M.; Manoli, S.; et al. Sequencing and assembly of low copy and genic regions of isolated Triticum aestivum chromosome arm 7DS. Plant Biotechnol. J. 2011, 9, 768–775.
69. Erwin, T.A.; Jewell, E.G.; Love, C.G.; Lim, G.A.C.; Li, X.; Chapman, R.; Batley, J.; Stajich, J.E.; Mongin, E.; Stupka, E.; et al. BASC: An integrated bioinformatics system for Brassica research. Nucl. Acid. Res. 2007, 35, D870–D873.

70. Love, C.G.; Robinson, A.J.; Lim, G.A.C.; Hopkins, C.J.; Batley, J.; Barker, G.; Spangenberg, G.C.; Edwards, D. Brassica ASTRA: An integrated database for Brassica genomic research. Nucl. Acid. Res. 2005, 33, D656–D659.
71. Stein, L.D.; Thierry-Mieg, J. Scriptable access to the Caenorhabditis elegans genome sequence and other ACEDB databases. Genome Res. 1998, 8, 1308–1315.
72. Berkman, P.J.; Lai, K.; Lorenc, M.T.; Edwards, D. Next generation sequencing applications for wheat crop improvement. Amer. J. Bot. 2012, 99, 365–371.
73. Edwards, D.; Batley, J., Plant Bioinformatics: From genome to phenome. Trends Biotech. 2004, 22, 232–237.
74. Duran, C.; Eales, D.; Marshall, D.; Imelfort, M.; Stiller, J.; Berkman, P.J.; Clark, T.; McKenzie, M.; Appleby, N.; Batley, J.; et al. Future tools for association mapping in crop plants. Genome 2010, 53, 1017–1023.

PART II

GOING BEYOND DNA VARIATIONS TO UNDERSTAND ENVIRONMENTAL APPLICATIONS

CHAPTER 8

UNCOVERING THE COMPLEXITY OF TRANSCRIPTOMES WITH RNA-SEQ

VALERIO COSTA, CLAUDIA ANGELINI, ITALIA DE FEIS, and ALFREDO CICCODICOLA

8.1 INTRODUCTION

It is commonly known that the genetic information is conveyed from DNA to proteins via the messenger RNA (mRNA) through a finely regulated process. To achieve such a regulation, the concerted action of multiple cis-acting proteins that bind to gene flanking regions—"core" and "auxiliary" regions—is necessary [1]. In particular, core elements, located at the exons' boundaries, are strictly required for initiating the pre-mRNA processing events, whereas auxiliary elements, variable in number and location, are crucial for their ability to enhance or inhibit the basal splicing activity of a gene.

Until recently—less than 10 years ago—the central dogma of genetics indicated with the term "gene" a DNA portion whose corresponding mRNA encodes a protein. According to this view, RNA was considered a "bridge" in the transfer of biological information between DNA and proteins, whereas the identity of each expressed gene, and of its transcriptional levels, were commonly indicated as "transcriptome" [2]. It was con-

This chapter was originally published under the Creative Commons Attribution License. Li, Z, Treyzon, L, Chen, S, Yan, E, Thames, G, and Carpenter, CL. Uncovering the Complexity of Transcriptomes with RNA-Seq. Journal of Biomedicine and Biotechnology ***2010*** *(2010), 19 pages. doi:10.1155/2010/853916.*

sidered to mainly consist of ribosomal RNA (80–90%, rRNA), transfer RNA (5–15%, tRNA), mRNA (2–4%) and a small fraction of intragenic (i.e., intronic) and intergenic noncoding RNA (1%, ncRNA) with undefined regulatory functions [3]. Particularly, both intragenic and intergenic sequences, enriched in repetitive elements, have long been considered genetically inert, mainly composed of "junk" or "selfish" DNA [4]. More recently it has been shown that the amount of noncoding DNA (ncDNA) increases with organism complexity, ranging from 0.25% of prokaryotes' genome to 98.8% of humans [5]. These observations have strengthened the evidence that ncDNA, rather than being junk DNA, is likely to represent the main driving force accounting for diversity and biological complexity of living organisms.

Since the dawn of genetics, the relationship between DNA content and biological complexity of living organisms has been a fruitful field of speculation and debate [6]. To date, several studies, including recent analyses performed during the ENCODE project, have shown the pervasive nature of eukaryotic transcription with almost the full length of nonrepeat regions of the genome being transcribed [7].

The unexpected level of complexity emerging with the discovery of endogenous small interfering RNA (siRNA) and microRNA (miRNA) was only the tip of the iceberg [8]. Long interspersed noncoding RNA (lincRNA), promoter- and terminator-associated small RNA (PASR and TASR, resp.), transcription start site-associated RNA (TSSa-RNA), transcription initiation RNA (tiRNA) and many others [8] represent part of the interspersed and crosslinking pieces of a complicated transcription puzzle. Moreover, to cause further difficulties, there is the evidence that most of the pervasive transcripts identified thus far, have been found only in specific cell lines (in most of cases in mutant cell lines) with particular growth conditions, and/or particular tissues. In light of this, discovering and interpreting the complexity of a transcriptome represents a crucial aim for understanding the functional elements of such a genome. Revealing the complexity of the genetic code of living organisms by analyzing the molecular constituents of cells and tissues, will drive towards a more complete knowledge of many biological issues such as the onset of disease and progression.

The main goal of the whole transcriptome analyses is to identify, characterize and catalogue all the transcripts expressed within a specific cell/tissue—at a particular stage—with the great potential to determine the correct splicing patterns and the structure of genes, and to quantify the differential expression of transcripts in both physio- and pathological conditions [9].

In the last 15 years, the development of the hybridization technology, together with the tag sequence-based approaches, allowed to get a first deep insight into this field, but, beyond a shadow of doubt, the arrival on the marketplace of the NGS platforms, with all their "Seq" applications, has completely revolutionized the way of thinking the molecular biology.

The aim of this paper is to give an overview of the RNA-Seq methodology, trying to highlight all the challenges that this application presents from both the biological and bioinformatics point of view.

8.2 NEXT GENERATION SEQUENCING TECHNOLOGIES

Since the first complete nucleotide sequence of a gene, published in 1964 by Holley [10] and the initial developments of Maxam and Gilbert [11] and Sanger et al. [12] in the 1970s (see Figure 1), the world of nucleic acid sequencing was a RNA world and the history of nucleic acid sequencing technology was largely contained within the history of RNA sequencing.

In the last 30 years, molecular biology has undergone great advances and 2004 will be remembered as the year that revolutionized the field; thanks to the introduction of massively parallel sequencing platforms, the Next Generation Sequencing-era, [13–15], started. Pioneer of these instruments was the Roche (454) Genome Sequencer (GS) in 2004 (http://www.454.com/), able to simultaneously sequence several hundred thousand DNA fragments, with a read length greater than 100 base pairs (bp). The current GS FLX Titanium produces greater than 1 million reads in excess of 400 bp. It was followed in 2006 by the Illumina Genome Analyzer (GA) (http://www.illumina.com/) capable to generate tens of millions of 32-bp reads. Today, the Illumina GAIIx produces 200 million 75–100 bp reads. The last to arrive in the marketplace was the Applied Biosystems

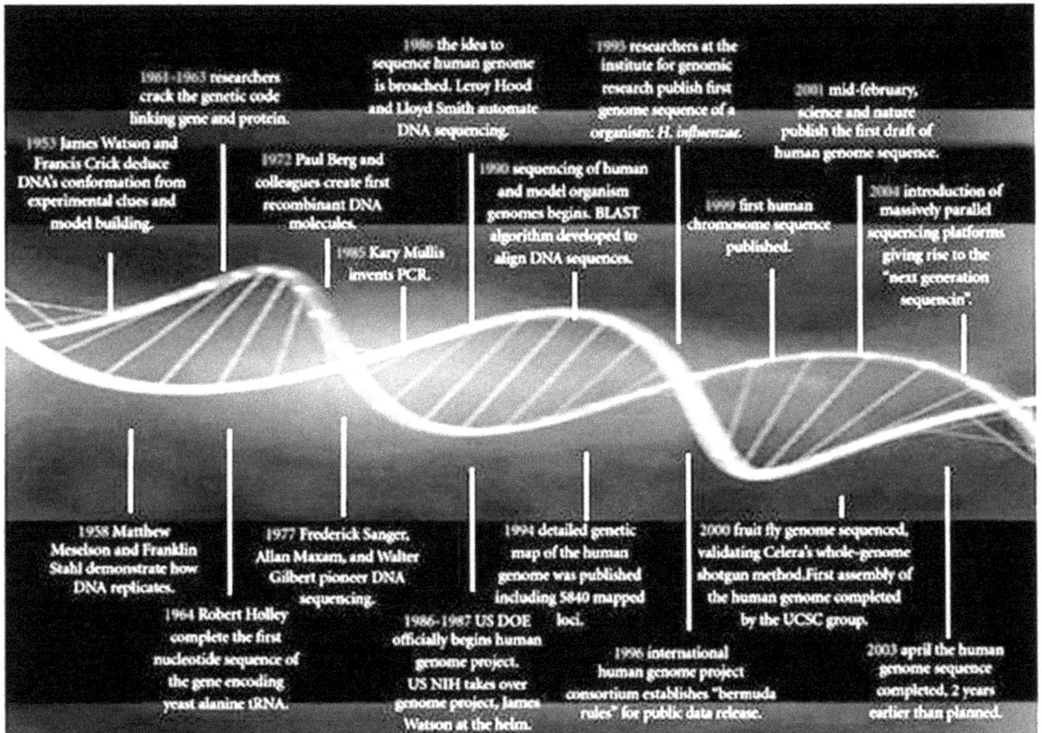

FIGURE 1: Evolution of DNA revolution.

platform based on Sequencing by Oligo Ligation and Detection (SOLiD) (http://www3.appliedbiosystems.com/AB_Home/index.htm), capable of producing 400 million 50-bp reads, and the Helicos BioScience HeliScope (http://www.helicosbio.com/), the first single-molecule sequencer that produces 400 millions 25–35 bp reads.

While the individual approaches considerably vary in their technical details, the essence of these systems is the miniaturization of individual sequencing reactions. Each of these miniaturized reactions is seeded with DNA molecules, at limiting dilutions, such that there is a single DNA molecule in each, which is first amplified and then sequenced. To be more precise, the genomic DNA is randomly broken into smaller sizes from which either fragment templates or mate-pair templates are created. A common theme among NGS technologies is that the template is attached to a solid surface or support (immobilization by primer or template) or indirectly immobilized (by linking a polymerase to the support). The immobilization of spatially separated templates allows simultaneous thousands to billions of sequencing reactions. The physical design of these instruments allows for an optimal spatial arrangement of each reaction, enabling an efficient readout by laser scanning (or other methods) for millions of individual sequencing reactions onto a standard glass slide. While the immense volume of data generated is attractive, it is arguable that the elimination of the cloning step for the DNA fragments to sequence is the greatest benefit of these new technologies. All current methods allow the direct use of small DNA/RNA fragments not requiring their insertion into a plasmid or other vector, thereby removing a costly and time-consuming step of traditional Sanger sequencing.

It is beyond a shadow of doubt that the arrival of NGS technologies in the marketplace has changed the way we think about scientific approaches in basic, applied and clinical research. The broadest application of NGS may be the resequencing of different genomes and in particular, human genomes to enhance our understanding of how genetic differences affect health and disease. Indeed, these platforms have been quickly applied to many genomic contexts giving rise to the following "Seq" protocols: RNA-Seq for transcriptomics, Chip-Seq for DNA-protein interaction, DNase-Seq for the identification of most active regulatory regions, CNV-Seq for copy number variation, and methyl-Seq for genome wide profiling of epigenetic marks.

8.3 RNA-SEQ

RNA-Seq is perhaps one of the most complex next-generation applications. Expression levels, differential splicing, allele-specific expression, RNA editing and fusion transcripts constitute important information when comparing samples for disease-related studies. These attributes, not readily available by hybridization-based or tag sequence-based approaches, can now be far more easily and precisely obtained if sufficient sequence coverage is achieved. However, many other essential subtleties in the RNA-Seq data remain to be faced and understood.

Hybridization-based approaches typically refer to the microarray platforms. Until recently, these platforms have offered to the scientific community a very useful tool to simultaneously investigate thousands of features within a single experiment, providing a reliable, rapid, and cost-effective technology to analyze the gene expression patterns. Due to their nature, they suffer from background and cross-hybridization issues and allow researchers to only measure the relative abundance of RNA transcripts included in the array design [16]. This technology, which measures gene expression by simply quantifying—via an indirect method—the hybridized and labeled cDNA, does not allow the detection of RNA transcripts from repeated sequences, offering a limited dynamic range, unable to detect very subtle changes in gene expression levels, critical in understanding any biological response to exogenous stimuli and/or environmental changes [9, 17, 18].

Other methods such as Serial, Cap Analysis of Gene Expression (SAGE and CAGE, resp.) and Polony Multiplex Analysis of Gene Expression (PMAGE), tag-based sequencing methods, measure the absolute abundance of transcripts in a cell/tissue/organ and do not require prior knowledge of any gene sequence as occurs for microarrays [19]. These analyses consist in the generation of sequence tags from fragmented cDNA and their following concatenation prior to cloning and sequencing [20]. SAGE is a powerful technique that can therefore be viewed as an unbiased digital microarray assay. However, although SAGE sequencing has been successfully used to explore the transcriptional landscape of various genetic disorders, such as diabetes [21, 22], cardiovascular diseases [23], and Downs syndrome [24, 25], it is quite laborious for the cloning and sequencing steps that have thus far limited its use.

TABLE 1: Selection of papers on mammalian RNA-Seq.

Reference	Organism	Cell type/tissue	NGS platform
Bainbridge et al., 2006 [27]	*Homo sapiens*	Prostate cancer cell line	Roche
Cloonan et al., 2008 [30]	*Mus musculus*	ES cells and Embryoid bodies	ABI
Core et al., 2008 [31]	*Homo sapiens*	Lung fibroblasts	Illumina
Hashimoto et al., 2008 [32]	*Homo sapiens*	HT29 cell line	ABI
Li et al., 2008 [33]	*Homo sapiens*	Prostate cancer cell line	Illumina
Marioni et al., 2008 [34]	*Homo sapiens*	Liver and kidney samples	Illumina
Morin et al., 2008 [35]	*Homo sapiens*	ES cells and Embryoid bodies	Illumina
Morin et al., 2008 [36]	*Homo sapiens*	HeLa S3 cell line	Illumina
Mortazavi et al., 2008 [37]	*Mus musculus*	Brain, liver and skeletal muscle	Illumina
Rosenkran et al., 2008 [38]	*Mus musculus*	ES cells	Illumina
Sugarbaker et al., 2008 [39]	*Homo sapiens*	Malignant pleural mesothelioma, adenocarcinoma and normal lung	Roche
Sultan et al., 2008 [40]	*Homo sapiens*	Human embryonic kidney and B cell line	Illumina
Asmann et al., 2009 [41]	*Homo sapiens*	Universal and brain human reference RNAs	Illumina
Chepelev et al., 2009 [42]	*Homo sapiens*	Jurkat and GD4+ T cells	Illumina
Levin et al., 2009 [43]	*Homo sapiens*	K562	Illumina
Maher et al., 2009 [44]	*Homo sapiens*	Prostate cancer cell lines	Roche
Illumina			
Parkhomchuk et al., 2009 [45]	*Mus musculus*	Brain	Illumina
Reddy et al., 2009 [46]	*Homo sapiens*	A549 cell line	Illumina
Tang et al., 2009 [47]	*Mus musculus*	Blastomere and oocyte	ABI
Blekhman et al., 2010 [48]	*Homo sapiens, Pan troglodytes, Rhesus macaca.*	Liver	Illumina
Heap et al., 2010 [49]	*Homo sapiens*	Primary GD4+ T cells	Illumina
Raha et al., 2010 [50]	*Homo sapiens*	K562 cell line	Illumina

In contrast, RNA-Seq on NGS platforms has clear advantages over the existing approaches [9, 26]. First, unlike hybridization-based technologies, RNA-Seq is not limited to the detection of known transcripts, thus allowing the identification, characterization and quantification of new splice isoforms. In addition, it allows researchers to determine the correct gene annotation, also defining—at single nucleotide resolution—the transcriptional boundaries of genes and the expressed Single Nucleotide Polymorphisms (SNPs). Other advantages of RNA-Seq compared to microarrays are the low "background signal," the absence of an upper limit for quantification and consequently, the larger dynamic range of expression levels over which transcripts can be detected. RNA-Seq data also show high levels of reproducibility for both technical and biological replicates.

Recent studies have clearly demonstrated the advantages of using RNA-Seq [27–50]. Table 1 provides a short description of recent and more relevant papers on RNA-Seq in mammals.

Many research groups have been able to precisely quantify known transcripts, to discover new transcribed regions within intronic or intergenic regions, to characterize the antisense transcription, to identify alternative splicing with new combinations of known exon sequences or new transcribed exons, to evaluate the expression of repeat elements and to analyze a wide number of known and possible new candidate expressed SNPs, as well as to identify fusion transcripts and other new RNA categories.

8.3.1 SAMPLE ISOLATION AND LIBRARY PREPARATION

The first step in RNA-Seq experiments is the isolation of RNA samples; further RNA processing strictly depends on the kind of analysis to perform. Indeed, as "transcriptome" is defined as the complete collection of transcribed elements in a genome (see [2]), it consists of a wide variety of transcripts, both mRNA and non-mRNA, and a large amount (90–95%) of rRNA species. To perform a whole transcriptome analysis, not limited to annotated mRNAs, the selective depletion of abundant rRNA molecules (5S, 5.8S, 18S and 28S) is a key step. Hybridization with rRNA sequence-specific 5'-biotin labeled oligonucleotide probes, and the following removal with streptavidin-coated magnetic beads, is the main procedure to selectively deplete large rRNA molecules from total isolated RNA. Moreover, since rRNA—but not capped mRNAs—is characterized by the presence of 5' phosphate, an useful approach for selective ribo-depletion is based on the use of an exonuclease able to specifically degrade RNA molecules bearing a phosphate (mRNA-ONLY kit, Epicentre). Compared to the polyadenylated (polyA+) mRNA fraction, the ribo-depleted RNA is enriched in non-polyA mRNA, preprocessed RNA, tRNA, regulatory molecules such as miRNA, siRNA, small ncRNA, and other RNA transcripts of yet unknown function (see review [8]).

How closely the RNA sequencing reflects the original RNA populations is mainly determined in the library preparation step, crucial in the whole transcriptome protocols. Although NGS protocols were first developed for the analysis of genomic DNA, these technical procedures have been rapidly and effectively adapted to the sequencing of double-strand (ds) cDNA for transcriptome studies [51].

A double-stranded cDNA library can be usually prepared by using: (1) fragmented double-stranded (ds) cDNA and (2) hydrolyzed or fragmented RNA.

The goal of the first approach is to generate high-quality, full-length cDNAs from RNA samples of interest to be fragmented and then ligated to an adapter for further amplification and sequencing. By the way, since the primer adaptor is ligated to a fragmented ds cDNA, any information on the transcriptional direction would completely be lost. Preserving the strandedness is fundamental for data analysis; it allows to determine the directionality of transcription and gene orientation and facilitates detection of opposing and overlapping transcripts. To take into account and thus to avoid this biologically relevant issue, many approaches, such as pretreating the RNA with sodium bisulphite to convert cytidine into uridine [52], have been so far developed. Other alternative protocols, differing in how the adaptors are inserted into ds cDNA, have been recently published: direct ligation of RNA adaptors to the RNA sample before or during reverse transcription [30, 31, 53], or incorporation of dUTP during second strand synthesis and digestion with uracil-Nglycosylase enzyme [45]. For instance, SOLiD Whole Transcriptome Kit contains two different sets of oligonucleotides with a single-stranded degenerate sequence at one end, and a defined sequence required for sequencing at the other end, constraining the orientation of RNA in the ligation reaction. The generation of ds cDNA from RNA involves a number of steps. First, RNA is converted into first-strand cDNA using reverse transcriptase with either random hexamers or oligo(dT) as primers. The resulting first-strand cDNA is then converted into double-stranded cDNA, further fragmented with DNAse I and then ligated to adapters for amplification and sequencing [54]. The advantage of using oligo dT is that the majority of cDNA produced should be polyadenylated mRNA, and hence more of the sequence obtained should be informative (nonribosomal). The significant disadvantage is that the reverse transcriptase enzyme will fall off of the template at a characteristic rate, resulting in a bias towards the end of transcripts. For long mRNAs this bias can be pronounced, resulting in an under representation (or worse in the absence) of the 5' end of the transcript in the data. The use of random primers would therefore be the preferred method to avoid this problem and to allow a better representation of the 5' end of long ORFs. However, when oligo dT primers are used for priming, the slope which is formed by the diminishing frequency of reads towards the 5' end of the ORF can, in some cases, be useful for determining the strand of origin for new transcripts if strand information has not been retained [28, 37].

Fragmenting RNA, rather than DNA, has the clear advantage of reducing possible secondary structures, particularly for tRNA and miRNA, resulting in a major heterogeneity in coverage and can also lead to a more comprehensive transcriptome analysis (Figure 2). In this case, the RNA sample is first fragmented by using controlled temperature or chemical/enzymatic hydrolysis, ligated to adapters and retro-transcribed by complementary primers. Different protocols have been so far developed. Indeed, the adaptor sequences may be directly ligated to the previously fragmented RNA molecules by using T4 RNA ligase, and the resulting library can be reverse transcribed with primer pairs specifically suited on the adaptor sequences, and then sequenced. Another approach, recently described in [55], consists in the in vitro polyadenilation of RNA fragments in order to have a template for the next step of reverse transcription using poly(dT) primers containing both adaptor sequences (linkers), separated back-to-back by an endonuclease site. The resulting cDNAs are circularized and then cleaved at endonuclease site in the adaptors, thus leaving ss cDNA with the adaptors at both ends [55]. A third protocol described by [33], named double random priming method, uses biotinylated random primers (a sequencing primer P1 at the 5' end, and a random octamer at the 3' end). After a first random priming reaction, the products are isolated by

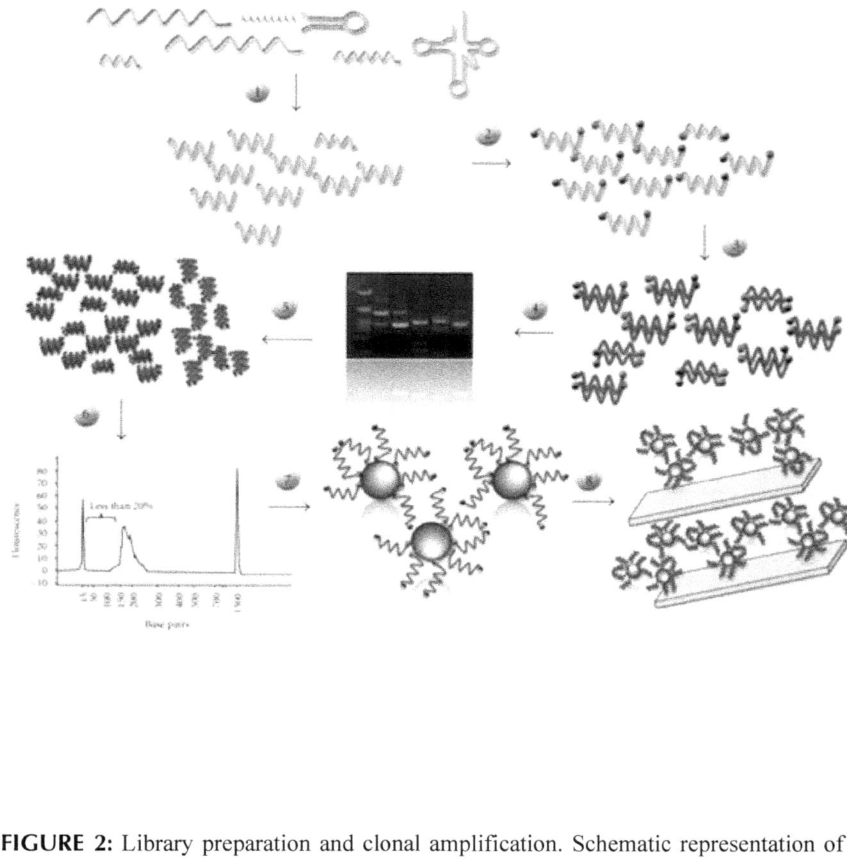

FIGURE 2: Library preparation and clonal amplification. Schematic representation of a workflow for library preparation in RNA-Seq experiments on the SOLiD platform. In the figure is depicted a total RNA sample after depletion of rRNA, containing both polyA and non-polyA mRNA, tRNAs, miRNAs and small noncoding RNAs. Ribo-depleted total RNA is fragmented (1), then ligated to specific adaptor sequences (2) and retro-transcribed (3). The resulting cDNA is size selected by gel electrophoresis (4), and cDNAs are PCR amplified (5). Then size distribution is evaluated (6). Emulsion PCR, with one cDNA fragment per bead, is used for the clonal amplification of cDNA libraries (7). Purified and enriched beads are finally deposited onto glass slides (8), ready to be sequenced by ligation.

using streptavidin beads and a second random priming reaction is performed on a solid phase with a random octamer carrying the sequencing primer P2. Afterwards, second random priming products are released from streptavidin beads by heat, PCR-amplified, gel-purified, and finally subjected to sequencing process from the P1 primer. Moreover, as already mentioned, in [45] the authors used dUTP—a surrogate for dTTP—during the second-strand synthesis to allow a selective degradation of second cDNA strand after adaptor ligation using a uracil-N-glycosylase. The use of engineered DNA adaptors, combined to the dUTP protocol, ensures that only the cDNA strand corresponding to the "real" transcript is used for library amplification and sequencing, reserving the strandedness of gene transcription [45].

However, independently on the library construction procedure, particular care should be taken to avoid complete degradation during RNA fragmentation.

The next step of the sequencing protocols is the clonally amplification of the cDNA fragments.

Illumina, 454 and SOLiD use clonally amplified templates. In particular, the last two platforms use an innovative procedure, emulsion PCR (emPCR), to prepare sequencing templates in a cell-free system. cDNA fragments from a fragment or paired-end library are separated into single strands and captured onto beads under conditions that favour one DNA molecule per bead. After the emPCR and beads enrichment, millions of them are chemically crosslinked to an amino-coated glass surface (SOLiD) or deposited into individual PicoTiterPlate (PTP) wells (454) in which the NGS chemistry can be performed. Solid-phase amplification (Illumina) can also be used to produce randomly distributed, clonally amplified clusters from fragment or mate-pair templates on a glass slide. High-density forward and reverse primers are covalently attached to the slide, and the ratio of the primers to the template defines the surface density. This procedure can produce up to 200 million spatially separated template clusters, providing ends for primer hybridization, needed to initiate the NGS reaction. A different approach is the use of single molecules templates (Helicos BioScience) usually immobilized on solid supports, in which PCR amplification is no more required, thus avoiding the insertion of possible confounding mutations in the templates. Furthermore, AT- and GC-rich sequences present amplification issues, with over- or under-representation bias in genome alignments and assemblies. Specific adaptors are bound to the fragmented templates, then hybridized to spatially distributed primers covalently attached to the solid support [56].

9.3.2 SEQUENCING AND IMAGING

NGS platforms use different sequencing chemistry and methodological procedures.

Illumina and HeliScope use the Cyclic Reversible Termination (CRT), which implies the use of reversible terminators (modified nucleotide) in a cyclic method. A DNA polymerase, bound to the primed template, adds one fluorescently modified nucleotide per cycle; then the remaining un-incorporated nucleotides are washed away and imaging capture is performed. A cleavage step precedes the next incorporation cycle to remove the terminating/inhibiting group and the fluorescent dye, followed by an additional washing. Although these two platforms use the same methodology, Illumina employs the four-colour CRT method, simultaneously incorporating all 4 nucleotides with different dyes; HeliScope uses the one-colour (Cy5 dye) CRT method.

Substitutions are the most common error type, with a higher portion of errors occurring when the previous incorporated nucleotide is a G base [57]. Under representation of AT-rich and GC-rich regions, probably due to amplification bias during template preparation [57–59], is a common drawback.

In contrast, SOLiD system uses the Sequencing by Ligation (SBL) with 1, 2-nucleotide probes, based on colour space, which is an unique feature of SOLiD. It has the main advantage to improve accuracy in colour and single nucleotide variations (SNV) calling, the latter of which requires an adjacent valid colour change. In particular, a universal primer is hybridized to the template beads, and a library of 1, 2-nucleotide probes is added. Following four-colour imaging, the ligated probes are chemically cleaved to generate a 5'-phosphate group. Probe hybridization and ligation, imaging, and probe cleavage is repeated ten times to yield ten colour calls spaced in five-base intervals. The extended primer is then stripped from the solid-phase-bound templates. A second ligation round is performed with a n - 1 primer, which resets the interrogation bases and the corresponding ten colour calls one position to the left. Ten ligation cycles ensue, followed by three rounds of ligation cycles. Colour calls from the five-ligation rounds are then ordered into a linear sequence (the csfasta colour space) and aligned to a reference genome to decode the sequence. The most common error type observed by using this platform are substitutions, and, similar to Illumina, SOLiD data have also revealed an under representation of AT- and GC-rich regions [58].

Another approach is pyrosequencing (on 454), a non-electrophoretic bioluminescence method, that unlike the above-mentioned sequencing approaches is able to measure the release of pyrophosphate by proportionally converting it into visible light after enzymatic reactions. Upon incorporation of the complementary dNTP, DNA polymerase extends the primer and pauses. DNA synthesis is reinitiated following the addition of the next complementary dNTP in the dispensing cycle. The enzymatic cascade generates a light recorded as a flowgram with a series of picks corresponding to a particular DNA sequence. Insertions and deletions are the most common error types.

An excellent and detailed review about the biotechnological aspects of NGS platforms can be found in [15].

8.3.3 FROM BIOLOGY TO BIOINFORMATICS

The unprecedented level of sensitivity in the data produced by NGS platforms brings with it the power to make many new biological observations, at the cost of a considerable effort in the development of new bioinformatics tools to deal with these massive data files.

First of all, the raw image files from one run of some next generation sequencers can require terabytes of storage, meaning that simply moving the data off the machine can represent a technical challenge for the computer networks of many research centers. Moreover, even when the data are transferred from the machine for subsequent processing, common desktop computer will be hopelessly outmatched by the volume of data from a single run. As a result, the use of a small cluster of computers is extremely beneficial to reduce computational bottleneck.

Another issue is the availability of software required to perform downstream analysis. Indeed after image and signal processing the output of a RNA-Seq experiment consists of 10–400 millions of short reads (together with their base-call quality values), typically of 30–400 bp, depending on the DNA sequencing technology used, its version and the total cost of the experiments.

NGS data analysis heavily relies on proper mapping of sequencing reads to corresponding reference genomes or on their efficient de novo assembly. Mapping NGS reads with high efficiency and reliability currently faces several challenges. As noticed by [60], differences between the sequencing platforms in samples preparation, chemistry, type and volume of raw data, and data formats are very large, implying that each platform produces data affected by characteristic error profiles. For example the 454 system can produce reads with insertion or deletion errors during homopolymer runs and generate fewer, but longer, sequences in fasta like format allowing to adapt classical alignment algorithms; the Illumina has an increased likelihood to accumulate sequence errors toward the end of the read and produce fasta reads, but they are shorter, hence requiring specific alignment algorithms; the SOLiD also tends to accumulate bias at the end of the reads, but uses di-base encoding strategy and each sequence output is encoded in a colour space csfasta format. Hence, some sequence errors are correctable, providing better discrimination between sequencing error and polymorphism, at the cost of requiring analysis tools explicitly built for handling this aspect of the data. It is not surprising that there are no "box standard" software available for end-users, hence the implementation of individualized data processing pipelines, combining third part packages and new computational methods, is the only advisable approach. While some existing packages are already enabling to solve general aspects of RNA-Seq analysis, they also require a time consuming effort due to the lack of clear documentation in most of the algorithms and the variety of the formats. Indeed, a much clear documentation of the algorithms is needed to ensure a full understanding of the processed data. Community adoption of input/output data formats for reference alignments, assemblies and detected variants is also essential for ease the data management problem. Solving these issues may simply shift the software gap from sequence processing (base-calling, alignment or assembly, positional counting and variant detection) to sequence analysis (annotation and functional impact).

8.3.4 GENOME ALIGNMENT AND READS ASSEMBLY

The first step of any NGS data analysis consists of mapping the sequence reads to a reference genome (and/or to known annotated transcribed sequences) if available, or de novo assembling to produce a genome-scale transcriptional map. (see Figure 3 for an illustration of a classical RNA-Seq computational pipeline). The decision to use one of strategies is mainly based on the specific application. However, independently on the followed approach, there is a preliminary step that can be useful to perform which involves the application of a quality filtering to remove poor quality reads and to reduce the computational time and the effort for further analysis.

Analyzing the transcriptome of organisms without a specific reference genome requires de novo assembling (or a guided assembly with the help of closely related organisms) of expressed sequence tags (ESTs) using short-read assembly programs such as [61, 62]. A reasonable strategy for improving the quality of the assembly is to increase the read coverage and to mix different reads types. However RNA-Seq experiments without a reference genome propose specific features and challenges that are out of the scope of the present paper; we refer the readers to [63, 64] for further details.

In most cases, the reference genome is available and the mapping can be carried out using either the whole genome or known transcribed sequences (see, e.g., [28–30, 32, 34, 37, 40, 46, 47]). In both cases, this preliminary but crucial step is the most computationally intensive of the entire process and strongly depends on the type of available sequences (read-length, error profile, amount of data and data format). It is not surprising that such nodal point still constitutes a very prominent area of research (see, e.g., [65–67] for a review) and has produced a great number of different algorithms in the last couple of years (e.g., [68–78]). Clearly, not all of them completely support the available platforms or are scalable for all amount

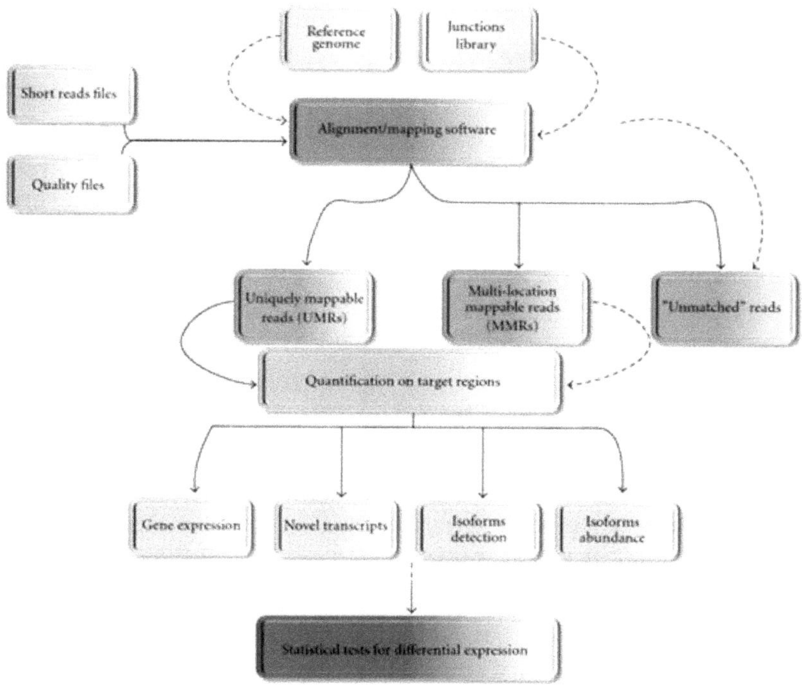

FIGURE 3: RNA-Seq computational pipeline.

of throughput or genome size. Nevertheless, the sequencing technologies are still in a developing phase with a very fast pace of increase in throughput, reads length and data formats after few months. Consequently, the already available mapping/assembly software are continuously under evolution in order to adapt themselves to the new data formats, to scale with the amount of data and to reduce their computational demand. New softwares are also continuously complementing the panorama. Moreover, the alignment phase of reads from RNA-Seq experiments presents many other subtleties to be considered; standard mapping algorithms are not able to fully exploit the complexity of the transcriptome, requiring to be modified or adapted in order to account for splicing events in eucaryotes.

The easiest way to handle such difficulty is to map the reads directly on known transcribed sequences, with the obvious drawback of missing new transcripts. Alternatively, the reads can be mapped continuously to the genome, but with the added opportunity of mapping reads that cross splice junctions. In this case, the algorithms differ from whether they require or not junctions's model. Algorithms such as Erange [37] or RNA-mate [79] require library of junctions constructed using known splice junctions extracted from data-bases and also supplemented with any set of putative splice junctions obtained, for instance, using a combinatorial approach on genes' model or ESTs sequences. Clearly, such approaches do not allow to map junctions not previously assembled in the junctions' library. On the other hand, algorithms like the WT [69], QPALMA [80], TopHat [81], G.Mo.R-Se [63], and PASS [78] potentially allow to detect new splice isoforms, since they use a more sophisticated mapping strategy. For instance, WT [69] splits the reads in left and right pieces, aligns each part to the genome, then attempts to extend each alignment on the other side to detect the junction. Whereas TopHat [81] first maps the reads against the whole reference genome using [77], second aggregates the mapped reads in islands of candidate exons on which compute a consensus measure, then generates potential donor/acceptor splice sites using neighboring exons, and finally tries to align the reads, unmapped to the genome, to these splice junction sequences.

Most of the RNA-Seq packages are built on top of optimized short read core mappers [68, 69, 72, 77] and the mapping strategy is carried out by performing multiple runs or cycles. At the end of each cycle the unmatched reads are trimmed from one extreme and another step of alignment is attempted (see, e.g., [79]). Specific tolerances can be set for each alignment in order to increase the amount of mappable data. Obviously the simplest core approach is to map the sequence reads across the genome allowing the user to specify only the number of tolerated mismatches, although other methods allow to use also gapped alignment. Such flexibility can be beneficial for the rest of the analysis since both sequencing errors, that usually increase with the length of the sequence, and SNPs may cause substitutions and insertion/deletion of nucleotides in the reads. On the other hand, increasing the mapping flexibility also introduces a higher level of noise in the data. The compromise between the number of mapped reads and the quality of the resulting mapping is a very time consuming process without an optimal solution.

At the end of the mapping algorithm one can distinguish between three types of reads: reads that map uniquely to the genome or to the splice junctions (Uniquely Mappable Reads, UMR), reads with multiple (equally or similarly likely) locations either to the genome or to the splice junctions (Multilocation Mappable Reads, MMR) and reads without a specific mapping location. MMRs arise predominantly from conserved domains of paralogous gene families and from repeats. The fraction of mappable reads that are MMRs depends on the length of the read, the genome under investigation, and the expression in the individual sample; however it is typically between 10–40% for mammalian derived libraries [30, 37]. Most of the studies [28, 34] usually discarded MMRs from further analysis, limiting the attention only to UMRs. Clearly, this omission introduces experimental bias, decreases the coverage and reduces the possibility of investigating expressed regions such as active retrotransposons and gene families. An alternative strategy for the removal of the MMRs is to probabilistically assign them to each genomic location they map to. The simplest assignment considers equal probabilities. However, far better results have been obtained using a guilt-by-association strategy that calculates the probability of a MMRs originating from a particular locus. In [82], the authors proposed to proportionally assign MMRs to each of their mapping locations based on unique coincidences with either UMRs and other MMRs. Such a technique was later adopted in [79]. By contrast, in [83], the authors computed the probability as the ratio between the number of UMRs occurring in a nominal window surrounding each locus occupied by the considered MMR and the total number of UMRs proximal to all loci associated with that MMR. Similarly, in [37] the MMRs were fractionally assigned to their different possible locations considering the expression levels of their respective gene models. All these rescue strategies lead to substantially higher transcriptome coverage and give expression estimates in better agreement with microarrays than those using only UMRs (see, [37, 83]). Very recently, a more sophisticated approach was proposed in [84]. The authors introduced latent random variables representing the true mappings, with the parameters of the graphical model corresponding to isoform expression levels, read distributions across transcripts, and

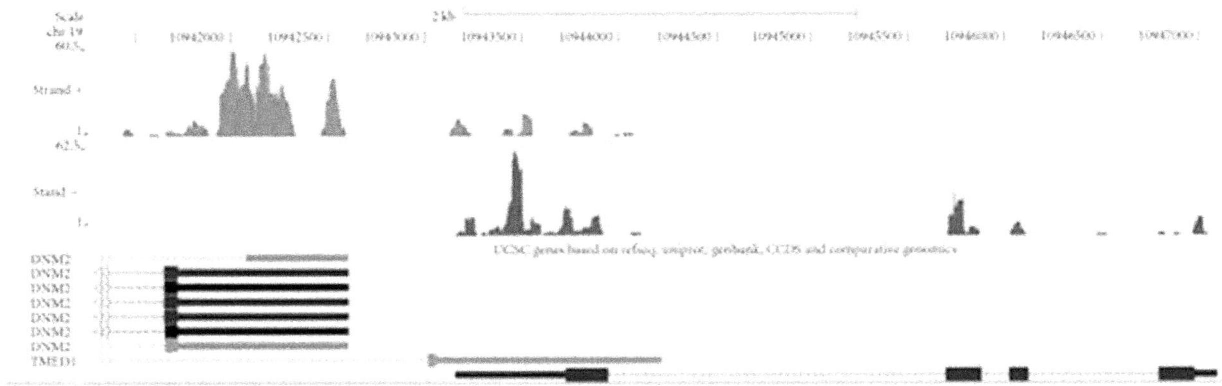

FIGURE 4: Strand-Specific Read Distribution in UCSC Genome Browser and IGV. (a) UCSC Genome Browser showing an example of stranded sequences generated by RNA-Seq experiment on NGS platform. In particular, the screenshot—of a characteristic "tail to tail" orientation of two human genes—clearly shows the specific expression in both strands where these two genes overlap, indicating that the strandedness of reads is preserved. (b) The same genomic location in the IGV browser, showing the reads (coloured blocks) distribution along TMED1 gene. The grey arrows indicate the sense of transcription. The specific expression in both strands where the genes overlap, indicates that the strandedness of reads is preserved. In (c) a greater magnification of the reads mapping to the same region at nucleotide level, useful to SNP analysis. The chromosome positions are shown at the top and genomic loci of the genes are shown at the bottom of each panel.

sequencing error. They allocated MMRs by maximizing the likelihood of the expression levels using an Expectation-Maximization (EM) algorithm. Additionally, they also showed that previous rescue methods introduced in [37, 82] are roughly equivalent to one iteration of EM. Independently on the specific proposal, we observe that all the above mentioned techniques work much better with data that preserve RNA strandedness. Alternatively, the use of paired-end protocols should help to alleviate the MMRs problem. Indeed, when one of the paired reads maps to a highly repetitive element in the genome but the second does not, it allows both reads to be unambiguously mapped to the reference genome. This is accomplished by first matching the first nonrepeat read uniquely to a genomic position and then looking within a size window, based on the known size range of the library fragments, for a match for the second read. The usefulness of this approach was demonstrated to improve read matching from 85% (single reads) to 93% (paired reads) [70], allowing a significant improvement in genome coverage, particularly in repeat regions. Currently, all of the next generation sequencing technologies are capable for generating data from paired-end reads, but unfortunately, till now only few RNA-Seq software support the use of paired-end reads in conjunction with the splice junctions mapping.

One of the possible reasons for reads not mapping to the genome and splice junctions is the presence of higher sequencing errors in the sequence. Other reasons can be identified in higher polymorphisms, insertion/deletion, complex exon-exon junctions, miRNA and small ncRNA: such situations could potentially be recovered by more sophisticated or combined alignment strategy.

Once mapping is completed, the user can display and explore the alignment on a genome browser (see Figure 4 for a screen-shot example) such as UCSC Genome Browser [85] (http://genome.ucsc.edu/) or the Integrative Genomics Viewer (IGV) (http://www.broadinstitute.org/igv), or on specifically devoted browsers such as EagleView [86], MapView [87] or Tablet [88], that can provide some highly informative views of the results at different levels of aggregations. Such tools allow to incorporate the obtained alignment with database annotations and other source of information, to observe specific polymorphism against sequence error, to identify well documented artifacts due to the DNA amplifications, as well as to detect other source of problems such as the not uniformity of the reads coverage across the transcript. Unfortunately, in many cases the direct visualization of the data is hampered by the lack of a common format for the alignment algorithm, causing a tremendous amount of extra work in format conversion for visualization purposes, feature extraction and other downstream analysis. Only recently, the SAM (Sequencing Alignment/Map) format [89] has been proposed as a possible standard for storing read alignment against reference sequences.

8.3.5 QUANTIFYING GENE EXPRESSION AND ISOFORMS' ABUNDANCE

Browser-driven analyses are very important for visualizing the quality of the data and to interpret specific events on the basis of the available annotations and mapped reads. However they only provide a qualitative picture of the phenomenon under investigation and the enormous amount of data does not allow to easily focus on the most relevant details. Hence, the second phase of most of the RNA-Seq pipeline consists of the automatic quantification of the transcriptional events across the entire genome (see Figure 4). From this point of view the interest is both quantifying known elements (i.e., genes or exons already annotated) and detecting new transcribed regions, defined as transcribed segments of DNA not yet annotated as exons in databases. The ability to detect these unannotated regions, even though biologically relevant, is one of the main advantages of the RNA-Seq over microarray technology. Usually, the quantification step is preliminary to any differential expression approach, see Figure 5.

In order to derive a quantitative expression for annotated elements (such as exons or genes) within a genome, the simplest approach is to provide the expression as the total number of reads mapping to the coordinates of each annotated element. In the classical form, such method weights all the reads equally, even though they map the genome with different stringency. Alternatively, gene expression can be calculated as the sum of the number of reads covering each base position of the annotated element; in this way the expression is provided in terms of base coverage. In both cases, the results depend on the accuracy of the used gene models

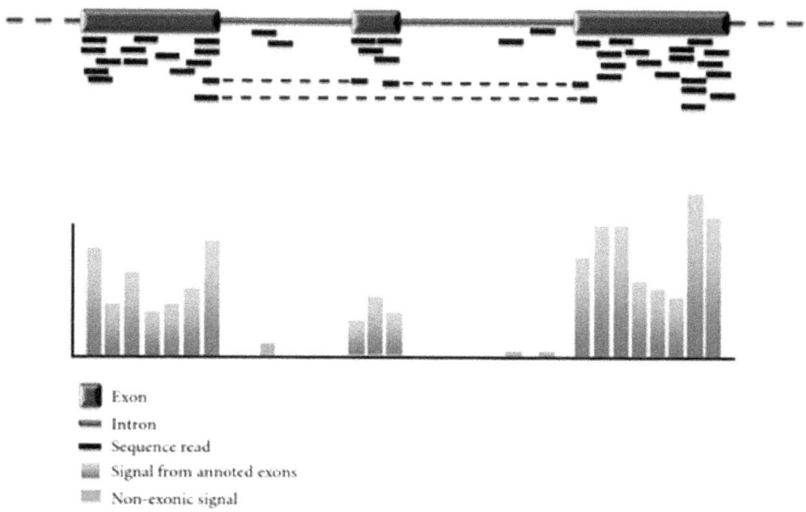

FIGURE 5: Mapping and quantification of the signal. RNA-seq experiments produce short reads sequenced from processed mRNAs. When a reference genome is available the reads can be mapped on it using efficient alignment software. Classical alignment tools will accurately map reads that fall within an exon, but they will fail to map spliced reads. To handle such problem suitable mappers, based either on junctions library or on more sophisticated approaches, need to be considered. After the mapping step annotated features can be quantified.

and the quantitative measures are a function of the number of mapped reads, the length of the region of interest and the molar concentration of the specific transcript. A straightforward solution to account for the sample size effect is to normalize the observed counts for the length of the element and the number of mapped reads. In [37], the authors proposed the Reads Per Kilobase per Million of mapped reads (RPKM) as a quantitative normalized measure for comparing both different genes within the same sample and differences of expression across biological conditions. In [84], the authors considered two alternative measures of relative expression: the fraction of transcripts and the fraction of nucleotides of the transcriptome made up by a given gene or isoform.

Although apparently easy to obtain, RPKM values can have several differences between software packages, hidden at first sight, due to the lack of a clear documentation of the analysis algorithms used. For example ERANGE [37] uses a union of known and new exon models to aggregate reads and determines a value for each region that includes spliced reads and assigned multireads too, whereas [30, 40, 81, 90] are restricted to known or prespecified exons/gene models. However, as noticed in [91], several experimental issues influence the RPKM quantification, including the integrity of the input RNA, the extent of ribosomal RNA remaining in the sample, the size selection steps and the accuracy of the gene models used.

In principle, RPKMs should reflect the true RNA concentration; this is true when samples have relatively uniform sequence coverage across the entire gene model. The problem is that all protocols currently fall short of providing the desired uniformity, see for example [37], where the Kolmogorov-Smirnov statistics is used to compare the observed reads distribution on each selected exon model with the theoretical uniform one. Similar conclusions are also illustrated in [57, 58], among others.

Additionally, it should be noted that RPKM measure should not be considered as the panacea for all RNA-Seq experiments. Despite the importance of the issue, the expression quantification did not receive the necessary attention from the community and in most of the cases the choice has been done regardless of the fact that the main question is the detection of differentially expressed elements. Regarding this point in [92] it is illustrated the inherent bias in transcript length that affect RNA-Seq experiments. In fact the total number of reads for a given transcript is roughly proportional to both the expression level and the length of the transcript. In other words, a long transcript will have more reads mapping to it compared to a short gene of similar expression. Since the power of an experiment is proportional to the sampling size, there will be more statistical power to detect differential expression for longer genes. Therefore, short transcripts will always be at a statistical disadvantage relative to long transcripts in the same sample. RPKM-type measures provide an expression level normalized by the length of the gene and this only apparently solves the problem; it gives an unbiased measure of the expression level, but also changes the variance of the data in a length dependent manner, resulting in the same bias to differential expression estimation. In order to account for such an inherent bias, in [92] the authors proposed to use a fixed length window approach, with a window size smaller than the smallest gene. This method can calculate aggregated tag counts for each window and consequently assess them for differential expression. However, since the analysis is performed at the window level some proportion of the data will be discarded; moreover such an approach suffers for a reduced power and highly expressed genes are more likely to be detected due to the fact that the sample variance decreases with the expression level. Indeed, it should be noticed that the sample variance depends on both the transcript length and the expression level.

Finally, we observe that annotation files are often inaccurate; boundaries are not always mapped precisely, ambiguities and overlaps among transcripts often occur and are not yet completely solved. Concerning this issue in [93] the authors proposed a method based on the definition of "union-intersection genes" to define the genomic region of interest and normalized absolute and relative expression measures within. Also, in this case we observe that all strategies work much better with data that preserve RNA strandedness, which is an extremely valuable information for transcriptome annotation, especially for regions with overlapping transcription from opposite directions.

The quantification methods described above do not account for new transcribed region. Although several studies have already demonstrated that RNA-Seq experiments, with their high resolution and sensitivity have great potentiality in revealing many new transcribed regions, unidentifiable by microarrays, the detection of new transcribed regions is mainly obtained by means of a sliding window and heuristic approaches. In [94] stretches of contiguous expression in intergenic regions are identified after removing all UTRs from the intergenic search space by using a combination of information arising from tiling-chip and sequence data and visual inspection and manual curation. The procedure is quite complex and is mainly due to the lack of strandedness information in their experiment. On the contrary, the hybridization data are less affected by these issues because they distinguish transcriptional direction and do not show any 5 bias (see [94] for further details). Then, new transcribed regions are required to have a length of at least 70 bp and an average sequence coverage of 5 reads per bp. A similar approach, with different choices of the threshold and the window, was proposed in [40], where the authors investigated either intergenic and intronic regions. The choices of the parameters are assessed by estimating noise levels by means of a Poisson model of the noncoding part of the genome. In [45] the whole genome is split into 50 bp windows (non-overlapping). A genomic region is defined as a new transcribed region if it results from the union of two consecutive windows, with at least two sequence reads mapped per window. Additionally, the gap between each new transcribed regions should be at least 50 bp, and the gap between a new transcribed region and an annotated gene (with the same strand) at least 100 bp. A slightly more sophisticated approach is used in ERANGE [37]. Reads that do not fall within known exons are aggregated into candidate exons by requiring regions with at least 15 reads, whose starts are not separated by more than 30 bp. Most of the candidate exons are assigned to neighboring gene models when they are within a specifiable distance of the model.

These studies, among others, reveal many of these new transcribed regions. Unfortunately, most of them do not seem to encode any protein, and hence their functions remain often to be determined. In any case, these new transcribed regions, combined with many undiscovered new splicing variants, suggest that there is considerably more transcript complexity than previously appreciated. Consequently further RNA-Seq experiments and more sophisticated analysis methods can disclose it.

The complexity of mammalian transcriptomes is also compounded by alternative splicing which allows one gene to produce multiple transcript isoforms. Alternative splicing includes events such as exon skipping, alternative 5' or 3' splicing, mutually exclusive exons, intron retention, and "cryptic" splice sites (see Figure 6). The frequency of occurrence of alternative splicing events is still underestimated. However it is well known that multiple transcript isoforms produced from a single gene can lead to protein isoforms with distinct functions, and that alternative splicing is widely involved in different physiological and pathological processes. One of the most important advantages of the RNA-Seq experiments is the possibility of understanding and comparing the transcriptome at the isoform level (see [95, 96]). In this context, two computational problems need to be solved: the detection of different isoforms and their quantification in terms of transcript abundance.

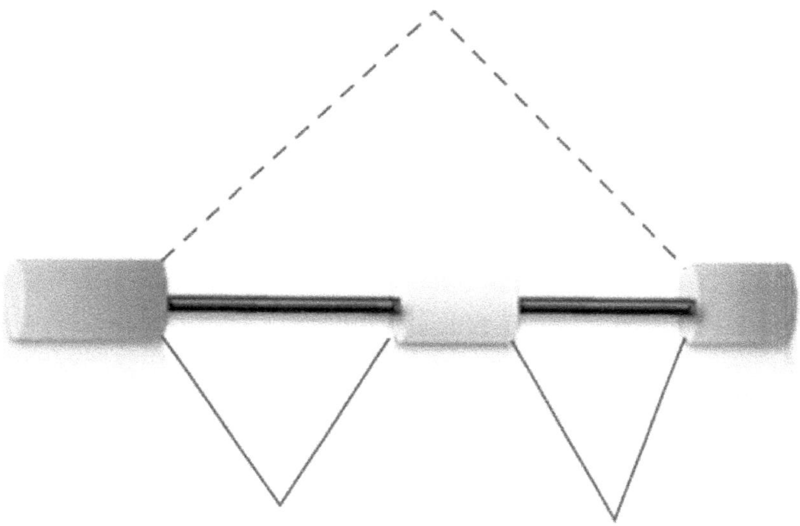

FIGURE 6: Alternative splicing. Schematic representation of the possible patterns of alternative splicing of a gene. Boxes are discrete exons that can be independently included or excluded from the mRNA transcript. Light blue boxes represent constitutive exons, violet and red boxes are alternatively spliced exons. Dashed lines represent alternative splicing events. (a) Canonical exon skipping; (b) 5 or (c) 3 alternative splicing; (d) Mutually exclusive splicing event involving the selection of only one from two or more exon variants; (e) Intra-exonic "cryptic" splice site causing the exclusion of a portion of the exon from the transcript; (f) Usage of new alternative 5 or (g) 3 exons; (h) Intron retention.

Initial proposals for solving these problems were essentially based on a gene-by-gene manual inspection usually focusing the attention to the detection of the presence of alternative splicing forms rather than to their quantification. For example, the knowledge of exon-exon junction reads and of junctions that fall into some isoform-specific regions can provide useful information for identifying different isoforms. The reliability of a splicing junction is usually assessed by counting features like the number of reads mapping to the junction, the number of mismatches on each mapped read, the mapping position on the junction and the mismatches location in a sort of heuristic approach. Unfortunately, these techniques cannot be scaled to the genome level and they are affected by a high false positive and false negative rate.

Following the above mentioned ideas, in [40] the authors detected junctions by computing the probability of a random hits for a read of length R on the splice junctions of length J with at most a certain number of mismatches. In [95], the authors used several information similar to those described above to train classifiers based on logistic regression for splicing junction detection. In [97], the authors introduced a new metric to measure the quality of each junction read. Then they estimated the distribution of such metric either with respect to known exon splice junctions and random splice junctions, and implemented an empirical statistical model to detect exon junctions evaluating the probability that an observed alignment distribution comes from a true junction.

The simple detection of specific isoforms does not provide useful information about their quantitative abundance. In principle, the quantification methods described above are equally applicable to quantify isoform expression. In practice, however, it is difficult to compute isoform-specific expression because most reads that are mapped to the genes are shared by more than one isoform and then it becomes difficult to assign each read only to a specific isoform. As a consequence, the assignment should rely on inferential methods that consider all data mapping to a certain region.

Several proposed methods for inferring isoforms' abundance are based on the preliminary knowledge of precise isoforms' annotation, on the assumption of uniform distribution of the reads across the transcript, on Poisson model for the reads' counts and equal weight for each read, regardless the quality of the match. The methods are often limited to handle only the cases where there is a relative small number of isoforms without confounding effects due to the overlap between genes. In particular in [98], the authors showed that the complexity of some isoform sets may still render the estimation problem nonidentifiable based on current RNA-Seq protocols and derived a mathematical characterization of identifiable isoform set. The main reason for such an effect is that current protocols with short single-end reads RNA-Seq are only able to asses local properties of a transcript. It is possible that the combination of short-read data with longer reads or paired-end reads will be able to go further in addressing such challenges.

Recently, in [90] the authors proposed a statistical method where, similar to [34], the count of reads falling into an annotated gene with multiple isoforms is modeled as a Poisson variable. They inferred the expression of each individual isoform using maximum likelihood approach, whose solution has been obtained by solving a convex optimization problem. In order to quantify the degree of uncertainty of the estimates, they carried out statistical inferences about the parameters from the posterior distribution by importance sampling. Interestingly, they showed that their method can be viewed as an extension of the RPKM concept and reduces to the RPKM index when there is only one isoform. An attempt to relax the assumption of uniform reads sampling is proposed in [84]. In this paper, the authors unified the notions of reads that map to multiple locations, that is, that could be potentially assigned to several genes, with those of reads that map to multiple isoforms through the introduction of latent random variables representing the true mappings. Then, they estimated the isoforms' abundance as the maximum likelihood expression levels using the EM algorithm. The Poisson distribution is also the main assumption in [99], where a comprehensive approach to the problem of alternative isoforms prediction is presented. In particular, the presence of alternative splicing event within the same sample is assessed by using Pearson's chi-square test on the parameter of a multinomial distribution and the EM algorithm is used to estimate the abundance of each isoform.

8.3.6 DIFFERENTIAL EXPRESSION

The final goal in the majority of transcriptome studies is to quantify differences in expression across multiple samples in order to capture differential gene expression, to identify sample-specific alternative splicing isoforms and their differential abundance.

Mimicking the methods used for microarray analysis, researchers started to approach such crucial question using statistical hypothesis' tests combined with multiple comparisons error procedures on the observed counts (or on the RPKM values) at the gene, isoform or exon level. Indeed, in [30] the authors applied the empirical Bayes moderated t-test proposed in [100] to the normalized RPKM. However in microarray experiments, the abundance of a particular transcript is measured as a fluorescence intensity, that can be effectively modeled as a continuous response, whereas for RNA-Seq data the abundance is usually a count. Therefore, procedures that are successful for microarrays do not seem to be appropriate for dealing with such type of data.

One of the pioneering works to handle such difference is [34], where the authors modeled the aggregated reads count for each gene using Poisson distribution. One can prove that the number of reads observed from a gene (or transcript isoform) follows a binomial distribution that can be approximated by a Poisson distribution, under the assumption that RNA-Seq reads follow a random sampling process, in which each read is sampled independently and uniformly from every possible nucleotide in the sample. In this set-up, in [34] the authors used a likelihood ratio test to test for significant differences between the two conditions. The Poisson model was also employed by [40], where the authors used the method proposed in [101] to determine the significance of differential expression. On the contrary, in [83], the authors simply estimated the difference in expression of a gene between two conditions through the difference of the count proportions p_1 and p_2 computed using a classical Z-test statistics. In [18], the authors employed the Fishers exact test to better weigh the genes with relatively small counts. Similarly in [99] the authors used Poisson model and Fishers exact test to detect alternative exon usage between conditions.

Recently, more sophisticated approaches have been proposed in [102, 103]. In [102], the authors proposed an empirical Bayesian approach, based on the negative binomial distribution; it results very flexible and reduces to the Poisson model for a particular choice of the hyperparameter. They carried out differential expression testing using a moderated Bayes approach similar in the spirit to the one described in [100], but adapted for data that are counts. We observed that the method is designed for finding changes between two or more groups when at least one of the groups has replicated measurements. In [103], the observed counts of reads mapped to a specific gene obtained from a certain sample was modeled using Binomial distribution. Under such assumption, it can be proved that the log ratio between the two samples conditioned to the intensity signal (i.e., the average of the two logs counts) follows an approximate normal distribution, that is used for assessing the significance of the test. All the above-mentioned methods assume that the quantification of the features of interest under the experimental conditions has been already done and each read has been assigned to only one elements, hence the methods are directly applicable to detect genes or exons differences provided that overlapping elements are properly filtered out. By contrast the above described methods are not directly suited for detecting isoforms' differences unless the quantification of the isoform abundance has been carried out using specific approaches. To handle such difficulties, in [104], the authors proposed a hierarchical Bayesian model to directly infer the differential expression level of each transcript isoform in response to two conditions. The difference in expression of each isoform is modeled by means of an inverse gamma model and a latent variable is introduced for guiding the isoform's selection. The model can handle the heteroskedasticity of the sequence read coverage and inference is carried out using Gibbs sampler.

It should be noticed that although these techniques already provide interesting biological insights, they have not been sufficiently validated on several real data-sets where different type of replicates are available, neither sufficiently compared each others in terms of advantages and disadvantages. As with any new biotechnology it is important to carefully study the different sources of variation that can affect measure of the biological effects of interest and to statistically asses the reproducibility of the biological findings in a rigorous way, and to date this has been often omitted. Indeed, it should be considered that there are a variety of experimental effects that could possibly increase the variability, the bias, or be confounded with sequencing-based measures, causing miss-understanding of the results. Unfortunately, such problems have received little of attention until now. In order to fill this gap, in [93] the authors presented a statistical inference framework for transcriptome analysis using RNA-Seq mapped read data. In particular, they proposed a new statistical method based on log-linear regression for investigating relationships between read counts and biological and experimental variables describing input samples as well as genomic regions of interest. The main advantage of the log-linear regression approach is that it allows to account both for biological effect and a variety of experimental effects. Their paper represents one of the few attempts of looking at the analysis of RNA-Seq data from a general point of view.

8.4 CHALLENGES AND PERSPECTIVE FOR NGS

From the development of the Sanger method to the completion of the HGP, genetics has made significant advances towards the understanding of gene content and function. Even though significant achievements were reached by Human Genome, HapMap and ENCODE Projects [7, 105, 106], we are far from an exhaustive comprehension of the genomic diversity among humans and across the species, and from understanding gene expression variations and its regulation in both physio and pathological conditions. Since the appearance of first NGS platforms in the 2004, it was clear that understanding this diversity at a cost of around \$5–10 million per genome sequence [107], placed it outside the real possibilities of most research laboratories, and very far from single individual economical potential. To date, we are in the "\$1,000 genome" era, and, although this important barrier has not yet been broken, its a current assumption that this target is going to be reached within the end of 2010. It is likely that the rapid evolution of DNA sequencing technology, able to provide researchers with the ability to generate data about genetic variation and patterns of gene expression at an unprecedented scale, will become a routine tool for researchers and clinicians within just a few years.

As we can see, the number of applications and the great amount of biological questions that can be addressed by "Seq" experiments on NGS platforms is leading a revolution in the landscape of molecular biology, but the imbalance between the pace at which technology innovations are introduced in the platforms and the biological discoveries derivable from them is growing up. The risk is the creation of a glut of "under-used" information that in few months becomes of no use because the new one is produced. It is necessary to invest in an equivalent development of new computational strategies and expertise to deal with the volumes of data created by the current generation of new sequencing instruments, to maximize their potential benefit.

These platforms are creating a new world to explore, not only in the definition of experimental/technical procedures of large-scale analyses, but also in the downstream computational analysis and in the bioinformatics infrastructures support required for high-quality data generation and for their correct biological interpretation. In practice, they have shifted the bottleneck from the generation of experimental data to their management and to their statistical and computational analysis. There are few key points to consider. The first one is the data management: downstream computational analysis becomes difficult without appropriate Information Technology (IT) infrastructure. The terabytes of data produced by each sequencing run requires conspicuous storage and backup capacity, which increases considerably the experimental costs. The second one regards the protocols used for the production of raw data: each platform has its peculiarity in both sample preparation and type and volume of raw data produced, hence they require individualized laboratory expertise and data processing pipelines. Third, beside vendor specific and commercial software, several other open-source analysis tools are continuously appearing. Unfortunately, there is often an incomplete documentation and it is easy to spend more time in evaluating software suites than in analyzing the output data. Whichever software is used, the most important question is to understand its limitations and assumptions. Community adoption of input/output data standards is also essential to efficiently handle the data management problem. Till now the effort has been mainly devoted to the technological development rather than to the methodological counterpart. The choice of a careful experimental design has been also not always adequately considered.

As regards the RNA-Seq, we have still to face several critical issues either from a biological and computational point of view. RNA-seq protocols are extremely sensitive and need a very careful quality control for each wet laboratory step. For instance, the contamination of reagents with RNAse and the degradation of RNA, even partial, must be avoided during all the technical procedures. The quality of total isolated RNA is the first, and probably the most crucial point for an RNA-Seq experiment. Poor yield of polyA enrichment or low efficiency of total RNA ribodepletion are also critical issues for preparing high-quality RNA towards the library construction. It is clear that, independently on the library construction procedure, particular care should be taken to avoid complete degradation of RNA during the controlled RNA fragmentation step. Furthermore, in order to correctly determine the directionality of gene transcription and to facilitate the detection of opposing and overlapping transcripts within gene-dense genomic regions, particular care should be taken to preserve the strandedness of RNA fragments during the library preparation. In addition, to provide a more uniform coverage throughout the transcript length, random priming for reverse transcription protocols, rather than oligo dT priming (with the bias of low coverage at the 5' ends), should be done after removal of rRNA. Finally, it should be considered that for the platforms based on CRT and SBL, substitutions and under representation of AT-rich and GC-rich regions, probably due to amplification bias during template preparation, are the most common error type. In contrast, for pyrosequencing platforms, insertions and deletions represent a common drawback.

For what concern the data analysis, to the above-mentioned points, we should note that most of the available software for read alignment are designed for genomic mapping hence they are not fully capable to discover exon junctions. The classical extension for handling RNA-Seq data involves the preconstruction of junction libraries reducing the possibility of discovering new junctions. It would be desirable to develop new methods that allow either new junction detection and also the use of paired-end reads, that are particularly promising for more accurate study. Additionally further developments are required to assess the significance of new transcribed regions, the construction of new putative genes and the precise quantification of each isoform, for which there is still a lack of statistical methodologies. For what concerns the detection of differential expression, existing techniques were not sufficiently validated on biological data and compared in terms of specificity and sensitivity. Moreover, of potentially great impact, is the lack of biological replicates which precludes gauging the magnitude of individual effects in relation to technical effects. Biological replicates is essential in a RNA-Seq experiment to draw generalized conclusions about the “real" differences observed between two or more biological groups.

Facing such multidisciplinary challenges will be the key point for a fruitful transfer from laboratory studies to clinical applications. Indeed, the availability of low-cost, efficient and accurate technologies for gene expression and genome sequencing will be useful in providing pathological gene expression profiles in a wide number of common genetic disorders including type II diabetes, cardiovascular disease, Parkinson disease and Downs syndrome. Moreover, the application of NGS to the emerging disciplines of pharmacogenomics and nutrigenomics will allow to understand drug response and nutrient-gene interactions on the basis of individual patient's genetic make-up, leading in turn to the development of targeted therapies for many human diseases or tailored nutrient supplementation [108].

REFERENCES

1. D. D. Licatalosi and R. B. Darnell, "RNA processing and its regulation: global insights into biological networks," Nature Reviews Genetics, vol. 11, no. 1, pp. 75–87, 2010.
2. V. E. Velculescu, L. Zhang, W. Zhou, et al., "Characterization of the yeast transcriptome," Cez,ll, vol. 88, no. 2, pp. 243–251, 1997.
3. J. Lindberg and J. Lundeberg, "The plasticity of the mammalian transcriptome," Genomics, vol. 95, no. 1, pp. 1–6, 2010.
4. W. F. Doolittle and C. Sapienza, "Selfish genes, the phenotype paradigm and genome evolution," Nature, vol. 284, no. 5757, pp. 601–603, 1980.
5. R. J. Taft, M. Pheasant, and J. S. Mattick, "The relationship between non-protein-coding DNA and eukaryotic complexity," BioEssays, vol. 29, no. 3, pp. 288–299, 2007.
6. T. Cavalier-Smith, "Cell volume and the evolution of eukaryote genome size," in The Evolution of Genome Size, T. Cavalier-Smith, Ed., pp. 105–184, John Wiley & Sons, Chichester, UK, 1985.
7. E. Birney, J. A. Stamatoyannopoulos, A. Dutta, et al., "Identification and analysis of functional elements in 1% of the human genome by the ENCODE pilot project," Nature, vol. 447, no. 7146, pp. 799–816, 2007.
8. A. Jacquier, "The complex eukaryotic transcriptome: unexpected pervasive transcription and novel small RNAs," Nature Reviews Genetics, vol. 10, no. 12, pp. 833–844, 2009.
9. Z. Wang, M. Gerstein, and M. Snyder, "RNA-Seq: a revolutionary tool for transcriptomics," Nature Reviews Genetics, vol. 10, no. 1, pp. 57–63, 2009.
10. R. W. Holley, "Alanine transfer RNA," in Nobel Lectures in Molecular Biology 1933–1975, pp. 285–300, Elsevier North Holland, New York, NY, USA, 1977.
11. A. M. Maxam and W. Gilbert, "A new method for sequencing DNA," Proceedings of the National Academy of Sciences of the United States of America, vol. 74, no. 2, pp. 560–564, 1977.
12. F. Sanger, S. Nicklen, and A. R. Coulson, "DNA sequencing with chain-terminating inhibitors," Proceedings of the National Academy of Sciences of the United States of America, vol. 74, no. 12, pp. 5463–5467, 1977.

13. E. R. Mardis, "Next-generation DNA sequencing methods," Annual Review of Genomics and Human Genetics, vol. 9, pp. 387–402, 2008.
14. J. Shendure and H. Ji, "Next-generation DNA sequencing," Nature Biotechnology, vol. 26, no. 10, pp. 1135–1145, 2008.
15. M. L. Metzker, "Sequencing technologies the next generation," Nature Reviews Genetics, vol. 11, no. 1, pp. 31–46, 2010.
16. R. A. Irizarry, D. Warren, F. Spencer, et al., "Multiple-laboratory comparison of microarray platforms," Nature Methods, vol. 2, no. 5, pp. 345–349, 2005.
17. P. A. C. 't Hoen, Y. Ariyurek, H. H. Thygesen, et al., "Deep sequencing-based expression analysis shows major advances in robustness, resolution and inter-lab portability over five microarray platforms," Nucleic Acids Research, vol. 36, no. 21, article e141, 2008.
18. J. S. Bloom, Z. Khan, L. Kruglyak, M. Singh, and A. A. Caudy, "Measuring differential gene expression by short read sequencing: quantitative comparison to 2-channel gene expression microarrays," BMC Genomics, vol. 10, article 221, 2009.
19. M. Harbers and P. Carninci, "Tag-based approaches for transcriptome research and genome annotation," Nature Methods, vol. 2, no. 7, pp. 495–502, 2005.
20. M. P. Horan, "Application of serial analysis of gene expression to the study of human genetic disease," Human Genetics, vol. 126, no. 5, pp. 605–614, 2009.
21. H. Misu, T. Takamura, N. Matsuzawa, et al., "Genes involved in oxidative phosphorylation are coordinately upregulated with fasting hyperglycaemia in livers of patients with type 2 diabetes," Diabetologia, vol. 50, no. 2, pp. 268–277, 2007.
22. T. Takamura, H. Misu, T. Yamashita, and S. Kaneko, "SAGE application in the study of diabetes," Current Pharmaceutical Biotechnology, vol. 9, no. 5, pp. 392–399, 2008.
23. D. V. Gnatenko, J. J. Dunn, S. R. McCorkle, D. Weissmann, P. L. Perrotta, and W. F. Bahou, "Transcript profiling of human platelets using microarray and serial analysis of gene expression," Blood, vol. 101, no. 6, pp. 2285–2293, 2003.
24. C. A. Sommer, E. C. Pavarino-Bertelli, E. M. Goloni-Bertollo, and F. Henrique-Silva, "Identification of dysregulated genes in lymphocytes from children with Down syndrome," Genome, vol. 51, no. 1, pp. 19–29, 2008.
25. W. Malagó Jr., C. A. Sommer, C. Del Cistia Andrade, et al., "Gene expression profile of human Down syndrome leukocytes," Croatian Medical Journal, vol. 46, no. 4, pp. 647–656, 2005.
26. B. T. Wilhelm and J.-R. Landry, "RNA-Seq-quantitative measurement of expression through massively parallel RNA-Sequencing," Methods, vol. 48, no. 3, pp. 249–257, 2009.
27. M. N. Bainbridge, R. L. Warren, M. Hirst, et al., "Analysis of the prostate cancer cell line LNCaP transcriptome using a sequencing-by-synthesis approach," BMC Genomics, vol. 7, article 246, 2006.
28. U. Nagalakshmi, Z. Wang, K. Waern, et al., "The transcriptional landscape of the yeast genome defined by RNA sequencing," Science, vol. 320, no. 5881, pp. 1344–1349, 2008.
29. T. T. Torres, M. Metta, B. Ottenwälder, and C. Schlötterer, "Gene expression profiling by massively parallel sequencing," Genome Research, vol. 18, no. 1, pp. 172–177, 2008.

30. N. Cloonan, A. R. R. Forrest, G. Kolle, et al., "Stem cell transcriptome profiling via massive-scale mRNA sequencing," Nature Methods, vol. 5, no. 7, pp. 613–619, 2008.
31. L. J. Core, J. J. Waterfall, and J. T. Lis, "Nascent RNA sequencing reveals widespread pausing and divergent initiation at human promoters," Science, vol. 322, no. 5909, pp. 1845–1848, 2008.
32. S.-I. Hashimoto, W. Qu, B. Ahsan, et al., "High-resolution analysis of the 5′-end transcriptome using a next generation DNA sequencer," PLoS ONE, vol. 4, no. 1, article e4108, 2009.
33. H. Li, M. T. Lovci, Y.-S. Kwon, M. G. Rosenfeld, X.-D. Fu, and G. W. Yeo, "Determination of tag density required for digital transcriptome analysis: application to an androgen-sensitive prostate cancer model," Proceedings of the National Academy of Sciences of the United States of America, vol. 105, no. 51, pp. 20179–20184, 2008.
34. J. C. Marioni, C. E. Mason, S. M. Mane, M. Stephens, and Y. Gilad, "RNA-Seq: an assessment of technical reproducibility and comparison with gene expression arrays," Genome Research, vol. 18, no. 9, pp. 1509–1517, 2008.
35. R. D. Morin, M. D. O'Connor, M. Griffith, et al., "Application of massively parallel sequencing to microRNA profiling and discovery in human embryonic stem cells," Genome Research, vol. 18, no. 4, pp. 610–621, 2008.
36. R. D. Morin, M. Bainbridge, A. Fejes, et al., "Profiling the HeLa S3 transcriptome using randomly primed cDNA and massively parallel short-read sequencing," BioTechniques, vol. 45, no. 1, pp. 81–94, 2008.
37. A. Mortazavi, B. A. Williams, K. McCue, L. Schaeffer, and B. Wold, "Mapping and quantifying mammalian transcriptomes by RNA-Seq," Nature Methods, vol. 5, no. 7, pp. 621–628, 2008.
38. R. Rosenkranz, T. Borodina, H. Lehrach, and H. Himmelbauer, "Characterizing the mouse ES cell transcriptome with Illumina sequencing," Genomics, vol. 92, no. 4, pp. 187–194, 2008.
39. D. J. Sugarbaker, W. G. Richards, G. J. Gordon, et al., "Transcriptome sequencing of malignant pleural mesothelioma tumors," Proceedings of the National Academy of Sciences of the United States of America, vol. 105, no. 9, pp. 3521–3526, 2008.
40. M. Sultan, M. H. Schulz, H. Richard, et al., "A global view of gene activity and alternative splicing by deep sequencing of the human transcriptome," Science, vol. 321, no. 5891, pp. 956–960, 2008.
41. Y. W. Asmann, E. W. Klee, E. A. Thompson, et al., "3′ tag digital gene expression profiling of human brain and universal reference RNA using Illumina Genome Analyzer," BMC Genomics, vol. 10, article 531, 2009.
42. I. Chepelev, G. Wei, Q. Tang, and K. Zhao, "Detection of single nucleotide variations in expressed exons of the human genome using RNA-Seq," Nucleic Acids Research, vol. 37, no. 16, article e106, 2009.
43. J. Z. Levin, M. F. Berger, X. Adiconis, et al., "Targeted next-generation sequencing of a cancer transcriptome enhances detection of sequence variants and novel fusion transcripts," Genome Biology, vol. 10, no. 10, article R115, 2009.
44. C. A. Maher, N. Palanisamy, J. C. Brenner, et al., "Chimeric transcript discovery by paired-end transcriptome sequencing," Proceedings of the National Academy of Sciences of the United States of America, vol. 106, no. 30, pp. 12353–12358, 2009.

45. D. Parkhomchuk, T. Borodina, V. Amstislavskiy, et al., "Transcriptome analysis by strand-specific sequencing of complementary DNA," Nucleic Acids Research, vol. 37, no. 18, article e123, 2009.
46. T. E. Reddy, F. Pauli, R. O. Sprouse, et al., "Genomic determination of the glucocorticoid response reveals unexpected mechanisms of gene regulation," Genome Research, vol. 19, no. 12, pp. 2163–2171, 2009.
47. F. Tang, C. Barbacioru, Y. Wang, et al., "mRNA-Seq whole-transcriptome analysis of a single cell," Nature Methods, vol. 6, no. 5, pp. 377–382, 2009.
48. R. Blekhman, J. C. Marioni, P. Zumbo, M. Stephens, and Y. Gilad, "Sex-specific and lineage-specific alternative splicing in primates," Genome Research, vol. 20, no. 2, pp. 180–189, 2010.
49. G. A. Heap, J. H. M. Yang, K. Downes, et al., "Genome-wide analysis of allelic expression imbalance in human primary cells by high-throughput transcriptome resequencing," Human Molecular Genetics, vol. 19, no. 1, pp. 122–134, 2010.
50. D. Raha, Z. Wang, Z. Moqtaderi, et al., "Close association of RNA polymerase II and many transcription factors with Pol III genes," Proceedings of the National Academy of Sciences of the United States of America, vol. 107, no. 8, pp. 3639–3644, 2010.
51. S. Marguerat and J. Bahler, "RNA-Seq: from technology to biology," Cellular and Molecular Life Sciences, vol. 67, no. 4, pp. 569–579, 2010.
52. Y. He, B. Vogelstein, V. E. Velculescu, N. Papadopoulos, and K. W. Kinzler, "The antisense transcriptomes of human cells," Science, vol. 322, no. 5909, pp. 1855–1857, 2008.
53. R. Lister, R. C. O'Malley, J. Tonti-Filippini, et al., "Highly integrated single-base resolution maps of the epigenome in Arabidopsis," Cell, vol. 133, no. 3, pp. 523–536, 2008.
54. B. T. Wilhelm, S. Marguerat, I. Goodhead, and J. Bahler, "Defining transcribed regions using RNA-Seq," Nature Protocols, vol. 5, no. 2, pp. 255–266, 2010.
55. N. T. Ingolia, S. Ghaemmaghami, J. R. S. Newman, and J. S. Weissman, "Genome-wide analysis in vivo of translation with nucleotide resolution using ribosome profiling," Science, vol. 324, no. 5924, pp. 218–223, 2009.
56. T. D. Harris, P. R. Buzby, H. Babcock, et al., "Single-molecule DNA sequencing of a viral genome," Science, vol. 320, no. 5872, pp. 106–109, 2008.
57. J. C. Dohm, C. Lottaz, T. Borodina, and H. Himmelbauer, "Substantial biases in ultra-short read data sets from high-throughput DNA sequencing," Nucleic Acids Research, vol. 36, no. 16, article e105, 2008.
58. O. Harismendy, P. C. Ng, R. L. Strausberg, et al., "Evaluation of next generation sequencing platforms for population targeted sequencing studies," Genome Biology, vol. 10, no. 3, article R32, 2009.
59. L. W. Hillier, G. T. Marth, A. R. Quinlan, et al., "Whole-genome sequencing and variant discovery in C. elegans," Nature Methods, vol. 5, no. 2, pp. 183–188, 2008.
60. J. D. McPherson, "Next-generation gap," Nature Methods, vol. 6, no. 11S, pp. S2–S5, 2009.
61. D. R. Zerbino and E. Birney, "Velvet: algorithms for de novo short read assembly using de Bruijn graphs," Genome Research, vol. 18, no. 5, pp. 821–829, 2008.
62. I. Birol, S. D. Jackman, C. B. Nielsen, et al., "De novo transcriptome assembly with ABySS," Bioinformatics, vol. 25, no. 21, pp. 2872–2877, 2009.

63. F. Denoeud, J.-M. Aury, C. Da Silva, et al., "Annotating genomes with massive-scale RNA sequencing," Genome Biology, vol. 9, no. 12, article R175, 2008.
64. M. Yassoura, T. Kaplana, H. B. Fraser, et al., "Ab initio construction of a eukaryotic transcriptome by massively parallel mRNA sequencing," Proceedings of the National Academy of Sciences of the United States of America, vol. 106, no. 9, pp. 3264–3269, 2009.
65. C. Trapnell and S. L. Salzberg, "How to map billions of short reads onto genomes," Nature Biotechnology, vol. 27, no. 5, pp. 455–457, 2009.
66. P. Flicek and E. Birney, "Sense from sequence reads: methods for alignment and assembly," Nature Methods, vol. 6, supplement 11, pp. S6–S12, 2009.
67. D. S. Horner, G. Pavesi, T. Castrignanò, et al., "Bioinformatics approaches for genomics and post genomics applications of next-generation sequencing," Briefings in Bioinformatics, vol. 11, no. 2, pp. 181–197, 2009.
68. A. Cox, "ELAND: efficient local alignment of nucleotide data," unpublished, http://bioit.dbi.udel.edu/howto/eland.
69. "Applied Biosystems mappread and whole transcriptome software tools," http://www.solidsoftwaretools.com/.
70. H. Li, J. Ruan, and R. Durbin, "Mapping short DNA sequencing reads and calling variants using mapping quality scores," Genome Research, vol. 18, no. 11, pp. 1851–1858, 2008.
71. A. D. Smith, Z. Xuan, and M. Q. Zhang, "Using quality scores and longer reads improves accuracy of Solexa read mapping," BMC Bioinformatics, vol. 9, article 128, 2008.
72. R. Li, Y. Li, K. Kristiansen, and J. Wang, "SOAP: short oligonucleotide alignment program," Bioinformatics, vol. 24, no. 5, pp. 713–714, 2008.
73. R. Li, C. Yu, Y. Li, et al., "SOAP2: an improved ultrafast tool for short read alignment," Bioinformatics, vol. 25, no. 15, pp. 1966–1967, 2009.
74. B. D. Ondov, A. Varadarajan, K. D. Passalacqua, and N. H. Bergman, "Efficient mapping of Applied Biosystems SOLiD sequence data to a reference genome for functional genomic applications," Bioinformatics, vol. 24, no. 23, pp. 2776–2777, 2008.
75. H. Jiang and W. H. Wong, "SeqMap: mapping massive amount of oligonucleotides to the genome," Bioinformatics, vol. 24, no. 20, pp. 2395–2396, 2008.
76. H. Lin, Z. Zhang, M. Q. Zhang, B. Ma, and M. Li, "ZOOM! Zillions of oligos mapped," Bioinformatics, vol. 24, no. 21, pp. 2431–2437, 2008.
77. B. Langmead, C. Trapnell, M. Pop, and S. L. Salzberg, "Ultrafast and memory-efficient alignment of short DNA sequences to the human genome," Genome Biology, vol. 10, no. 3, article R25, 2009.
78. D. Campagna, A. Albiero, A. Bilardi, et al., "PASS: a program to align short sequences," Bioinformatics, vol. 25, no. 7, pp. 967–968, 2009.
79. N. Cloonan, Q. Xu, G. J. Faulkner, et al., "RNA-MATE: a recursive mapping strategy for high-throughput RNA-sequencing data," Bioinformatics, vol. 25, no. 19, pp. 2615–2616, 2009.
80. F. De Bona, S. Ossowski, K. Schneeberger, and G. Rätsch, "Optimal spliced alignments of short sequence reads," Bioinformatics, vol. 24, no. 16, pp. i174–i180, 2008.

81. C. Trapnell, L. Pachter, and S. L. Salzberg, "TopHat: discovering splice junctions with RNA-Seq," Bioinformatics, vol. 25, no. 9, pp. 1105–1111, 2009.
82. G. J. Faulkner, A. R. R. Forrest, A. M. Chalk, et al., "A rescue strategy for multi-mapping short sequence tags refines surveys of transcriptional activity by CAGE," Genomics, vol. 91, no. 3, pp. 281–288, 2008.
83. T. Hashimoto, M. J. L. de Hoon, S. M. Grimmond, C. O. Daub, Y. Hayashizaki, and G. J. Faulkner, "Probabilistic resolution of multi-mapping reads in massively parallel sequencing data using MuMRescueLite," Bioinformatics, vol. 25, no. 19, pp. 2613–2614, 2009.
84. B. Li, V. Ruotti, R. M. Stewart, J. A. Thomson, and C. N. Dewey, "RNA-Seq gene expression estimation with read mapping uncertainty," Bioinformatics, vol. 26, no. 4, pp. 493–500, 2009.
85. W. J. Kent, C. W. Sugnet, T. S. Furey, et al., "The human genome browser at UCSC," Genome Research, vol. 12, no. 6, pp. 996–1006, 2002.
86. W. Huang and G. Marth, "EagleView: a genome assembly viewer for next-generation sequencing technologies," Genome Research, vol. 18, no. 9, pp. 1538–1543, 2008.
87. H. Bao, H. Guo, J. Wang, R. Zhou, X. Lu, and S. Shi, "MapView: visualization of short reads alignment on a desktop computer," Bioinformatics, vol. 25, no. 12, pp. 1554–1555, 2009.
88. I. Milne, M. Bayer, L. Cardle, et al., "Tablet-next generation sequence assembly visualization," Bioinformatics, vol. 26, no. 3, pp. 401–402, 2010.
89. H. Li, B. Handsaker, A. Wysoker, et al., "The sequence alignment/map format and SAMtools," Bioinformatics, vol. 25, no. 16, pp. 2078–2079, 2009.
90. H. Jiang and W. H. Wong, "Statistical inferences for isoform expression in RNA-Seq," Bioinformatics, vol. 25, no. 8, pp. 1026–1032, 2009.
91. S. Pepke, B. Wold, and A. Mortazavi, "Computation for ChIP-Seq and RNA-Seq studies," Nature Methods, vol. 6, no. 11S, pp. S22–S32, 2009.
92. A. Oshlack and M. J. Wakefield, "Transcript length bias in RNA-Seq data confounds systems biology," Biology Direct, vol. 4, article 14, 2009.
93. J. H. Bullard, E. A. Purdom, K. D. Hansen, S. Durinck, and S. Dudoit, "Statistical inference in mRNA-Seq: exploratory data analysis and differential expression," Tech. Rep. 247/2009, University of California, Berkeley, 2009.
94. B. T. Wilhelm, S. Marguerat, S. Watt, et al., "Dynamic repertoire of a eukaryotic transcriptome surveyed at single-nucleotide resolution," Nature, vol. 453, no. 7199, pp. 1239–1243, 2008.
95. Q. Pan, O. Shai, L. J. Lee, B. J. Frey, and B. J. Blencowe, "Deep surveying of alternative splicing complexity in the human transcriptome by high-throughput sequencing," Nature Genetics, vol. 40, no. 12, pp. 1413–1415, 2008.
96. E. T. Wang, R. Sandberg, S. Luo, et al., "Alternative isoform regulation in human tissue transcriptomes," Nature, vol. 456, no. 7221, pp. 470–476, 2008.
97. L. Wang, Y. Xi, J. Yu, L. Dong, L. Yen, and W. Li, "A statistical method for the detection of alternative splicing using RNA-Seq," PLoS ONE, vol. 5, no. 1, article e8529, 2010.
98. D. Hiller, H. Jiang, W. Xu, and W. H. Wong, "Identifiability of isoform deconvolution from junction arrays and RNA-Seq," Bioinformatics, vol. 25, no. 23, pp. 3056–3059, 2009.

99. H. Richard, M. H. Schulz, M. Sultan, et al., "Prediction of alternative isoforms from exon expression levels in RNA-Seq experiments," Nucleic Acids Research, vol. 38, no. 10, p. e112, 2010.
100. G. K. Smyth, "Linear models and empirical Bayes methods for assessing differential expression in microarray experiments," Statistical Applications in Genetics and Molecular Biology, vol. 3, no. 1, article 3, 2004.
101. S. Audic and J.-M. Claverie, "The significance of digital gene expression profiles," Genome Research, vol. 7, no. 10, pp. 986–995, 1997.
102. M. D. Robinson, D. J. McCarthy, and G. K. Smyth, "edgeR: a bioconductor package for differential expression analysis of digital gene expression data," Bioinformatics, vol. 26, no. 1, pp. 139–140, 2010.
103. L. Wang, Z. Feng, X. Wang, X. Wang, and X. Zhang, "DEGseq: an R package for identifying differentially expressed genes from RNA-Seq data," Bioinformatics, vol. 26, no. 1, pp. 136–138, 2009.
104. S. Zheng and L. Chen, "A hierarchical Bayesian model for comparing transcriptomes at the individual transcript isoform level," Nucleic Acids Research, vol. 37, no. 10, article e75, 2009.
105. F. S. Collins, E. S. Lander, J. Rogers, and R. H. Waterson, "Finishing the euchromatic sequence of the human genome," Nature, vol. 431, no. 7011, pp. 931–945, 2004.
106. International Human Genome Sequencing Consortium, "A haplotype map of the human genome," Nature, vol. 437, no. 7063, pp. 1299–1320, 2005.
107. E. R. Mardis, "Anticipating the 1,000 dollar genome," Genome Biology, vol. 7, no. 7, article 112, 2006.
108. V. Costa, A. Casamassimi, and A. Ciccodicola, "Nutritional genomics era: opportunities toward a genome-tailored nutritional regimen," The Journal of Nutritional Biochemistry, vol. 21, no. 6, pp. 457–467, 2010.

CHAPTER 9

EFFICIENT EXPERIMENTAL DESIGN AND ANALYSIS STRATEGIES FOR THE DETECTION OF DIFFERENTIAL EXPRESSION USING RNA-SEQUENCING

JOSÉ A. ROBLES, SUMAIRA E. QURESHI, STUART J. STEPHEN, SUSAN R. WILSON, CONRAD J. BURDEN, and JENNIFER M. TAYLOR

9.1 BACKGROUND

RNA sequencing (RNA-Seq) allows an entire transcriptome to be surveyed at single-base resolution whilst concurrently profiling gene expression levels on a genome scale [1]. RNA-Seq is an attractive approach as it profiles the transcriptome directly through sequencing and therefore does not require prior knowledge of the transcriptome under consideration. An example of the use of RNA-Seq as a high-resolution exploratory tool is the discovery of thousands of additional novel coding and noncoding genes, transcripts and isoforms of known genes despite the prior extensive annotation of the mouse [2-4] and human genomes [5,6].

Arguably, the most popular use of RNA-Seq is profiling of gene expression or transcript abundance between samples or differential expression (DE). The efficiency, resolution and cost advantages of using RNA-

This chapter was originally published under the Creative Commons Attribution License. Robles JA, Qureshi SE, Stephen SJ, Wilson SR, Burden CJ, and Taylor JM. Efficient Experimental Design and Analysis Strategies for the Detection of Differential Expression using RNA-Sequencing. BMC Genomics ***13****,484 (2012), 14 pages. doi:10.1186/1471-2164-13-484.*

Seq as a tool for profiling DE has prompted many biologists to abandon microarrays in favour of RNA-Seq [7,8].

Despite the advantages of using RNA-Seq for DE analysis, there are several sources of sequencing bias and systematic noise that need to be considered when using this approach. Clearly, RNA-Seq analysis is vulnerable to the general biases and errors inherent in the next-generation sequencing (NGS) technology upon which it is based. These errors and biases include: sequencing errors (wrong base calls), biases in sequence quality, nucleotide composition and error rates relative to the base position in the read [9,10], variability in sequence depth across the transcriptome due to preferential sites of fragmentation, variable primer and transcript nucleotide composition effects [11] and finally, differences in the coverage and composition of raw sequence data generated from technical and biological replicate samples [1,12].

Recently, there have been several investigations [13-15] into the biases that affect the accuracy with which RNA-Seq represents the absolute abundance of a given transcript as measured by high precision approaches such as Taqman RT-PCR [16]. It has been shown that these abundance measures are prone to biases correlated with the nucleotide composition [14,17] and length of the transcript [1,18]. Several within and between sample correction and normalisation procedures have recently been developed to address these biases either as nucleotide composition effects [17] or various combinations of nucleotide, length or library preparation biases [14,15]. These approaches all yield improvements in the correspondence of RNA-Seq read counts with expression estimates gained by other experimental approaches.

Despite the known biases, RNA-Seq continues to be widely and successfully used to profile relative transcript abundances across samples to identify differentially expressed transcripts [19]. The profile of a given transcript across a biological population would be hoped to be less prone to nucleotide composition and length biases as these variables remain constant. Nevertheless, to accurately detect DE across samples it is necessary to understand the sources of variation across technical and biological replication and where possible respond to these with an appropriate experimental design and statistically robust analysis [17,20]. To date, there has been little discussion in the literature of efficient experimental designs for the detection of DE and a lack of consensus about a standard and comprehensive approach to counter the many sources of noise and biases present in RNA-Seq has meant that some of the biological community remain sceptical about its reliability and unsure of how to design cost-efficient RNA-Seq experiments (see [19]).

Good experimental design and appropriate analysis is integral to maximising the power of any NGS study. With regard to RNA-Seq, important experimental design decisions include the choice of sequencing depth and number of technical and/or biological replicates to use. For researchers with a fixed budget, often a critical design question is whether to increase the sequencing depth at the cost of reduced sample numbers or to increase the sample size with limited sequencing depth for each sample [20].

9.1.2 SEQUENCING DEPTH

Sequencing depth is usually referenced to be the expected mean coverage at all loci over the target sequence(s), in the case of RNA-seq experiments assuming all transcripts having similar levels of expression. Without the benefit of extensive previous RNA-Seq studies, it is difficult in most cases to estimate prior to data generation the optimal sequencing depth or amount of sequencing data required to adequately power the detection of DE in the transcriptome of interest. Pragmatically, RNA-seq sequencing depth is typically chosen based on an estimation of total transcriptome length (bases) and the expected dynamic range of transcript abundances. Given the dynamic nature of the transcriptome, the suitability of these estimates could vary substantially across organisms, tissues, time points and biological contexts.

Wang et al. [21] found a significant increase in correlation between gene transcripts observed and number of sequence reads generated when increasing sequencing depth from 1.6 to 10 million reads after which the gains plateau – 10 million reads detected about 80% of the annotated chicken transcripts. Despite the expectation of continuous sequencing depth increases in the near future, Łabaj et al. [22] argue that most of the additional reads will align to the subset of already extensively sampled transcripts. As a result, transcripts with low to moderate expression levels will remain difficult to quantify with good precision using current RNA-Seq protocols even at higher read depths. Greater sequencing depth will also increase sensitivity to detect smaller changes in relative expression, however this does not guarantee that these changes have functional impact in the biological system under study as opposed to tolerated fluctuations in transcript abundance [20]. Ideally, an efficient experimental design will be informed by an understanding of when increasing sequencing depth begins to provide rapidly diminishing returns with regard to transcript detection and DE testing.

9.1.2 REPLICATION

Replication is vital for robust statistical inference of DE. In the context of RNA sequencing, multiple nested levels of technical replication exist depending upon whether it is the sequence data generation, library preparation or RNA extraction technical processes that are being replicated from the same biological sample. Several published studies have incorporated technical replicates into their RNA-Seq experimental designs [23-25]. The degree of technical variation present in these datasets appears to vary and the main source of technical variation appears to be library preparation [15]. Biological replication measures variation within the target population and simultaneously can counteract random technical variation as part of independent sample preparation [20].

It has been shown that power to detect DE improves when the number of biological replicates n is increased from n = 2 to n = 5 [26], however, to date few studies have incorporated extensive biological replication and extensive testing of the effects of replication on power is needed. More recently with the increasing utility and availability of multiplex experimental designs, the incorporation of biological replicates with decreased sequencing depth is becoming a much more attractive and cost-effective strategy. The relative merits of sacrificing sequencing depth for increased replication has not been rigorously explored.

9.1.3 EFFICIENT EXPERIMENTAL DESIGN

Multiplexing is an increasingly popular approach that allows the sequencing of multiple samples in a single sequencing lane or reaction and consequently the reduction in sequencing costs per sample [27,28]. Multiplexing uses indexing tags, "barcodes" or short ($\leq$ 20 bp) stretches of sequence that are ligated to the start of sample sequence fragments during the library preparation step. Barcodes are distinct between sample libraries and allow pooling for sequencing followed by allocation of reads back to individual samples after sequencing by analysis of the sequenced barcode. Multiplex barcode designs are routinely available with up to 12 samples in the same lane, recently up to 96 yeast DNA samples were profiled in single lane [28]. Novel methods are continuing to emerge for low-cost strategies to multiplex RNA-Seq samples [29]. With the dramatic increases in sequencing yields being achieved with current chemistries and new platforms, multiplexing is becoming the method of choice to increase sample throughput. These designs have direct impacts on sequencing depth generated that need to be considered in the power of the experimental design. Also, when multiplex strategies are used, biologists need to be mindful of potential systematic variations between sequencing lanes. These variations can be addressed through randomisation or blocking designs to distribute samples across lanes, see [30] for a discussion of barcoding bias in multiplex sequencing, and [31] for an alternative to barcoding. In a comparison between microarray and NGS technologies in synthetic pools of small RNA, Willenbrock et al. [13] found that multiplexing resulted in decreased sensitivity due to a reduction of sequencing depth and a loss of reproducibility; however the authors did not investigate power for detection of DE in their study.

9.1.4 APPROACH

Improving detection of DE requires not only an appropriate experimental design but also a suitably powered analysis approach. Several algorithms have recently been developed specifically to appropriately handle expected technical and biological variation arising from RNA-Seq experiments. A non-exhaustive list of these algorithms is: edgeR [32], DESeq [25], NBPSeq [33], BBSeq [34], FDM [35], RSEM [36], NOISeq [37], Myrna [38], Cuffdiff [2]. A thorough comparison of these packages' performance with datasets of different properties falls beyond the scope of this study, however before considering issues relating to power and experimental design, it is important to investigate whether packages for DE analysis give the correct type I error rate under the null hypothesis of no DE. To do this evaluation we considered three popular packages for DE analysis of RNA-sequencing data. These packages are based on a negative binomial distribution model of read counts [39] and include edgeR [32], DESeq [25] and NBPSeq [33].

To quantify the effects of different sequencing depths and replication choices we compared a range of realistic experimental designs for their ability to robustly detect DE. Using simulated data with known DE transcripts allowed us to estimate the false positive rate (FPR) and true positive rate (TPR) of DE calls. The changes of these rates were used to compare the detection power yielded by each choice of number of biological replicates and sequencing depth.

In the Methods section, we outline the definitions used for FPR and TPR as well as explaining the method used for the construction of the synthetic data; which includes induced differential expression, simulates the variations that biological replicates introduce and simulates loss of sequencing depth.

In our study, we test a wide range of real-world experimental design scenarios for performance under the null hypothesis and in the presence of DE. In these scenarios both the numbers of biological replicates n and the sequencing depth are varied. This provides a comprehensive quantitative comparison of different experimental design strategies and is particularly informative for those accessing modern multiplex approaches.

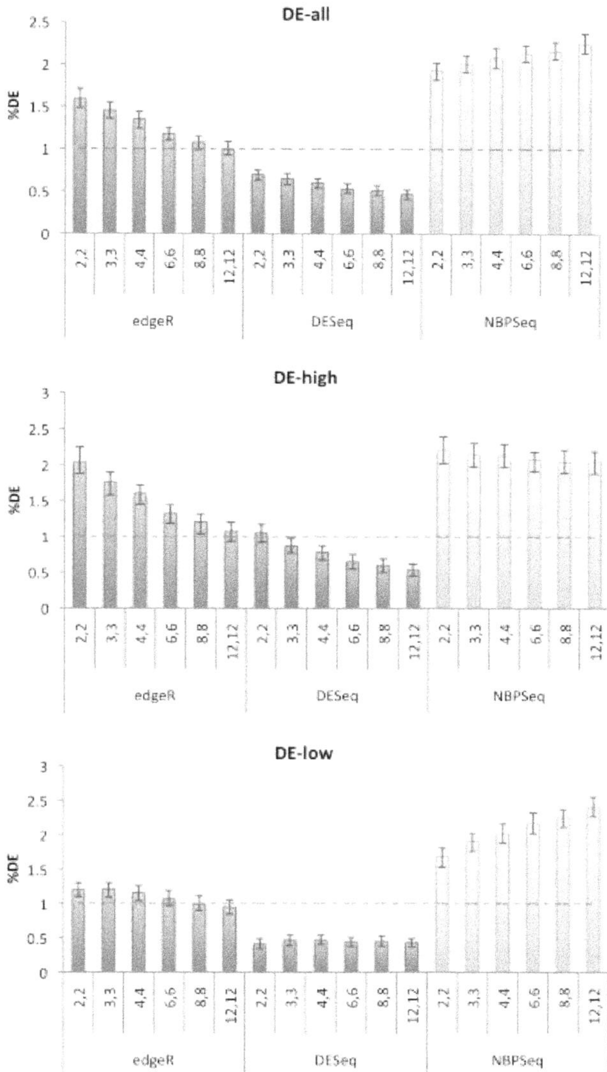

FIGURE 1: The percentage of transcripts reported differentially expressed, FPR defined by Eq. 4 by three software packages for synthetic data generated under the null hypothesis of no DE between two conditions. In the lower two panels the set of transcripts has been divided into those with greater than 100 counts (DE-high) and those with less than or equal to 100 counts (DE-low) averaged over biological replicates. The number of biological replicates in each condition was varied over the range n = 2, 3, …12. The experiment was repeated for 100 independently generated datasets. The top of each bar is the median value obtained and its 90% confidence interval.

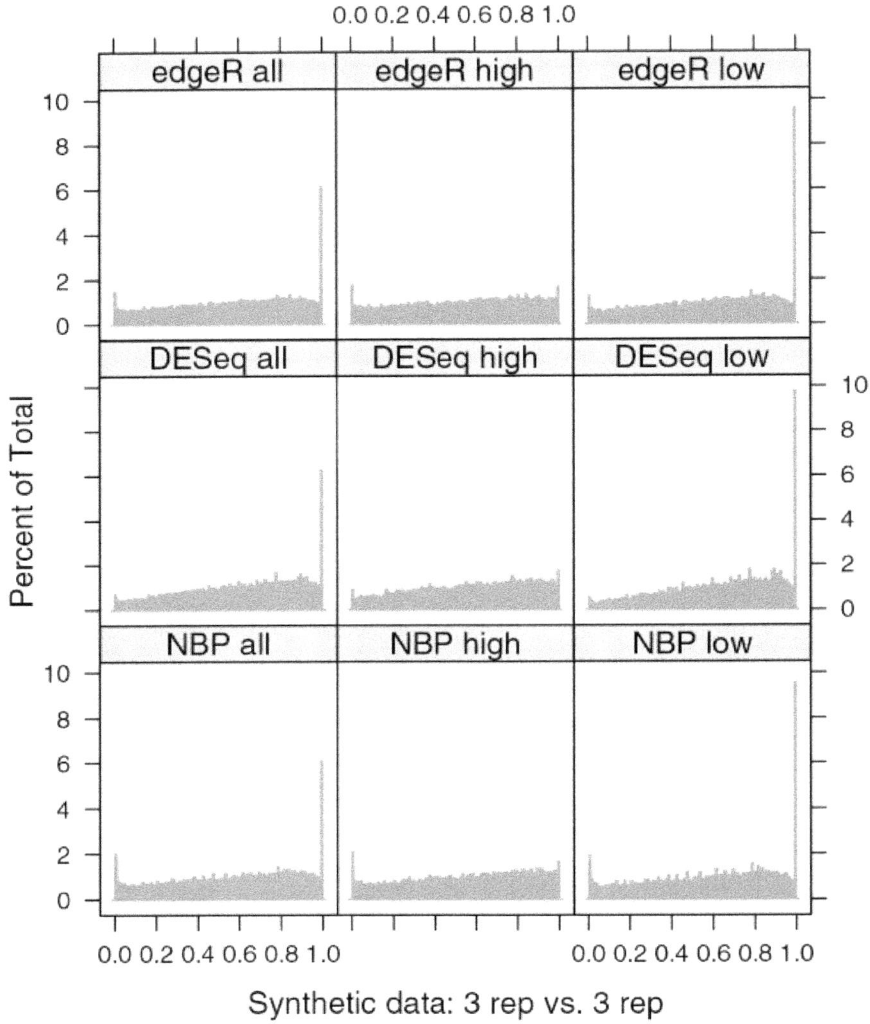

FIGURE 2: Histograms of p-values calculated by three software packages for one particular example of synthetic data generated under the null hypothesis for the case n = 3. In the two right hand columns the set of transcripts has been divided into high-count transcripts (> 100 counts) and low-count transcripts (≤ 100 counts) respectively. 'Percentage of total' is the percentage of p-values falling within each of 100 bins in each histogram.

9.2 RESULTS

9.2.1 COMPARISONS OF STATISTICAL METHODS: EDGER, DESEQ, AND NBPSEQ USING SIMULATED DATA UNDER THE NULL

To test the performance of each package under the null hypothesis, we simulated sets of n "control" and n "treatment" lanes of counts in accordance with the procedure described in the Methods section, for a range of values of n and with no DE between treatments. For each value of n and for each package the simulation and testing were repeated 100 times. Figure 1 shows the percentage of transcripts reported as differentially expressed at the 1% significance level by each of the three packages for a range of values of n. The height of each bar is the median value obtained from 100 repetitions of the synthetic data generation with its associated 90% confidence intervals. Under the null hypothesis, the percentage reported is the false positive rate (FPR) defined by Eq. 4, and should match the significance level of $\alpha = 1\%$ if the package is performing correctly. Also shown are FPRs for high-count transcripts (> 100 counts averaged across biological replicates) and low-count transcripts ($\leq$ 100 counts averaged across biological replicates). Figure 2 shows an example of the p-value distribution obtained for one experiment at n = 3 biological replicates. Ideally, p-values should have a uniform distribution in the interval [0,1] if the package is performing correctly.

Immediately noticeable in the p-value histogram is a sharp spike in the right hand bin for low count transcripts, which is observed to be present in general for all values of n and all packages. This is a known artifact of calculating p-values for discrete random variables using the method described in [40] and summarised in our Methods subsection 'Under the null hypothesis': when count sums in both conditions are equal the computed p-value is exactly 1. The situation is most likely to occur for transcripts with extremely low counts, in which case it is difficult to draw meaningful conclusions regarding DE via any method. The behaviour at the left hand end of the histogram, which drives the FPRs plotted in Figure 1, varies considerably between packages and numbers of biological replicates. It is affected mainly by the method used for estimating a dispersion parameter ϕi for each transcript i (see Methods section).

The package edgeR performs well for large numbers of biological replicates (n = 12), for which squeezing of the dispersion estimate towards the common dispersion is minimal, and a tagwise estimate is appropriate. For small numbers of biological replicates, because the dispersion cannot be estimated accurately on a per-transcript basis, information is borrowed from the complete set of transcripts to squeeze the estimate towards a common dispersion estimate. For the high-count transcripts in particular, the squeezing causes the dispersion of the most highly dispersed transcripts to be underestimated, causing too many transcripts to be deemed differentially expressed, leading to an inflated FPR.

In an effort to be conservative, DESeq chooses as its estimate of dispersion the maximum of a per-transcript estimate and the functional form Eq. 2 which is fitted to the per-transcript estimates for all transcripts. Our results indicate that the method performs well for the high-count transcripts when the number of biological replicates is small (n = 2 or 3), but is otherwise over-conservative. This is generally to be preferred to an inflated FPR, as one has more evidence that what is called DE is truly DE.

The package NBPSeq imposes the functional relationship Eq. 3, which appears to be too restrictive for a number of relatively highly dispersed transcripts. For those transcripts the dispersion parameter is underestimated, leading to an overestimate of significance and hence an inflated FPR irrespective of the number of biological replicates.

Based on these results we selected DESeq (v1.6.1) and edgeR (v2.4.0) for use in subsequent experimental design testing. Throughout these tests, results obtained using DESeq and edgeR are mostly compatible with each other. However, our comparison revealed a slightly inflated FPR from edgeR while DESeq behaves more conservatively throughout. Therefore, in the following section we will focus on the results obtained using DESeq.

9.2.2 COMPARISON OF STATISTICAL METHODS: DESEQ AND EDGER USING SIMULATED DATA WITH 15% DE TRANSCRIPTS

To test the performance of packages in the presence of an alternate hypothesis, we simulated sets of n "control" and n "treatment" lanes of counts with 15% of the transcripts either up- or down-regulated according to the procedure described in the Methods section. All results presented from this point on are derived from DESeq.

9.2.3 DETECTION OF DE AS A FUNCTION OF NUMBER OF BIOLOGICAL REPLICATES N

With an increase in replication we saw a steady increase in the percentage DE calls by DESeq (call rate), increasing from 0.44% to 5.12% as n increased from 2 to 12 (at 100% depth). As n increased, the FPR, defined by Eq. 5 at a significance level of $\alpha = 1\%$, remained below 0.1% for all values of n, and the TPR, defined by Eq. 6 with $\alpha = 1\%$, increased substantially from 3.26% to 41.57% (see Table 1).

TABLE 1: Effects of biological replication on power to detect DE using DESeq

%	n = 2	n = 3	n = 4	n = 6	n = 8	n = 12
call rate %	0.44	1.15	1.76	3.03	4.08	5.12
FPR %	0.04	0.06	0.06	0.06	0.05	0.04
TPR %	3.26	8.95	13.95	24.30	32.72	41.57

Effects of biological replication on power to detect DE using DESeq. FPR and TPR are defined in Eqs. 5 & 6 respectively at 1%. "call rate" is the total number of reported positives / the total number of transcripts. These values are also represented in Figure 3 at 100% sequencing depth.

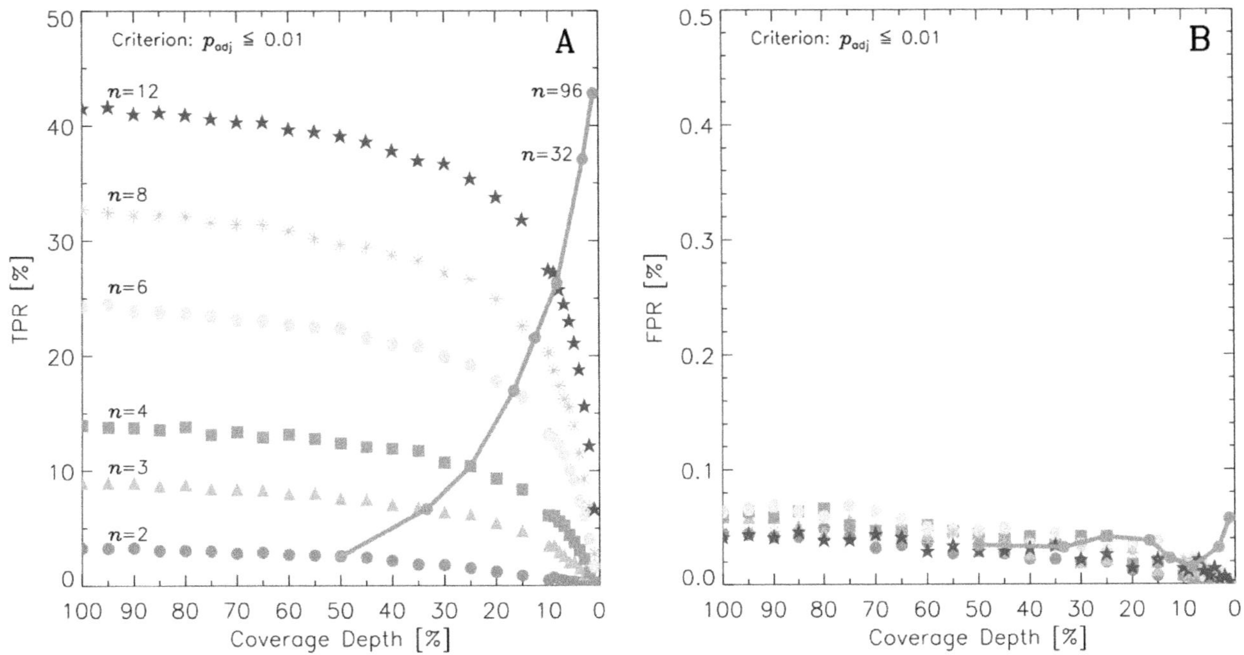

FIGURE 3: TPR and FPR detected by DESeq as a function of sequencing depth and replication. Different symbols represent the number n of control vs. treatment samples (n = 2, 3, 4, 6, 8, and 12) across sequence depths [100%→1%]. A: TPR (Eq. 6 at $\alpha = 1\%$) padj ≤ 0.01. B: FPR (Eq. 5 at $\alpha = 1\%$) padj ≤ 0.01. The solid grey line ("multiplex line") connecting the TPR values of n biological replicates at $1/n \times 100\%$ sequencing depth shows the increase of TPR as more biological replicates n are used despite the loss power due to the sequencing depth reduction required by the multiplexing of lanes. This trend remains true even for the n = 32 and n = 96 cases.

9.2.4 DETECTION OF DE AS A FUNCTION OF SEQUENCING DEPTH

Figure 3 represents the combined results of decreasing sequencing depth for all values of n. It can be seen that as sequencing depth decreases the TPR generated by DESeq decreases monotonically across all n while the FPR remains below 0.1%.

Table 2 shows the FPR for all biological replicates n and a subset of sequencing depths: 25%, 50%, 75% and 100%, the FPR remains below 0.1% at all sequencing depths. Table 3 shows the TPR reported by DESeq for the same subset of sequencing depths, here the TPR increases strongly as sequencing depth increases for any number of biological replicates n.

TABLE 2: Effects of sequencing depth on FPR at different n and depths

Depth	n = 2	n = 3	n = 4	n = 6	n = 8	n = 12
25%	0.02	0.02	0.04	0.03	0.03	0.03
50%	0.03	0.03	0.04	0.05	0.04	0.03
75%	0.04	0.06	0.05	0.07	0.04	0.04
100%	0.04	0.06	0.06	0.06	0.05	0.04

Effects of sequencing depth on FPR values for a subset of our tested depths = 25%, 50%, 75% & 100%.

TABLE 3: Effects of sequencing depth on TPR at different n and depths

Depth	n = 2	n = 3	n = 4	n = 6	n = 8	n = 12
25%	1.57	6.24	10.40	19.18	26.08	35.41
50%	2.58	7.63	12.40	22.34	29.66	39.16
75%	3.01	8.47	13.16	23.44	31.57	40.65
100%	3.26	8.95	13.95	24.30	32.72	41.57

Effects of sequencing depth on TPR values for a subset of our tested depths = 25%, 50%, 75% & 100%.

9.2.5 DETECTION OF DE ACROSS MULTIPLEX EXPERIMENTAL DESIGN STRATEGIES

We simulated various scenarios of multiplexing n-control samples vs. n-treatment samples into two sequencing lanes – each control and treatment sample at a sequencing depth $=1/n\times100\%$. In Figures 3 and 4, a solid grey line connecting every value of n at its corresponding sequencing depth provides a summary of the performance of these multiplexing scenarios. We call this trend the "multiplex line" and it provides an insight into the results obtained by increasing the number of biological replicates used into a fixed number of sequencing lanes, in this case 2 sequencing lanes.

The multiplex line in Figure 3 shows a clear increase in TPR as replication increases despite the loss of detection power that decreasing sequencing depth induces. It can also be seen that the FPR remains below 0.1% for all multiplex scenarios tested (Figure 3B). Note that for completeness we also added multiplex scenarios for $n = 32$ & $n = 96$, whose results follow the trends well. The multiplex line strongly favours adding more biological replicates despite the inherent loss of sequencing depth as shown by its dramatic positive slope for the TPR while maintaining a roughly constant, low FPR.

9.2.6 FOLD-CHANGES AS INDICATORS OF DE

It is common practice among biologists to use fold-change, rather than p-values, as an indicator of DE. Figure 4 shows results analogous to those of Figure 3 when the criterion of fold-change ≥ 2 (instead of p-values) is used to detect DE: as replication n increases, both TPR and FPR decrease because more biological replicates have the effect of averaging out differences between control and treatment lanes. Note that, as sequencing depth decreases, the FPR increases owing to the growing number of transcripts with very low numbers of counts (Figure 4B), in which case the Poisson shot noise of the sequencer can easily induce a spurious doubling or halving of counts. This effect is ameliorated by adding 1 count to all transcripts

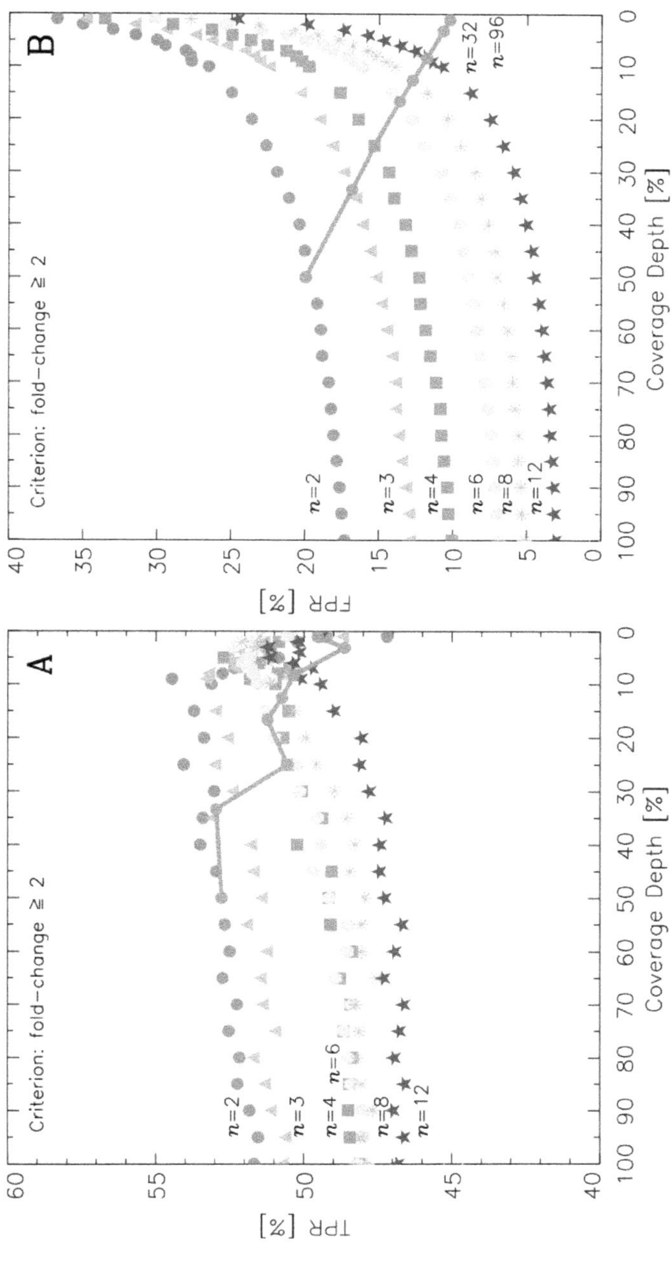

FIGURE 4: Same as Figure3but using 2-fold-changes as the criterion for FPR and TPR instead of padj ≤ 0.01. A: TPR fold-change ≥ 2. B: FPR fold-change ≥ 2. The "multiplex line" connects the TPR and and FPR values of n biological replicates at $1/n \times 100\%$ sequencing depth.

prior to DE analysis – doing so, does not affect the calculation of p-values (data not shown).

9.3 DISCUSSION

9.3.1 COMPARISONS OF DE ALGORITHMS: EDGER, DESEQ AND NBPSEQ

Our comparison of these three DE detection algorithms under the null hypothesis revealed different performances (measured by their FPR) when different numbers of biological replicates n, are used. DESeq consistently performed more conservatively across the different n biological replicates scenarios. DESeq's performance was closest to the expected significance level when only using high-count (counts > 100) transcripts while for only low-count (counts $\leq$ 100) transcripts over-conservative behaviour is shown. edgeR overestimates DE detection for small values of n while its performance improves as n increases. edgeR's level of detection is constant over n when only low-count transcripts are used while overestimation increases when only high-count transcripts are used. NBPSeq overestimated detection across n for the three scenarios (all-transcripts, high-counts and low-counts).

This comparison led us to use both DESeq and edgeR throughout our replication and sequencing depth simulations. We ultimately chose DESeq's results as this package behaved slightly more conservatively and appeared less sensitive to changes in replication (see Figure 1). In a study by Tarazona et al. [37], it is argued that negative-binomial based DE analysis packages like DESeq and edgeR are highly sensitive to sequence depth increases and are therefore unable to control the FPR as sequencing depth increases. Tarazona et al. propose a non-parametric algorithm (NOISeq) to calculate DE based on a noise distribution created with fold-changes and the absolute differences between the transcript's control and experiment lane counts. However, Kvam and Liu [26] argue that due to the small number of replicates typically used for RNA-Seq experiments, non-parametric methods do not offer enough detection power and suggest that current statistical methods to detect DE genes based on parametric models for RNA-Seq data (e.g. DESeq and edgeR) remain a more adequate approach. In our study we also find that both DESeq and edgeR tend to slightly increase the FPR as sequencing depth increases – as higher depths induce DESeq and edgeR to assign smaller p-values to transcripts with small fold-changes. However in no instance do we obtain a FPR larger than 1% for DESeq (2% for edgeR) – (see Figures 1 and 3A). It is worth mentioning that the latest updates to DESeq (v1.6.1) and edgeR (v2.4.0) – released after the studies [37] and [26] reduce the number of false positive calls by about 50% (data not shown).

9.3.2 EFFECTS OF REPLICATION FOR DETECTION OF DE

To quantify the effects of replication in RNA-Seq DE experiments, we tested n-control vs. n-treatment biological replicates (2, 3, 4, 6, 8 and 12) while maintaining sequencing depth constant. We find that as n increases both algorithms increase the call rate and TPR while the FPR remains unchanged (Table 1).

Our results clearly support the simple message that more biological replicates are not only desirable but needed to improve the quality and reliability of DE detection using RNA-Seq, however, due to the costs associated with RNA-Seq, many experiments are likely to need to use multiplex designs to achieve this level of replication.

This study is concerned with the simulation of overdispersion effects due to biological variability and it is implied that overdispersion due to technical variability is nested within this estimation (see Methods section). It is worth mentioning that, while biological variability is important, the contribution to overdispersion by technical variation is not negligible, and disagreements between estimates of expression can occur at all levels of coverage [41]. Ideally, RNA-Seq experimental design with biological replication should also aim to block sources of technical variation, such as between lane variations, to constrain the dispersion of RNA-Seq experiments.

9.3.3 EFFECTS OF SEQUENCING DEPTH FOR DETECTION OF DE

To quantify the effects of sequencing depth in RNA-Seq DE experiments, we simulated an extensive sequencing depth range (100% to 1%) for every case of n-control vs. n-treatment biological replicates. As the amount of available sequencing data is decreased, both packages decrease the call rate and TPR while the FPR remains low. TPR decreases very slowly as sequencing depth decreases, suggesting that sequencing depth can be reduced to -15% without much impact on TPR.

We conclude that DE analysis with RNA-Seq is robust to substantial loss of sequencing data as indicated by a slow decline in TPR as sequencing depth is lost accompanied by no increase in FPR. These findings seem consistent with the results reported by Bashir et al. [42] who observed that lower levels of transcriptome sequencing had sufficient information to estimate the distribution of expression values arising from observed transcripts. Bashir et al. did not directly test power to detect DE, however as testing for DE relies on good concordance with the expected distribution, it follows that DE is reasonably robust to loss of sequencing data.

9.3.4 MULTIPLEXING EXPERIMENTAL DESIGNS

To quantify the effects of varying both n and sequencing depth, we simulated multiplexing n-control vs. n-treatment lanes into two sequencing lanes. We observed a steady increase in TPR with the increase in n, despite the corresponding decrease in sequencing data per transcript by 1/n. Similarly, for both DESeq and edgeR the number of DE calls and the TPR increases with n, as we observed previously and is unaffected by the decrease in data. For DESeq, the FPR remains roughly constant and always below 0.1%, while for edgeR, the FPR decreases slowly as n increases.

Our simulations strongly support that the benefits of multiplexing n-biological replicates into one sequencing lane (two lanes for a n-control vs. n-treatment DE experiment), far outweigh the decrease of available data per sample by 1/n. These multiplexing experimental designs improve TPR and FPR while greatly reducing the cost of the experiment.

While the detection of DE appears robust to available sequence data, there remains the question of how multiplexing affects coverage of the transcriptome and detection of low abundant or rare transcripts. This coverage issue will increasingly be counterbalanced by rapid increases in data generation capacity from a single sequencing experiment. In a detailed study of the Marioni [23] human (liver and kidney) dataset, Banshir et al. [42], reported that over 90% of the total observed transcripts were sampled with 1 million reads. This should be considered in the context of the quickly evolving sequencing technologies like HiSeq 2000 and HiSeq 2500 which can produce up to 300 million reads per sequencing lane. In an evaluation of coverage depth of the chicken transcriptome, Wang et al. [21] find that while 10 million reads allow detection of 80% of the annotated genes, an increase from 10 to 20 million reads does not have a significant effect on transcriptome coverage or reliability of mRNA measurements. That said, current estimates of transcriptome coverage and the impacts of multiplexing strategies analysed in this paper assume unbiased sampling of transcripts. It is highly likely that the power to detect DE varies across transcripts with their sequence content, isoform complexity and abundance. Fang and Cui [20] warn against and discuss several sequencing biases that could create the need for high sequencing coverage to accurately estimate transcript abundance and variation. The authors mention the importance of choosing whether to increase the sequencing depth per sample or to increase the number of biological replicates when planning an experiment. Here, we quantitatively argue that given a fixed budget, the benefits of increasing the number of biological replicates outweigh the corresponding decrease of sequencing depth. This suggestion is backed by the patterns in Figures 3 and 4 in which for a given number of n-biological replicates TPR drops very slowly as depth decreases, FPR remains low when sequencing depth is decreased. In the light of new sequencing technologies rapidly increasing the available sequencing depth per lane, the inforrmation provided by biological replicates' variation is likely to become a priority over sequencing depth.

9.4 CONCLUSIONS

Not surprisingly, our results indicate that more biological replicates are needed to improve the quality and reliability of DE detection using RNA-Seq. Importantly however, we also find that DE analysis with RNA-Seq is robust to substantial loss of sequencing data as indicated by a slow decline in TPR accompanied by no increase in FPR. Our simulations strongly support that multiplexing experimental designs improve TPR and FPR while greatly reducing the cost of the experiment, as the benefits of multiplexing n-biological replicates far outweigh the decrease of available data per sample by 1/n.

As many available packages for DE analysis are increasingly becoming faster and easier to use, our recommendation for most RNA-Seq DE experiments is to use 2 different packages for DE testing. The combined use of packages based on different distribution statistics or a different set of assumptions could generate useful information about a possible bias susceptibility of a given package particular to the specific dataset of interest.

To our knowledge, this is the most up-to-date comparison of DESeq and edgeR's performance relative to ability to detect DE in a range of experimental designs. It directly tests the efficiency of modern multiplex experimental design strategies. Our study informs important experimental design decisions now relevant when trying to maximise an RNA-Seq study to reliably detect DE.

9.5 METHODS

9.5.1 NEGATIVE BINOMIAL MODEL AND BIOLOGICAL VARIATION SIMULATION

Our synthetic data is based on a negative binomial (NB) model of read counts assumed by [39] and used in edgeR [32], DESeq [25] and NBPSeq [33]. The model is a hierarchical model which takes into account sources of variability in the molar concentration of each transcript isoform in the prepared cDNA library due to i) library preparation steps and, in the case of biological replicates, ii) biological variation. This variation is compounded by an additional Poisson shot-noise arising from the sequencing step. Assuming the molar concentration in the prepared cDNA library to have a Gamma distribution, one arrives at a NB distribution for the number of counts K mapped onto a particular transcript of interest in a given lane of the sequencer:

$$K \sim NB\,(\text{mean} = \mu, \text{var} = \mu(1 + \phi\mu)). \tag{1}$$

The mean μ is proportional to the concentration of the transcript of interest in the original biological sample, up to a normalisation factor specific to the lane of the sequencer. A suitable model for this normalisation factor is the Robinson-Oshlack TMM factor [32]. The quantity ϕ is called the dispersion parameter [39], and is specific to the transcript isoform and the library preparation.

9.5.2 R PACKAGES FOR DE IN RNA-SEQ

All three packages considered are based on a NB model, and differ principally in the way the dispersion parameter is estimated. Unless otherwise stated, tests of these packages used herein use default settings.

9.5.3 EDGER (VERSION 2.4.0, BIOCONDUCTOR)

To begin with, edgeR [43] calculates for each transcript a quantile adjusted log conditional likelihood function for the dispersion ϕ[39]. Here, "quantile adjusted" refers to an adjustment of the number of counts to adjust for the total number of counts across all transcripts in each biological replicate, and "conditional" means conditioning on the sum of counts for the given transcript across biological replicates. The "common dispersion" estimate defined by edgeR assumes $\phi$to be a constant over all transcripts in one lane of the sequencer, and is obtained by maximising the log-likelihood summed over transcripts. However, edgeR recommends a "tagwise dispersion" function, which estimates the dispersion on a gene-by-gene basis, and implements an empirical Bayes strategy for squeezing the estimated dispersions towards the common dispersion. Under the default setting, the degree of squeezing is adjusted to suit the number of biological replicates within each condition: more biological replicates will need to borrow less information from the complete set of transcripts and require less squeezing.

9.5.4 DESEQ (VERSION 1.6.1, BIOCONDUCTOR)

In previous versions of the package DESeq [25], ϕ was assumed to be a function of μ determined by nonparametric regression. The recent version used in this paper follows a more versatile procedure. Firstly, for each transcript, an estimate of the dispersion is made, presumably using maximum likelihood. Secondly, the estimated dispersions for all transcripts are fitted to the functional form:

$$\phi = a + b/\mu \qquad \text{(DESeq parametric fit)} \qquad (2)$$

using a gamma-family generalised linear model. The per-transcript estimate is considered to be more appropriate when large numbers of replicates (≥ 4) are present, while the functional form is considered to be more appropriate when small numbers of replicates (≤ 2) are present, in which case information is borrowed from the general trend of all transcripts. Recognising that the dispersion may be underestimated by the functional fit, leading to an overestimate of significance in detecting DE, DESeq by default chooses the maximum of the two methods for each transcript. Also by default, DESeq assumes a model in which the mean μ differs between conditions, but the dispersion ϕ is common across all conditions.

9.5.5 NBPSEQ (VERSION 0.1.4, CRAN)

As for edgeR, the package NBPSeq [33] considers per-transcript log likelihood conditioned on the sum of counts across replicates. However, NBPSeq further imposes the following functional relationship between ϕ and μ:

$$\phi = c\mu^{\alpha - 2} \qquad \text{(NBPSeq model)} \qquad (3)$$

that is, a linear relationship between $\log\phi$ and $\log\mu$. The cases $\alpha = 1$ and $\alpha = 2$ (equivalent to common dispersion) of this function are referred to as NB1 and NB2 respectively. The global parameters α and c are estimated by maximising the log conditional likelihood summed over all replicates and transcripts.

9.5.6 CONSTRUCTION OF THE SYNTHETIC DATASETS

Each of our synthetic datasets consists of a 'control' dataset of read counts K_{ij}^{contr} and a 'treatment' dataset of read counts K_{ij}^{treat}, for i = 1,…,t transcript isoforms sequenced from j = 1,…,n biological replicate cDNA libraries.

For each transcript isoform, we begin by providing a pair of NB parameters. A read count K_{ij}^{contr} for each isoform in each biological replicate is then generated by sampling randomly from a NB distribution with these estimated parameters to from the control dataset. To create the treatment dataset, the set of isoforms is first divided into a non-regulated subset, an up-regulated subset and a down-regulated subset. A regulating factor θ_i = 1,…,t, which is equal to 1 (non-regulated), > 1 (up-regulated) or < 1 (down-regulated) is then chosen from a suitable distribution. A treatment read count K_{ij}^{treat} is then generated for each isoform in each biological replicate from a NB distribution with the mean $\theta_i\mu_i$ and unchanged dispersion ϕ_i.

The basis for the parameters μ_i and ϕ_i is a subset of the Pickrell [24] dataset of sequenced cDNA libraries generated form mRNA from 69 lymphoblastoid cell lines derived from Nigerian individuals as part of the International HapMap Project. For each individual, a library prepared for the Illumina GA2 platform was split into two portions, with one portion sequenced at the Argonne sequencing centre and the other at the Yale sequencing centre. For 12 of the individuals a second library was also prepared, split, and sequenced at both centres. Only data from the initial 69 libraries sequenced at Argonne was used for the current study. The raw reads were re-aligned onto the human transcriptome (hg18, USCS) using the KANGA aligner [44]. The total number of reads mapped to annotated genes per lane varied substantially from 2 × 106 to 20 × 106. To provide a uniform set of biological replicates from which to estimate μ_i and ϕ_i, a subset of 44 libraries for which the total number of mappings to the transcriptome per lane was in the range 10×10^6 to 16×10^6 was chosen. Finally, any transcript for which the total number of reads was less than 44, i.e. an average of less than one transcript per lane, was culled from the dataset to leave a list of 46,446 transcripts. The resulting subset of the Pickrell dataset is considered to exhibit overdispersion due to both library preparation and biological variation.

Note that for generation of synthetic data it is not necessary to provide an accurate estimate of μ_i and ϕ_i for each isoform in the reduced Pickrell dataset, but simply to provide a plausible distribution of values of these parameters over the transcriptome representing typical isoform abundances and their variation due to technical and/or biological overdispersion. Parameter values μ_i and ϕ_i, were obtained from the reduced Pickrell dataset as follows. The total number of counts from each of the 44 lanes was first reduced to that of the lane with the smallest number of counts by sampling from the counts in each lane while keeping track of the transcript to which each count is mapped. This forms a normalised set of counts for the ith transcript in the jth lane.

For each transcript a maximum likelihood estimate (MLE) μ_i and ϕ_i, was then made from the n = 44 biological replicates. Details of the construction of this estimate are given in the Additional file 3. For each simulation described herein, a synthetic dataset was constructed consisting of n biological replicates of 'control data' generated from NB distributions with the estimated μ_i and ϕ_i, and a further n biological replicates of treatment data generated from NB distributions with means $\theta i\mu\hat{}i$ and unchanged dispersion $\phi\hat{}i$.

Two sets of simulations were performed:

1. To test performance under the null hypothesis, the regulating factor was set to $\theta i = 1$ for all transcripts.
2. To test ability to detect DE in the presence of an alternative hypothesis, the regulating factor θi was set to $1 + Xi$ for a randomly chosen 7.5% of the transcripts (up-regulated), $(1 + Xi)-1$ for a further 7.5% (down-regulated) and 1 for the remaining 85% of the transcripts, where the Xi are identically and independently distributed exponential random variables with mean 1.

9.5.7 CALCULATION OF TRUE AND FALSE POSITIVE RATES

9.5.7.1 UNDER THE NULL HYPOTHESIS

All three packages test for DE in single-factor experiments by calculating p-values using the method described in [25]. For each transcript i, a probability is calculated for the number of counts in each of two conditions control and treatment, conditional on the sum of the counts in both conditions assuming the NB model described above. The p-value is the sum of the probabilities of all ways of apportioning the sum of counts between the two conditions, which have a lower probability than the observed counts.

To test the performance of each package under the null hypothesis, we simulated sets of n-control and n-treatment lanes of counts for a range of values of n. The FPR, quoted as a percentage, was calculated at a given significance level α as:

$$\text{FPR} = (\text{number of transcripts with } 100 \times \text{p-value} < \alpha) / (\text{total number of transcripts} \times 100\%) \quad (4)$$

Ideally, the FPR should match the significance level of α if the package is performing correctly.

9.5.7.2 IN THE PRESENCE OF AN ALTERNATIVE HYPOTHESIS

All three packages provide an adjusted p-value, p_{adj}, to correct for multiple hypothesis testing with the Benjamini-Hochberg procedure using the R function p.adjust(). All calculations herein of true and false positive rates in the presence of an alternative hypothesis use adjusted p-values.

From the 6,966/46,446 (15%) of the transcripts induced with a regulating factor other than 1, we selected the 5,726 (12%) with a regulation factor satisfying either $\theta_i \leq 0.83$ or $\theta_i \geq 1.20$. We define these as "effectively DE" transcripts. This additional filter on minimal fold-change is designed to quantify the performance of algorithms and experimental designs for detection of DE that might be considered more biologically relevant by researchers. Likewise we define the remaining transcripts, those satisfying $0.83 < \theta i < 1.20$, as "effectively non-DE". These definitions were used to estimate the FPR and TPR at significance level α via the following formulae:

$$\text{FPR} = \frac{\text{number of effectively non-DE transcripts with } 100\times p_{adj} < \alpha}{\text{total number of effectively non-DE transcripts}} \times 100\% \quad (5)$$

$$\text{TPR} = \frac{\text{number of effectively DE transcripts with } 100\times p_{adj} < \alpha}{\text{total number of effectively DE transcripts}} \times 100\% \quad (6)$$

Apart from the use of adjusted p-values, the formula for FPR reduces to Eq. 4 if the number of simulated DE transcripts is set to zero, since in this case all transcripts are, by definition, "effectively non-DE". The quantities 1−FPR and TPR are commonly referred to in the literature as "specificity" and "sensitivity" respectively.

9.5.8 SIMULATING VARIABLE LEVELS OF SEQUENCE DATA AND REPLICATION

Simulating variations in available sequencing data is a fundamental part of investigating the impacts of multiplex experimental design strategies. Variability in the amount of sequence data amongst samples can occur for reasons such as restrictions on available resources, machine error, or sequencing reads sequestered by pathogen transcriptome fractions present in the sample. To simulate loss of sequencing depth, we randomly sub-sampled without replacement counts from the original table of counts simulated in the presence of an alternative hypothesis for each biological replicate. Sequencing depth was decreased in both control and treatment samples over a range of 100% (a full lane of sequence) to 1% of the original data. After this sub-sampling, the resulting table of counts was analysed in DESeq (edgeR) and the total number of effectively-DE calls, TPR, FPR and fold-changes were recorded for every n scenario. We simulated experimental choices of n-controls vs. n-treatments biological replicates at n = 2, 3, 4, 6, 8 and 12.

9.5.9 MULTIPLEXING EXPERIMENTAL DESIGNS

Multiplexing various samples into one sequencing lane reduces the monetary cost of RNA-Seq DE analysis, albeit by dividing the available sequencing depth over various samples. Our strategy consisted of simulating multiplexing n-control samples vs. n treatment samples into two sequencing lanes. This way, the amount of total sequenced data is constrained and each control and treatment sample is expected to be represented at an average depth $1/n \times 100\%$. The absolute value of reads produced in a lane of sequence (i.e. 100% depth) has increased as RNA-Seq technologies evolve, currently this value can be up to 100 million reads. The multiplex experimental setups we tested are:

- 2 vs. 2 biological replicates at 50% sequencing depth
- 3 vs. 3 biological replicates at 33% sequencing depth
- 4 vs. 4 biological replicates at 25% sequencing depth

- 6 vs. 6 biological replicates at 17% sequencing depth
- 8 vs. 8 biological replicates at 13% sequencing depth
- 12 vs. 12 biological replicates at 8% sequencing depth
- 32 vs. 32 biological replicates at 3% sequencing depth
- 96 vs. 96 biological replicates at 1% sequencing depth

REFERENCES

1. Mortazavi A, Williams BA, McCue K, Schaeffer L, Wold B: Mapping and quantifying mammilian transcriptomes by RNA-seq. Nat Methods 2008, 5(7):621-628.
2. Trapnell C, Williams BA, Pertea G, Mortazavi A, Kwan G, van Baren MJ, Salzberg SL, Wold BJ, Pachter L: Transcript assembly and quantification by RNA-Seq reveals unannotated transcripts and isoform switching during cell differentiation. Nat Biotechnol 2010, 28:511-515.
3. Guttman M, Garber M, Levin JZ, Donaghey J, Robinson J, Adiconis X, Fan L, Koziol MJ, Gnirke A, Nusbaum C, Rinn JL, Lander ES, Regev A: Ab initio reconstruction of cell type-specific transcriptomes in mouse reveals the conserved multi-exonic structure of lincRNAs. Nat Biotechnol 2010, 28:503-510.
4. Haas BJ, Zody MC: Advancing RNA-Seq analysis. Nat Biotechnol 2010, 28:421-423.
5. Pan Q, Shai O, Lee LJ, Frey BJ, Blencowe BJ: Deep surveying of alternative splicing complexity in the human transcriptome by high-throughput sequencing. Nat Genet 2008, 40:1413-1415.
6. Lovci MT, Li HR, Fu XD, Yeo GW: RNA-seq analysis of gene expression and alternative splicing by double-random priming strategy. Methods Mol Biol 2011, 729:247-255.
7. Bullard JH, Purdom E, Hansen KD, Dudoit S: Evaluation of statistical methods for normalization and differentail expression in mRNA-seq experiments. BMC Bioinf 2010, 11:94.
8. Oshlack A, Robinson MD, Young MD: From RNA-seq reads to differential expression results. Genome Biol 2010, 11:220.
9. Dohm JC, Lottaz C, Borodina T, Himmelbauer H: Substantial biases in ultra-short read data sets from high-throughput DNA sequencing. Nucleic Acids Res 2008, 36:e105.
10. Hansen KD, Brenner SE, Dudoit S: Biases in Illumina transcriptome sequencing caused by random hexamer priming. Nucleic Acids Res 2010, 38:e131.
11. Sendler E, Johnson GD, Krawetz SA: Local and global factors affecting RNA sequencing analysis. Anal Biochem 2011, 419:317-322.
12. Lü B, Yu J, Xu J, Chen J, Lai M: A novel approach to detect differentially expressed genes from count-based digital databases by normalizing with housekeeping genes. Genomics 2009, 94:211-216.

13. Willenbrock H, Salomon J, Søkilde R, Barken KB, Hansen TN, Nielsen FC, Møller S, Litman T: Quantitative miRNA expression analysis: comparing microarrays with next-generation sequencing. RNA 2009, 15:2028-2034.
14. Zheng W, Chung LM, Zhao H: Bias detection and correction in RNA-Sequencing data. BMC Bioinf 2011, 12:290.
15. Roberts A, Trapnell C, Donaghey J, Rinn JL, Pachter L: Improving RNA-Seq expression estimates by correcting for fragment bias. Genome Biol 2011, 12:R22.
16. Canales RD, Luo Y, Willey JC, Austermiller B, Barbacioru CC, Boysen C, Hunkapiller K, Jensen RV, Knight CR, Lee KY, Ma Y, Maqsodi B, Papallo A, Peters EH, Poulter K, Ruppel PL, Samaha RR, Shi L, Yang W, Zhang L, Goodsaid FM: Evaluation of DNA microarray results with quantitative gene expression platforms. Nat Biotechnol 2006, 24:1115-1122.
17. Risso D, Schwartz K, Sherlock G, Dudoit S: GC-Content normalization for RNA-seq data. BMC Bioinf 2011, 12:480.
18. Oshlack A, Wakefield MJ: Transcript length bias in RNA-seq data confounds systems biology. Biol Direct 2009, 4:14.
19. Auer PL, Srivastava S, Doerge RW: Differential expression–the next generation and beyond. Brief Funct Genomics 2011.
20. Fang Z, Cui X: Design and validation issues in RNA-seq experiments. Brief Bioinf 2011, 12:280-287.
21. Wang Y, Ghaffari N, Johnson CD, Braga-Neto UM, Wang H, Chen R, Zhou H: Evaluation of the coverage and depth of transcriptome by RNA-seq in chickens. BMC Bioinf 2011, 12(Suppl 10):S5.
22. Łabaj PP, Leparc GG, Linggi BE, Markillie LM, Wiley HS, Kreil DP: Characterization and improvement of RNA-Seq precision in quantitative transcript expression profiling. Bioinformatics 2011, 27:i383-391.
23. Marioni JC, Mason C, Mane SM, Stephens S, Gilad Y: RNA-seq: an assessment of technical reproducability and comparison with gene expression arrays. Genome Res 2008, 18:1509-1517.
24. Pickrell JK, Marioni JC, Pai AA, Degner JF, Engelhardt BE, Nkadori E, Veyrieras JB, Stephens M, Gilad Y, Pritchard JK: Understanding mechanisms underlying human gene expression variation with RNA sequencing. Nature 2010, 464:768-72.
25. Anders S, Huber W: Differential expression analysis for sequence count data. Genome Biol 2010, 11(10):R106.
26. Kvam VM, Liu P, Si Y: A comparison of statistical methods for detecting differentially expressed genes from RNA-seq data. Am J Bot 2012, 99(2):248-256.
27. Porreca GJ, Zhang K, Li JB, Xie B, Austin D, Vassallo SL, LeProust EM, Peck BJ, Emig CJ, Dahl F, Gao Y, Church GM, Shendure J: Multiplex amplification of large sets of human exons. Nat Methods 2007, 4:931-936.
28. Smith AM, Heisler LE, St Onge RP, Farias-Hesson E, Wallace IM, Bodeau J, Harris AN, Perry KM, Giaever G, Pourmand N, Nislow C: Highly-multiplexed barcode sequencing: an efficient method for parallel analysis of pooled samples. Nucleic Acids Res 2010, 38:e142.

29. Wang L, Si Y, Dedow LK, Shao Y, Liu P, Brutnell TP: A low-cost library construction protocol and data analysis pipeline for Illumina-based strand-specific multiplex RNA-seq. PLoS ONE 2011, 6:e26426.
30. Alon S, Vigneault F, Eminaga S, Christodoulou DC, Seidman JG, Church GM, Eisenberg E: Barcoding bias in high-throughput multiplex sequencing of miRNA. Genome Res 2011, 21:1506-1511.
31. Timmermans MJ, Dodsworth S, Culverwell CL, Bocak L, Ahrens D, Littlewood DT, Pons J, Vogler AP: Why barcode? High-throughput multiplex sequencing of mitochondrial genomes for molecular systematics. Nucleic Acids Res 2010, 38:e197.
32. Robinson MD, Oshlack A: A scaling normalization method for differential expression analysis of RNA-seq data. Genome Biol 2010, 11:R25.
33. Di Y, Schafer D, Cumbie J, Chang J: The NBP negative binomial model for assessing differential gene expression from RNA-seq. Stat Appl in Genet and Mol Biol 2011, 10:Article 24.
34. Zhou YH, Xia K, Wright FA: A powerful and flexible approach to the analysis of RNA sequence count data. Bioinformatics 2011, 27:2672-2678.
35. Singh D, Orellana CF, Hu Y, Jones CD, Liu Y, Chiang DY, Liu J, Prins JF: FDM: a graph-based statistical method to detect differential transcription using RNA-seq data. Bioinformatics 2011, 27:2633-2640.
36. Li B, Dewey CN: RSEM: accurate transcript quantification from RNA-Seq data with or without a reference genome. BMC Bioinformatics 2011, 12:323.
37. Tarazona S, Garcia-Alcalde F, Dopazo J, Ferrer A, Conesa A: Differential expression in RNA-seq: a matter of depth. Genome Res 2011, 21:2213-2223.
38. Langmead B, Hansen KD, Leek JT: Cloud-scale RNA-sequencing differential expression analysis with MYRNA. Genome Biol 2010, 11:R83.
39. Robinson M, Smyth G: Moderated statistical tests for assessing differences in tag abundance. Bioinformatics 2007, 23(21):2881-2887.
40. Anders S: Analysing RNA-Seq data with the DESeq Package. 2010. [http://www.bioconductor.org/help/course-materials/2011/BioC2011/LabStuff/DESeq.pdf]
41. McIntyre LM, Lopiano KK, Morse AM, Amin V, Oberg AL, Young LJ, Nuzhdin SV: RNA-seq: technical variability and sampling. BMC Genomics 2011, 12:293.
42. Bashir A, Bansal V, Bafna V: Designing deep sequencing experiments: detecting structural variation and estimating transcript abundance. BMC Genomics 2010, 11:385.
43. Robinson M, McCarthy D, Smyth G: edgeR: a Bioconductor package for differential expression analysis of digital gene expression data. Bioinformatics 2010, 26:139-140.
44. Stephen S, Cullerne D, Spriggs A, Helliwell C, Lovell D, Taylor JM: BioKanga: a suite of high performance bioinformatics applications. in preparation 2012, [http://code.google.com/p/biokanga/]
45. JabRef Development Team: JabRef . JabRef Development Team 2010. [http://jabref.sourceforge.net/faq.php]

46. Muller A: TeXMed – a BibTeX interface for PubMed 2002–2012. [http://www.bioinformatics.org/texmed/]
47. Chen H, Boutros PC: VennDiagram: a package for the generation of highly-customizable Venn and Euler diagrams in R. BMC Bioinf 2011, 12:35.

CHAPTER 10

ANALYZING THE MicroRNA TRANSCRIPTOME IN PLANTS USING DEEP SEQUENCING DATA

XIAOZENG YANG and LEI LI

10.1 INTRODUCTION

One of the most exciting biological finding in recent years is the discovery of many functional small RNA species that regulate diverse spatial and temporal function of the genome [1–4]. After the initial discovery of miRNAs in the worm *C. elegans* [5,6], they are emerging as an important class of endogenous gene regulators acting at the post-transcriptional level in both animals and plants. In plants, much of the effort to identify, experimentally validate, and functionally characterize miRNAs has been directed toward the model plant *Arabidopsis thaliana*. Consequently, dozens of miRNA-target pairs have been identified and studied [7–9]. It is now well-established that these gene circuits are crucial for many plant development processes as well as responses to environmental challenges [4,9,10].

Although hundreds of miRNAs have been predicted in a broad range of plant lineages [11], there are two indications that current miRNA

This chapter was originally published under the Creative Commons Attribution License. Yang X and Li L. Analyzing the microRNA Transcriptome in Plants Using Deep Sequencing Data. Biology **1** *(2012), 297–310. doi:10.3390/biology1020297. doi:10.3390/biology1020297.*

collections in many model plants and important crop species are far from completion. First of all, the numbers of predicted miRNAs among different plant species are conspicuously uneven. As shown in Figure 1, the well-annotated *Arabidopsis* and rice (*Oryza sativa*) genomes contain approximately one miRNA for every 100 protein-coding genes. In other plant species, the relative density of miRNAs is only half or even less than that in *Arabidopsis* and rice (Figure 1). Because it is highly unlikely that these species indeed encode a smaller complement of miRNA genes, the only explanation is that most miRNAs in species other than *Arabidopsis* and rice still await discovery.

In plants, various studies have established that there are about 20 families of conserved miRNAs [14,15]. In many plant species, only miRNAs belonging to conserved families are identified. However, the sequencing of small RNA populations is increasingly revealing miRNAs that are not conserved between species, suggesting a recent evolutionary origin [8,14,16–18]. In fact, there is increasing evidence that species-specific or subfamily-specific miRNAs are functional constituents of the miRNA-mediated regulatory networks and underscore the dynamic nature of these networks [19–21]. Thus, it is highly desirable to elucidate the full spectrum of miRNAs in diverse plant lineages to gain a comprehensive understanding of miRNA origin, evolution and function.

10.2 BRIEF OVERVIEW OF miRNA BIOGENESIS IN PLANTS

There is no question that any systematic effort to identify miRNAs will depend on a clear understanding of the miRNA biogenesis pathways. Since miRNA biogenesis is the subject of numerous reviews, we only provide a brief overview here (Figure 2). Like protein-coding genes, miRNAs are encoded by class II genes and transcribed by RNA polymerase II [22,23]. Although mature miRNAs are typically 20- to 24-nucleotides (nt) in length, their precursor transcripts can be much longer. As shown in Figure 2, after initial transcription by Pol II, splicing and further processing of pri-miRNAs are carried out in the nucleus and involve the interactive functions of HYL1, DDL, TGHand SE, as well as the cap-binding proteins CBP20 and CBP80 [24–26].

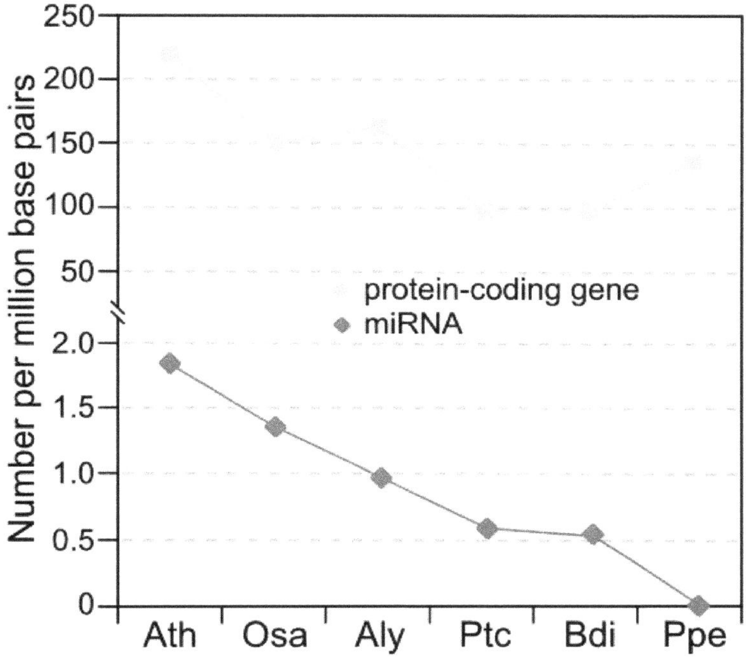

FIGURE 1: Comparison of the density of protein-coding and MIR genes in six plant lineages. Vertical axis indicates the density of protein-coding and MIR genes per million genomic base pairs. The six plants are Ath (*Arabidopsis thaliana*), Osa (*Oryza sativa*), Aly (*Arabidopsis lyrata*), Ptc (*Populus trichocarpa*), Bdi (*Brachypodium distachyon*) and Ppe (Prunus persica). The numbers of protein-coding genes are obtained from TAIR10 (Ath), RGAP6.1 (Osa) [12] and Phytozome7.0 (Aly, Ptc, Bdi and Ppe) [13] while all the numbers of MIR genes are from miRBase17 [11].

A characteristic of pri-miRNAs is that they contain internally complementary sequences that fold back to form a hairpin structure, which is called pre-miRNAs [2,4,27,28]. Pri-miRNAs and pre-miRNAs are sequentially processed by Dicer to yield one or several phased miRNA/miRNA* duplexes. Unique to higher plants, pri-miRNA and pre-miRNA processing are both carried out in the nucleus [29]. The duplexes are stabilized through end methylation catalyzed by HEN1 [25] and transported to the cytoplasm by HST1 [26]. Only the mature miRNA is integrated into the AGO1-containing RNA-induced silencing complex (RISC) and accumulated with DCP1 and DCP2 whereas the passenger strand, called RNA*, is degraded as a RISC substrate [2,30–32]. After loading into RISC, miRNAs base pair with their targets and direct either cleavage [33,34] or translational repression [35] of the target transcripts. Recently, silencing of target genes by miRNA-directed DNA methylation at the target loci has also been reported [36].

As a consequence of miRNA maturation, a series of RNA intermediates are generated in addition to the mature miRNAs, which include the stem-loop-structured pre-miRNAs, miRNA* and sliced RNA fragments derived from other parts of the precursors. Detection, quantification, and reconstruction of these RNA intermediates are the goal of essentially all available methods to identify and profile miRNAs.

10.3 COMPREHENSIVE IDENTIFICATION OF miRNAS FROM NEXT-GENERATION SEQUENCING DATA

Direct sequencing of specifically prepared low-molecular-weight RNA has long been recognized as powerful approach to sample the small RNA species [37]. Typically, small RNA species in the 18–35 nucleotides range are isolated and ligated to the 5' and 3' RNA adapters. The ligated RNA molecules are reverse-transcribed into cDNA using a primer specific to the 5' adapter and amplified by PCR with two primers that anneal to the ends of the adapters. Quality controlled cDNA libraries are then sequenced [8,37]. In one of the earliest studies, Fahlgren et al. [8] sequenced small RNA populations from wild-type *Arabidopsis* as well several mutants defective in miRNA biogenesis using the 454 technology. A total of 48 non-conserved miRNA families were identified by a computational analysis of sequence composition and secondary structure based on knowledge of annotated miRNAs at that time [8].

Encouraged by success in *Arabidopsis* [8,17], deep sampling of small RNA libraries by next-generation sequencing has become a popular approach to identify miRNAs for functional and evolutionary studies in diverse plant species [18,38,39]. An advantage of the sequencing methods

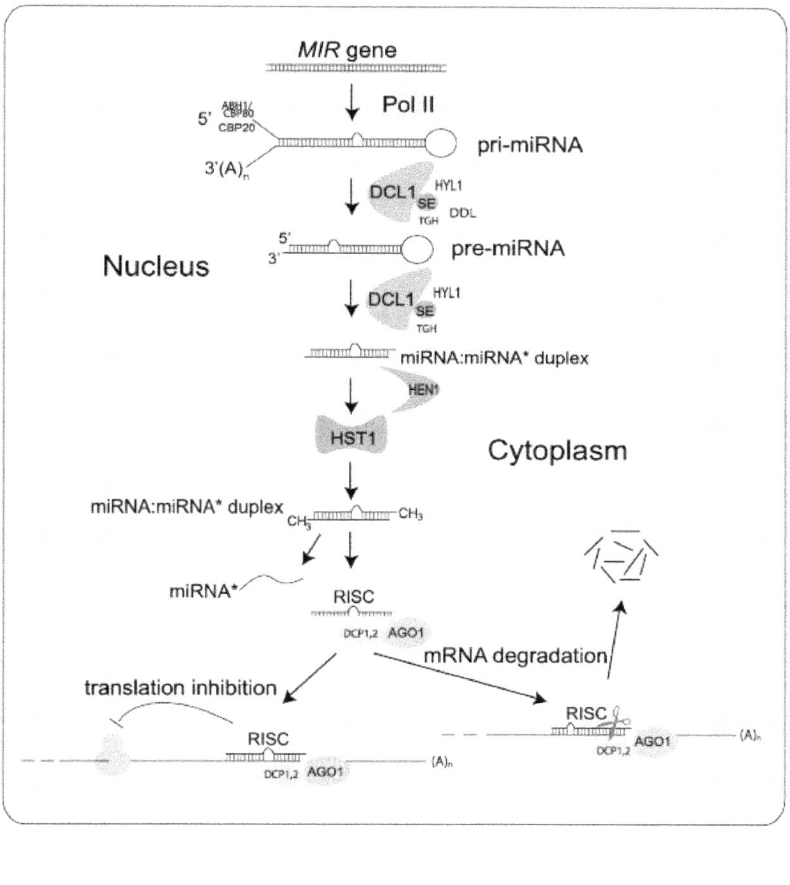

FIGURE 2: Simplified model of miRNA biogenesis in plants. MIR genes are initially transcribed by Pol II into pri-miRNAs that fold back to form hairpin structure. Splicing and further processing in nuclear involve the interactive functions of HYL1, DDL, TGH and SE and of the cap-binding proteins CBP20 and CBP80. Pri-miRNAs and pre-miRNAs are sequentially processed by DCL1 to yield one or several phased miRNA/miRNA* duplexes, which are methylated by HEN1 and transported to the cytoplasm by HST1. The miRNA is selected and incorporated into dedicated AGO1-containing RISC that directs translation inhibition or cleavage of the target mRNA transcript.

is their sensitivity in detecting even poorly expressed or species-specific miRNAs [39]. The potential of deep sequencing to provide quantitative information on the expression pattern of known miRNAs has been explored [8,40]. In addition to validating annotated miRNAs, large numbers of putative new miRNAs have been identified. Following we discuss various recently developed algorithms and programs to profile miRNAs from deep sequencing data, with an emphasis on their applications in plants.

10.4 AVAILABLE TOOLS FOR ANALYZING miRNAS FROM DEEP SEQUENCING DATA

A number of computational tools for identifying and profiling miRNAs from deep sequencing data have been developed. With an easy-to-use graphical interface, most of these tools are web-based while a few, such as miRDeep [41] and miRNAKey [42], are also packaged into a stand-alone version (Table 1). A common module employed by these tools is sequence similarity search to detect miRNAs cross multiple species based on the fact that many miRNAs are evolutionarily conserved. Meanwhile, other algorithms are also introduced to detect new miRNAs based on different models in terms of the pre-miRNA hairpin structures and the duplex of miRNA and miRNA* (Table 1). The challenge now is to separate miRNAs from the pool of other sequenced small RNAs or mRNA degradation products. Further, as most of the miRNA-detecting methods focus exclusively on the mature miRNAs, a drawback is the failure to collect and quantify information on the precursors, which could result in limitations in elucidating the miRNA transcriptome.

10.4.1 mi*RANALYZER*

Utilizing a machine learning algorithm, Hackenberg et al. [46] developed miRanalyzer, a web server tool for analyzing results from deep-sequencing experiments on small RNAs. This program requires a simple input file containing a list of unique reads and their copy numbers. Application of this program in seven animal model species (human, mouse, rat, fruit fly, round-worm, zebra fish and dog) not only detected known miRNA sequences annotated in miRBase, but led to prediction of new miRNAs. The core algorithm of miRanalyzer is based on the random forest classifier and was trained on experimental data, which could accurately predict novel miRNAs with a low false positive rate in animals. Later, miRanalyzer was updated to include a module on miRNA prediction in plants by taking into account differences between plant and animal miRNAs. Currently, 31 genomes, including 6 plant genomes, have been analyzed by the updated miRanalyzer [47].

TABLE 1: Tools for analyzing deeply sequenced small RNA data.

Name	Designed Model	Algorithm	Availability
UEA sRNA Toolkit*	animal & plant	based on criteria of miRNA	web-based
miRDeep	animal	probabilistic model	stand-alone
miRanalyzer**	animal	machine learning	web-based
SeqBuster [43]	animal	sequence similarity	web-based
DSAP [44]	animal	sequence similarity	web-based
mirTools [45]	animal sequence	similarity & miRDeep	web-based
miRNAKey	animal	sequence similarity	stand-alone
miRNEST	animal & plant	sequence similarity	web-based

**Only the miRCat component is for detecting new miRNAs; ** Updated miRanalyzer could also predict new miRNAs in plants, and it has a new stand-alone version.*

10.4.2 UEA SRNA TOOLKIT

UEA sRNA Toolkit [48] combines two integrated parts, miRCat and SiLoCo, to analyze miRNAs using deep sequencing data. miRCat, the package for detecting miRNAs, adopts a number of empirical and published criteria for bona fide miRNA loci to mine miRNAs from deeply sequenced small RNA data. In brief, the program accepts a FASTA file of small RNA sequences as input, which are mapped to a plant genome using PatMaN [49] and grouped into discrete loci. Then it obtains miRNA candidates by searching for a two-peak alignment pattern of sequence reads on one strand of the locus and assessing the secondary structures of a series of putative precursor transcripts using RNAfold [50] and randfold [51]. On the other hand, SiLoCo is the tool to compare the miRNA expression between different samples. It weighs each small RNA hit by its repetitiveness in the genome acquired from mapping by PatMaN [49]. For each locus, the log2 ratio and the average of the normalized small RNA hit counts are used to calculate the miRNA expression difference [48].

10.4.3 mi*RDEEP*

Maturation of miRNAs from the stem-loop structured pre-miRNAs results in three species of small RNAs: mature miRNA, miRNA* and RNA fragments derived from other parts of the precursors. Typically the mature miRNAs are very stable, which results in uneven abundance of the different small RNA species derived from the same pre-miRNA (Figure 3A). miRDeep, developed by Friedländer et al., employs a novel algorithm based on a probabilistic model of miRNA biogenesis to score compatibility of the nucleotide position and frequency of sequenced small RNA reads with the secondary structure of miRNA precursors [41]. When using miRDeep, the small RNA sequence reads are first aligned to the genome. The genomic DNA bracketing these alignments are extracted and computed for secondary RNA structure. Plausible miRNA precursor sequences based on a model for Dicer-mediated miRNA processing are identified. Finally, miRDeep scores the likelihood of the putative miRNA precursors and outputs a scored list of known and new miRNA precursors and mature miRNAs in the deep-sequencing samples [41].

10.4.4 mi*RNAKEY*

miRNAkey [36] is a software pipeline designed to be used as a base-station for the analysis of deep sequencing data on miRNAs. The package implements a common set of steps generally taken to analyze deep sequencing data including the use of similarity search to detect known or conserved miRNAs as well as adding miRDeep [41] to predict new miRNAs. This

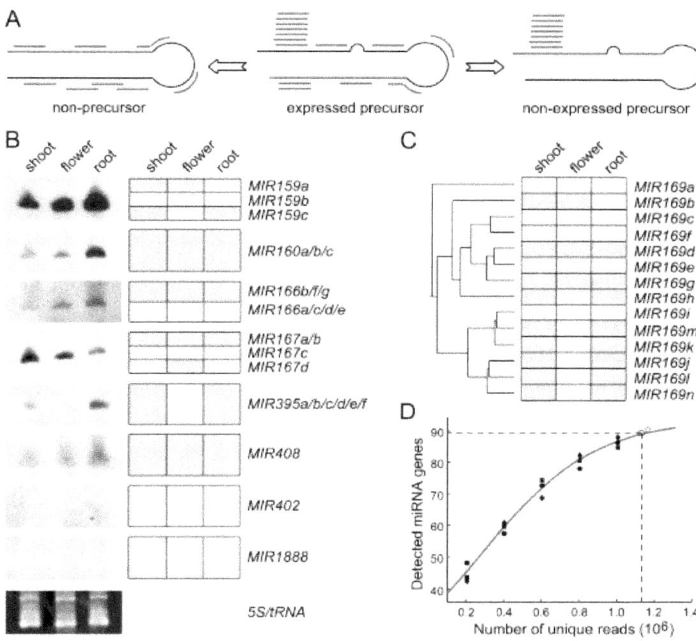

FIGURE 3: Using miRDeep-P to identify and profile miRNAs from deep sequencing data. (A) The core algorithm is based on a miRNA biogenesis model in which the small RNAs derived from a pre-miRNA are considered to have certain probabilities of being sequenced [41]. This model distinguishes an expressed pre-miRNA from a non-expressed pre-miRNA or a genomic locus with the potential to form a hairpin but is not processed by Dicer. (B) Validation of miRDeep-P results in *Arabidopsis* by Northern blotting. RNA blots are shown on the left. Ethidium bromide staining of the 5S/tRNA is used as a loading control. The expression pattern (represented by the green color) of individual genes deduced from the sequencing data is shown on the right with genes with identical pattern combined. (C) Relating miRNA expression to pre-miRNA phylogeny. The annotated pre-miRNAs of the miR169 family in *Arabidopsis* were used to construct a phylogenetic tree. The gene-level expression profile of family members is depicted to the right of the tree. (D) Simulation of the relation between miRNA detection rate and sequencing depth in *Arabidopsis*. Perfectly mapped unique reads from the shoot library were randomly retrieved to create five different simulated sequencing depths. The number of expressed miRNAs was determined at each depth. The scatter plot represents results from three independent simulations and was used for curve fitting. The star sign indicates the actual data from the shoot library. Dashed lines indicate a 95% detection rate of the theoretic maximal number of expressed miRNAs and the corresponding sequencing depth. Adapted from Yang et al. [52].

program also includes unique features such as data statistics and multiple mapping levels to generate a comprehensive platform for the analysis of miRNA expression. Based on a statistic analysis of small RNA sequencing data, it could generate measurement of differentially expressed miRNAs in paired samples by a tabular and graphical output format.

It should be noted that most of the programs listed in Table 1 are originally designed for analyzing miRNAs in animals. Several considerations prevent their direct application to profile miRNAs in plants. First, many available methods highly depend on sequence similarity search to detect known and new miRNAs, which is not sufficient to uncover species-specific miRNAs. In fact, only a minority of annotated miRNAs in plants is conserved between different lineages, suggesting that most unknown miRNAs would not be discovered through sequence similarity search [14]. Second, for those programs that do consider other features their models are usually based on the animal systems. However, it is well studied that some aspects of miRNA biogenesis in plants and animals are critically different. For instance, pre-miRNAs in animals possess a rather uniform length at ~80 nucleotides, which is a key to the success of miRDeep, miRCat, and miRNAKey [41,42,48]. In plants, the precursor length is longer and more variable. Thus, it is not feasible to simply employ tools developed for animals to detect plant miRNAs. Third, considering the easy-to-use feature, most tools in Table 1 are available as a web-based version, which could easily handle small size sequencing data. However, at present it is standard to generate tens of millions of reads per sample, resulting in increased difficulty of processing the large amount of data on line and reinforcing the desirability of developing plant-specific tools.

10.5 miRDEEP-P IS A PROGRAM FOR COMPREHENSIVE IDENTIFICATION OF MiRNAS IN PLANTS

miRDeep-P was modified from miRDeep [41] to specifically retrieve and quantify miRNA related information from deep sequencing data in plants [52,53]. Similar to miRDeep, this program maps the small RNA reads to a reference genome and extracts the sequence flanking each anchored read for predicting RNA secondary structure and quantifying the compatibility of the distribution of reads with Dicer-mediated processing. After progressively processing all mapped reads, candidate miRNAs as well pre-miRNAs are scored based on the core miRDeep algorithm [41] and filtered with plant-specific criteria [52–54]. It thus provides reliable information on the transcription and processing of the pre-miRNAs (Figure 3A). Using training data from both *Arabidopsis* and rice, it was demonstrated that miRDeep-P works effectively for deep sequencing data in plants [52]. miRDeep-P is freely available as a stand-alone package that runs in a command line environment [53].

miRDeep-P was tested utilizing annotated miRNAs in *Arabidopsis* and available deep sequencing data from three independent small RNA libraries prepared from shoot, root and inflorescence [52]. By retrieving the signature small RNA distribution from each of the 199 annotated pre-miRNAs (miRBase release 15), the tissue-specific expression pattern of individual miRNAs was determined. In shoot, root and flower, 81, 70 and 55 expressed pre-miRNAs were detected, respectively, indicating that only 40% of the annotated pre-miRNAs are expressed in major organ types. Northern blotting was performed and the results were clearly consistent with expression determined from the deep sequencing data for miRNAs of single member families. For the multiple member families, the cumulated expression levels combining individual gene-level expression also showed strong agreement with that from the Northern analysis (Figure 3B).

Gene specific expression patterns generated by miRDeep-P revealed the transcriptional relationship of paralogous members. For example, the MIR169 family has 14 members in *Arabidopsis*. According to the neighbor-joining tree constructed from pre-miRNA sequences, this family could be grouped into three major clusters (Figure 3C). The smallest clade only consisting of MIR169a was not expressed according to miRDeep-P analysis. By contrast, the clade consisting of MIR169i/j/k/l/m/n was expressed simultaneously but only in root and shoot. The only exception for this clade was MIR169i, which was not expressed. Meanwhile, the MIR169b/c/d/e/g/f/h branch was detected in flower as well (Figure 3C). These results indicate that the tissue specific expression determined from deep sequencing data by miRDeep-P is consistent with the phylogenetic relationship of paralogous MIR genes.

Reliable estimation of gene level miRNA expression makes it possible to determine the relationship between miRNA detection rate and the sequencing depth. A simulation approach was taken in which sequence reads in the shoot library were randomly selected to simulate six different sequencing depths. These subsets were processed using miRDeep-P and the number of expressed miRNAs at each sequencing depth calculated. After reiterating this process for three times, the mathematical relationship between the number of detected miRNAs and the number of unique sequence reads was determined by curve fitting based on a logistic function (Figure 3D). This analysis indicates that, when sampling of the RNA population is unbiased, there would be a finite number of reads needed to reach a saturated detection rate of expressed miRNAs. Further, the maximal number of expressed miRNAs in shoot was estimated to be 94. Accordingly, 1.13 million unique, perfectly mapped sequence reads were required to detect 95% of the expressed miRNAs in this simulation (Figure 3D).

10.6 PLANT miRNA DATABASES WITH INTEGRATED DEEP SEQUENCING DATA

Increasing accumulation and mining of deep sequencing data have resulted in the development of comprehensive databases to facilitate miRNA annotation in a variety of species or experimental systems. The earliest and most comprehensive miRNA database, miRBase, combines deep sequencing data with miRNA annotation in chromalveolata, metazoan, mycetozoa, viridiplantae, and viruses [55]. On the other hand, genome annotation databases for plant species such as TAIR and TFGD [56] also include miRNA loci. The online databases most relevant to plant miRNAs are summarized in Table 2. ASRP, the *Arabidopsis* Small RNA Project Database, is the first database providing a repository for sequences of miRNAs from various *Arabidopsis* genotypes and tissues [57]. PmiRKB, currently focusing on the two model plants *Arabidopsis* and rice was developed to emphasize on single nucleotide polymorphisms regarding miRNAs in these two species supported by deep sequencing data [58]. miRNEST is a newly released comprehensive miRNA database including miRNA sequences from more than 200 plants, and those annotated miRNAs from some model plants are supported by deep sequencing data from different samples [59].

Transcriptome profiling has become indispensable in biology, which now includes not only mRNA but also other regulatory RNA species and intermediates of RNA metabolism. To fully decipher the transcriptome, systems based approaches are highly desirable to integrate the expression profiles with data characterizing other functional elements of the cell. Toward this goal, databases integrating miRNA annotation and deep sequencing data represent an important step forward. Combining sequencing data from various genotypes, diverse tissues, different developmental stages, or in response to different environmental challenges, could elaborate the expression patterns of miRNAs. Further, these databases should prove useful to integrate deep sequencing data of small RNA with other types of high throughput data such as those for mRNA transcriptome and degradome. The integrated data, in conjugation with modeling and model testing, will provide important clues to the regulatory networks that ultimately elucidate genome transcription, function, and adaptation.

TABLE 2: Databases for plant miRNAs with integrated deep sequencing data.

Database	Description	URL
miRBase	Including miRNA sequences and annotations in more than 50 plants.	http://www.mirbase.org/
miRNEST	Combining miRNA sequences in more than 200 plant organisms.	http://lemur.amu.edu.pl/share/php/mirnest/home.php
PmiRKB	*Arabidopsis* and Rice miRNA knowledge base.	http://bis.zju.edu.cn/pmirkb/
ASRP	Database of *Arabidopsis* small RNA sequences.	http://asrp.cgrb.oregonstate.edu/

REFERENCES

1. Huttenhofer, A.; Schattner, P.; Polacek, N. Non-coding RNAs: Hope or hype? Trends Genet. 2005, 21, 289–297.
2. Bartel, D.P. MicroRNAs: Target recognition and regulatory functions. Cell 2009, 136, 215–233.
3. Carthew, R.W.; Sontheimer, E.J. Origins and Mechanisms of miRNAs and siRNAs. Cell 2009, 136, 642–655.
4. Voinnet, O. Origin, biogenesis, and activity of plant microRNAs. Cell 2009, 136, 669–687.

5. Lee, R.C.; Feinbaum, R.L.; Ambros, V. The C. elegans heterochronic gene lin-4 encodes small RNAs with antisense complementarity to lin-14. Cell 1993, 75, 843–854.
6. Wightman, B.; Ha, I.; Ruvkun, G. Posttranscriptional regulation of the heterochronic gene lin-14 by lin-4 mediates temporal pattern formation in C. elegans. Cell 1993, 75, 855–862.
7. Alves, L., Jr.; Niemeier, S.; Hauenschild, A.; Rehmsmeier, M.; Merkle, T. Comprehensive prediction of novel microRNA targets in *Arabidopsis thaliana*. Nucleic Acids Res. 2009, 37, 4010–4021.
8. Fahlgren, N.; Howell, M.D.; Kasschau, K.D.; Chapman, E.J.; Sullivan, C.M.; Cumbie, J.S.; Givan, S.A.; Law, T.F.; Grant, S.R.; Dangl, J.L.; et al. High-throughput sequencing of *Arabidopsis* microRNAs: Evidence for frequent birth and death of MIRNA genes. PLoS One 2007, 2, e219.
9. Jones-Rhoades, M.W.; Bartel, D.P.; Bartel, B. MicroRNAS and their regulatory roles in plants. Annu. Rev. Plant. Biol. 2006, 57, 19–53.
10. Garcia, D. A miRacle in plant development: Role of microRNAs in cell differentiation and patterning. Semin. Cell Dev. Biol. 2008, 19, 586–595.
11. Kozomara, A.; Griffiths-Jones, S. miRBase: Integrating microRNA annotation and deep-sequencing data. Nucleic Acids Res. 2011, 39, D152–D157.
12. Ouyang, S.; Zhu, W.; Hamilton, J.; Lin, H.; Campbell, M.; Childs, K.; Thibaud-Nissen, F.; Malek, R.L.; Lee, Y.; Zheng, L.; et al. The TIGR rice genome annotation resource: Improvements and new features. Nucleic Acids Res. 2007, 35, D883–D887.
13. Goodstein, D.M.; Shu, S.; Howson, R.; Neupane, R.; Hayes, R.D.; Fazo, J.; Mitros, T.; Dirks, W.; Hellsten, U.; Putnam, N.; et al. Phytozome: A comparative platform for green plant genomics. Nucleic Acids Res. 2012, 40, D1178–D1186.
14. Cuperus, J.T.; Fahlgren, N.; Carrington, J.C. Evolution and functional diversification of MIRNA genes. Plant Cell 2011, 23, 431–442.
15. Nobuta, K.; Venu, R.C.; Lu, C.; Belo, A.; Vemaraju, K.; Kulkarni, K.; Wang, W.; Pillay, M.; Green, P.J.; Wang, G.L.; et al. An expression atlas of rice mRNAs and small RNAs. Nat. Biotechnol. 2007, 25, 473–477.
16. Molnar, A.; Schwach, F.; Studholme, D.J.; Thuenemann, E.C.; Baulcombe, D.C. miRNAs control gene expression in the single-cell alga Chlamydomonas reinhardtii. Nature 2007, 447, 1126–1129.
17. Rajagopalan, R.; Vaucheret, H.; Trejo, J.; Bartel, D.P. A diverse and evolutionarily fluid set of microRNAs in *Arabidopsis thaliana*. Genes Dev. 2006, 20, 3407–3425.
18. Zhu, Q.H.; Spriggs, A.; Matthew, L.; Fan, L.; Kennedy, G.; Gubler, F.; Helliwell, C. A diverse set of microRNAs and microRNA-like small RNAs in developing rice grains. Genome Res. 2008, 18, 1456–1465.
19. Lu, S.; Yang, C.; Chiang, V.L. Conservation and diversity of microRNA-associated copper-regulatory networks in Populus trichocarpa. J. Integr. Plant Biol. 2011, 53, 879–891.
20. Ng, D.W.; Zhang, C.; Miller, M.; Palmer, G.; Whiteley, M.; Tholl, D.; Chen, Z.J. cis- and trans-Regulation of miR163 and target genes confers natural variation of secondary metabolites in two *Arabidopsis* species and their allopolyploids. Plant Cell 2011, 23, 1729–1740.

21. Wang, Y.; Itaya, A.; Zhong, X.; Wu, Y.; Zhang, J.; van der Knaap, E.; Olmstead, R.; Qi, Y.; Ding, B. Function and evolution of a microRNA that regulates a Ca2+-ATPase and triggers the formation of phased small interfering RNAs in tomato reproductive growth. Plant Cell 2011, 23, 3185–3203.
22. Lee, Y.; Kim, M.; Han, J.; Yeom, K.H.; Lee, S.; Baek, S.H.; Kim, V.N. MicroRNA genes are transcribed by RNA polymerase II. EMBO J. 2004, 23, 4051–4060.
23. Tam, W. Identification and characterization of human BIC, a gene on chromosome 21 that encodes a noncoding RNA. Gene 2001, 274, 157–167.
24. Laubinger, S.; Sachsenberg, T.; Zeller, G.; Busch, W.; Lohmann, J.U.; Ratsch, G.; Weigel, D. Dual roles of the nuclear cap-binding complex and SERRATE in pre-mRNA splicing and microRNA processing in *Arabidopsis thaliana.* Proc. Natl. Acad. Sci. USA 2008, 105, 8795–8800.
25. Yu, B.; Bi, L.; Zheng, B.; Ji, L.; Chevalier, D.; Agarwal, M.; Ramachandran, V.; Li, W.; Lagrange, T.; Walker, J.C.; et al. The FHA domain proteins DAWDLE in *Arabidopsis* and SNIP1 in humans act in small RNA biogenesis. Proc. Natl. Acad. Sci. USA 2008, 105, 10073–10078.
26. Ren, G.; Xie, M.; Dou, Y.; Zhang, S.; Zhang, C.; Yu, B. Regulation of miRNA abundance by RNA binding protein TOUGH in *Arabidopsis*. Proc. Natl. Acad. Sci. USA 2012, 109, 12817–12821.
27. Bartel, D.P. MicroRNAs: Genomics, biogenesis, mechanism, and function. Cell 2004, 116, 281–297.
28. Chen, X. MicroRNA biogenesis and function in plants. FEBS Lett. 2005, 579, 5923–5931.
29. Papp, I.; Mette, M.F.; Aufsatz, W.; Daxinger, L.; Schauer, S.E.; Ray, A.; van der Winden, J.; Matzke, M.; Matzke, A.J. Evidence for nuclear processing of plant micro RNA and short interfering RNA precursors. Plant Physiol. 2003, 132, 1382–1390.
30. Khvorova, A.; Reynolds, A.; Jayasena, S.D. Functional siRNAs and miRNAs exhibit strand bias. Cell 2003, 115, 209–216.
31. Schwarz, D.S.; Hutvagner, G.; Du, T.; Xu, Z.; Aronin, N.; Zamore, P.D. Asymmetry in the assembly of the RNAi enzyme complex. Cell 2003, 115, 199–208.
32. Motomura, K.; Le, Q.T.; Kumakura, N.; Fukaya, T.; Takeda, A.; Watanabe, Y. The role of decapping proteins in the miRNA accumulation in *Arabidopsis thaliana.* RNA Biol. 2012, 9, 644–652.
33. Llave, C.; Xie, Z.; Kasschau, K.D.; Carrington, J.C. Cleavage of scarecrow-like mRNA targets directed by a class of *Arabidopsis* miRNA. Science 2002, 297, 2053–2056.
34. Reinhart, B.J.; Weinstein, E.G.; Rhoades, M.W.; Bartel, B.; Bartel, D.P. MicroRNAs in plants. Genes Dev. 2002, 16, 1616–1626.
35. Brodersen, P.; Sakvarelidze-Achard, L.; Bruun-Rasmussen, M.; Dunoyer, P.; Yamamoto, Y.Y.; Sieburth, L.; Voinnet, O. Widespread translational inhibition by plant miRNAs and siRNAs. Science 2008, 320, 1185–1190.
36. Wu, L.; Zhou, H.; Zhang, Q.; Zhang, J.; Ni, F.; Liu, C.; Qi, Y. DNA methylation mediated by a microRNA pathway. Mol. Cell 2010, 38, 465–475.
37. Lu, J.; Getz, G.; Miska, E.A.; Alvarez-Saavedra, E.; Lamb, J.; Peck, D.; Sweet-Cordero, A.; Ebert, B.L.; Mak, R.H.; Ferrando, A.A.; et al. MicroRNA expression profiles classify human cancers. Nature 2005, 435, 834–838.

38. Moxon, S.; Jing, R.; Szittya, G.; Schwach, F.; Rusholme Pilcher, R.L.; Moulton, V.; Dalmay, T. Deep sequencing of tomato short RNAs identifies microRNAs targeting genes involved in fruit ripening. Genome Res. 2008, 18, 1602–1609.
39. Sunkar, R.; Zhou, X.; Zheng, Y.; Zhang, W.; Zhu, J.K. Identification of novel and candidate miRNAs in rice by high throughput sequencing. BMC Plant Biol. 2008, 8, 25.
40. Creighton, C.J.; Reid, J.G.; Gunaratne, P.H. Expression profiling of microRNAs by deep sequencing. Brief Bioinform. 2009, 10, 490–497.
41. Friedlander, M.R.; Chen, W.; Adamidi, C.; Maaskola, J.; Einspanier, R.; Knespel, S.; Rajewsky, N. Discovering microRNAs from deep sequencing data using miRDeep. Nat. Biotechnol. 2008, 26, 407–415.
42. Ronen, R.; Gan, I.; Modai, S.; Sukacheov, A.; Dror, G.; Halperin, E.; Shomron, N. miRNAkey: A software for microRNA deep sequencing analysis. Bioinformatics 2010, 26, 2615–2616.
43. Pantano, L.; Estivill, X.; Marti, E. SeqBuster, a bioinformatic tool for the processing and analysis of small RNAs datasets, reveals ubiquitous miRNA modifications in human embryonic cells. Nucleic Acids Res. 2010, 38, e34.
44. Huang, P.J.; Liu, Y.C.; Lee, C.C.; Lin, W.C.; Gan, R.R.; Lyu, P.C.; Tang, P. DSAP: Deep-sequencing small RNA analysis pipeline. Nucleic Acids Res. 2010, 38, W385–W391.
45. Zhu, E.; Zhao, F.; Xu, G.; Hou, H.; Zhou, L.; Li, X.; Sun, Z.; Wu, J. mirTools: microRNA profiling and discovery based on high-throughput sequencing. Nucleic Acids Res. 2010, 38, W392–W397.
46. Hackenberg, M.; Sturm, M.; Langenberger, D.; Falcon-Perez, J.M.; Aransay, A.M. miRanalyzer: A microRNA detection and analysis tool for next-generation sequencing experiments. Nucleic Acids Res. 2009, 37, W68–W76.
47. Hackenberg, M.; Rodriguez-Ezpeleta, N.; Aransay, A.M. miRanalyzer: An update on the detection and analysis of microRNAs in high-throughput sequencing experiments. Nucleic Acids Res. 2011, 39, W132–W138.
48. Moxon, S.; Schwach, F.; Dalmay, T.; Maclean, D.; Studholme, D.J.; Moulton, V. A toolkit for analysing large-scale plant small RNA datasets. Bioinformatics 2008, 24, 2252–2253.
49. Prufer, K.; Stenzel, U.; Dannemann, M.; Green, R.E.; Lachmann, M.; Kelso, J. PatMaN: Rapid alignment of short sequences to large databases. Bioinformatics 2008, 24, 1530–1531.
50. Denman, R.B. Using RNAFOLD to predict the activity of small catalytic RNAs. Biotechniques 1993, 15, 1090–1095.
51. Bonnet, E.; Wuyts, J.; Rouze, P.; van de Peer, Y. Evidence that microRNA precursors, unlike other non-coding RNAs, have lower folding free energies than random sequences. Bioinformatics 2004, 20, 2911–2917.
52. Yang, X.; Zhang, H.; Li, L. Global analysis of gene-level microRNA expression in *Arabidopsis* using deep sequencing data. Genomics 2011, 98, 40–46.
53. Yang, X.; Li, L. miRDeep-P: A computational tool for analyzing the microRNA transcriptome in plants. Bioinformatics 2011, 27, 2614–2615.

54. Meyers, B.C.; Axtell, M.J.; Bartel, B.; Bartel, D.P.; Baulcombe, D.; Bowman, J.L.; Cao, X.; Carrington, J.C.; Chen, X.; Green, P.J.; et al. Criteria for annotation of plant MicroRNAs. Plant Cell 2008, 20, 3186–3190.
55. Griffiths-Jones, S.; Grocock, R.J.; van Dongen, S.; Bateman, A.; Enright, A.J. miRBase: microRNA sequences, targets and gene nomenclature. Nucleic Acids Res. 2006, 34, D140–D144.
56. Fei, Z.; Joung, J.G.; Tang, X.; Zheng, Y.; Huang, M.; Lee, J.M.; McQuinn, R.; Tieman, D.M.; Alba, R.; Klee, H.J.; et al. Tomato functional genomics database: A comprehensive resource and analysis package for tomato functional genomics. Nucleic Acids Res. 2011, 39, D1156–D1163.
57. Gustafson, A.M.; Allen, E.; Givan, S.; Smith, D.; Carrington, J.C.; Kasschau, K.D. ASRP: The *Arabidopsis* Small RNA Project database. Nucleic Acids Res. 2005, 33, D637–D640.
58. Meng, Y.; Gou, L.; Chen, D.; Mao, C.; Jin, Y.; Wu, P.; Chen, M. PmiRKB: A plant microRNA knowledge base. Nucleic Acids Res. 2011, 39, D181–D187.
59. Szczesniak, M.W.; Deorowicz, S.; Gapski, J.; Kaczynski, L.; Makalowska, I. miRNEST database: An integrative approach in microRNA search and annotation. Nucleic Acids Res 2012, 40, D198–D204.

CHAPTER 11

MAIZE (*Zea Mays* L.) GENOME DIVERSITY AS REVEALED BY RNA-SEQUENCING

CANDICE N. HANSEY, BRIEANNE VAILLANCOURT, RAJANDEEP S. SEKHON, NATALIA DE LEON, SHAWN M. KAEPPLER, and C. ROBIN BUELL

11.1 INTRODUCTION

Maize is genetically and phenotypically diverse. In populations containing only a small proportion of the variation present in the entirety of maize, progress based on phenotypic selection is being realized. This is exemplified by consistent gains in hybrid yields [1], [2] and through progress in long-term selection experiments, such as the Illinois long-term selection study for grain oil and protein content [3], selection for prolificacy [4], and selection for seed size [5], [6]. Phenotypic diversity in maize for yield, composition, and morphological traits has also been documented in multiple diversity panels [7], [8], [9], and in analysis of structured populations such as the nested association mapping (NAM) population, which represents diverse maize types [10], [11], [12], [13], [14]. Understanding the genetic factors that underlie this extensive phenotypic diversity and allow for continual improvement in populations is essential for efficient

*This chapter was originally published under the Creative Commons Attribution License. Hansey CN, Vaillancourt B, Sekhon RS, de Leon N, Kaeppler SM, and Buell CR. Maize (*Zea mays L.*) Genome Diversity as Revealed by RNA-Sequencing.* PLoS ONE *7,3 (2012), 10 pages. doi:10.1371/journal.pone.0033071.*

manipulation of maize to meet the demands of the increasing human population and the need to adapt to global climate changes.

A wide range of sequence level variation exists in maize including single nucleotide polymorphisms (SNPs), small insertions/deletions, presence/absence variation (PAV), and copy number variation (CNV) [15], [16], [17], [18], [19]. Furthermore, epigenetic mechanisms also generate variation via differences in transcript abundance [16], [18], [19], [20], [21]. With the recent release of the maize B73 reference genome [17] and the availability of high-output, affordable sequencing technologies, a greater understanding of this diversity, particularly structural variation, is being realized.

The concept of separate core and dispensable portions of genomes was first described in prokaryotes. The core genome is defined as the portion present in all sequenced strains and the dispensable genome is defined as sequence that is present in one or more but not all strains [22], [23]. This phenomenon has been subsequently observed in many bacterial species [23], [24], [25]. In addition to understanding environmental adaptation, characterizing microbial pan-genomes has led to the development of universal vaccine cocktails using genes from the dispensable genome [26], [27], the elucidation of virulence factors [28], and improved pathogen diagnostics [29]. The presence of a pan-genome has also been demonstrated in eukaroyote species, including human (*Homo sapiens*) [30], maize (*Zea mays* L.) [16], [18], [19], [31], [32], and *Arabidopsis thaliana* [33], [34], [35]; however, the size and effect of the dispensable genome in eukaryotes is not well understood.

The goal of this study was to explore sequence and expression-based variation, and to identify novel transcripts in a common set of diverse maize lines using RNA based sequencing (RNA-seq). RNA-seq permits all of these types of variation to be evaluated simultaneously in a cost effective manner. It also has a major advantage in large complex genomes of reducing the effective genome size substantially. For example, the 2.3 Gb haploid maize genome is reduced to a 97 Mb transcriptome assuming all genes are expressed. In this study, we generated 17.8 Gb of seedling RNA-seq data to explore a set of 21 diverse maize lines that are representative of North American as well as exotic germplasm to improve our understanding of the maize genome and the genetic relationships between lines with diverse pedigrees.

TABLE 1: Read mapping, expression, and single nucleotide polymorphism (SNP) summary for 21 diverse maize lines.

Inbred Line	Group	Purity Filtered Reads	% Reads Mapped FPKM	% Reads Mapped SNP	% Reads Mapped with Assembled Transcripts	Max FPKM	% Genes Expressed	% SNPs with Coverage	% Genes with SNP Coverage
B14A	SSS	11.1M	77.85%	57.79%	78.34%	37,727	61.48%	64.8%	50.0%
B37	SSS	20.SM	82.65%	61.07%	83.19%	36,901	64.73%	81.3%	53.7%
B73	SSS	19.4M	87.81%	64.17%	88.31%	37,978	64.51%	83.9%	53.1%
B97	NSS	19.7M	82.00%	60.07%	82.57%	41,010	64.17%	79.8%	53.3%
CML103	Exotic	10.0M	80.47%	58.01%	81.05%	40,148	61.02%	63.9%	49.4%
CML322	Exotic	20.7M	82.10%	59.44%	82.64%	40,302	64.69%	79.5%	49.4%
CML333	Exotic	8.3M	77.80%	55.30%	78.34%	49,670	60.23%	49.8%	46.3%
H99	NA	19.7M	80.55%	59.74%	81.10%	32,713	65.05%	76.2%	53.3%
M37W	NA	17.4M	83.10%	59.76%	83.71%	50,022	63.49%	75.7%	52.2%
Mo17	NSS	9.3M	74.57%	56.35%	75.10%	27,418	61.35%	60.7%	49.6%
MoG	NA	14.8M	77.32%	57.31%	77.87%	40,860	62.91%	69.8%	51.6%
MS71	NSS	6.2M	78.07%	55.88%	78.69%	26,667	60.11%	45.9%	45.7%
NC350	Exotic	7.2M	81.50%	60.99%	82.05%	34,546	57.08%	61.1%	48.9%
NC358	Exotic	17.9M	78.61%	58.44%	79.14%	45,634	63.84%	75.0%	52.1%

TABLE 1: *Cont.*

Inbred Line	Group	Purity Filtered Reads	% Reads Mapped FPKM	% Reads Mapped SNP	% Reads Mapped with Assembled Transcripts	Max FPKM	% Genes Expressed	% SNPs with Coverage	% Genes with SNP Coverage
Oh43	NSS	39.7M	82.84%	57.89%	83.48%	48,470	64.39%	86.0%	54.4%
Oh7B	NA	13.5M	83.38%	60.60%	83.94%	32,622	64.28%	73.1%	52.2%
PHG47	NSS	11.SM	75.77%	56.40%	76.30%	35,519	62.28%	64.3%	50.4%
PHNIl	lodent	10.7M	80.47%	57.42%	81.10%	34,790	61.76%	66.1%	50.2%
PHW65	NSS	19.IM	83.42%	59.80%	83.99%	53,606	64.44%	78.6%	53.1%
W605S	NSS	6.4M	78.48%	55.61%	79.12%	28,518	59.12%	45.2%	45.0%
W64A	NSS	29.SM	80.67%	59.94%	81.24%	39,564	65.96%	86.3%	54.6%

Reads were mapped requiring a unique hit for the SNPs and multiple hits for fragments per kilobase of exon model per million fragments mapped (FPKM) and mapping to the pseudomolecules plus the Velvet and Oases [45] assembled transcripts. FPKM values were calculated using Cufflinks [56]. Genes with a FPKM 95% confidence interva l lower boundary greater than zero were considered expressed. For each inbred line, a gene was considered to have SNP coverage if there was at least one polymorphic locus with coverage in the gene. SSS= Stiff Stalk Synthetic, NSS =Non-Stiff Stalk Synthetic, NA= Not applicable.

11.2 RESULTS AND DISCUSSION

11.2.1 DESCRIPTION OF GERMPLASM AND DATASETS

The inbred lines used in this study were selected to provide a representative sampling of genetic and phenotypic variation in maize [7], [8], [15], [16], [17], [18], [19] by inclusion of diverse exotic germplasm, and germplasm from two major heterotic groups involved in United States grain hybrids, Stiff Stalk Synthetic (SSS) and Non-Stiff Stalk Synthetic (NSS) [36]. We generated between 6.2 and 39.7 million single end (36–74 bp) RNA-seq reads for each of the 21 inbred lines (Table 1). This dataset allows for in-depth evaluation of sequence and expression variation, as well as the identification of novel expressed maize sequences in a common set of 21 maize lines.

Using the maize B73 5b pseudomolecules (http://ftp.maizesequence.org/), between 74.57% and 87.81% of the reads could be mapped to the reference sequence with B73 having the highest percentage of reads mapped (Table 1). The range in percent of sequences aligning to the reference sequence is reflective of the diversity of the germplasm explored in this study relative to the reference sequence. While there is a broad range in percentage of mapped reads across the 20 non-B73 lines (Table 1), for all genotypes there is still a relatively high and substantial percentage of reads mapping. In addition, the consistent proportions of reads that mapped uniquely and to multiple locations, suggests little bias attributable to use of the maize B73 reference sequence in downstream identification of sequence, expression, and novel expressed sequence variants among the other 20 lines.

11.2.2 SINGLE NUCLEOTIDE POLYMORPHISM VARIATION

There are limitations to studying genetic diversity using RNA-seq rather than genomic sequencing, namely that a gene/allele must be expressed in the tissue being sequenced in order to detect variants. Because the germplasm in this study is nearly homozygous, we do not expect multiple alleles at any given locus within an inbred line. Using 511 SNP makers from an Illumina Golden Gate SNP assay, the range in observed heterozygosity across these 21 lines ranged from 0.0% to 0.4% [8]. Thus, bias due to allele specific expression is not a concern. In addition, a previous study, which included 60 maize tissue types, showed that seedling tissue had the highest number of expressed genes [37]. Thus, use of seedling RNA-seq will allow for identification of a large portion of the total SNPs in the maize transcriptome.

A major advantage of RNA-seq in SNP calling relative to whole genome sequencing is the reduction in the effective size of the genome. This reduction in effective genome size can also be accomplished with exome capture and subsequent sequencing [38]. Exome capture is advantageous in allowing for the detection of unexpressed alleles in heterozygous species as well as SNP detection in unexpressed genes. However, exome capture does not allow for simultaneous quantification of transcript abundance, another major advantage of RNA-seq. Using RNA-seq we were able to identify 351,710 high confidence SNP polymorphisms between this set of 21 diverse inbred lines, of which, 329,027 SNPs were within annotated gene models in the version 5b annotation (http://ftp.maizesequence.org/). Congruent SNP positions with sequence coverage in at least 15 of the inbred lines were abundant within our dataset (197,720 SNPs in 17,149 genes). Utilization of RNA-seq requires much less sequence depth to identify the majority of the variants in medium to highly expressed transcripts relative to whole genome sequencing. Without the reduction in effective genome size afforded by RNA-seq, significantly more reads or imputation following skim sequencing of the whole genome to permit consistent comparison of congruent positions across the genotypes would be required [15], [39].

To assess the error rate associated with the RNA-seq based SNP calls, genotype scores were compared to an independent SNP data set from this same set of 21 inbred lines obtained by Illumina sequencing of genomic DNA using a modified version of the genotyping-by-sequencing (GBS) protocol [39]. The B73 sequences from the RNA-seq and GBS datasets were first compared to the B73 5b reference pseudomolecules. From the

RNA-seq data, 32,341,048 nucleotide positions were compared to the 2.3 Gb reference sequence, with 99.86% of the loci concordant to the reference sequence. Similarly, for the GBS data, of the 47,727,306 nucleotides compared, 99.92% matched the reference sequence. This error rate may be slightly inflated due to errors in the B73 reference sequence; however, for the 6,202,079 positions in common between the RNA-seq and GBS datasets, only 147 were different from the reference sequence and concordant between the two sequencing methods. This suggests a low error rate within the reference B73 genome sequence. To further test the quality of these datasets, data from both sequencing methods were compared across all 21 lines. Between the RNA-seq and GBS methods, there were 147,857 data points (nucleotide position-by-inbred line) in common representing 16,915 positions in the reference sequence. Of these data points, 97.25% (143,796 SNPs) were concordant between the two methods, with only 40 of the 16,915 positions having more than two inbred lines with contradicting nucleotides. Both of these comparisons indicate a very low error rate in the SNP calling methods used in this study and provide strong support for the quality of this data as well as the quality of the 5b maize reference pseudomolecule sequences.

The 350,710 SNPs identified in this study were distributed throughout the genome, with the number of SNPs per 1 Mb window coincidental with gene density and number of expressed genes per window. There were, however, windows with relatively high or low SNP density compared to the number of genes and expressed genes, such as on the long arm of chromosome 2. On a single gene basis, RNA-seq-derived SNPs ranged from zero SNPs to a maximum of 170 SNPs per gene, with 22,831 genes having at least one SNP (Figure 1A).

After normalizing the number of SNPs per gene for length of the gene, less than 0.5 SNPs per 100 bp were identified for the majority of the genes expressed in seedling tissue (Figure 1B). Based on a study that analyzed more than 32 Gb of sequence in the low copy region of the genome across 27 diverse lines, maize haplotypes are comprised of over 3 million SNPs and insertions/deletions with a polymorphism on average every 44 bp [15]. This discrepancy in observed SNP density is likely due to the inclusion of non-coding sequence in the genomic sequencing, which is less conserved than genic regions, consistent with observations from whole-genome sequencing of rice inbred lines [40]. Genomic sequencing of six elite Chinese inbred lines identified 468,966 SNPs in the 97 Mb gene space, averaging to approximately one SNP every 207 bp [16]. This is similar to the density observed in this study and consistent with the observation of decreased diversity in the coding sequence. There were, however, a subset of the genes with high SNP density after normalizing for gene length, with 384 genes containing 2–3 SNPs per 100 bp and 66 genes with greater than 3 SNPs per 100 bp (Figure 1B).

Reads were mapped against the 5b pseudomolecules (http://ftp.maizesequence.org/) with Bowtie version 0.12.7 [50] and TopHat version 1.2.0 [51] requiring a unique hit for the SNP mapping. Gene assignment was determined based on the 5b annotation (http://ftp.maizesequence.org/), and not all SNPs identified were assigned to a gene model. (A) Distribution of the number of SNPs per gene. (B) Distribution of the average number of SNPs per 100 bp window per gene.

Cluster analysis using allele frequency based distances from the 350,710 SNPs revealed the expected grouping of SSS and NSS type germplasm, and of exotic lines (Figure 2). Studies using 511 random SNP markers [8] and 100 simple sequence repeat markers [41] in larger diversity panels have also demonstrated that pedigree based groups cluster together using genomic markers. Higher marker density in this study markedly improved the clustering, although fewer lines were included.

11.2.3 TRANSCRIPTOME PROFILES

In addition to the high allelic variation among these diverse lines, the transcriptome profiles were variable between the 21 inbred lines. The total number of expressed genes ranged from 22,522 to 26,026 (57.08% to 65.96%; Table 1), but the maximum observed expression value (fragments per kilobase of exon model per million fragments mapped, FPKM) varied substantially from 26,667 to 53,606 in M37W and PHW65, respectively (Table 1). In every inbred line except NC358, the gene with the maximum FPKM value encoded Chlorophyll a-b binding protein, as expected in photosynthetic seedling tissue.

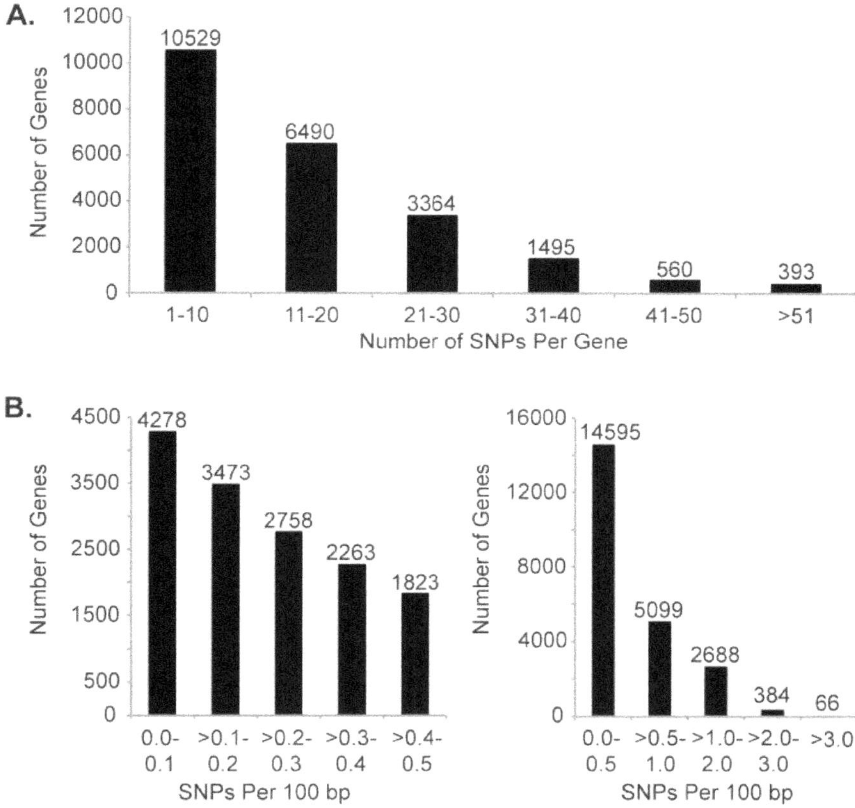

FIGURE 1: Distribution of the number of single nucleotide polymorphisms (SNPs) and SNP density per gene.

Transcriptional variation on both a quantitative and qualitative level has been implicated in important agricultural traits [42]. Using a quantitative approach, we binned the genes as present or absent in the core maize seedling transcriptome. A gene was considered transcribed and included in the core transcriptome of a given inbred line if the FPKM 95% confidence interval lower boundary was greater than zero. Using this classification, 19,225 (48.7%) genes were expressed in all 21 inbred lines and 11,011 (27.9%) genes were expressed in one to 20 of the inbred lines in a fairly

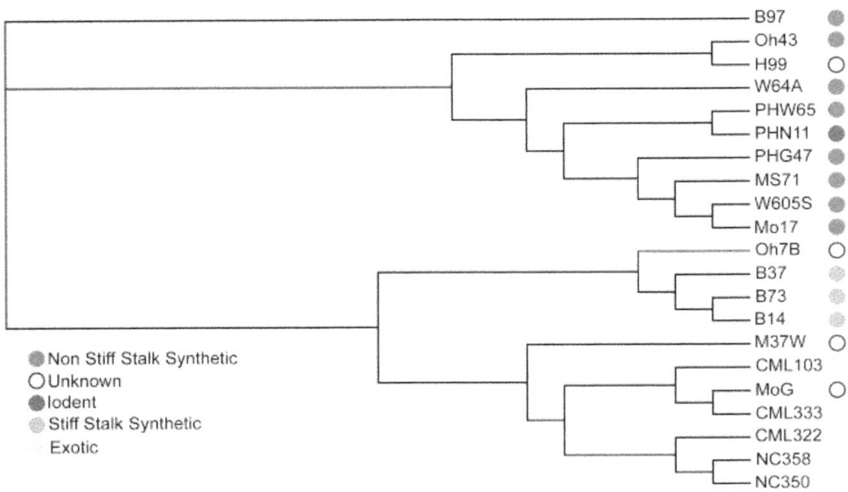

FIGURE 2: Neighbor-Joining tree of 21 diverse maize lines based on 351,710 single nucleotide polymorphisms (SNPs). Frequency based distances were calculated as pair-wise Rogers distances [53]. PowerMarker version 3.25 [54] was used to construct the tree.

symmetric distribution. The 11,011 genes, which were expressed in only a subset of the lines, comprise the seedling variable transcriptome and may contribute to the phenotypic diversity observed in maize. However, sampling limitations and the range of sequence depth in this study are limitations in consistently detecting lowly expressed genes, and consequently, very lowly expressed genes may not be detected as expressed in a given genotype.

Transcriptome profile variation can extend beyond PAV for each transcript. Using a semi-quantitative approach where inbred lines were categorized as having no, low, medium, or high expression for each gene, we also observed variation in transcript abundance between the lines (Figure 3). In this classification approach, a gene could have constitutive expression across all 21 lines within any one of the four categories. Alternatively, a gene could have variable expression, with inbred lines categorized into multiple expression level categories. Using this method, the no, medium, and high expression categories had a similar distribution to that observed in the expressed/not expressed based analysis described above (Figure 3; Figure S3), where a large number of genes (24,378) had constitutive expression across all 21 inbred lines and the remainder of genes were variable in their expression across the lines. For the low expression category, the majority of the genes had only 5 or fewer lines with low expression, and the other lines were predominantly no or medium expression. A mere 12 genes had low expression in all 21 lines. It is possible that lowly expressed genes are less frequently expressed across all 21 genotypes, or that there is erroneous transcription in a small number of lines. However, this altered distribution is most likely a technical limitation attributable to sampling limitations. While there are a large number of genes with constitutive expression, there are also many genes with variable transcript abundance, both quantitatively and qualitatively that may contribute to observed phenotypic diversity.

11.2.3 NOVEL TRANSCRIPT DISCOVERY AND PRESENCE/ ABSENCE VARIATION

Presence/absence variation on a genome level may be a contributing factor to the phenotypic diversity and heterosis observed in maize [16], [18], [19], [43], [44]. Whole genome array-based comparative genome hybridization experiments across 33 maize and teosinte lines have identified nearly four thousand genes exhibiting either CNV or PAV [18], [19]. Using 83.7 Gb of sequence across six elite inbreds related by pedigree and a combination of whole genome shotgun sequencing, mapping onto the B73 reference genome, and de novo assembly approaches, 296 genes in the B73 filtered gene set were identified as being absent in at least one of the non-B73 lines and at least 157 novel non-B73 genes were identified [16].

We used de novo assembly of the unmapped RNA-seq reads from all 21 lines to identify additional novel expressed genes that are either not present in the B73 inbred line or are missing from the B73 5b pseudomolecules altogether. Assembly of the unmapped RNA-seq reads using Velvet/Oases [45] with multiple k-mers indicated that 23 was the optimal

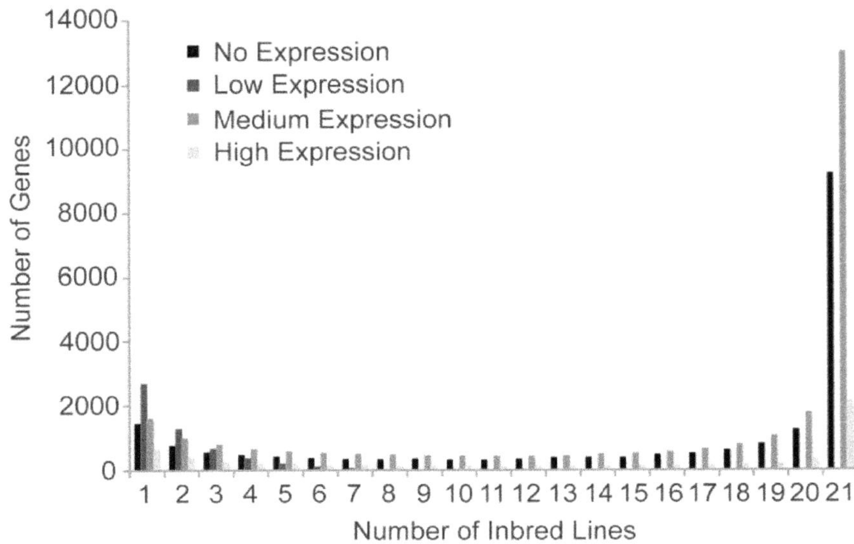

FIGURE 3: Distribution of genes in the maize seedling core and dispensable transcriptomes determined using a semi-qualitative approach. Reads were mapped to the 5b pseudomolecules (http://ftp.maizesequence.org/) using Bowtie version 0.12.7 [50] and TopHat version 1.2.0 [51], and fragments per kilobase of exon model per million fragments mapped (FPKM) were determined with Cufflinks version 0.9.3 [56] and the 5b annotation (http://ftp.maizesequence.org/).For each gene, a line was considered not expressed if the low confidence FPKM value was equal to zero, low expressed if the low confidence interval was greater than zero and the FPKM value was less than 5, medium expressed if the low confidence interval was greater than zero and the FPKM value was greater than or equal to 5 and less than or equal to 200, and high expressed if the low confidence interval was greater than zero and the FPKM value was greater than 200.

k-mer (data not shown). The combined de novo assembly used 7,919,942 (17.5%) of the total unmapped reads to generate 8,595 transcripts across 4,701 loci, with an N50 of 725 bp. The average transcript size was 752 bp, with some transcripts greater than 3,000 bp. To identify truly novel sequences and remove those that were either alleles or paralogs with similar sequences in the B73 reference genome, the representative transcript for each locus was aligned to the B73 5b pseudomolecules. Multiple percent identity and percent coverage cutoffs were used to establish the optimal parameters for determining if a transcript should be retained. For both parameters, there was a large decrease in the number of transcripts that could be mapped between 85% and 90%, therefore the percent coverage and identity cutoffs were set to 85%. Using this filtering, 1,321 high confidence novel representative transcripts were retained, significantly more than previously identified using de novo assembly of genomic sequence [16], yet similar to that observed using comparative genome hybridization [18], [19]. Of these representative transcripts, 365 did not align to the 5b pseudomolecules at any alignment criteria, and 956 had alignments below the coverage and/or identity cutoff criteria. To confirm the novel nature of these transcripts, we re-mapped all of the RNA-seq reads from all 21 lines to an amended set of B73 5b pseudomolecules in which we added a supplemental pseudomolecule constructed by concatenating our novel representative transcripts. Using the same mapping parameters as described previously, the total number of mapped reads for every inbred line increased with the inclusion of the de novo assembled transcripts (Table 1), whereas the proportion of reads that mapped to multiple locations remained relatively constant.

RT-PCR was conducted on 21 of the filtered representative transcripts across eight inbred lines to confirm the predicted size and transcript PAV across the lines. The 21 transcripts were selected to represent transcripts with computationally predicted expression in one, four, or eight of the lines. The use of transcripts within these three categories allowed for more accurate experiment-wide inferences to be made. A transcript was predicted to be present in a given line if the FPKM 95% low confidence interval was greater than zero, and absent if the 95% confidence interval was equal to zero. Seven transcripts were concordant for both the de novo assembly length and PAV predictions. Another seven were concordant with the de novo assembly length but not with the FPKM predicted PAV, and the remaining seven transcripts did not amplify under the PCR conditions used. In total, of the 14 transcripts that were amplified in the eight surveyed lines, only 14 of the 112 inbred line-transcript pairs (12.5%) did not match the computational PAV predictions. Additionally, all of the genes with predicted expression in each inbred line based on FPKM values were confirmed experimentally with RT-PCR. The lack of concordance in expression, i.e., positive experimentally with RT-PCR, but FPKM values equal to zero, are due to either insufficient sequence depth or sequence divergence resulting in the reads failing to map to the reference genome.

To further test the accuracy of the assembled transcript sequences, RT-PCR reactions with a visible band (75 reactions across the 14 novel transcripts) were Sanger sequenced from the 5′ and 3′ end of the fragment. The assembled novel transcripts are the product of a hybrid assembly across the 21 maize lines. Because of this, we do not expect the assembly to represent a single inbred line in its entirety, but to be reflective of lines with higher expression levels, as reads from those lines will be more abundant. For all 14 novel transcripts, the sequence from the 5′ and 3′ ends contained complete coverage of the predicted sequence indicating that the assembled transcripts were likely not chimeras and that there were not large exons missing in the assembly. For all but one of the inbred line-transcript pairs, the percent identity for the bases that aligned was greater than 90%, indicating conservation of these gene sequences, while still reflecting some diversity between the lines. We also evaluated the percent coverage in each of the pair wise alignments between the assembled sequences and the Sanger sequences. For 12 of the 14 novel transcripts, the coverage was greater than 90% for all of the sequenced lines. The lower coverage for the remaining two transcripts is reflective of lower sequence conservation between the lines for some portion of the gene. The high validation rate of the RT-PCR and subsequent Sanger sequencing demonstrates the robust nature of the computational pipeline used to generate the de novo assembled transcripts and to predict presence/absence status of the transcript.

The novel transcripts that did not map to the B73 reference sequence may be the result of non-maize sequence contamination. To test for the presence of contamination, the high confidence assembled transcripts were searched against UniRef100 [46]. Of the 1,321 de novo assembled transcripts, 531 could be aligned to the UniRef100 database at greater than 70% coverage and 70% identity, and the top hit for these transcripts were all to *Poaceae* sequences. An additional 638 transcripts had alignments below this cutoff; 610 of these were to *Plantae* sequences and the remaining 28 were to non-*Plantae* sequences, indicating a very low contamination frequency. Similarly, alignments to the *Oryza sativa* and *Sorghum bicolor* protein sequences revealed 379 and 409 transcripts with alignments greater than 70% coverage and 70% identity, respectively, and 621 sequences with alignments below this cutoff to *Oryza sativa* and 616 to *Sorghum bicolor* protein sequences. Interestingly, only 225 transcripts of our de novo assembled transcripts aligned to the 181,717 maize PlantGDB-assembled Unique Transcripts (PUTs) [47] at greater than 85% coverage and 85% identity, and 214 of the transcripts did not align at all to the PUTs. This indicates that there is a breadth of maize sequence that has not yet been sampled.

Across all 1,321 novel transcripts, a range in the number of lines with read support was observed (Figure 4A). Using a stringent criterion of at least one read mapping uniquely to the transcript, 42 transcripts across all 21 lines did not have support. However, reducing the stringency criterion such that reads were permitted to map to multiple locations, all of the de novo novel transcripts had read support. These transcripts with read support only when reads were allowed to map to multiple locations most likely represent paralogous genes or genes with CNV at a diverged locus, such that they cannot be aligned at greater than 85% coverage and 85% identity. A total of 564 transcripts had read support across all 21 lines and likely represent essential maize genes in the core genome that are missing from the current B73 reference assembly. There were 270 additional transcripts with read support in B73 yet missing in the current B73 reference assembly, but not present in one or more of the other lines. In total, when requiring reads to map uniquely, there were 715 transcripts with read support in one to 20 of the lines. These transcripts present in only a portion of the 21 lines are part of the dispensable transcriptome, and could be due to PAV on the genomic level or transcriptional variation due to promoter sequence variation or epiallelic variation [20], [21]. Of these 715 transcripts, 24 were specific to one of the pedigree groups, and comparison of only the SSS and NSS type germplasm, 145 were present in only one of the heterotic groups (Figure 4B). Expanding this to all genes in the 5b annotation and the assembled transcripts, 1,529 were specific to one of the pedigree groups (SSS, NSS, exotic, and other), and 2,114 were heterotic group specific in the SSS and NSS germplasm comparison (Figure 4C).

We have shown RNA-seq to be a robust, rapid, and inexpensive method to identify SNPs in genic regions in maize, an important crop species with a large, complex, repetitive genome. Furthermore, we have identified

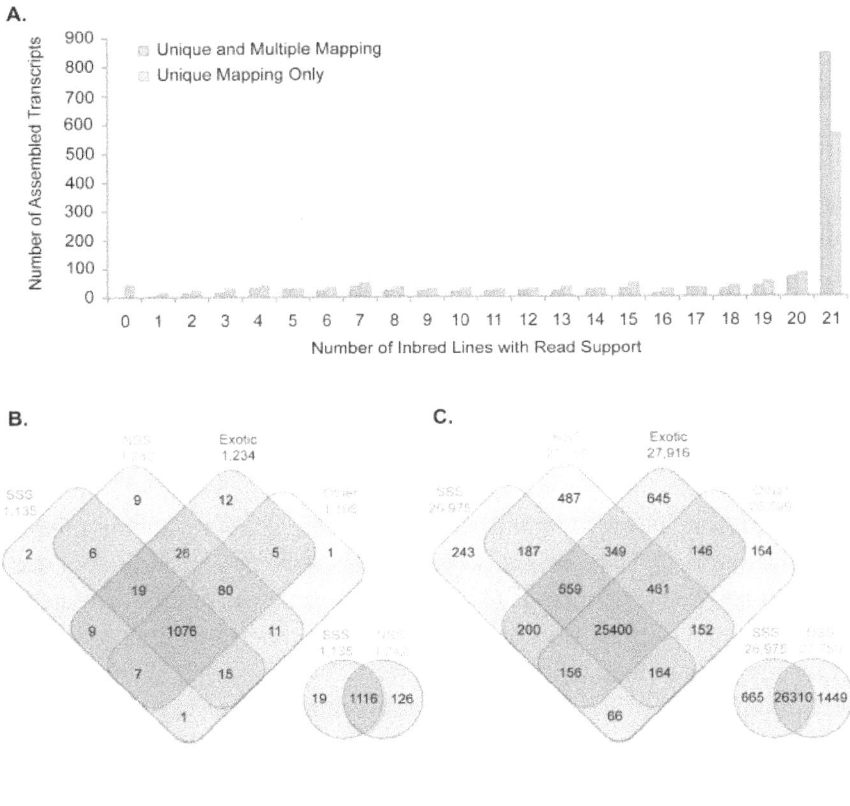

FIGURE 4: Frequency of novel de novo assembled transcripts across lines and heterotic groups. Reads were mapped to the 5b pseudomolecules plus assembled novel transcripts with Bowtie version 0.12.7 [50] and TopHat version 1.2.0 [51]. Novel transcripts from unmapped reads were assembled using Velvet version 1.0.17 and Oases version 0.1.18 [45]. (A) Distribution of the number of inbred lines with read support for each novel de novo assembled transcript requiring unique alignments and allowing for multiple mapping. (B) Venn diagram of shared and group specific novel de novo assembled transcripts. (C) Venn diagram of shared and group specific novel de novo assembled transcripts and transcripts from the 5b annotated genes (http://ftp.maizesequence.org/).

a core set of essential genes, as well as a set of genes that may be dispensable to the maize seedling transcriptome. Using de novo assembly, we have discovered transcripts previously unidentified in maize. We have also shown that the CNV and PAV observed at the genome level between lines in opposite heterotic groups extends to the transcriptome for the annotated maize genes as well as novel transcripts identified through assembly of RNA-seq within this set of inbred lines. In addition to these utilities, RNA-seq can also be used to analyze simple sequence repeat diversity in the gene space as well as the presence of short insertions/deletions using similar methods to those consistently used with genomic sequence. In contrast, due to the confounding of CNV in the genome and transcription rate variation, it is not practical to use RNA-seq data to infer CNV as can be determined with genomic sequence.

Heterosis, defined as the superior performance of F1 hybrids relative to their inbred parents, is a complex phenomenon that is likely underlined by multiple molecular mechanisms. This phenomenon has been highly exploited for economic gain through hybrid seed production, and in maize these gains are estimated to provide a 15% yield advantage to hybrids [48]. Understanding the molecular basis of heterosis can have major implications for improving yield beyond what has been realized to date with our limited understanding of the mechanisms underlying heterosis. It has been hypothesized that allelic complementation, structural variation, and epiallelic variation may contribute to this phenomenon [16], [18], [19], [20], [21], [43], [44]. The systematic divergence of germplasm by breeders for decades has generated relatively distinct heterotic groups within maize [36]. In this study, we showed tight clustering of genotypes within heterotic groups using genome wide SNP markers. It is likely that the heterosis observed from this allelic divergence is the product of complementation of deleterious alleles present within each heterotic group. Genomic level PAV between heterotic groups represents an extreme case of complementation (dominance model of heterosis) and has been shown in this study to extend to the transcriptome. The SSS and NSS heterotic groups in maize are large and diverse sets of germplasm, and because of such, it is necessary to test the extent to which the observations made in this study extend beyond these 21 inbred lines. While additional research is needed to definitively implicate allelic, structural, and transcriptome level variation in heterosis, this study provides growing evidence to the involvement of all of these levels of variation in heterosis.

11.3 MATERIALS AND METHODS

11.3.1 PLANT MATERIALS AND RNA ISOLATION

A set of 21 diverse maize inbred lines was evaluated in this study representing the SSS heterotic group (B14A, PI 550461; B37, PI 550467; B73, PI 550473), the NSS heterotic group (B97, PI 564682; Mo17, PI 558532; MS71, PI 587137; Oh43, Ames 19288; PHG47, PI 601318; PHW65, PI 601501; W605S, Ames 30557; W64A, NSL 30058), Iodent (PHN11, PI 601497), exotic (CML 103, Ames 27081; CML 322, Ames 27096; CML 333, Ames 27101; NC350, Ames 27171; and NC358, Ames 27175), and other (H99, PI 587129; M37W, Ames 27133; MoG, Ames 27136; and Oh7B, Ames 19323) germplasm. Plants were grown under greenhouse conditions (27°C/24°C day/night and 16 h/8 h light/dark) with five plants per pot (30 cm top diameter, 28 cm height, 14.5 L volume) in Metro-Mix 300 (Sun Gro Horticulture, http://www.sungro.com/) with no additional fertilizer. Whole above ground seedling tissue from three plants per inbred line was pooled 6–7 days after planting at the vegetative 1 stage [49], and immediately frozen in liquid nitrogen. Total RNA was extracted with TRIZOL (Invitrogen, http://www.invitrogen.com) and subsequently purified with the RNeasy MinElute Cleanup kit (Qiagen, http://www.qiagen.com), both according to the manufacturer's protocol.

11.3.2 RNA-SEQ LIBRARY CONSTRUCTION AND SEQUENCING

mRNA was isolated from approximately 5 µg of total RNA, fragmented, converted to cDNA, and PCR amplified according to the Illumina RNA-seq protocol (Illumina, Inc. San Diego, CA, Cat # RS-100-0801), and sequenced using the Illumina Genome Analyzer II (San Diego, CA) at the University of Wisconsin Biotechnology Center (Madison, WI). Illumina barcodes were used to pool three samples per lane. Two technical replicates were conducted for each library to generate 36 and 74 bp single-end sequence reads. Sequences are available in the Sequence Read Archive at the National Center for Biotechnology Information (study accession number SRP006703). For all subsequent analysis, reads from the two technical replicates were pooled.

11.3.3 SINGLE NUCLEOTIDE POLYMORPHISM DIVERSITY ANALYSIS

The FASTX toolkit (http://hannonlab.cshl.edu/fastx_toolkit/index.html) was used to clean reads prior to mapping. The fastx_clipper program was used to remove the Illumina adapter sequences requiring a minimum sequence length of 20 bp after clipping, and the fastq_quality_trimmer was used to remove low quality bases from the ends of reads requiring a minimum Phred score of 20 and a minimum length of 20 bp after trimming.

Cleaned RNA-seq reads were mapped to the maize B73 5b pseudomolecules (http://ftp.maizesequence.org/) [17] using Bowtie version 0.12.7 [50] and TopHat version 1.2.0 [51], requiring a minimum intron size of 5 bp and a maximum intron size of 60,000 bp. Alignments for reads that mapped uniquely to the pseudomolecules were processed using the sort, index, and pileup programs within SAMtools version 0.1.7 [52] to generate unfiltered pileup files. A custom Perl script was used to provide counts of reads for each nucleotide (A, T, C, and G) with a Phred score greater than 20 at each position. To determine the genotype for each of the inbred lines at each position, a nucleotide had to be present in greater than 5.0% of the filtered reads and a minimum of two of the filtered reads supporting that nucleotide to be included. A genotype containing more than one allele was defined as missing data as it is not possible to resolve if this is a true heterozygous locus or the product of a duplication not represented in the reference sequence with sequence divergence appearing heterozygous. To confirm the presence of an allele at each nucleotide position, we required data from at least two inbred lines to support the presence of that allele, and furthermore at least two alleles had to be present to call a SNP. Thus, with these criteria, SNPs could only be detected at a nucleotide position for which there was sequence coverage in at least four of the 21 genotypes. For example, to identify a biallelic SNP, data had to be present in at least two inbred lines for one allele and at least two other inbred lines for the alternative allele.

Genomic DNA sequence reads were obtained using genotyping-by-sequencing [39], with an additional size selection step to restrict the fragment size to approximately 300 bp. The multiplexed library was parsed into individual lines and barcode sequences removed using a custom Perl script that required a perfect match to the barcode and ApeKI cut site (GC[A/T]GC). The reads were cleaned using the same method as the RNA-seq reads and subsequently mapped to the 5b pseudomolecules using Bowtie version 0.12.7 [50] permitting up to two mismatches. The alignments were compiled using SAMtools version 0.1.7 [52] as described above. Genotype calls were determined with a custom perl script requiring at least 70% of the reads at a given loci to support a single nucleotide.

Functional annotation of genes with a high number of SNPs and high SNP density was obtained from maizesequence.org. Pair-wise allele frequency based distances between inbred lines were calculated as Rogers distances [53] using the 351,710 SNPs identified from the RNA-seq reads. PowerMarker version 3.25 [54] was used to construct the neighbor-joining tree, and TreeView version 1.6.6 was used to generate the tree image [55].

11.3.4 TRANSCRIPTOME PROFILE ANALYSIS

To evaluate variation in gene expression, purity filtered reads were mapped to the 5b pseudomolecules (http://ftp.maizesequence.org/) using Bowtie version 0.12.7 [50] and TopHat version 1.2.0 [51], which utilize a quality and splice site aware alignment algorithm. The minimum and maximum intron length was set to 5 bp and 60,000 bp respectively; all other parameters were set to the default values. Normalized gene expression values were determined using Cufflinks version 0.9.3 [56], data provided in Table S5. The maximum intron length was set to 60,000 bp, the quartile normalization option was used, the 5b annotation (http://ftp.maizesequence.org/) was provided for the reference annotation, and the v2 pseduomolecules were provided for the bias detection and correction algorithms. The default settings were used for all other parameters.

The Cufflinks program provides the upper and lower bound FPKM values for a 95% confidence interval. For the qualitative categorization of genes based on expression status, if the low confidence interval value was greater than zero the gene was considered expressed. For the semi-quantitative categorization, a gene was considered lowly expressed if the low confidence interval was greater than zero and the FPKM value was less than 5, moderately expressed if the low confidence interval was greater than zero and the FPKM value was greater than or equal to 5 and less than or equal to 200, and highly expressed if the low confidence interval was greater than zero and the FPKM value was greater than 200.

11.3.5 IDENTIFICATION AND CHARACTERIZATION OF NOVEL MAIZE TRANSCRIPTS

RNA-seq reads were mapped to the 5b pseudomolecules (http://ftp.maizesequence.org/) using the same methods as described above for the transcriptome analysis. Unmapped reads were cleaned with the FASTX toolkit (http://hannonlab.cshl.edu/fastx_toolkit/index.html), using the fastx_clipper requiring a minimum read length of 30 bp after clipping, the fastx_artifacts_filter, and the fastq_quality_trimmer requiring a minimum Phred score of 20 and a minimum read length of 30 bp. Cleaned RNA-seq reads from all 21 lines were de novo assembled in a joint assembly using Velvet version 1.0.17 and Oases version 0.1.18 [45] requiring a minimum transcript length of 500 bp. Four assemblies were generated using a k-mer of 21, 23, 25, and 27. The optimal k-mer size based on the number of reads incorporated, number of loci/transcripts, and the N50 contig size was 23. For all future analyses only the representative transcript, defined as the longest transcript, for each locus was used. Representative transcripts were aligned to the 5b pseudomoleucles using GMAP version 2010-07-27 [57], a splice site aware aligner designed for mRNA and expressed sequence tag sequences. Transcripts with alignments greater than 85% coverage and 85% identity were removed from future analysis. The representative transcripts that did not align to the reference sequence were compiled into a single chromosome and added to the 5b pseudomolecules hereafter referred to as the reference plus assembly. All of the RNA-seq reads were aligned to the reference plus assembly and FPKM values determined using the methods described in the transcript analysis section. The reads were also aligned to the reference plus assembly requiring a unique alignment.

Primers were designed to confirm the transcript assembly and PAV for 21 representative transcripts from the assembly across eight of the inbred lines (B14A, B37, B73, B97, CML103, CML322, CML333, and H99; Table S3). The transcripts selected were computationally predicted to have expression in one, four, or eight of these lines. The same RNA used for the RNA-seq library construction was converted to cDNA using the SuperScript One-Step RT-PCR kit (Invitrogen, http://www.invitrogen.com) according to the manufacturer's protocol and PCR amplified using the following program: 94°C for 15 sec, 58°C for 30 sec, and 70°C for 1.5 min, for 40 cycles, followed by 72°C for 8 min. RT-PCR products with a visible band were cleaned and prepared for sequencing using FastAP Thermosensitive Alkaline Phosphatase (Thermo Scientific, http://www.fermentas.com) according to the manufacturer's suggested protocol. Sanger dideoxy-termination sequencing was performed at the Michigan State University Research Technology Support Facility (East Lansing, MI). Sequences were aligned to the computationally predicted sequences in pair-wise alignments with NCBI BLASTN [58] using the longest high scoring pair and in multiple sequence alignments with CodonCode Aligner version 3.7.1.1 (CodonCode Corporation, Dedham, MA, USA).

Assembled transcripts were searched against the UniRef100 database release 2011_04 [46] with WU BLASTX [59], [60] requiring a minimum E-value of 1e-5, a seed word length of 4, a neighborhood word score of 1000, and the shortqueryok option. The transcripts were also searched against the rice version 6.1 proteins [61] and the sorghum version 1 proteins [62] using WU BLASTX [59], [60], and against the maize PUTs excluding sequences less than 250 bases or with 10 or more N's [47] using WU BLASTN [59], all requiring a minimum E-value of 1e-5. For all database searches only the best hit was used.

REFERENCES

1. Troyer AF (1990) A Retrospective View of Corn Genetic Resources. Journal of Heredity 81: 17–24.
2. Duvick DN (2005) The Contribution of Breeding to Yield Advances in maize (Zea mays L.). In: Donald LS, editor. Advances in Agronomy: Academic Press. pp. 83–145.
3. Dudley JW, Lambert RJ (1992) Ninety generations of selection for oil and protein in maize. Maydica 37: 1–7.
4. de Leon N, Coors JG (2002) Twenty-Four Cycles of Mass Selection for Prolificacy in the Golden Glow Maize Population. Crop Sci 42: 325–333. doi: 10.2135/cropsci2002.0325.
5. Russell WK (2006) Registration of KLS_30 and KSS_30 Populations of Maize. Crop Sci 46: 1405–1406. doi: 10.2135/cropsci2005.08-0253.
6. Odhiambo MO, Compton WA (1987) Twenty Cycles of Divergent Mass Selection for Seed Size in Corn1. Crop Sci 27: 1113–1116. doi: 10.2135/cropsci1987.0011183X002700060005x.
7. Flint-Garcia SA, Thuillet AC, Yu J, Pressoir G, Romero SM, et al. (2005) Maize association population: a high-resolution platform for quantitative trait locus dissection. Plant J 44: 1054–1064. doi: 10.1111/j.1365-313X.2005.02591.x.
8. Hansey CN, Johnson JM, Sekhon RS, Kaeppler SM, de Leon N (2011) Genetic Diversity of a Maize Association Population with Restricted Phenology. Crop Sci 51: 704–715. doi: 10.2135/cropsci2010.03.0178.
9. Yan J, Shah T, Warburton ML, Buckler ES, McMullen MD, et al. (2009) Genetic Characterization and Linkage Disequilibrium Estimation of a Global Maize Collection Using SNP Markers. PLoS ONE 4: e8451. doi: 10.1371/journal.pone.0008451.
10. Buckler ES, Holland JB, Bradbury PJ, Acharya CB, Brown PJ, et al. (2009) The genetic architecture of maize flowering time. Science 325: 714–718. doi: 10.1126/science.1174276.
11. Kump KL, Bradbury PJ, Wisser RJ, Buckler ES, Belcher AR, et al. (2011) Genome-wide association study of quantitative resistance to southern leaf blight in the maize nested association mapping population. Nat Genet 43: 163–168. doi: 10.1038/ng.747.
12. Poland JA, Bradbury PJ, Buckler ES, Nelson RJ (2011) Genome-wide nested association mapping of quantitative resistance to northern leaf blight in maize. Proc Natl Acad Sci U S A 108: 6893–6898. doi: 10.1073/pnas.1010894108.
13. Tian F, Bradbury PJ, Brown PJ, Hung H, Sun Q, et al. (2011) Genome-wide association study of leaf architecture in the maize nested association mapping population. Nat Genet 43: 159–162. doi: 10.1038/ng.746.

14. Yu J, Holland JB, McMullen MD, Buckler ES (2008) Genetic design and statistical power of nested association mapping in maize. Genetics 178: 539–551. doi: 10.1534/genetics.107.074245.
15. Gore MA, Chia JM, Elshire RJ, Sun Q, Ersoz ES, et al. (2009) A first-generation haplotype map of maize. Science 326: 1115–1117. doi: 10.1126/science.1177837.
16. Lai J, Li R, Xu X, Jin W, Xu M, et al. (2010) Genome-wide patterns of genetic variation among elite maize inbred lines. Nat Genet 42: 1027–1030. doi: 10.1038/ng.684.
17. Schnable PS, Ware D, Fulton RS, Stein JC, Wei F, et al. (2009) The B73 Maize Genome: Complexity, Diversity, and Dynamics. Science 326: 1112–1115. doi: 10.1126/science.1178534.
18. Springer NM, Ying K, Fu Y, Ji T, Yeh CT, et al. (2009) Maize inbreds exhibit high levels of copy number variation (CNV) and presence/absence variation (PAV) in genome content. PLoS Genet 5: e1000734. doi: 10.1371/journal.pgen.1000734.
19. Swanson-Wagner RA, Eichten SR, Kumari S, Tiffin P, Stein JC, et al. (2010) Pervasive gene content variation and copy number variation in maize and its undomesticated progenitor. Genome Res 20: 1689–1699. doi: 10.1101/gr.109165.110.
20. Makarevitch I, Stupar RM, Iniguez AL, Haun WJ, Barbazuk WB, et al. (2007) Natural variation for alleles under epigenetic control by the maize chromomethylase zmet2. Genetics 177: 749–760. doi: 10.1534/genetics.107.072702.
21. Eichten SR, Swanson-Wagner RA, Schnable JC, Waters AJ, Hermanson PJ, et al. (2011) Heritable epigenetic variation among maize inbreds. PLoS Genet 7: e1002372. doi: 10.1371/journal.pgen.1002372.
22. Medini D, Donati C, Tettelin H, Masignani V, Rappuoli R (2005) The microbial pan-genome. Curr Opin Genet Dev 15: 589–594. doi: 10.1016/j.gde.2005.09.006.
23. Tettelin H, Masignani V, Cieslewicz MJ, Donati C, Medini D, et al. (2005) Genome analysis of multiple pathogenic isolates of Streptococcus agalactiae: implications for the microbial "pan-genome". Proc Natl Acad Sci U S A 102: 13950–13955. doi: 10.1073/pnas.0506758102.
24. Tettelin H, Riley D, Cattuto C, Medini D (2008) Comparative genomics: the bacterial pan-genome. Curr Opin Microbiol 11: 472–477. doi: 10.1016/j.mib.2008.09.006.
25. Hogg JS, Hu FZ, Janto B, Boissy R, Hayes J, et al. (2007) Characterization and modeling of the Haemophilus influenzae core and supragenomes based on the complete genomic sequences of Rd and 12 clinical nontypeable strains. Genome Biol 8: R103. doi: 10.1186/gb-2007-8-6-r103.
26. Maione D, Margarit I, Rinaudo CD, Masignani V, Mora M, et al. (2005) Identification of a universal Group B Streptococcus vaccine by multiple genome screen. Science 309: 148–150. doi: 10.1126/science.1109869.
27. Muzzi A, Masignani V, Rappuoli R (2007) The pan-genome: towards a knowledge-based discovery of novel targets for vaccines and antibacterials. Drug Discov Today 12: 429–439. doi: 10.1016/j.drudis.2007.04.008.
28. Lauer P, Rinaudo CD, Soriani M, Margarit I, Maione D, et al. (2005) Genome analysis reveals pili in Group B Streptococcus. Science 309: 105. doi: 10.1126/science.1111563.

29. Castellanos E, Aranaz A, Gould KA, Linedale R, Stevenson K, et al. (2009) Discovery of stable and variable differences in the Mycobacterium avium subsp. paratuberculosis type I, II, and III genomes by pan-genome microarray analysis. Appl Environ Microbiol 75: 676–686. doi: 10.1128/AEM.01683-08.
30. Li R, Li Y, Zheng H, Luo R, Zhu H, et al. (2010) Building the sequence map of the human pan-genome. Nat Biotechnol 28: 57–63. doi: 10.1038/nbt.1596.
31. Brunner S, Fengler K, Morgante M, Tingey S, Rafalski A (2005) Evolution of DNA sequence nonhomologies among maize inbreds. Plant Cell 17: 343–360. doi: 10.1105/tpc.104.025627.
32. Morgante M, De Paoli E, Radovic S (2007) Transposable elements and the plant pan-genomes. Curr Opin Plant Biol 10: 149–155. doi: 10.1016/j.pbi.2007.02.001.
33. Ossowski S, Schneeberger K, Clark RM, Lanz C, Warthmann N, et al. (2008) Sequencing of natural strains of Arabidopsis thaliana with short reads. Genome Res 18: 2024–2033. doi: 10.1101/gr.080200.108.
34. Weigel D, Mott R (2009) The 1001 genomes project for Arabidopsis thaliana. Genome Biol 10: 107. doi: 10.1186/gb-2009-10-5-107.
35. Cao J, Schneeberger K, Ossowski S, Gunther T, Bender S, et al. (2011) Whole-genome sequencing of multiple Arabidopsis thaliana populations. Nat Genet.
36. Tracy W, Chandler M (2006) The Historical and Biological Basis of the Concept of Heterotic Patterns in Corn Belt Dent Maize. Plant Breeding: The Arnel R Hallauer International Symposium: Blackwell Publishing. pp. 219–233.
37. 37. Sekhon RS, Lin H, Childs KL, Hansey CN, Robin Buell C, et al. (2011) Genome-wide atlas of transcription during maize development. Plant J 66: 553–563. doi: 10.1111/j.1365-313X.2011.04527.x.
38. Clark MJ, Chen R, Lam HY, Karczewski KJ, Euskirchen G, et al. (2011) Performance comparison of exome DNA sequencing technologies. Nat Biotechnol 29: 908–914. doi: 10.1038/nbt.1975.
39. Elshire RJ, Glaubitz JC, Sun Q, Poland JA, Kawamoto K, et al. (2011) A Robust, Simple Genotyping-by-Sequencing (GBS) Approach for High Diversity Species. PLoS ONE 6: e19379. doi: 10.1371/journal.pone.0019379.
40. Subbaiyan GK, Waters DL, Katiyar SK, Sadananda AR, Vaddadi S, et al. (2012) Genome-wide DNA polymorphisms in elite indica rice inbreds discovered by whole-genome sequencing. Plant Biotechnol J.
41. Liu K, Goodman M, Muse S, Smith JS, Buckler E, et al. (2003) Genetic structure and diversity among maize inbred lines as inferred from DNA microsatellites. Genetics 165: 2117–2128.
42. Zheng J, Fu J, Gou M, Huai J, Liu Y, et al. (2010) Genome-wide transcriptome analysis of two maize inbred lines under drought stress. Plant Mol Biol 72: 407–421. doi: 10.1007/s11103-009-9579-6.
43. Springer NM, Stupar RM (2007) Allelic variation and heterosis in maize: how do two halves make more than a whole? Genome Res 17: 264–275. doi: 10.1101/gr.5347007.
44. Fu H, Dooner HK (2002) Intraspecific violation of genetic colinearity and its implications in maize. Proc Natl Acad Sci U S A 99: 9573–9578. doi: 10.1073/pnas.132259199.

45. Zerbino DR, Birney E (2008) Velvet: algorithms for de novo short read assembly using de Bruijn graphs. Genome Res 18: 821–829. doi: 10.1101/gr.074492.107.
46. Suzek BE, Huang H, McGarvey P, Mazumder R, Wu CH (2007) UniRef: comprehensive and non-redundant UniProt reference clusters. Bioinformatics 23: 1282–1288. doi: 10.1093/bioinformatics/btm098.
47. Duvick J, Fu A, Muppirala U, Sabharwal M, Wilkerson MD, et al. (2008) PlantGDB: a resource for comparative plant genomics. Nucleic Acids Res 36: D959–965. doi: 10.1093/nar/gkm1041.
48. Duvick DN (1999) Heterosis: Feeding People and Protecting Natural Resources. In: Coors JG, Pandey S, editors. The Genetics and Exploitation of Heterosis in Crops. Madison, WI: American Society of Agronomy, Inc., Crop Science Society of America, Inc., Soil Science Society of America, Inc. pp. 19–29.
49. Abendroth LJ, Elmore RW, Boyer MJ, Marlay SK (2011) Corn growth and development. PMR 1009 Iowa State University Extension, Ames Iowa.
50. Langmead B, Trapnell C, Pop M, Salzberg SL (2009) Ultrafast and memory-efficient alignment of short DNA sequences to the human genome. Genome Biol 10: R25. doi: 10.1186/gb-2009-10-3-r25.
51. Trapnell C, Pachter L, Salzberg SL (2009) TopHat: discovering splice junctions with RNA-Seq. Bioinformatics 25: 1105–1111. doi: 10.1093/bioinformatics/btp120.
52. Li H, Handsaker B, Wysoker A, Fennell T, Ruan J, et al. (2009) The Sequence Alignment/Map format and SAMtools. Bioinformatics 25: 2078–2079. doi: 10.1093/bioinformatics/btp352.
53. Rogers JS (1972) Measure of genetic similarity and genetic distance. Studies in Genomics VII Univ Tex Publ 7213: 145–153.
54. Liu K, Muse SV (2005) PowerMarker: an integrated analysis environment for genetic marker analysis. Bioinformatics 21: 2128–2129. doi: 10.1093/bioinformatics/bti282.
55. Page RD (1996) TreeView: an application to display phylogenetic trees on personal computers. Comput Appl Biosci 12: 357–358. doi: 10.1093/bioinformatics/12.4.357.
56. Trapnell C, Williams BA, Pertea G, Mortazavi A, Kwan G, et al. (2010) Transcript assembly and quantification by RNA-Seq reveals unannotated transcripts and isoform switching during cell differentiation. Nat Biotechnol 28: 511–515. doi: 10.1038/nbt.1621.
57. Wu TD, Watanabe CK (2005) GMAP: a genomic mapping and alignment program for mRNA and EST sequences. Bioinformatics 21: 1859–1875. doi: 10.1093/bioinformatics/bti310.
58. Johnson M, Zaretskaya I, Raytselis Y, Merezhuk Y, McGinnis S, et al. (2008) NCBI BLAST: a better web interface. Nucleic Acids Res 36: W5–9. doi: 10.1093/nar/gkn201.
59. Altschul SF, Gish W, Miller W, Myers EW, Lipman DJ (1990) Basic local alignment search tool. J Mol Biol 215: 403–410. doi: 10.1006/jmbi.1990.9999.
60. Gish W, States DJ (1993) Identification of protein coding regions by database similarity search. Nat Genet 3: 266–272. doi: 10.1038/ng0393-266.

61. Ouyang S, Zhu W, Hamilton J, Lin H, Campbell M, et al. (2007) The TIGR Rice Genome Annotation Resource: improvements and new features. Nucleic Acids Res 35: D883–887. doi: 10.1093/nar/gkl976.
62. Paterson AH, Bowers JE, Bruggmann R, Dubchak I, Grimwood J, et al. (2009) The Sorghum bicolor genome and the diversification of grasses. Nature 457: 551–556. doi: 10.1038/nature07723.

CHAPTER 12

MOLECULAR MECHANISMS OF EPIGENETIC VARIATION IN PLANTS

RYO FUJIMOTO, TAKU SASAKI, RYO ISHIKAWA, KENJI OSABE, TAKAHIRO KAWANABE, and ELIZABETH S. DENNIS

12.1 INTRODUCTION

Variation in DNA sequence can cause variation in gene expression, which influences quantitative phenotypic variation in organisms and is an important factor in natural variation. Gene expression regulatory networks are comprised of *cis*- and *trans*-acting factors, and differences in gene expression are attributable to genetic variation. In eukaryotes, the genome is compacted into chromatin, and the chromatin structure plays an important role in gene expression: Gene expression can be controlled by changes in the structure of chromatin without changing the DNA sequence, and this phenomenon is termed "epigenetic" control. Recently, there have been many reports indicating that epigenetic change can cause phenotypic variation, and thus epigenetic change can be considered as an important factor in understanding phenotypic change. DNA methylation and histone modifications are well known epigenetic modifications. DNA methylation refers to an addition of a methyl group at the fifth carbon position of a cytosine ring, and in plants it is observed not only in the symmetric CG context but also in sequence contexts of CHG and CHH (where H is A, C, or T) [1–3]. DNA methylation is enriched in heterochromatic regions, such as in centromeric and pericentromeric regions, predominantly consisting of transposons [3–7]. Most transposons are immobile to protect genome integrity and are silenced via DNA methylation [3,8–12]. DNA methylation is also observed in euchromatic regions such as gene-coding regions (gene body methylation), and it is widely seen in eukaryotes [3,13,14].

Nucleosomes are formed by a histone octamer containing two of each of the core histones H2A, H2B, H3, and H4, and 147bp of DNA is wrapped around this core. The N-terminal regions of histone proteins are subject to various chemical modifications such as methylation or acetylation, and these histone modifications are associated with gene transcription [15,16]. In plants, DNA methylation, histone deacetylation and histone methylation in H3K9 (9th lysine of H3) and H3K27 are associated with gene repression, and DNA demethylation, histone acetylation and histone methylation in H3K4 and H3K36 are associated with gene activation [15,17]. Histone lysine residues are able to be mono-, di-, or tri-methylated and each methylation state is associated with different functions [15,17]. Epigenetic modifications play important roles in various aspects of the plant life cycle such as genome integrity, transgene silencing, nucleosome arrangement, nucleolar dominance, paramutation, flowering, and parent of origin-specific gene expression (imprinting) [15,18–21].

Genome-wide profiles of epigenetic information (the epigenome) are available in plants using new technologies such as tiling arrays or high-throughput next generation sequencing [15]. High-resolution maps of epigenetic features have been obtained from bisulfite sequencing (bisulfite converted DNA is directly sequenced) or a combination of chromatin immunoprecipitation (ChIP) technology and genomic tiling arrays (ChIP on chip) or ChIP and high-throughput sequencing (ChIP-seq) [15]. Using these technologies, effects of epigenetic modifications in mutants and variations of DNA methylation status between accessions in *Arabidopsis thaliana*, rice and maize have been shown at the whole genome level [22–26].

In general, heritable variation is a consequence of differences of nucleotide sequence. However, more studies are reporting heritable variation caused by epigenetic variation [27,28]. These epigenetic variations were categorized "obligatory", "facilitated", or "pure epialleles" by Richards [27]. "Obligatory" epigenetic variation is entirely dependent on DNA sequence changes, "facilitated" epigenetic variation is caused by stochastic variation in epigenetic status associated with a DNA sequence change, while "pure" epigenetic variation is generated stochastically and is completely independent of DNA sequence [27]. The "pure" epigenetic variations are subcategorized as stably or metastably inherited [29]. Sometimes these heritable epigenetic changes with or without genetic changes accompany phenotypic change, and there is evidence that spontaneous epigenetic changes generate new plant phenotypes in nature or in cultivars [30–32]. In addition, abnormalities in DNA hypo-methylated mutants have been characterized and some of them are due to the change of DNA methylation status without any difference in nucleotide sequence [33]. Increased knowledge about heritable epigenetic change associated with phenotypic variation suggests that heritable epigenetic changes may become a resource in plant breeding or play a role in plant adaptation [34].

In this review, we describe instances of naturally occurring epigenetic variants and how these can affect plant phenotype. We speculate on the possible causes and analyze the molecular basis of many of these variants and where possible, we elaborate on the resulting phenotypes. Most of our examples are from *A. thaliana*, as its genomics resources are most advanced. We conclude that epigenetic variation is widespread and contributes significantly to the generation of natural variation probably in most species, not just *A. thaliana*.

12.2 EPIGENETIC VARIATION INDUCED BY MUTATIONS OF GENES INVOLVED IN EPIGENETIC MODIFICATION IN *A. THALIANA*

Epigenetic variation can arise in a number of ways. One way is through mutations in the genes responsible for maintaining epigenetic modifications such as DNA methylation. In *A. thaliana*, DNA methylation in the CG context is maintained by *Met1* (METHYLTRANSFERASE 1), while non-CG contexts are maintained by *Drm* (DOMAINS REARRANGED METHYLTRANSFERASE) and *Cmt3* (CHROMOMETHYLASE 3) [3,35]. In addition to DNA methyltransferase, a chromatin remodeling factor| *Ddm1* (DECREASE IN DNA METHYLATION 1), histone methyltransferase| SUVH4/KYP (SU(VAR)3-9 HOMOLOG 4/KRYPTONITE) (hereafter KYP) and SUVH5/6, or SRA-domain methylcytosine-binding protein VIM1/2/3 (VARIANT IN METHYLATION 1/2/3) are also involved in the maintenance of DNA methylation [35]. The process of de novo DNA methylation is triggered by 24-nt siRNAs produced by the RNAi (RNA interference) pathway, termed RdDM (RNA-directed DNA methylation) [36]. Two plant specific RNA polymerases, Pol IV and Pol V, RDR2 (RNA-DEPENDENT RNA POLYMERASE 2), DCL3 (DICER-LIKE 3), and AGO4 (ARGONAUTE 4) proteins function in this RNAi pathway [36,37].

Plant developmental abnormalities have been detected in mutants with disturbed epigenetic modifications, some of which are heritable. An allele of a heritable variant, which is caused by a change in an epigenetic modification without a change in the DNA sequence, is termed an "epiallele". In *A. thaliana*, *ddm1* hypo-methylated mutants showed only slight morphological changes in the early generations, but morphological abnormalities increased after repeated self-pollination over several generations [38,39]. Some developmental abnormalities are heritable and are not linked to the *Ddm1* gene [40]. Some of these mutants are a consequence of the mobilization of transposons due to removal of DNA methylation from the transposon. Genes responsible for these abnormal phenotypes have been able to be identified by map-based cloning, because these phenotypes are heritable and indistinguishable from genetic mutations [33]. One of them is *clm* (*clam*), which showed a lack of elongation in shoots and petioles. This *clm* mutant is caused by an insertion of a *CACTA1* transposon in the *DWF4* gene, which encodes 22-α-hydroxylase in the brassinosteroid biosynthetic pathway. In wild type, *CACTA1* is silent, but it can transpose in *ddm1* [8]. This transposition has also been observed in *met1 cmt3* double mutants, indicating that DNA methylation is important for the silencing of *CACTA1* [41]. Another mutant, *wvs* (*wavy-sepal*), is also caused by an insertion of a transposon into the *FASCIATA1* gene. This transposon is a member of LTR (Long-terminal repeat) retrotransposon class, *AtGP3-1*. *AtGP3-1* is silent in wild type, but it can transpose in *ddm1* [11]. *clm* and *wvs* are genetic mutants caused by epigenetic changes. Another transposon is mobilized in the hypo-methylated mutants, *ddm1* or *met1*, and has the potential to generate a new genic mutant [11,42].

Other mutants can be caused directly by changes in DNA methylation affecting transcription of the gene. The late flowering mutant *FWA* (*FLOWERING WAGENINGEN*) caused by ectopic expression of the *FWA* gene, encodes a homeodomain-containing transcription factor. In wild type, the promoter region of *FWA* is DNA methylated and *FWA* is not expressed in vegetative tissues, this DNA methylation is removed in the *ddm1* mutant and *FWA* is expressed in vegetative tissues (Figure 1) and causes late flowering [43]. This late flowering phenotype is also observed in the *met1* mutant [44,45], but not in the *drm1 drm2 cmt3* triple mutant [46], suggesting that silencing of *FWA* is mainly dependent on CG methylation. The DNA hypo-methylation in the promoter region of *FWA* and the late flowering phenotype are stable in the normal *Ddm1* background, indicating that *FWA* is a gain of function epigenetic mutant.

Another mutant phenotype seen in the *ddm1* background, change of plant structure (short and compact inflorescence with reduced plant height), is *BNS* (BONSAI), which is unstably inherited in the presence of the *Ddm1* gene. The *BNS* gene encodes a protein with similarity to the mammalian cell cycle regulator Swm1/Apc13. In wild type, the *BNS* gene is normally expressed and not methylated (Figure 1). However, in a self-pollinated *ddm1* mutant, the *BNS* gene is methylated and stochastically silenced (Figure 1), indicating that *BNS* is a loss of function epigenetic mutant. The *BNS* gene is flanked by a LINE (Long interspersed repeated element) sequence in a tail-to-tail orientation, and in the *ddm1* mutant DNA methylation in the *BNS* coding region spreads from the LINE (Figure 1). In *ddm1*, the DNA methylation level in *BNS* gradually increases over generations and a phenotype develops. There are two types (with or without LINE sequence) of variation in the *BNS* gene among 96 accessions of *A. thaliana*, 70 of these 96 accessions have LINE sequences at the *BNS* locus. Cvi that lacks the LINE sequence does not show DNA methylation at the *BNS* locus even in a *ddm1* background, indicating that the LINE is essential for the spread of DNA methylation in the *ddm1* mutant background [47]. This shows there is ectopic local DNA hyper-methylation of a specific locus in the global DNA hypo-methylation mutant, *ddm1*. Although small RNAs corresponding to the *BNS* locus accumulate in the *ddm1* mutant, ectopic induction of de novo DNA methylation at the *BNS*

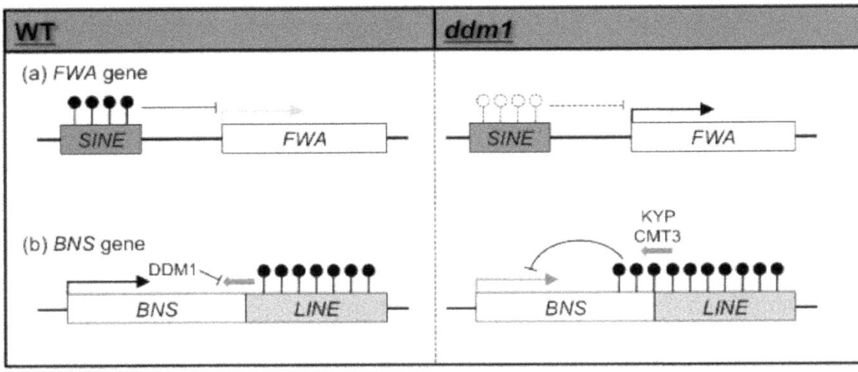

FIGURE 1: Epialleles in the *ddm1* mutant. (a) In WT (wild type) plants, expression of the FWA gene is repressed by DNA methylation of a promoter region-harboring short interspersed nuclear element (SINE) (left). In *ddm1* mutants, decreased DNA methylation in the SINE element induces ectopic expression of the *FWA* gene (right); (b) *BONSAI* (*BNS*) gene is flanked by LINE sequences, which are hyper-methylated, in tail-to-tail manner. In ddm1 mutants, DNA methylation spreads into the *BNS* gene from the LINE sequence in a CMT3-KYP dependent manner, and stochastically induces silencing of the *BNS* gene.

locus in the *ddm1* background was independent of the RdDM pathway because mutations in RdDM components such as RDR2, DCL3, AGO4, PolIV, and PolV did not affect *ddm1*-induced DNA methylation at the *BNS* locus [48]. However, KYP and *Cmt3* were essential for this ectopic DNA hyper-methylation at the *BNS* locus. In addition, meDIP (methylated DNA immnoprecipitation)-chip analysis revealed that *BNS*-like loci were widespread within the *A. thaliana* genome, and that they are DNA hyper-methylated in the *ddm1* mutant background in a *Cmt3*-KYP-dependent manner. Although *Cmt3* is known for the maintenance of DNA methylation in the CHG context, *Cmt3*-KYP dependent alternative de novo DNA methylation was found in all three contexts [48].

The *met1* mutants in *A. thaliana* also show developmental abnormalities such as reduced apical dominance, alterations in flowering time, floral abnormalities, curled leaves, embryogenesis, and formation of viable seeds [44,45,49,50], some of which are inherited even when the wild type allele is present. Genome-wide inheritance of hypo-methylation status even in the presence of the *Met1* wild type locus has been observed in an F8 population derived from hybrids between *met1* and wild type [51]. The floral abnormalities in the *met1* mutant or *Met1* antisense lines are due to DNA hyper-methylation and silencing of *SUP* (*SUPERMAN*) and/or *AG* (*AGAMOUS*) [52,53]. DNA hyper-methylation occurs at CT-rich repeats in the promoter of *SUP* or in the promoter and second intron of *AG*. This shows global DNA hypo-methylation by the *met1* mutation, which causes local DNA hyper-methylation: Stochastic non-CG methylation has been observed in the *met1* mutant [54].

The *drm1*, *drm2* or *cmt3* single mutants did not show any apparent phenotypes, but *drm1 drm2 cmt3* triple mutants showed pleiotropic phenotypes including developmental retardation, reduced plant size, and partial sterility [46,55]. Unlike *ddm1* or *met1*, the *drm1 drm2 cmt3* phenotype is completely recessive: Pleiotropic phenotypes are not inherited independently of the *drm* and *cmt3* mutations [55]. The misexpression of *SDC* (supressor of *drm1 drm2 cmt3*) was observed in the *drm1 drm2 cmt3* triple mutant, and it is sufficient for pleiotropic phenotypes in *drm1 drm2 cmt3* triple mutants. The promoter region harboring tandem repeat regions is densely methylated in all contexts in wild type, but DNA methylation in the promoter region is eliminated in *drm1 drm2 cmt3* triple mutant. F1 progeny between *drm1 drm2 cmt3* triple mutant and wild type show reversion of developmental phenotypes and the promoter region of *SDC* becomes methylated and *SDC* expression is lost. siRNAs corresponding to tandem repeat regions are expressed in wild type leading to DNA methylation of the *SDC* promoter region dependent on the RdDM pathway. Taken together, the pleiotropic phenotypes in the *drm1 drm2 cmt3* triple mutant are due to *SDC* misexpression caused by the elimination of DNA methylation in its promoter region [56].

12.3 NATURAL VARIATION OF EPIGENETIC STATUS

It is well known that DNA sequence polymorphisms at a single locus or multiple loci cause phenotypic variation, and that they are important sources of variation in plants during evolution. In addition to DNA sequence polymorphisms, epigenetic variation has the potential to contribute to the natural variation of plant traits. Epigenome analysis aids in explaining how natural epigenetic variation causes phenotypic differences in plants. The DNA methylation status at the whole genome level has been examined in several species [1,2,4–7,13,14]. Accessions of a species may have been sourced from different environments and different epigenetic modifications selected over time to ensure optimum adaptation to specific environments. Variation of DNA methylation between accessions in *A. thaliana* occurs with gene-body methylation being more variable than DNA methylation of transposable elements among 96 natural accessions [22,25,26,57,58]. Differentially methylated regions have been detected in a comparison of whole genome DNA methylation statuses between two lines of rice or maize [23,24]. In the case of maize, differentially DNA methylated regions were generally observed in intergenic regions. Stable inheritance of DNA methylation was exhibited using near-isogenic lines of maize, though *trans*-acting control of DNA methylation was detected at a few regions [24]. These differences in DNA methylation could have consequences for differential expressions of genes.

A comparison of DNA methylation statuses between parental lines and their progenies generated from single seed descent over 30 generations showed that larger regions of DNA methylation were stable and changes of DNA methylation accumulated through generations [59,60]. The rate of spontaneous changes of DNA methylation is higher than the rate of spontaneous genetic mutations [59–61], suggesting that sequence-independent epialleles play important roles in phenotypic diversity (Figure 2) [59,60]. To identify loci causing phenotypic variation, populations of epigenetic recombinant inbred lines (epi-RILs) between parents, which differed only in epigenetic marks, have been established in *A. thaliana*, and plant complex traits caused by epigenetic variation are observed [29,51,62]. In *A. thaliana*, two sets of epi-RILs were generated from *ddm1* or *met1* mutants that were crossed with wild type [29,48]. Stable inheritance of complex traits such as flowering time and plant height has been observed in these epi-RIL populations, providing important evidence that epigenetic variation can contribute to complex traits [29,51]. Heritable variation that was segregating in epi-RILs is similar to the phenotypic diversity observed in natural populations, suggesting that epigenetic variation in complex traits may drive some portion of natural variation (Figure 2) [63].

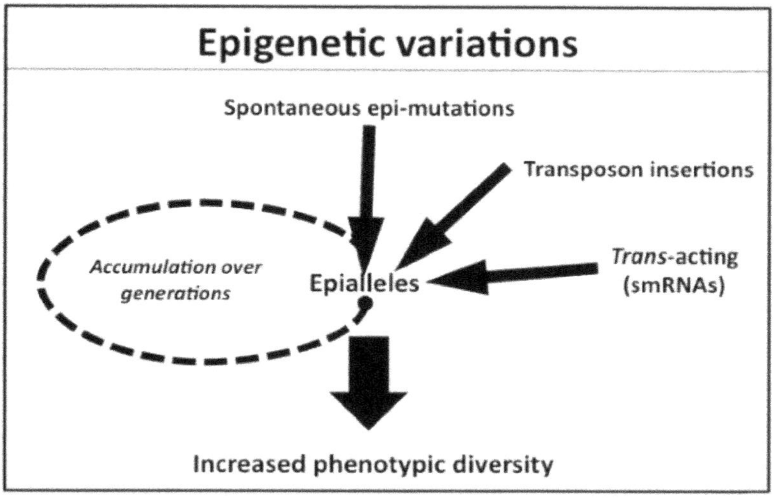

FIGURE 2: Factors that can lead to epigenetic variation in plants. Spontaneous epi-mutations, transposon insertions, and trans-acting (small RNAs) factors can contribute to the generation of epialleles. Epialleles can change gene expression and lead to phenotypic changes, and heritable epialleles can accumulate over generations and increase phenotypic diversity.

12.3.1 SPONTANEOUS EPIGENETIC MUTANTS OCCURRING AT SINGLE LOCI

Examples of spontaneous epi-mutants at single-loci, which influence plant traits, have been reported (Figure 2). Such epi-mutants are a change of flower structure from fundamental symmetry to radial symmetry in Linaria vulgaris (peloric) [30] and a *Cnr* (*colorless non-ripening*) mutant in tomato [31]. The peloric mutation is recessive and prevents expression of *Lcyc* (*Linaria cycloidea-like gene*) [30], and non-ripening of tomato fruit is due to the silencing of the *LeSPL-CNR*, which encodes an SBP-box (Squamosa promoter binding protein-like) transcription factor [31]. In these two cases, there is no sequence polymorphism between mutant and wild type, but high levels of DNA methylation of the causative genes were detected [30,31]. Occasionally some branches, which showed flowers near identical to wild type, were produced in the peloric plant population, and the flowers showed partial DNA demethylation in the *Lcyc* gene [30]. Similarly, the non-ripening phenotype in tomato is stable, but is reversible (showing normal ripening) at a low frequency [31]. In rice, the spontaneous dwarf mutant, *Epi-D1*, shows a metastable inheritance, and has been maintained for more than 90 years as breeding material like in the case of *LeSPL-CNR*. *Epi-D1* plants varied from dwarf to normal. The responsible gene, *D1* (*Dwarf1*), of *Epi-D1* encodes the α-subunit of a GTP-binding protein that is expressed differently between normal (active) and dwarf (inactive) plants, and this differential gene expression is not due to DNA sequence polymorphism. The silencing of the *D1* gene in *Epi-D1* is associated with H3K9 di-methylation in the genic region and DNA methylation in the *D1* promoter region. The promoter region harbors repeat regions, which show DNA methylation, and the repeat region is required for dwarf phenotypic metastability [32]. Tandem repeats are associated with paramutation at the b1 locus of maize. Paramutation refers to the process where alleles interact in trans to establish meiotically heritable expression states [19], but *Epi-D1* did not show a paramutation-like phenotype [32]. These three examples reveal that spontaneous epigenetic changes can be metastably heritable for hundreds of years in nature or during domestication.

12.3.2 TRANSPOSON INSERTION CAN GENERATE EPIGENETIC ALLELES

Transposon insertion in a coding region normally abolishes protein function, and there are some reports of insertion of transposons in a flanking region or intron of protein-coding genes, which can change the expression level of nearby genes [64–69]. Sometimes genetic variation such as transposon insertion drives spontaneous epialleles (Figure 2). Two cases in melon and *A. thaliana* showed that transposon insertion causes phenotypic change through the heritable epialleles.

Uni-sexual females (gynoecy) arise in melon by the action of a recessive g allele, which leads to a transition from male to female flowers. A 1.4

kb region was mapped at the g locus, which harbors a DNA transposon of the hAT family, termed *Gyno-hAT*. The insertion of *Gyno-hAT* downstream of *CmWIP*1, which encodes a C2H2 zinc-finger transcription factor of the WIP protein subfamily, induces DNA methylation in its promoter region, suggesting that DNA methylation caused by *Gyno-hAT* insertion *SUP*presses *CmWIP* expression [70].

A. thaliana accessions can be categorized into early- and late-flowering, which is largely dependent on the allelic variation at two loci, *FRI* (*FRIGIDA*) and *FLC* (*FLOWERING LOCUS C*). Landsberg *erecta* (L*er*) accession is early flowering and shows low-level *FLC* expression [71,72]. The L*er* *FLC* allele (*FLC*-L*er*) has a non-autonomous *Mutator*-like transposable element insertion in the first intron, which may cause low-level *FLC* expression [71]. siRNAs corresponding to the inserted transposable element (TE) sequence accumulate, and HEN1 (HUA ENHANCER 1), SDE4 (SILENCING MOVEMENT DEFICIENT 2)/ NRP*D1* (Nuclear RNA polymerase *D1*A), and AGO4 are involved in this accumulation. High-level *FLC* expression with a late-flowering phenotype was observed in the *hen1-1* mutant, but *FLC* expression level or flowering time did not change in *ago4-1*. The TE in *FLC*-Ler is DNA methylated, but surrounding regions were not. This DNA methylation of the TE was reduced in the *hen1-1* and *ago4-1* mutants, indicating that DNA methylation of the TE is not associated with *FLC* expression. However H3K9 di-methylation was detected in Ler or *ago4-1*, but not in *hen1-1*, indicating that the level of H3K9 di-methylation inversely correlated with the level of *FLC* expression. This suggests that TE in *FLC*-Ler results in low level of *FLC* expression through H3K9 di-methylation triggered by siRNA [73].

These two examples suggest that transposable insertion can drive the generation of new epialleles via changing the epigenetic modifications of nearby genes. In the epiallele, *bsn*, caused by hyper-methylation, DNA methylation in the *BNS* locus is dependent on the existence of a LINE transposable element (Chapter 2) [47]. Transposon insertion sites, number of transposons, and activity of transposons vary among accession of *A. thaliana* [74] and between *A. thaliana* and the related species, Arabidopsis lyrata [75,76], suggesting that distribution of transposable elements may drive natural variation via epigenetic changes in the nearby genes.

12.3.3 *TRANS*-ACTING EPIGENETIC MODIFICATIONS

In addition to transposable element insertion, structural differences such as tandem repeats between accessions may trigger *trans*-acting DNA methylation and silencing through small RNAs (Figure 2). One example of *trans*-acting DNA methylation is the *PAI* (*Phosphoribosylanthranilate isomerase*) gene, which is involved in catalyzing the third step of the tryptophan biosynthetic pathway. The majority of *A. thaliana* accessions have three unlinked *PAI* genes, while in Ws and several other accessions, one of the *PAI* loci is rearranged as a tail-to-tail inverted repeat (IR) of two genes, *PAI*1-*PAI*4 [77,78]. In Ws-type accessions, all four *PAI* genes are DNA methylated, while there is no DNA methylation in the three *PAI* genes in Col-type accessions [77,78]. The *PAI* mutant in Ws, which lacks *PAI*1-*PAI*4 IR, showed blue florescence under UV light and *PAI*2 expression without DNA methylation [77]. The Col *PAI* genes were DNA methylated in the hybrid between Col and Ws [77], and transformation of Ws *PAI*1-*PAI*4 IR into Col induced DNA hyper-methylation of *PAI* genes [79]. From these results, the IR structure triggers DNA methylation not only at *PAI*1-*PAI*4 but also at the unlinked singlet genes *PAI*2 and *PAI*3. A *cmt3* mutant or a *suvh4 suvh5 suvh6* triple mutant showed reduction of non-CG methylation at Ws *PAI* genes, and the *cmt3 met1* double mutant showed depletion of both CG and non-CG methylation [80–82]. Non-CG methylation at Ws *PAI* genes was reduced in the *dcl2 dcl3 dcl4* triple mutant, while DNA methylation did not change in *drm2* or *dcl3* mutants, indicating that a new pathway involving *DCL*-dependent small RNAs and the SUVH/*Cmt3* pathway but not involving the RdDM pathway controls DNA methylation at the Ws *PAI* genes [83]. RdDM independent but *Cmt3*-KYP dependent de novo DNA methylation is observed in many loci in plants derived from *ddm1* [48], suggesting that the *Cmt3*-KYP pathway is also involved in DNA methylation in trans by an uncharacterized mechanism.

DNA methylation in trans is also involved in plant reproduction. Sometimes hybrids between intra-specific accessions are unviable, which is known as hybrid incompatibility. The hybrid incompatibility caused by the genotypic combination of Col at the K4 locus and Sha (Shahdara) at the K5 locus is due to the lack of AtFOLT transcripts. In Col, *AtFOLT1* is expressed, but there is no *AtFOLT2* gene. In Sha, *AtFOLT2* is expressed, but *AtFOLT1* is not expressed. In Sha, lack of *AtFOLT1* expression was due to the high level of DNA methylation in its promoter region and there are siRNA transcripts corresponding to the promoter and first exon regions of *AtFOLT1*. The K4 locus in Sha comprises two additional rearranged truncated sequences homologous to parts of *AtFOLT2*, suggesting that siRNAs are produced from these rearranged gene copies and they can trigger de novo DNA methylation in *AtFOLT1* [84]. Further study will reveal which pathways, RdDM, *Cmt3*-KYP, or others, are involved in de novo DNA methylation in trans via siRNAs.

DNA methylation in trans affects the expression not only of protein coding genes but also of transposable elements. The *MuK* (Mu Killer) locus dominantly silences an active *MuDR* [85,86]. As *MuK* results from an inverted duplication of a partially deleted autonomous *MuDR* element, it forms a perfect 2.4 kb hairpin RNA, which is processed into siRNAs [86]. *MuK* triggers DNA methylation of the terminal inverted repeats of *MuDR*. Once exposed to *MuK*, silencing of *MuDR* is heritable even in the absence of *MuK*, but *MuDR* elements can occasionally be reactivated with DNA demethylation when they are in a particular chromosomal position [85–87]. The *mop1* (*Mediator of paramutation 1*) mutant does not prevent the establishment of silencing of *MuDR* by *MuK*, but the *NAP1* (*Nucleosome assembly protein 1*) knockdown mutant can. *Mop1* encodes a RNA-dependent RNA polymerase, which is an ortholog of RDR2 in *A. thaliana*, and is involved in the production of 24-nt siRNAs, and *NAP1* has been implicated as a histone chaperon [88,89]. The NAPs are required to establish a form of heritable silencing, perhaps by recruiting specific histone variants, but they are not required once the silencing state is established. By contrast, *Mop1* is not required for the establishment of heritable silencing, but maintenance of *MuDR* silencing is assisted by *Mop1* through the RdDM pathway [88].

Another type of small RNAs, which can trigger de novo DNA methylation in trans, has been identified [90]. The dominance-relationship in the male determinant of self-incompatibility in *Brassica* is controlled by de novo DNA methylation in the promoter region of the recessive S determinant gene, *SP11/SCR* (*S locus protein 11/S locus cystein rich*), through small RNAs, *SMI* (*SP11 methylation inducer*). Self-incompatibility is controlled by one locus, the S locus, and the female-determinant gene of self-incompatibility, *SRK* (*S receptor kinase*), and *SP11/SCR* are located at the S locus [91,92]. As these two genes are inherited without recombination, they are called S haplotypes. As the self-incompatibility of Brassica is sporophytically controlled, there are dominant relationships between S haplotypes in the heterozygous plants on both the pollen and stigma side [92,93]. In Brassica, there are two types of S haplotypes, Class-I and Class-II, which are sequence based, and Class-I S haplotypes are dominant over Class-II S haplotypes in the Class-I/Class-II S heterozygote plants of pollen [92,93]. In Class-I/Class-II S heterozygotes, expression of Class-II *SP11/SCR* is suppressed and the promoter region of Class-II *SP11/SCR* is DNA methylated (Figure 3) [94]. The Class-I S haplotypes have the *SMI* (*SP11-methylation-inducing region*) located in the S locus, and its sequence has homology to the promoter region of Class-II S haplotypes (Figure 3). The 24nt-small RNAs, *SMI*, are expressed from *SMI*, and these small RNAs can trigger the de novo DNA methylation of the promoter region of Class-II *SP11/SCR* (Figure 3), indicating that Class-I derived *SMI* induces silencing of the recessive *SP11* allele by *trans*-acting de novo DNA methylation in the Class-I/Class-II S heterozygote plants [90]. Models of the molecular mechanism of *SMI* dependent de novo DNA methylation in trans have been suggested [95,96].

12.4 NATURAL VARIATION OF IMPRINTED *FWA* GENES IN THE GENUS ARABIDOPSIS

FWA is responsible for a late flowering phenotype in *A. thaliana* that is caused by the inhibition of FT function by protein-protein interaction between ectopically expressed FWA and FT [97,98]. *FWA* is expressed only in the central cell and endosperm in *A. thaliana* and reciprocal crosses between Col and Ler have shown that only the maternal allele of *FWA* is expressed in the endosperm, indicating that *FWA* is an imprinted gene in *A. thaliana* [99]. The maternal allele is demethylated in the central cell by

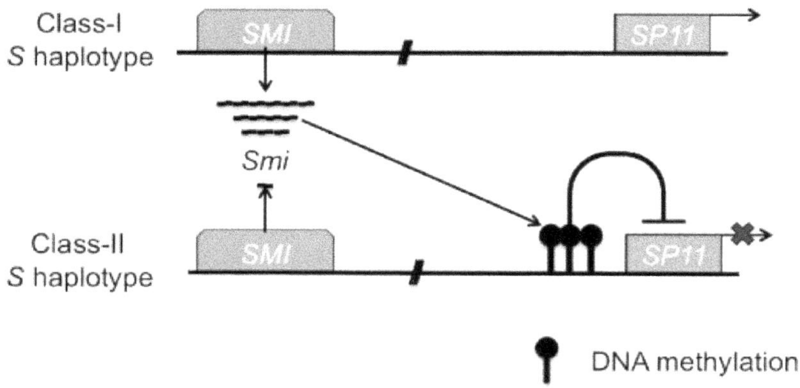

FIGURE 3: Dominance relationship in pollen. Smi derived from Class-I S locus can induce the de novo DNA methylation in the promoter region of Class-II *SP11*. These four examples have revealed that genetic changes generating small RNAs can trigger de novo DNA methylation in trans during plant development. There are two types of trans-acting de novo DNA methylation, heritable as in paramutation or non-heritable, and several different molecular mechanisms induce de novo DNA methylation via small RNAs. These molecular mechanisms are generally plant specific and may be one factor generating natural variation.

the demethylase, *DME* (*DEMETER*), which also acts on other imprinted genes in *A. thaliana* [99,100]. In vegetative tissues, DNA methylation of *FWA* occurs in the promoter region, which harbors two pairs of tandem repeats and a SINE (short interspersed nuclear element). This DNA methylation is reduced in the endosperm of *A. thaliana*, suggesting that methylation of this region participates in silencing of *FWA* [43,99]. Small RNAs are produced from the promoter region of *FWA*, suggesting that DNA methylation in this region is mediated by the RdDM pathway [101,102]. Indeed, DNA methylation of the promoter region of a "transgene" of *FWA* is dependent on the function of DRM2, RDR2, DCL3, and AGO4 [103]. Transformation of a double stranded RNA construct, which can cause de novo DNA methylation directed to a target region, into the *FWA* mutant has shown that DNA methylation in the region harboring the two pairs of tandem repeats and SINE region is sufficient for the silencing of *FWA* expression in vegetative tissues of *A. thaliana* [104].

Using species related to *A. thaliana*, the structures that cause DNA methylation, imprinting, and vegetative silencing of *FWA* have been examined [105]. *FWA* genes are conserved in the genus in *Arabidopsis*, as there is high sequence homology not only in exon regions but also in the intron and promoter regions among species (Figure 4a). The SINE sequence is found in all species examined, *A. arenosa, A. halleri, A. lyrata, A. suecica* (allotetraploid between *A. thaliana* and *A. arenosa*) and *A. kamchatica* (allotetraploid between *A. halleri* and *A. lyrata*), suggesting that the SINE insertion is an ancient event (Figure 4a). In contrast, the structure of the tandem repeats is different among species: *A. halleri* and *A. halleri* allele of *A. kamchatica* have no tandem repeat in the SINE region, while *A. arenosa, A. lyrata, A. arenosa* and *A. thaliana* alleles of *A. suecica*, and the *A. lyrata* allele of *A. kamchatica* have tandem repeats like *A. thaliana* (Figure 4a). The sizes of the repeated and duplicated regions are different between species, suggesting that duplications occurred after speciation (Figure 4a) [105]. The ancient species, *Arabis glabra*, has a SINE region but no tandem repeat, supporting the hypothesis that the tandem repeat was not in the original structure (Figure 4). The *FWA* genes of *A. lyrata* and *A. halleri* show imprinted expression in immature seeds. DNA methylation of *FWA* in vegetative tissues in the SINE region is observed in all species. In *A. halleri* subsp. *gemmifera*, which lacks the tandem repeat structure, *FWA* shows imprinted expression, silencing in vegetative tissues, and DNA methylation in the SINE region, suggesting that the SINE sequence per se is important for epigenetic regulation of the *FWA* gene and *FWA* may have evolved silencing mechanism for transposable elements [101,105]. Transposable elements are extensively demethylated in endosperm, and the flanking regions of imprinted genes involving repetitive sequences are also demethylated, suggesting that imprinted genes evolved from targeted DNA methylation of transposable elements in *A. thaliana* [106,107].

Vegetative silencing of *FWA* varies not only between species but also within species [105,108]. In *A. thaliana*, 93 out of 96 accessions have two *PAI*rs of tandem repeats (termed Type-A), and three have large tandem repeats but not short tandem repeats (termed Type-B). All 96 natural accessions have DNA methylation in the SINE region [22]. *FWA* is not expressed in all 21 accessions of Type-A that we selected randomly from 93 accessions, but two of three accessions of type-B, Fab-4, Var2-1, and Var2-6 showed a low level of *FWA* expression. However the DNA methylation level in the SINE region is almost the same among Type-B accessions (Figure 5). Though it is still unknown what the difference in the silencing stability among the three Type-B accessions is, two pairs of tandem repeats stabilize *FWA* silencing. Indeed, both large and small tandem repeats are involved in silencing *FWA* [105,108]. In *A. lyrata*, *FWA* is expressed in two strains of subsp. *lyrata* that has three tandem repeats and *FWA* is not expressed in subsp. *petraea* that has four tandem repeats. The *FWA*

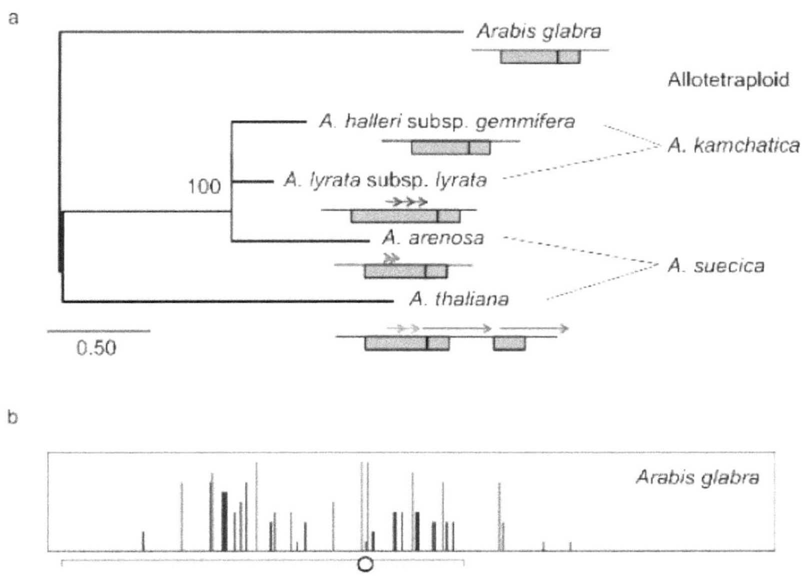

FIGURE 4: (a) Neighbor-joining tree of amino acid sequences of the *FWA* in the genus *Arabidopsis*. Bootstrap values with 1,000 replicates are indicated at the node of the neighbor-joining trees. *Arabis glabra* is used as out-group. Schematic views show the structure of the tandem repeats in the *FWA* promoter. Gray boxes reveal the SINE region, and vertical lines in the gray box show the transcription start site. Tandem repeats covering different regions are shown by different shades. *A. kamchatica* and *A. suecica* are allotetraploids between *A. halleri* and *A. lyrata* and between *A. thaliana* and *A. arenosa*, respectively; (b) Cytosine methylation status of the FWA promoter in *Arabis glabra*. Ten clones from bisulfite-treated templates were examined for each sample. Bars represent methylation in CG, CHG, and asymmetric sites, respectively. Gray bars show the SINE-related sequences. The circle shows the transcription start site.

expression level tended to be inversely correlated with the DNA methylation level of the SINE in *A. lyrata* [105]. Another repeat in the subsp. petraea enlarged the DNA methylated region, suggesting that more tandem duplications might lead to greater stabilization of *FWA* silencing, similar to the indications from *A. thaliana*. In *A. halleri*, *FWA* was expressed in vegetative tissues of subsp. *halleri, tatlica*, and *ovirensis* and in several strains of subsp. *gemmifera*, while *FWA* was silenced in the majority of strains of subsp. *gemmifera*. One strain, IK, showed variation of *FWA* expression level among ten individual plants in spite of a perfect match of the promoter sequences, and there is a negative correlation between *FWA* expression level and DNA methylation level, especially with the non-CG methylation level in the region just upstream of the TSS (transcription start site) [105,108]. From these results, silencing of *FWA* is stable in Type-A of *A. thaliana* and *A. lyrata* subsp. *petraea*, but unstable in other species.

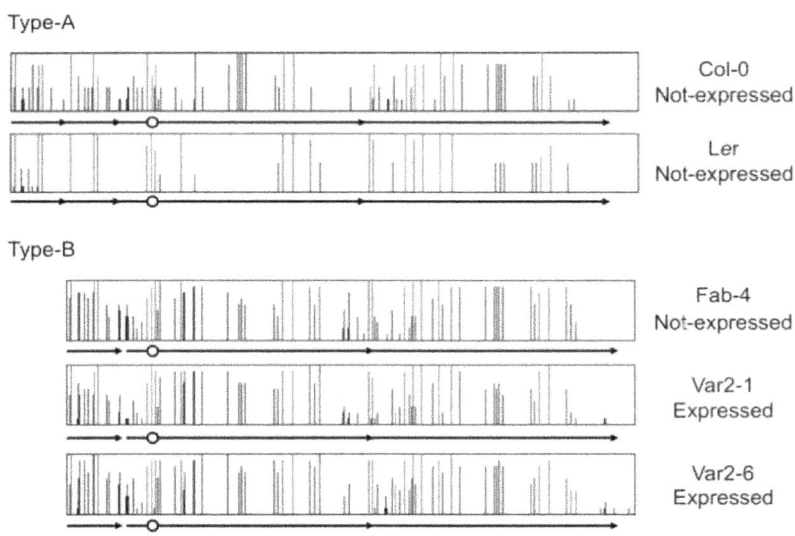

FIGURE 5: Cytosine methylation status of the *FWA* promoter in two types of accessions of *A. thaliana*. Type-A has short and large tandem repeats (shown by arrows), while Type-B has only large repeat. Ten clones from bisulfite-treated templates were examined for each sample. Bars represent methylation in CG, CHG, and asymmetric sites. The circle shows the transcription start site. *FWA* is not expressed in vegetative tissues of Col, Ler, and Fab-4, while being expressed in Var2-1 and Var2-6.

This difference might be due to the number of tandem repeats, which can expand the DNA methylated region.

The results from inter-specific hybridization support this suggestion. In the inter-specific hybrid between *A. thaliana* (Col or L*er*) and *A. lyrata* subsp. *lyrata* (pn3 or MN47), only the *A. lyrata* allele of *FWA* was expressed in vegetative tissues and this expression level was higher than the expression level of the parent *A. lyrata* subsp. *lyrata* (Figure 6). The DNA methylation level in the SINE region of the *A. lyrata* allele in the inter-specific hybrid was reduced in vegetative tissues [105]. In the inter-specific hybrid between *A. thaliana* (Col or L*er*) and *A. halleri* subsp. *gemmifera*, only the *A. halleri* allele of *FWA* was expressed in vegetative tissues in spite of no *FWA* expression in the parents (Figure 6). The DNA methylation level in the SINE region of the *A. halleri* allele of the inter-specific hybrids was also reduced in vegetative tissues, especially in the non-CG methylation of the region upstream of TSS (Figure 7). Though up-regulation of the *A.*

		Parent	Hybrid between At and Al	Hybrid between At and Ah
A. thaliana	Col	-	-	-
→→ → →	L*er*	-	-	-
A. lyrata subsp. *lyrata*	pn3	+	++	
→→→	MN47	+	++	
A. lyrata subsp. *petraea* →→→→	Mue-1	-	-	
A. halleri subsp. *gemmifera* →	Tada	-		+

FIGURE 6: Summary of the vegetative FWA expression in inter-specific hybrids between *A. thaliana* and *A. lyrata* or between *A. thaliana* and *A. halleri.* Arrows show the tandem repeats in the *FWA* promoter. −; Absence of vegetative *FWA* expression, +; low level *FWA* expression in vegetative tissues, ++; More vegetative *FWA* expression. At; *A. thaliana*, Al; *A. lyrata*, Ah; *A. halleri.*

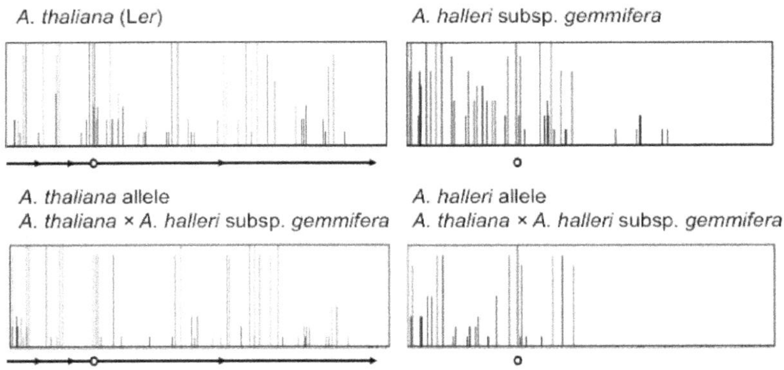

FIGURE 7: Cytosine methylation of the *FWA* promoter in *A. halleri* allele is reduced in the inter-specific hybrids between *A. thaliana* and *A. halleri* subsp. *gemmifera,* relative to direct parent. Ten clones from bisulfite-treated templates were examined for each sample. Bars represent methylation in CG, CHG, and asymmetric sites. The circle and arrows show the transcription start site and tandem repeats, respectively.

lyrata or A. halleri allele in inter-specific hybrids was detected by RT-PCR, this up-regulation could not be detected by microarray analysis using ATH1 [109]. The *A. thaliana FWA* allele in two inter-specific hybrids was silenced and non-CG DNA methylation was slightly reduced in vegetative tissues (Figures 6 and 7). In the inter-specific hybrid between *A. thaliana* and *A. lyrata* subsp. *petraea*, there was no *FWA* expression in vegetative tissues as in their parents (Figure 6). The DNA methylation level in the *A. lyrata* allele of the inter-specific hybrid did not change (Figure 8). These results suggest that silencing of *FWA* might be affected by inter-specific hybridization, if the silencing level of the parent is unstable. These results also support the possibility of enhancement of silencing by tandem duplications.

From these results, two possibilities arise. (1) Tandem duplications stabilize *FWA* silencing, especially in Type-A of *A. thaliana*; (2) Non-CG DNA methylation in the region upstream of TSS is important for *FWA* silencing in the species related to *A. thaliana*. To confirm these possibilities,

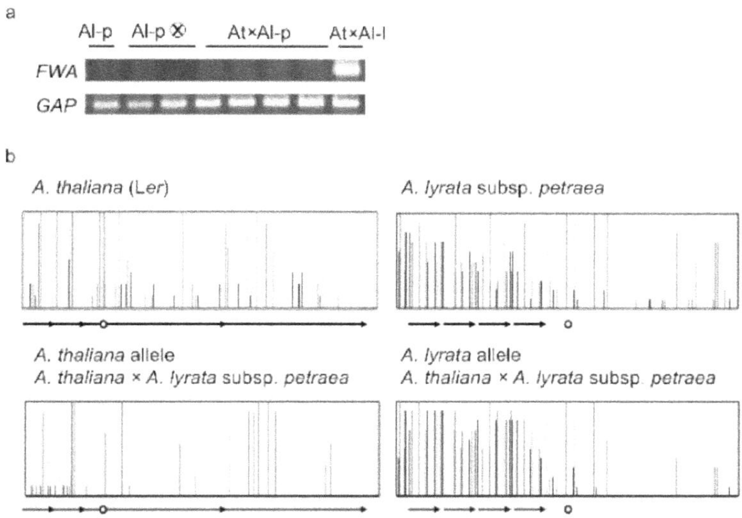

FIGURE 8: No alteration of vegetative *FWA* expression in the inter-specific hybrid between *A. thaliana* and *A. lyrata* subsp. *petraea.* (a) *FWA* transcripts in an inter-specific hybrid between *A. thaliana* and *A. lyrata* subsp. *lyrata* (Al-p). Al-l; *A. lyrata* subsp. lyrata; (b) Cytosine methylation of the *FWA* promoter in the inter-specific hybrid between *A. thaliana* and *A. lyrata* subsp. *petraea.* Ten clones from bisulfite-treated templates were examined for each sample. Bars represent methylation in CG, CHG, and asymmetric sites. The circle and arrows show the transcription start site and tandem repeats, respectively.

critical methylated residues controlling *FWA* silencing were examined using a double-stranded RNA to direct DNA methylation to target regions. In *A. thaliana*, DNA methylation in both short (region upstream of the TSS) and large tandem repeats (region downstream of the TSS) played a role in *FWA* silencing. In contrast, DNA methylation in the region upstream of the TSS played a role in *FWA* silencing in *A. lyrata* and *A. halleri,* but DNA methylation in the region downstream of the TSS was not sufficient for *FWA* silencing in *A. lyrata.* In *A. thaliana*, expression of small RNAs corresponding to the SINE region with two pairs of tandem repeats was confirmed, but few small RNAs were detected in *A. lyrata*, suggesting that DNA methylation in the SINE region is independent of the RdDM pathway in *A. lyrata*, unlike *A. thaliana* [101,102,104,108]. From these results, the critical methylated region for *FWA* silencing is different between *A. thaliana* and *A. lyrata*/*A. halleri*, and tandem duplications in *A. thaliana* enlarged the critical DNA methylated regions, which can stabilize the *FWA* silencing.

There is the question why the silencing mechanism is different between *A. thaliana* and species related to *A. thaliana*. This could be due to the ability of *FWA* to inhibit flowering in *A. thaliana* but not in *A. lyrata*. Over-expression of *FWA* from *A. lyrata* does not cause late flowering in an *A. thaliana* background, suggesting that *A. lyrata FWA* cannot inhibit FT function. Over-expression of both *A. thaliana FWA* and *A. thaliana FT* did not show any obvious developmental abnormality in flowers, but over-expression of both *A. thaliana FWA* and *A. lyrata FT* reveal occasional floral defects, which are due to misexpression of *AP1* (*APETALA1*) and *LFY* (*LEAFY*) [110]. *A. thaliana* shows amino acid changes in the C-terminal region of *FWA* close to the region important for binding of FT [98,108], suggesting that *A. thaliana FWA* might have gained the ability to interact with the FT protein after speciation. Thus ectopic *FWA* expression caused by DNA demethylation might be disadvantageous for both summer and winter annual natural accessions of *A. thaliana*, so *FWA* is stably silenced in *A. thaliana*. Tajima's D test showed negative selection against mutations in the C/G site, suggesting that silencing of *FWA* mediated by DNA methylation plays an important role in adaptation of *A. thaliana* [105]. In *A. thaliana*, a more stable *FWA* silencing mechanism (spreading of critical methylated regions by tandem duplications) has been selected during the process of evolution. This can prevent late flowering caused by a newly generated *FWA* function involving spontaneous substitutions, which enable *FWA* to interact with FT and inhibit FT function.

12.5 CONCLUSIONS

Increasing numbers of epialleles are being reported in various species, and it is clear that epi-mutations can affect plant phenotypes. Some naturally occurring epialleles affect genes involved in plant fitness; some epialleles are stably or metastably inherited [34]. There are multiple causes of epi-mutations such as change of epigenetic status without genetic changes or via genetic changes such as transposon insertion or tandem repeat formations [34]. As plants are sessile organisms, they rely on adaptation mechanisms to withstand environmental stress. Phenotypic modifications by DNA sequence changes cannot respond quickly to environmental stresses. Metastable inheritance may be more useful in adaptation than genetic mutations because metastable epigenetic changes are more flexible and may contribute to phenotypic plasticity under environmental stress conditions [111,112]. Natural variation of epigenetic status has been found among accessions in several plant species, and this variation might be a consequence of the different growing condition in nature [15,20,112]. The higher epi-mutation rate has the potential to contribute to natural variation [59,60], and results using epi-RILs support the idea that complex epigenetic variations are one of the factors of natural variation [29,51]. More research focusing on naturally occurring epigenetic changes will increase our understanding of how epigenetic variation has contributed to natural variation.

REFERENCES

1. Cokus, S.J.; Feng, S.; Zhang, X.; Chen, Z.; Merriman, B.; Haudenschild, C.D.; Pradhan, S.; Nelson, S.F.; Pellegrini, M.; Jacobsen, S.E. Shotgun bisulphite sequencing of the Arabidopsis genome reveals DNA methylation patterning. Nature 2008, 452, 215–219.
2. Lister, R.; O'Malley, R.C.; Tonti-Filippini, J.; Gregory, B.D.; Berry, C.C.; Millar, A.H.; Ecker, J.R. Highly integrated single-base resolution maps of the epigenome in Arabidopsis. Cell 2008, 133, 523–536.
3. Law, J.A.; Jacobsen, S.E. Establishing, maintaining and modifying DNA methylation patterns in plants and animals. Nat. Rev. Genet. 2010, 11, 204–220.
4. Zhang, X.; Yazaki, J.; Sundaresan, A.; Cokus, S.; Chan, S.W.; Chen, H.; Henderson, I.R.; Shinn, P.; Pellegrini, M.; Jacobsen, S.E.; et al. Genome-Wide high-resolution mapping and functional analysis of DNA methylation in Arabidopsis. Cell 2006, 126, 1189–1201.
5. Zilberman, D.; Gehring, M.; Tran, R.K.; Ballinger, T.; Henikoff, S. Genome-Wide analysis of Arabidopsis thaliana DNA methylation uncovers an interdependence between methylation and transcription. Nat. Genet. 2007, 39, 61–69.
6. Li, X.; Wang, X.; He, K.; Ma, Y.; Su, N.; He, H.; Stolc, V.; Tongprasit, W.; Jin, W.; Jiang, J.; et al. High-resolution mapping of epigenetic modifications of the rice genome uncovers interplay between DNA methylation, histone methylation, and gene expression. Plant Cell 2008, 20, 259–276.
7. Wang, X.; Elling, A.A.; Li, X.; Li, N.; Peng, Z.; He, G.; Sun, H.; Qi, Y.; Liu, X.S.; Deng, X.W. Genome-Wide and organ-specific landscapes of epigenetic modifications and their relationships to mRNA and small RNA transcriptomes in maize. Plan Cell 2009, 21, 1053–1069.

8. Miura, A.; Yonebayashi, S.; Watanabe, K.; Toyama, T.; Shimada, H.; Kakutani, T. Mobilization of transposons by a mutation abolishing full DNA methylation in Arabidopsis. Nature 2001, 411, 212–214.
9. Singer, T.; Yordan, C.; Martienssen, R.A. Robertson's Mutator transposons in *A. thaliana* are regulated by the chromatin-remodeling gene Decrease in DNA methylation (*Ddm1*). Genes Dev. 2001, 15, 591–602.
10. Fujimoto, R.; Sasaki, T.; Inoue, H.; Nishio, T. Hypomethylation and transcriptional reactivation of retrotransposon-like sequences in *ddm1* transgenic plants of Brassica rapa. Plant Mol. Biol. 2008, 66, 463–473.
11. Tsukahara, S.; Kobayashi, A.; Kawabe, A.; Mathieu, O.; Miura, A.; Kakutani, T. Bursts of retrotransposition reproduced in Arabidopsis. Nature 2009, 461, 423–426.
12. Sasaki, T.; Fujimoto, R.; Kishitani, S.; Nishio, T. Analysis of target sequences of *Ddm1*s in Brassica rapa by MSAP. Plant Cell Rep. 2011, 30, 81–88.
13. Feng, S.; Cokus, S.J.; Zhang, X.; Chen, P.Y.; Bostick, M.; Goll, M.G.; Hetzel, J.; Jaine, J.; Strauss, S.H.; Halpern, M.E.; et al. Conservation and divergence of methylation patterning in plants and animals. Proc. Natl. Acad. Sci. USA 2010, 107, 8689–8694.
14. Zemach, A.; McDaniel, I.E.; Silva, P.; Zilberman, D. Genome-Wide evolutionary analysis of eularyotic DNA methylation. Science 2010, 328, 916–919.
15. He, G.; Elling, A.A.; Deng, X.W. The epigenome and plant development. Annu. Rev. Plant Biol. 2011, 62, 411–435.
16. Lauria, M.; Rossi, V. Epigenetic control of gene regulation in plants. Biochim. Biophys. Acta 2011, 1809, 369–378.
17. Fuchs, J.; Demidov, D.; Houben, A.; Schubert, I. Chromosomal histone modification patterns–from conservation to diversity. Trends Plant Sci. 2006, 11,199–208.
18. Preuss, S.; Pikaard, C.S. rRNA gene silencing and nucleolar dominance: Insights into a chromosome-scale epigenetic on/off switch. Biochim. Biophys. Acta 2007, 1769, 383–392.
19. Erhard, K.F., Jr.; Hollick, J.B. Paramutation: A process for acquiring *trans*-generational regulatory states. Curr. Opin. Plant Biol. 2011, 14, 210–216.
20. Groszmann, M.; Greaves, I.K.; Albert, N.; Fujimoto, R.; Helliwell, C.A.; Dennis, E.S.; Peacock, W.J. Epigenetics in plants-vernalisation and hybrid vigour. Biochem. Biophys. Acta 2011, 1809, 427–437.
21. Wollmann, H.; Berger, F. Epigenetic reprogramming during pant reproduction and seed development. Curr. Opin. Plant Biol. 2012, 15, 63–69.
22. Vaughn, M.W.; Tanurdžić, M.; Lippman, Z.; Jiang, H.; Carrasquillo, R.; Rabinowicz, P.D.; Dedhia, N.; McCombie, W.R.; Agier, N.; Bulski, A.; et al. Epigenetic natural variation in Arabidopsis thaliana. PLoS Biol. 2007, 5, e174.
23. He, G.; Zhu, X.; Elling, A.A.; Chen, L.; Wang, X.; Guo, L.; Liang, M.; He, H.; Zhang, H.; Chen, F.; et al. Global epigenetic and transcriptional trends among two rice subspecies and their reciprocal hybrids. Plant Cell 2010, 22, 17–33.
24. Eichten, S.R.; Swanson-Wagner, R.A.; Schnable, J.C.; Waters, A.J.; Hermanson, P.J.; Liu, S.; Yeh, C.T.; Jia, Y.; Gendler, K.; Freeling, M.; et al. Heritable epigenetic variation among maize inbreds. PLoS Genet. 2011, 7, e1002372.

25. Greaves, I.K.; Groszmann, M.; Ying, H.; Taylor, J.M.; Peacock, W.J.; Dennis, E.S. Trans chromosomal methylation in Arabidopsis hybrids. Proc. Natl. Acad. Sci. USA 2012, 109, 3570–3575.
26. Shen, H.; He, H.; Li, J.; Chen, W.; Wang, X.; Guo, L.; Peng, Z.; He, G.; Zhong, S.; Qi, Y.; et al. Genome-Wide analysis of DNA methylation and gene expression changes in two Arabidopsis ecotypes and their reciprocal hybrids. Plant Cell 2012, 24, 875–892.
27. Richards, E.J. Inherited epigenetic variation—Revisiting soft inheritance. Nat. Rev. Genet. 2006, 7, 395–401.
28. Paszkowski, J.; Grossniklaus, U. Selected aspects of transgenerational epigenetic inheritance and resetting in plants. Curr. Opin. Plant Biol. 2011, 14, 195–203.
29. Johannes, F.; Porcher, E.; Teixeira, F.K.; Saliba-Colombani, V.; Simon, M.; Agier, N.; Bulski, A.; Albuisson, J.; Heredia, F.; Audigier, P.; et al. Assessing the impact of transgenerational epigenetic variation on complex traits. PLoS Genet. 2009, 5, e1000530.
30. Cubas, P.; Vincent, C.; Coen, E. An epigenetic mutation responsible for natural variation in floral symmetry. Nature 1999, 401, 157–161.
31. Manning, K.; Tör, M.; Poole, M.; Hong, Y.; Thompson, A.J.; King, G.J.; Giovannoni, J.J.; Seymour, G.B. A naturally occurring epigenetic mutation in a gene encoding an SBP-box transcription factor inhibits tomato fruit ripening. Nat. Genet. 2006, 38, 948–952.
32. Miura, K.; Agetsuma, M.; Kitano, H.; Yoshimura, A.; Matsuoka, M.; Jacobsen, S.E.; Ashikari, M. A metastable DWARF1 epigenetic mutant affecting plant stature in rice. Proc. Natl. Acad. Sci. USA 2009, 106, 11218–11223.
33. Kakutani, T. Epi-Alleles in plants: Inheritance of epigenetic information over generations. Plant Cell Physiol. 2002, 43, 1106–1111.
34. Richards, E.J. Natural epigenetic variation in plant species: A view from the field. Curr. Opin. Plant Biol. 2011, 14, 204–209.
35. Saze, H. Epigenetic memory tran*SMI*ssion through mitosis and meiosis in plants. Semin. Cell Dev. Biol. 2008, 19, 527–536.
36. Matzke, M.; Kanno, T.; Daxinger, L.; Huettel, B.; Matzke, A.J.M. RNA-Mediated chromatin-based silencing in plants. Curr. Opin. Cell Biol. 2009, 21, 367–376.
37. Haag, J.R.; Pikaard, C.S. Multisubunit RNA polymerases IV and V: Purveyors of non-coding RNA for plant gene silencing. Nat. Rev. Mol. Cell Biol. 2011, 12, 483–492.
38. Kakutani, T.; Jeddeloh, J.A.; Flowers, S.K.; Munakata, K.; Richards, E.J. Developmental abnormalities and epimutations associated with DNA hypomethylation mutations. Proc. Natl. Acad. Sci. USA 1996, 93, 12406–12411.
39. Kakutani, T. Genetic characterization of late-flowering traits induced by DNA hypomethylation mutation in Arabidopsis thaliana. Plant J. 1997, 12, 1447–1451.
40. Kakutani, T.; Munakata, K.; Richards, E.J.; Hirochika, H. Meiotically and mitotically stable inheritance of DNA hypomethylation induced by *ddm1* mutation of Arabidopsis thaliana. Genetics 1999, 151, 831–838.
41. Kato, M.; Miura, A.; Bender, J.; Jacobsen, S.E.; Kakutani, T. Role of CG and non-CG methylation in immobilization of transposons in Arabidopsis. Curr. Biol. 2003, 13, 421–426.

42. Mirouze, M.; Reinders, J.; Bucher, E.; Nishimura, T.; Schneeberger, K.; Ossowski, S.; Cao, J.; Weigel, D.; Paszkowski, J.; Mathieu, O. Selective epigenetic control of retrotransposition in Arabidopsis. Nature 2009, 461, 427–430.
43. Soppe, W.J.; Jacobsen, S.E.; Alonso-Blanco, C.; Jackson, J.P.; Kakutani, T.; Koornneef, M.; Peeters, A.J.M. The late flowering phenotype of *FWA* mutants is caused by gain-of-function epigenetic alleles of a homeodomain gene. Mol. Cell 2000, 6, 791–802.
44. Kankel, M.W.; Ramsey, D.E.; Stokes, T.L.; Flowers, S.K.; Haag, J.R.; Jeddeloh, J.A.; Riddle, N.C.; Verbsky, M.L.; Richards, E.J. Arabidopsis *Met1* cytosine methyltransferase mutants. Genetics 2003, 163, 1109–1122.
45. Saze, H.; Scheid, O.M.; Paszkowski, J. Maintenance of CpG methylation is essential for epigenetic inheritance during plant gametogenesis. Nat. Genet. 2003, 34, 65–69.
46. Cao, X.; Jacobsen, S.E. Locus-specific control of asymmetric and CpNpG methylation by the *Drm* and *Cmt3* methyltransferase genes. Proc. Natl. Acad. Sci. USA 2002, 99, 16491–16498.
47. Saze, H.; Kakutani, T. Heritable epigenetic mutation of a transposon-flanked Arabidopsis gene due to lack of the chromatin-remodeling factor *Ddm1*. EMBO J. 2007, 26, 3641–3652.
48. Sasaki, T.; Kobayashi, A.; Saze, H.; Kakutani, T. RNAi-independent de novo DNA methylation revealed in Arabidopsis mutants of chromatin remodeling gene *Ddm1*. Plant J. 2012, 70, 750–758.
49. Finnegan, E.J.; Peacock, W.J.; Dennis, E.S. Reduced DNA methylation in Arabidopsis thaliana results in abnormal plant development. Proc. Natl. Acad. Sci. USA 1996, 93, 8449–8454.
50. Ronemus, M.J.; Galbiati, M.; Ticknor, C.; Chen, J.; Dellaporta, S.L. Demethylation-induced developmental pleiotropy in Arabidopsis. Science 1996, 273, 654–657.
51. Reinders, J.; Wulff, B.B.H.; Mirouze, M.; Marí-Ordóñez, A.; Dapp, M.; Rozhon, W.; Bucher, E.; Theiler, G.; Paszkowski, J. Compromised stability of DNA methylation and transposon immobilization in mosaic Arabiopsis epigenomes. Genes Dev. 2009, 23, 939–950.
52. Jacobsen, S.E.; Meyerowitz, E.M. Hypermethylated *SUP*ERMAN epigenetic alleles in Arabidopsis. Science 1997, 277, 1100–1103.
53. Jacobsen, S.E.; Sakai, H.; Finnegan, E.J.; Cao, X.; Meyerowitz, E.M. Ectopic hypermethylation of flower-specific genes in Arabidopsis. Curr. Biol. 2000, 10, 179–186.
54. Mathieu, O.; Reinders, J.; Caikovski, M.; Smathajitt, C.; Paszkowski, J. Transgenerational stability of the Arabidopsis epigenome is coordinated by CG methylation. Cell 2007, 130, 851–862.
55. Chan, S.W.; Henderson, I.R.; Zhang, X.; Shah, G.; Chien, J.S.; Jacobsen, S.E. RNAi, DR*D1*, and histone methylation actively target developmentally important non-CG DNA methylation in Arabidopsis. PLoS Genet. 2006, 2, e83.
56. Henderson, I.R.; Jacobsen, S.E. Tandem repeats upstream of the Arabidopsis endogene *SDC* recruit non-CG DNA methylation and initiate siRNA spreading. Genes Dev. 2008, 22, 1597–1606.
57. Zhang, X.; Shiu, S.H.; Cal, A.; Borevitz, J.O. Global analysis of genetic, epigenetic and transcriptional polymorphisms in Arabidopsis thaliana using whole genome tiling arrays. PLoS Genet. 2008, 4, e1000032.

58. Zhai, J.; Liu, J.; Liu, B.; Li, P.; Meyers, B.C.; Chen, X.; Cao, X. Small RNA-directed epigenetic natural variation in Arabidopsis thaliana. PLoS Genet. 2008, 4, e1000056.
59. Becker, C.; Hagmann, J.; Müller, J.; Koenig, D.; Stegle, O.; Borgwardt, K.; Weigel, D. Spontaneous epigenetic variation in the Arabidopsis thaliana methylome. Nature 2011, 480, 245–249.
60. Schmitz, R.J.; Schultz, M.D.; Lewsey, M.G.; O'Malley, R.C.; Urich M.A.; Libiger, O.; Schork, N.J.; Ecker, J.R. Transgenerational epigenetic instability is a source of novel methylation variants. Science 2011, 334, 369–373.
61. Ossowski, S.; Schneeberger, K.; Lucas-Lledó, J.I.; Warthmann, N.; Clark, R.M.; Shaw, R.G.; Weigel, D.; Lynch, M. The rate and molecular spectrum of spontaneous mutations in Arabidopsis thaliana. Science 2010, 327, 92–94.
62. Teixeira, F.K.; Heredia, F.; Sarazin, A.; Roudier, F.; Boccara, M.; Ciaudo, C.; Cruaud, C.; Poulain, J.; Berdasco, M.; Fraga, M.F.; et al. A role for RNAi in the selective correction of DNA methylation defects. Science 2009, 323, 1600–1604.
63. Roux, F.; Colomé-Tatché, M.; Edelist, C.; Wardenaar, R.; Guerche, P.; Hospital, F.; Colot, V.; Jansen, R.C.; Johannes, F. Genome-Wide epigenetic perturbation jump-starts patterns of heritable variation found in nature. Genetics 2011, 88, 1015–1017.
64. Kobayashi, S.; Goto-Yamamoto, N.; Hirochika, H. Retrotransposon-induced mutations in grape skin color. Science 2004, 304, 982.
65. Fujimoto, R.; Sugimura, T.; Fukai, E.; Nishio, T. *SUP*pression of gene expression of a recessive *SP11/SCR* allele by an untranscribed *SP11/SCR* allele in Brassica self-incompatibility. Plant Mol. Biol. 2006, 61, 577–587.
66. Naito, K.; Zhang, F.; Tsukiyama, T.; Saito, H.; Hancock, C.N.; Richardson, A.O.; Okumoto, Y.; Tanisaka, T.; Wessler, S.R. Unexpected consequences of a sudden and massive transposon amplification on rice gene expression. Nature 2009, 461, 1130–1134.
67. Fernandez, L.; Torregrosa, L.; Segura, V.; Bouquet, A.; Martinez-Zapater, J.M. Transposon-induced gene activation as a mechanism generating cluster shape somatic variation in grapevine. Plant J. 2010, 61, 545–557.
68. Hollister, J.D.; *SMI*th, L.M.; Guo, Y.L.; Ott, F.; Weigel, D.; Gaut, B.S. Transposable elements and small RNAs contribute to gene expression divergence between Arabidopsis thaliana and Arabidopsis lyrata. Proc. Natl. Acad. Sci. USA 2011, 108, 2322–2327.
69. Butelli, E.; Licciardello, C.; Zhang, Y.; Liu, J.; Mackay, S.; Bailey, P.; Reforgiato-Recupero, G.; Martin, C. Retrotransposons control fruit-specific, cold-dependent accumulation of anthocyanins in blood oranges. Plant Cell 2012, 24, 1242–1255.
70. Martin, A.; Troadec, C.; Boualem, A.; Rajab, M.; Fernandez, R.; Morin, H.; Pitrat, M.; Dogimont, C.; Bendahmane, A. A transposon-induced epigenetic change leads to sex determination in melon. Nature 2009, 461, 1135–1138.
71. Michaels, S.D.; Ditta, G.; Gustafson-Brown, C.; Pelaz, S.; Yanofsky, M.; Amasino, R.M. AGL24 acts as a promoter of flowering in Arabidopsis and is positively regulated by vernalization. Plant J. 2003, 33, 867–874.
72. Gazzani, S.; Gendall, A.R.; Lister, C.; Dean, C. Analysis of the molecular basis of flowering time variation in Arabidopsis accessions. Plant Physiol. 2003, 132, 1107–1114.

73. Liu, J.; He, Y.; Amasino, R.; Chen, X. siRNAs targeting an intronic transposon in the regulation of natural flowering behavior in Arabidopsis. Genes Dev. 2004, 18, 2873–2878.
74. Rutter, M.T.; Cross, K.V.; van Woert, P.A. Birth, death and subfunctionalization in the Arabidopsis genome. Trends Plant Sci. 2012, 17, 204–212.
75. Hollister, J.D.; *SMI*th, L.M.; Guo, Y.L.; Ott, F.; Weigel, D.; Gaut, B.S. Transposable elements and small RNAs contribute to gene expression divergence between Arabidopsis thaliana and Arabidopsis lyrata. Proc. Natl. Acad. Sci. USA 2011, 108, 2322–2327.
76. Kawanabe, T.; Fujimoto, R.; Taku Sasaki, T.; Taylor, J.M.; Dennis, E.S. A comparison of transcriptome and epigenetic status between closely related species in the genus Arabidopsis. Gene 2012, 506, 301–309.
77. Bender, J.; Fink, G.R. Epigenetic control of an endogenous gene family is revealed by a novel blue fluorescent mutant of Arabidopsis. Cell 1995, 83, 725–734.
78. Melquist, S.; Luff, B.; Bender, J. Arabidopsis *PAI* gene arrangements, cytosine methylation and expression. Genetics 1999, 153, 401–413.
79. Luff, B.; Pawlowski, L.; Bender, J. An inverted repeat triggers cytosine methylation of identical sequences in Arabidopsis. Mol. Cell 1999, 3, 505–511.
80. Ebbs, M.L.; Bartee, L.; Bender, J. H3 Lysine 9 methylation is maintained on a transcribed inverted repeat by combined action of SUVH6 and SUVH4 methyltransferases. Mol. Cell Biol. 2005, 25, 10507–10515.
81. Ebbs, M.L.; Bender, J. Locus-Specific control of DNA methylation by the Arabidopsis SUVH5 histone methyltransferase. Plant Cell 2006, 18, 1166–1176.
82. Melquist, S.; Bender, J. Transcription from an upstream promoter controls methylation signaling from an inverted repeat of endogenous genes in Arabidopsis. Genes Dev. 2003, 17, 2036–2047.
83. Enke, R.A.; Dong, Z.; Bender, J. Small RNAs prevent transcription-coupled loss of Histone H3 Lysine 9 methylation in Arabidopsis thaliana. PLoS Genet. 2011, 7, e1002350.
84. Durand, S.; Bouché, N.; Strand, E.P.; Loudet, O.; Camilleri, C. Rapid establishment of genetic incompatibility through natural epigenetic variation. Curr. Biol. 2012, 22, 326–331.
85. Slotkin, R.K.; Freeling, M.; Lisch, D. Mu killer causes the heritable inactivation of the Mutator family of transposable elements in Zea mays. Genetics 2003, 165, 781–797.
86. Slotkin, R.K.; Freeling, M.; Lisch, D. Heritable transposon silencing initiated by a naturally occurring transposon inverted duplication. Nat. Genet. 2005, 37, 641–644.
87. Singh, J.; Freeling, M.; Lisch, D. A position effect on the heritability of epigenetic silencing. PLoS Genet. 2008, 4, e1000216.
88. Woodhouse, M.R.; Freeling, M.; Lisch, D. Initiation, establishment, and maintenance of heritable *MuDR* transposon silencing in maize are mediated by distinct factors. PLoS Biol. 2006, 4, e339.
89. Nobuta, K.; Lu, C.; Shrivastava, R.; Pillay, M.; de Paoli, E.; Accerbi, M.; Arteaga-Vazquez, M.; Sidorenko, L.; Jeong, D.H.; Yen, Y.; et al. Distinct size distribution

of endogenous siRNAs in maize: Evidence from deep sequencing in the *mop1*-1 mutant. Proc. Natl. Acad. Sci. USA 2008, 105, 14958–14963.
90. Tarutani, Y.; Shiba, H.; Ito, T.; Kakizaki, T.; Suzuki, G.; Watanabe, M.; Isogai, A.; Takayama, S. *Trans*-acting small RNA determines dominance relationships in Brassica self-incompatibility. Nature 2010, 466, 983–986.
91. Takayama, S.; Isogai, A. Self-incompatibility in plants. Annu. Rev. Plant Biol. 2005, 56, 231–251.
92. Fujimoto, R.; Nishio, T. Self-incompatibility. Adv. Bot. Res. 2007, 45, 139–154.
93. Watanabe, M.; Takayama, S.; Isogai, A.; Hinata, K. Recent progresses on self-incompatibility research in Brassica species. Breed. Sci. 2003, 53, 199–208.
94. Shiba, H.; Kakizaki, T.; Iwano, M.; Tarutani, Y.; Watanabe, M.; Isogai, A.; Takayama, S. Dominance relationships between self-incompatibility alleles controlled by DNA methylation. Nat. Genet. 2006, 38, 297–299.
95. Tarutani, Y.; Takayama, S. Monoallelic gene expression and its mechanisms. Curr. Opin. Plant Biol. 2011, 14, 608–613.
96. Finnegan, E.J.; Liang, D.; Wang, M.B. Self-incompatibility: *SMI* silences through a novel sRNA pathway. Trends Plant Sci. 2011, 16, 238–241.
97. Koornneef, M.; Hanhart, C.J.; van der Veen, J.H. A genetic and physiological analysis of late flowering mutants in Arabidopsis thaliana. Mol. Gen. Genet. 1991, 229, 57–66.
98. Ikeda, Y.; Kobayashi, Y.; Yamaguchi, A.; Abe, M.; Araki, T. Molecular basis of late-flowering phenotype caused by dominant epi-alleles of the *FWA* locus in Arabidopsis. Plant Cell Physioly 2007, 48, 205–220.
99. Kinoshita, T.; Miura, A.; Choi, Y.; Kinoshita, Y.; Cao, X.; Jacobsen, S.E.; Fischer, R.L.; Kakutani, T. One-way control of *FWA* imprinting in Arabidopsis endosperm by DNA methylation. Science 2004, 303, 521–523.
100. Choi, Y.; Gehring, M.; Johnson, L.; Hannon, M.; Harada, J.J.; Goldberg, R.B.; Jacobsen, S.E.; Fischer, R.L. DEMETER, a DNA Glycosylase Domain Protein, is required for endosperm gene imprinting and seed viability in Arabidopsis. Cell 2002, 110, 33–42.
101. Lippman, Z.; Gendrel, A.V.; Black, M.; Vaughn, M.W.; Dedhia, N.; McCombie, W.R.; Lavine, K.; Mittal, V.; May, B.; Kasschau, K.D.; et al. Role of transposable elements in heterochromatin and epigenetic control. Nature 2004, 430, 471–476.
102. Chan, S.W.; Zhang, X.; Bernatavichute, Y.V.; Jacobsen, S.E. Two-Step recruitment of RNA-directed DNA methylation to tandem repeats. PLoS Biol. 2006, 4, e363.
103. Chan, S.W.; Zilberman, D.; Xie, Z.; Johansen, L.K.; Carrington, J.C.; Jacobsen, S.E. RNA silencing genes control de novo DNA methylation. Science 2004, 303, doi: 10.1126/science.109598.
104. Kinoshita, Y.; Saze, H.; Kinoshita, T.; Miura, A.; Soppe, W.J.; Koornneef, M.; Kakutani, T. Control of *FWA* gene silencing in Arabidopsis thaliana by SINE-related direct repeats. Plant J. 2007, 49, 38–45.
105. Fujimoto, R.; Kinoshita, Y.; Kawabe, A.; Kinoshita, T.; Takashima, K.; Nordborg, M.; Nasrallah, M.E.; Shimizu, K.K.; Kudoh, H.; Kakutani, T. Evolution and metastable epigenetic states of imprinted *FWA* genes in the genus Arabidopsis. PLoS Genet. 2008, 4, e1000048.

106. Gehring, M.; Bubb, K.L.; Henikoff, S. Extensive demethylation of repetitive elements during seed development underlies gene imprinting. Science 2009, 324, 1447–1451.
107. Hsieh, T.F.; Ibarra, C.A.; Silva, P.; Zemach, A.; Eshed-Williams, L.; Fischer, R.L.; Zilberman, D. Genome-wide demethylation of Arabidopsis endosperm. Science 2009, 324, 1451–1454.
108. Fujimoto, R.; Sasaki, T.; Kudoh, H.; Taylor, J.M.; Kakutani, T.; Dennis, E.S. Epigenetic variation in the *FWA* gene within the genus Arabidopsis. Plant J. 2011, 66, 831–843.
109. Fujimoto, R.; Taylor, J.M.; Sasaki, T.; Kawanabe, T.; Dennis, E.S. Genome wide gene expression in artificially synthesized amphidiploids of Arabidopsis. Plant Mol. Biol. 2011, 77, 419–431.
110. Kawanabe, T.; Fujimoto, R. Inflorescence abnormalities occur with overexpression of Arabidopsis lyrata FT in the *FWA* mutant of Arabidopsis thaliana. Plant Sci. 2011, 181, 496–503.
111. Angers, B.; Castonguay, E.; Massicotte, R. Environmentally induced phenotypes and DNA methylation: How to deal with unpredictable conditions until the next generation and after. Mol. Ecol. 2010, 19, 1283–1295.
112. Grativol, C.; Hemerly, A.S.; Ferreira, P.C. Genetic and epigenetic regulation of stress responses in natural plant populations. Biochim. Biophys. Acta 2012, 1819, 176–185.

CHAPTER 13

METAGENOMICS: A GUIDE FROM SAMPLING TO DATA ANALYSIS

TORSTEN THOMAS, JACK GILBERT, and FOLKER MEYER

13.1 INTRODUCTION

Arguably, one of the most remarkable events in the field of microbial ecology in the past decade has been the advent and development of metagenomics. Metagenomics is defined as the direct genetic analysis of genomes contained with an environmental sample. The field initially started with the cloning of environmental DNA, followed by functional expression screening [1], and was then quickly complemented by direct random shotgun sequencing of environmental DNA [2,3]. These initial projects not only showed proof of principle of the metagenomic approach, but also uncovered an enormous functional gene diversity in the microbial world around us [4].

Metagenomics provides access to the functional gene composition of microbial communities and thus gives a much broader description than phylogenetic surveys, which are often based only on the diversity of one gene, for instance the 16S rRNA gene. On its own, metagenomics gives genetic information on potentially novel biocatalysts or enzymes, genomic linkages between function and phylogeny for uncultured organisms, and

This chapter was originally published under the Creative Commons Attribution License. Thomas T, Gilbert J, and Meyer F. Metagenomics: A Guide from Sampling to Data Analysis. Microbial Informatics and Experimentation, *2,3 (2012), 12 pages. doi:10.1186/2042-5783-2-3.*

evolutionary profiles of community function and structure. It can also be complemented with metatranscriptomic or metaproteomic approaches to describe expressed activities [5,6]. Metagenomics is also a powerful tool for generating novel hypotheses of microbial function; the remarkable discoveries of proteorhodopsin-based photoheterotrophy or ammonia-oxidizing Archaea attest to this fact [7,8].

The rapid and substantial cost reduction in next-generation sequencing has dramatically accelerated the development of sequence-based metagenomics. In fact, the number of metagenome shotgun sequence datasets has exploded in the past few years. In the future, metagenomics will be used in the same manner as 16S rRNA gene fingerprinting methods to describe microbial community profiles. It will therefore become a standard tool for many laboratories and scientists working in the field of microbial ecology.

This review gives an overview of the field of metagenomics, with particular emphasis on the steps involved in a typical sequence-based metagenome project (Figure 1). We describe and discuss sample processing, sequencing technology, assembly, binning, annotation, experimental design, statistical analysis, and data storage and sharing. Clearly, any kind of metagenomic dataset will benefit from the rich information available from other metagenome projects, and it is hoped that common, yet flexible, standards and interactions among scientists in the field will facilitate this sharing of information. This review article summarizes the current thinking in the field and introduces current practices and key issues that those scientists new to the field need to consider for a successful metagenome project.

13.2 SAMPLING AND PROCESSING

Sample processing is the first and most crucial step in any metagenomics project. The DNA extracted should be representative of all cells present in the sample and sufficient amounts of high-quality nucleic acids must be obtained for subsequent library production and sequencing. Processing requires specific protocols for each sample type, and various robust methods for DNA extraction are available (e.g. [3,9,10]). Initiatives are also under way to explore the microbial biodiversity from tens of thou-

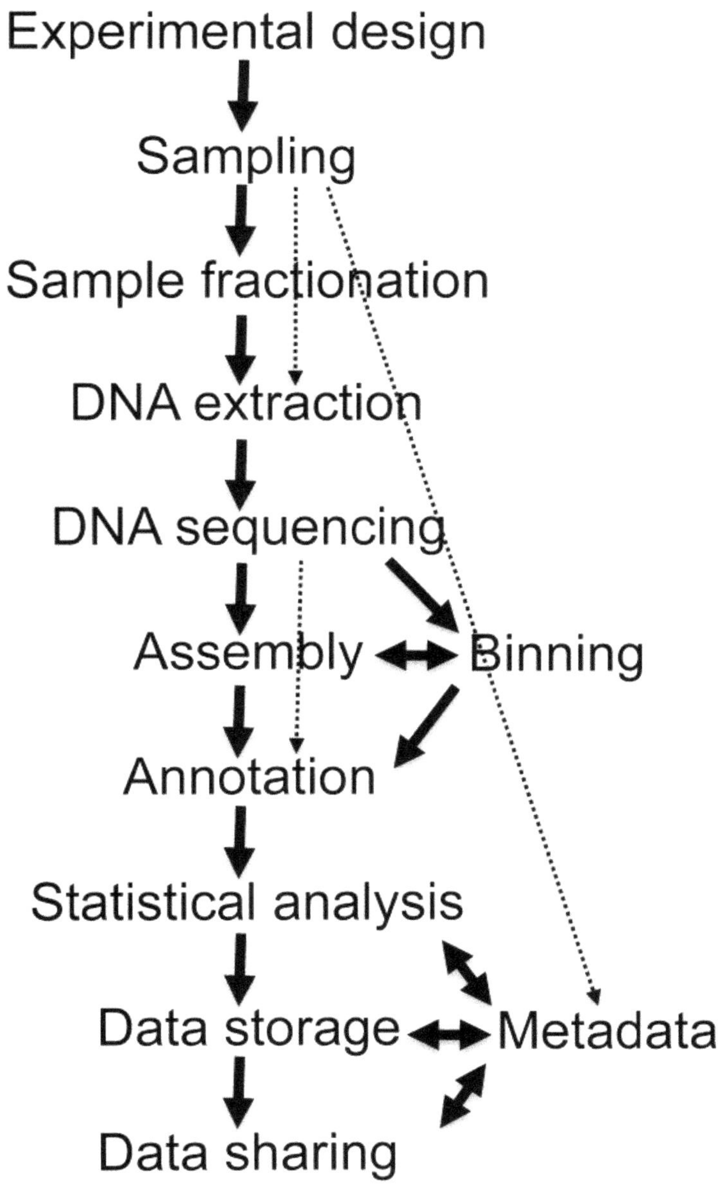

FIGURE 1: Flow diagram of a typical metagenome projects. Dashed arrows indicate steps that can be omitted.

sands of ecosystems using a single DNA extraction technology to ensure comparability [11].

If the target community is associated with a host (e.g. an invertebrate or plant), then either fractionation or selective lysis might be suitable to ensure that minimal host DNA is obtained (e.g. [9,12]). This is particularly important when the host genome is large and hence might "overwhelm" the sequences of the microbial community in the subsequent sequencing effort. Physical fractionation is also applicable when only a certain part of the community is the target of analysis, for example, in viruses seawater samples. Here a range of selective filtration or centrifugation steps, or even flow cytometry, can be used to enrich the target fraction [3,13,14]. Fractionation steps should be checked to ensure that sufficient enrichment of the target is achieved and that minimal contamination of non-target material occurs.

Physical separation and isolation of cells from the samples might also be important to maximize DNA yield or avoid coextraction of enzymatic inhibitors (such as humic acids) that might interfere with subsequent processing. This situation is particularly relevant for soil metagenome projects, and substantial work has been done in this field to address the issue ([10] and references therein). Direct lysis of cells in the soil matrix versus indirect lysis (i.e. after separation of cells from the soil) has a quantifiable bias in terms of microbial diversity, DNA yield, and resulting sequence fragment length [10]. The extensive work on soil highlights the need to ensure that extraction procedures are well benchmarked and that multiple methods are compared to ensure representative extraction of DNA.

Certain types of samples (such as biopsies or ground-water) often yield only very small amounts of DNA [15]. Library production for most sequencing technologies require high nanograms or micrograms amounts of DNA (see below), and hence amplification of starting material might be required. Multiple displacement amplification (MDA) using random hexamers and phage phi29 polymerase is one option employed to increase DNA yields. This method can amplify femtograms of DNA to produce micrograms of product and thus has been widely used in single-cell genomics and to a certain extent in metagenomics [16,17]. As with any amplification method, there are potential problems associated with reagent contaminations, chimera formation and sequence bias in the amplification, and their impact will depend on the amount and type of starting material and the required number of amplification rounds to produce sufficient amounts of nucleic acids. These issues can have significant impact on subsequent metagenomic community analysis [15], and so it will be necessary to consider whether amplification is permissible.

13.3 SEQUENCING TECHNOLOGY

Over the past 10 years metagenomic shotgun sequencing has gradually shifted from classical Sanger sequencing technology to next-generation sequencing (NGS). Sanger sequencing, however, is still considered the gold standard for sequencing, because of its low error rate, long read length (> 700 bp) and large insert sizes (e.g. > 30 Kb for fosmids or bacterial artificial chromosomes (BACs)). All of these aspects will improve assembly outcomes for shotgun data, and hence Sanger sequencing might still be applicable if generating close-to-complete genomes in low-diversity environments is the objective [18]. A drawback of Sanger sequencing is the labor-intensive cloning process in its associated bias against genes toxic for the cloning host [19] and the overall cost per gigabase (appr. USD 400,000).

Of the NGS technologies, both the 454/Roche and the Illumina/Solexa systems have now been extensively applied to metagenomic samples. Excellent reviews of these technologies are available [20,21], but a brief summary is given here with particular attention to metagenomic applications.

The 454/Roche system applies emulsion polymerase chain reaction (ePCR) to clonally amplify random DNA fragments, which are attached to microscopic beads. Beads are deposited into the wells of a picotitre plate and then individually and in parallel pyrosequenced. The pyrosequencing process involves the sequential addition of all four deoxynucleoside triphosphates, which, if complementary to the template strand, are incorporated by a DNA polymerase. This polymerization reaction releases pyrophosphate, which is converted via two enzymatic reactions to produce light. Light production of ~ 1.2 million reactions is detected in parallel via a charge-coupled device (CCD) camera and converted to the actual sequence of the template. Two aspects are important in this process with respect to metagenomic applications. First, the ePCR has been shown to produce artificial replicate sequences, which will impact any estimates of gene abundance. Understanding the amount of replicate sequences is crucial for the data quality of sequencing runs, and replicates can be identified and filtered out with bioinformatics tools [22,23]. Second, the intensity of light produced when the polymerase runs through a homopolymer is often difficult to correlate to the actual number of nucleotide positions. Typically, this results in insertion or deletion errors in homopolymers and can hence cause reading frameshifts, if protein coding sequences (CDSs) are called on a single read. This type of error can however be incorporated into models of CDS prediction thus resulting in high, albeit not perfect, accuracy [24]. Despite these disadvantages, the much cheaper cost of ~ USD 20,000 per gigabase pair has made 454/Roche pyrosequencing a popular choice for shotgun-sequencing metagenomics. In addition, the 454/Roche technology produces an average read length between 600-800 bp, which is long enough to cause only minor loss in the number of reads that can be annotated [25]. Sample preparation has also been optimized so that tens of nanograms of DNA are sufficient for sequencing single-end libraries [26,27], although pair-end sequencing might still require micrograms quantities. Moreover, the 454/Roche sequencing platform offers multiplexing allowing for up to 12 samples to be analyzed in a single run of ~500 Mbp.

The Illumina/Solexa technology immobilizes random DNA fragments on a surface and then performs solid-surface PCR amplification, resulting in clusters of identical DNA fragments. These are then sequenced with reversible terminators in a sequencing-by-synthesis process [28]. The cluster density is enormous, with hundreds of millions of reads per surface channel and 16 channels per run on the HiSeq2000 instrument. Read length is now approaching 150 bp, and clustered fragments can be sequenced from both ends. Continuous sequence information of nearly 300 bp can be obtained from two overlapping 150 bp paired-reads from a single insert. Yields of ~60 Gbp can therefore be typically expected in a single channel. While Illumina/Solexa has limited systematic errors, some datasets have shown high error rates at the tail ends of reads [29]. In general, clipping reads has proven to be a good strategy for eliminating the error in "bad" datasets, however, sequence quality values should also be used to detect "bad" sequences. The lower costs of this technology (~ USD 50 per Gbp) and recent success in its application to metagenomics, and even the generation of draft genomes from complex dataset [30,31], are currently making the Illumina technology an increasingly popular choice. As with 454/Roche sequencing, starting material can be as low as a 20 nanograms, but larger amounts (500-1000 ng) are required when matepair-libraries for longer insert libraries are made. The limited read length of the Illumina/Solexa technology means that a greater proportion of unassembled reads might be too short for functional annotation than are with 454/Roche technology [25]. While assembly might be advisable in such a case, potential bias, such as the suppression of low-abundance species (which can not be assembled) should be considered, as should the fact that some current software packages (e.g. MG-RAST) are capable of analyzing unassembled Illumina reads of 75 bp and longer. Multiplexing of samples is also available for individual sequencing channels, with more than 500 samples multiplexed per lane. Another important factor to consider is run time, with a 2 × 100 bp paired-end sequencing analysis taking approx. 10 days HiSeq2000 instrument time, in contrast to 1 day for the 454/ Roche technology. However, faster runtime (albeit at higher cost per Gbp of approx. USD 600) can be achieved with the new Illumina MiSeq instrument. This smaller version of Illumina/Solexa technology can also be used to test-run sequencing libraries, before analysis on HiSeq instrument for deeper sequencing.

A few additional sequencing technologies are available that might prove useful for metagenomic applications, now or in the near future. The Applied Biosystems SOLiD sequencer has been extensively used, for example, in genome resequencing [32]. SOLiD arguably provides the lowest error rate of any current NGS sequencing technology, however it does not achieve reliable read length beyond 50 nucleotides. This will limit its applicability for direct gene annotation of unassembled reads or for assembly of large contigs. Nevertheless, for assembly or mapping of metagenomic data against a reference genome, recent work showed encouraging outcomes [33]. Roche is also marketing a smaller-scale sequencer based on pyrosequencing with about 100 Mbp output and low per run costs. This system might be useful, because relatively low coverage of metagenomes can establish meaningful gene profiles [34]. Ion Torrent (and more recently Ion Proton) is another emerging technology and is based on the principle that protons released during DNA polymerization can detect nucleotide incorporation. This system promises read lengths of > 100 bp and throughput on the order of magnitude of the 454/Roche sequencing systems. Pacific Biosciences (PacBio) has released a sequencing technology based on single-molecule, real-time detection in zero-mode waveguide wells. Theoretically, this technology on its RS1 platform should provide much greater read lengths than the other technologies mentioned, which would facilitate annotation and assembly. In addition, a process called strobing will mimic pair-end reads. However, accuracy of single reads with PacBio is currently only at 85%, and random reads are "dropped," making the instrument unusable in its current form for metagenomic sequencing [35]. Complete Genomics is offering a technology based on sequencing DNA nanoballs with combinatorial probe-anchor ligation [36]. Its read length of 35 nucleotides is rather limited and so might be its utility for de novo assemblies. While none of the emerging sequencing technologies have been thoroughly applied and tested with metagenomics samples, they offer promising alternatives and even further cost reduction.

13.4 ASSEMBLY

If the research aims at recovering the genome of uncultured organisms or obtain full-length CDS for subsequent characterization rather than a functional description of the community, then assembly of short read fragments will be performed to obtain longer genomic contigs. The majority of current assembly programs were designed to assemble single, clonal genomes and their utility for complex pan-genomic mixtures should be approached with caution and critical evaluation.

Two strategies can be employed for metagenomics samples: reference-based assembly (co-assembly) and de novo assembly.

Reference-based assembly can be done with software packages such as Newbler (Roche), AMOS http://sourceforge.net/projects/amos/ webcite, or MIRA [37]. These software packages include algorithms that are fast and memory-efficient and hence can often be performed on laptop-sized machines in a couple of hours. Reference-based assembly works well, if the metagenomic dataset contains sequences where closely related reference genomes are available. However, differences in the true genome of the sample to the reference, such as a large insertion, deletion, or polymorphisms, can mean that the assembly is fragmented or that divergent regions are not covered.

De novo assembly typically requires larger computational resources. Thus, a whole class of assembly tools based on the de Bruijn graphs was specifically created to handle very large amounts of data [38,39]. Machine requirements for the de Bruijn assemblers Velvet [40] or SOAP [41] are still significantly higher than for reference-based assembly (co-assembly), often requiring hundreds of gigabytes of memory in a single machine and run times frequently being days.

The fact that most (if not all) microbial communities include significant variation on a strain and species level makes the use of assembly algorithms that assume clonal genomes less suitable for metagenomics. The "clonal" assumptions built into many assemblers might lead to suppression of contig formation for certain heterogeneous taxa at specific parameter settings. Recently, two de Bruijn-type assemblers, MetaVelvet and Meta-IDBA [42] have been released that deal explicitly with the nonclonality of natural populations. Both assemblers aim to identify within the entire de Bruijn graph a sub-graph that represents related genomes. Alternatively, the metagenomic sequence mix can be partition into "species bins" via k-mer binning (Titus Brown, personal communications). Those subgraphs or subsets are then resolved to build a consensus sequence of the genomes. For Meta-IDBA a improvement in terms of N50 and maximum contig length has been observed when compared to "classical" de Bruijn assembler (e.g. Velvet or SOAP; results from the personal experience of the authors; data not shown here). The development of "metagenomic assemblers" is however still at an early stage, and it is difficult to access their accuracy for real metagenomic data as typically no references exist to compare the results to. A true gold standard (i.e. a real dataset for a diverse microbial community with known reference sequences) that assemblers can be evaluated against is thus urgently required.

Several factors need to be considered when exploring the reasons for assembling metagenomic data; these can be condensed to two important questions. First, what is the length of the sequencing reads used to generate the metagenomic dataset, and are longer sequences required for annotation? Some approaches, e.g. IMG/M, prefer assembled contigs, other pipelines such as MG-RAST [43] require only 75 bp or longer for gene prediction or similarity analysis that provides taxonomic binning and functional classification. On the whole, however, the longer the sequence information, the better is the ability to obtain accurate information. One obvious impact is on annotation: the longer the sequence, the more information provided, making it easier to compare with known genetic data (e.g. via homology searches [25]). Annotation issues will be discussed in the next section. Binning and classification of DNA fragments for phylogenetic or taxonomic assignment also benefits from long, contiguous sequences and certain tools (e.g. Phylopythia) work reliably only over a specific cut-off point (e.g. 1 Kb) [44]. Second, is the dataset assembled to reduce data-processing requirements? Here, as an alternative to assembling reads into contigs, clustering near-identical reads with cd-hit [45] or uclust [46] will provide clear benefits in data reduction. The MG-RAST pipeline also uses clustering as a data reduction strategy.

Fundamentally, assembly is also driven by the specific problem that single reads have generally lower quality and hence lower confidence in accuracy than do multiple reads that cover the same segment of genetic information. Therefore, merging reads increases the quality of information. Obviously in a complex community with low sequencing depth or coverage, it is unlikely to actually get many reads that cover the same fragment of DNA. Hence assembly may be of limited value for metagenomics.

Unfortunately, without assembly, longer and more complex genetic elements (e.g., CRISPRS) cannot be analyzed. Hence there is a need for metagenomic assembly to obtain high-confidence contigs that enable the study of, for example, major repeat classes. However, none of the current assembly tools is bias-free. Several strategies have been proposed to increase assembly accuracy [38], but strategies such as removal of rare k-mers are no longer considered adequate, since rare k-mers do not represent sequence errors (as initially assumed), but instead represent reads from less abundant pan-genomes in the metagenomic mix.

13.5 BINNING

Binning refers to the process of sorting DNA sequences into groups that might represent an individual genome or genomes from closely related organisms. Several algorithms have been developed, which employ two types of information contained within a given DNA sequence. Firstly, compositional binning makes use of the fact that genomes have conserved nucleotide composition (e.g. a certain GC or the particular abundance distribution of k-mers) and this will be also reflected in sequence fragments of the genomes. Secondly, the unknown DNA fragment might encode for a gene and the similarity of this gene with known genes in a reference database can be used to classify and hence bin the sequence.

Compositional-based binning algorithms include Phylopythia [44], S-GSOM [47], PCAHIER [48,49] and TACAO [49], while examples of purely similarity-based binning software include IMG/M [50], MG-RAST [43], MEGAN [51], CARMA [52], SOrt-ITEMS [53] and MetaPhyler [54]. There is also number of binning algorithms that consider both composition and similarity, including the programs PhymmBL [55] and MetaCluster [56]. All these tools employ different methods of grouping sequences, including self-organising maps (SOMs) or hierarchical clustering, and are operated in either an unsupervised manner or with input from the user (supervised) to define bins.

Important considerations for using any binning algorithm are the type of input data available and the existence of a suitable training datasets or reference genomes. In general, composition-based binning is not reliable for short reads, as they do not contain enough information. For example, a 100 bp read can at best possess only less than half of all 256 possible

4-mers and this is not sufficient to determine a 4-mer distribution that will reliably relate this read to any other read. Compositional assignment can however be improved, if training datasets (e.g. a long DNA fragment of known origin) exist that can be used to define a compositional classifier [44]. These "training" fragments can either be derived from assembled data or from sequenced fosmids and should ideally contain a phylogenetic marker (such as a rRNA gene) that can be used for high-resolution, taxonomic assignment of the binned fragments [57].

Short reads may contain similarity to a known gene and this information can be used to putatively assign the read to a specific taxon. This taxonomic assignment obviously requires the availability of reference data. If the query sequence is only distantly related to known reference genomes, only a taxonomic assignment at a very high level (e.g. phylum) is possible. If the metagenomic dataset, however, contains two or more genomes that would fall into this high taxon assignment, then "chimeric" bins might be produced. In this case, the two genomes might be separated by additional binning based on compositional features. In general, however this might again require that the unknown fragments have a certain length.

Binning algorithm will obviously in the future benefit from the availability of a greater number and phylogenetic breadth of reference genomes, in particular for similarity-based assignment to low taxonomic levels. Post-assembly the binning of contigs can lead to the generation of partial genomes of yet-uncultured or unknown organisms, which in turn can be used to perform similarity-based binning of other metagenomic datasets. Caution should however been taken to ensure the validity of any newly created genome bin, as "contaminating" fragments can rapidly propagate into false assignments in subsequent binning efforts. Prior to assembly with clonal assemblers binning can be used to reduce the complexity of an assembly effort and might reduce computational requirement.

As major annotation pipelines like IMG/M or MG-RAST also perform taxonomic assignments of reads, one needs to carefully weigh the additional computational demands of the particular binning algorithm chosen against the added value they provide.

13.6 ANNOTATION

For the annotation of metagenomes two different initial pathways can be taken. First, if reconstructed genomes are the objective of the study and assembly has produced large contigs, it is preferable to use existing pipelines for genome annotation, such as RAST [58] or IMG [59]. For this approach to be successful, minimal contigs length of 30,000 bp or longer are required. Second, annotation can be performed on the entire community and relies on unassembled reads or short contigs. Here the tools for genome annotation are significantly less useful than those specifically developed for metagenomic analyses. Annotation of metagenomic sequence data has in general two steps. First, features of interest (genes) are identified (feature prediction) and, second, putative gene functions and taxonomic neighbors are assigned (functional annotation).

Feature prediction is the process of labeling sequences as genes or genomic elements. For completed genome sequences a number of algorithms have been developed [60,61] that identify CDS with more than 95% accuracy and a low false negative ratio. A number of tools were specifically designed to handle metagenomic prediction of CDS, including FragGeneScan [24], MetaGeneMark [62], MetaGeneAnnotator (MGA)/ Metagene [63] and Orphelia [64,65]. All of these tools use internal information (e.g. codon usage) to classify sequence stretches as either coding or non-coding, however they distinguish themselves from each other by the quality of the training sets used and their usefulness for short or error-prone sequences. FragGeneScan is currently the only algorithm known to the authors that explicitly models sequencing errors and thus results in gene prediction errors of only 1-2%. True positive rates of FragGeneScan are around 70% (better than most other methods), which means that even this tool still misses a significant subset of genes. These missing genes can potentially be identified by BLAST-based searches, however the size of current metagenomic datasets makes this computational expensive step often prohibitive.

There exists also a number of tools for the prediction of non-protein coding genes such as tRNAs [66,67], signal peptides [68] or CRISPRs

[69,70], however they might require significant computational resources or long contiguous sequences. Clearly subsequent analysis depends on the initial identification of features and users of annotation pipelines need to be aware of the specific prediction approaches used. MG-RAST uses a two-step approach for feature identification, FGS and a similarity search for ribosomal RNAs against a non-redundant integration of the SILVA [71], Greengenes [72] and RDP [73] databases. CAMERA's RAMCAPP pipeline [74] uses FGA and MGA, while IMG/M employs a combination of tools, including FGS and MGA [58,59].

Functional annotation represents a major computational challenge for most metagenomic projects and therefore deserves much attention now and over the next years. Current estimates are that only 20 to 50% of a metagenomic sequences can be annotated [75], leaving the immediate question of importance and function of the remaining genes. We note that annotation is not done de novo, but via mapping to gene or protein libraries with existing knowledge (i.e., a non-redundant database). Any sequences that cannot be mapped to the known sequence space are referred to as ORFans. These ORFans are responsible for the seemingly never-ending genetic novelty in microbial metagenomics (e.g. [76]. Three hypotheses exist for existence of this unknown fraction. First, ORFans might simply reflect erroneous CDS calls caused by imperfect detection algorithms. Secondly, these ORFans are real genes, but encode for unknown biochemical functions. Third, ORFan genes have no sequence homology with known genes, but might have structural homology with known proteins, thus representing known protein families or folds. Future work will likely reveal that the truth lies somewhere between these hypotheses [77]. For improving the annotation of ORFan genes, we will rely on the challenging and labor-intensive task of protein structure analysis (e.g. via NMR and x-ray crystallography) and on biochemical characterization.

Currently, metagenomic annotation relies on classifying sequences to known functions or taxonomic units based on homology searches against available "annotated" data. Conceptually, the annotation is relatively simple and for small datasets (< 10,000 sequences) manual curation can be used increase the accuracy of any automated annotation. Metagenomic datasets are typically very large, so manual annotation is not possible. Automated annotation therefore has to become more accurate and computationally inexpensive. Currently, running a BLASTX similarity search is computationally expensive; as much as ten times the cost of sequencing [78]. Unfortunately, computationally less demanding methods involving detecting feature composition in genes [44] have limited success for short reads. With growing dataset sizes, faster algorithms are urgently needed, and several programs for similarity searches have been developed to resolve this issue [46,79-81].

Many reference databases are available to give functional context to metagenomic datasets, such as KEGG [82], eggNOG [83], COG/KOG [84], PFAM [85], and TIGRFAM [86]. However, since no reference database covers all biological functions, the ability to visualize and merge the interpretations of all database searches within a single framework is important, as implemented in the most recent versions of MG-RAST and IMG/M. It is essential that metagenome analysis platforms be able to share data in ways that map and visualize data in the framework of other platforms. These metagenomic exchange languages should also reduce the burden associated with re-processing large datasets, minimizing, the redundancy of searching and enabling the sharing of annotations that can be mapped to different ontologies and nomenclatures, thereby allowing multifaceted interpretations. The Genomic Standards Consortium (GSC) with the M5 project is providing a prototypical standard for exchange of computed metagenome analysis results, one cornerstone of these exchange languages.

Several large-scale databases are available that process and deposit metagenomic datasets. MG-RAST, IMG/M, and CAMERA are three prominent systems [43,50,74]. MG-RAST is a data repository, an analysis pipeline and a comparative genomics environment. Its fully automated pipeline provides quality control, feature prediction and functional annotation and has been optimized for achieving a trade-off between accuracy and computational efficiency for short reads using BLAT {Kent, 2002 #64}. Results are expressed in the form of abundance profiles for specific taxa or functional annotations. Supported are the comparison of NCBI taxonomies derived from 16S rRNA gene or whole genome shotgun data and the comparison of relative abundance for KEGG, eggNOG, COG and SEED subsystems on multiple levels of resolution. Users can also download all data products generated by MG-RAST, share them and publish within the portal. The MG-RAST web interface allows comparison using a number of statistical techniques and allows for the incorporation of metadata into the statistics. MG-RAST has more than 7000 users, > 38,000 uploaded and analyzed metagenomes (of which 7000 are publicly accessible) and 9 Terabases analyzed as of December 2011. These statistics demonstrate a move by the scientific community to centralize resources and standardize annotation.

IMG/M also provides a standardized pipeline, but with "higher" sensitivity as it performs, for example, hidden Markov model (HMM) and BLASTX searches at substantial computational cost. In contrast to MG-RAST, comparisons in IMG/M are not performed on an abundance table level, but are based on an all vs. all genes comparison. Therefore IMG/M is the only system that integrates all datasets into a single protein level abstraction. Both IMG/M and MG-RAST provide the ability to use stored computational results for comparison, enabling comparison of novel metagenomes with a rich body of other datasets without requiring the end-user to provide the computational means for reanalysis of all datasets involved in their study. Other systems, such as CAMERA [74], offer more flexible annotation schema but require that individual researchers understand the annotation of data and analytical pipelines well enough to be confident in their interpretation. Also for comparison, all datasets need to be analyzed using the same workflow, thus adding additional computational requirements. CAMERA allows the publication of datasets and was the first to support the Genomic Standards Consortium's Minimal Information checklists for metadata in their web interface [87].

MEGAN is another tool used for visualizing annotation results derived from BLAST searches in a functional or taxonomic dendrogram [51]. The use of dendrograms to display metagenomic data provides a collapsible network of interpretation, which makes analysis of particular functional or taxonomic groups visually easy.

13.7 EXPERIMENTAL DESIGN AND STATISTICAL ANALYSIS

Owing to the high costs, many of the early metagenomic shotgun-sequencing projects were not replicated or were focused on targeted exploration of specific organisms (e.g. uncultured organisms in low-diversity acid mine drainage [2]). Reduction of sequencing cost (see above) and a much wider appreciation of the utility of metagenomics to address fundamental questions in microbial ecology now require proper experimental designs with appropriate replication and statistical analysis. These design and statistical aspects, while obvious, are often not properly implemented in the field of microbial ecology [88]. However, many suitable approaches and strategies are readily available from the decades of research in quantitative ecology of higher organisms (e.g. animals, plants). In a simplistic way, the data from multiple metagenomic shotgun-sequencing projects can be reduced to tables, where the columns represent samples and the rows indicate either a taxonomic group or a gene function (or groups thereof) and the fields containing abundance or presence/absence data. This is analogous to species-sample matrices in ecology of higher organisms, and hence many for the statistical tools available to identify correlations and statistically significant patterns are transferable. As metagenomic data however often contain many more species or gene functions then the number of samples taken, appropriate corrections for multiple hypothesis testing have to be implemented (e.g. Bonferroni correction for t-test based analyses).

The Primer-E package [89] is a well-established tool, allowing for a range of multivariate statistical analyses, including the generation of multidimensional scaling (MDS) plots, analysis of similarities (ANOSIM), and identification of the species or functions that contribute to the difference between two samples (SIMPER). Recently, multivariate statistics was also incorporated in a web-based tools called Metastats [90], which revealed with high confidence discriminatory functions between the replicated metagenome dataset of the gut microbiota of lean and obese mice [91]. In addition, the ShotgunFunctionalizeR package provides several statistical procedures for assessing functional differences between samples, both for individual genes and for entire pathways using the popular R statistical package [92].

Ideally, and in general, experimental design should be driven by the question asked (rather than technical or operational restriction). For example, if a project aims to identify unique taxa or functions in a particular habitat, then suitable reference samples for comparison should be taken and processed in consistent manner. In addition, variation between sample types can be due to true biological variation, (something biologist would be most interested in) and technical variation and this should be carefully considered when planning the experiment. One should also be aware that many microbial systems are highly dynamic, so temporal aspects of sampling can have a substantial impact on data analysis and interpretation. While the question of the number of replicates is often difficult to predict prior to the final statistical analysis, small-scale experiments are often useful to understand the magnitude of variation inherent in a system. For example, a small number of samples could be selected and sequenced to shallower depth, then analyzed to determine if a larger sampling size or greater sequencing effort are required to obtain statistically meaningful results [88]. Also, the level at which replication takes place is something that should not lead to false interpretation of the data. For example, if one is interested in the level of functional variation of the microbial community in habitat A, then multiple samples from this habitat should be taken and processed completely separately, but in the same manner. Taking just one sample and splitting it up prior to processing will provide information only about technical, but not biological, variation in habitat A. Taking multiple samples and then pooling them will lose all information on variability and hence will be of little use for statistical purposes. Ultimately, good experimental design of metagenomic projects will facilitate integration of datasets into new or existing ecological theories [93].

As metagenomics gradually moves through a range of explorative biodiversity surveys, it will also prove itself extremely valuable for manipulative experiments. These will allow for observation of treatment impact on the functional and phylogenetic composition of microbial communities. Initial experiments already showed promising results [94]. However, careful experimental planning and interpretations should be paramount in this field.

One of the ultimate aims of metagenomics is to link functional and phylogenetic information to the chemical, physical, and other biological parameters that characterize an environment. While measuring all these parameters can be time-consuming and cost-intensive, it allows retrospective correlation analysis of metagenomic data that was perhaps not part of the initial aim of the project or might be of interest for other research questions. The value of such metadata cannot be overstated and, in fact, has become mandatory or optional for deposition of metagenomic data into some databases [50,74].

13.8 SHARING AND STORAGE OF DATA

Data sharing has a long tradition in the field of genome research, but for metagenomic data this will require a whole new level of organization and collaboration to provide metadata and centralized services (e.g., IMG/M, CAMERA and MG-RAST) as well as sharing of both data and computational results. In order to enable sharing of computed results, some aspects of the various analytical pipelines mentioned above will need to be coordinated - a process currently under way under the auspices of the GSC. Once this has been achieved, researchers will be able to download intermediate and processed results from any one of the major repositories for local analysis or comparison.

A suite of standard languages for metadata is currently provided by the Minimum Information about any (x) Sequence checklists (MIxS) [95]. MIxS is an umbrella term to describe MIGS (the Minimum Information about a Genome Sequence), MIMS (the Minimum Information about a Metagenome Sequence) and MIMARKS (Minimum Information about a MARKer Sequence)[87] and contains standard formats for recording environmental and experimental data. The latest of these checklists, MIMARKS builds on the foundation of the MIGS and MIMS checklists, by including an expansion of the rich contextual information about each environmental sample.

The question of centralized versus decentralized storage is also one of "who pays for the storage," which is a matter with no simple answer. The US National Center for Biotechnology Information (NCBI) is mandated to store all metagenomic data, however, the sheer volume of data being generated means there is an urgent need for appropriate ways of storing vast amounts of sequences. As the cost of sequencing continues to drop while the cost for analysis and storing remains more or less constant, selection of data storage in either biological (i.e. the sample that was sequenced) or digital form in (de-) centralized archives might be required. Ongoing work and successes in compression of (meta-) genomic data [96], however, might mean that digital information can still be stored cost-efficiently in the near future.

13.9 CONCLUSION

Metagenomics has benefited in the past few years from many visionary investments in both financial and intellectual terms. To ensure that those investments are utilized in the best possible way, the scientific community should aim to share, compare, and critically evaluate the outcomes of metagenomic studies. As datasets become increasingly more complex and comprehensive, novel tools for analysis, storage, and visualization will be required. These will ensure the best use of the metagenomics as a tool to address fundamental question of microbial ecology, evolution and diversity and to derive and test new hypotheses. Metagenomics will be employed as commonly and frequently as any other laboratory method, and "metagenomizing" a sample might become as colloquial as "PCRing." It is therefore also important that metagenomics be taught to students and young scientists in the same way that other techniques and approaches have been in the past.

REFERENCES

1. Handelsman J, Rondon MR, Brady SF, Clardy J, Goodman RM: Molecular biological access to the chemistry of unknown soil microbes: a new frontier for natural products. Chem Biol 1998, 5(10):R245-249.
2. Tyson GW, Chapman J, Hugenholtz P, Allen EE, Ram RJ, Richardson PM, Solovyev VV, Rubin EM, Rokhsar DS, Banfield JF: Community structure and metabolism through reconstruction of microbial genomes from the environment. Nature 2004, 428(6978):37-43.
3. Venter JC, Remington K, Heidelberg JF, Halpern AL, Rusch D, Eisen JA, Wu D, Paulsen I, Nelson KE, Nelson W, Fouts DE, Levy S, Knap AH, Lomas MW, Nealson K, White O, Peterson J, Hoffman J, Parsons R, Baden-Tillson H, Pfannkoch C, Rogers YH, Smith HO: Environmental genome shotgun sequencing of the Sargasso Sea. Science 2004, 304(5667):66-74.
4. Simon C, Daniel R: Metagenomic analyses: past and future trends. Appl Environ Microbiol 2011, 77(4):1153-1161.

5. Wilmes P, Bond PL: Metaproteomics: studying functional gene expression in microbial ecosystems. Trends Microbiol 2006, 14(2):92-97.
6. Gilbert JA, Field D, Huang Y, Edwards R, Li W, Gilna P, Joint I: Detection of large numbers of novel sequences in the metatranscriptomes of complex marine microbial communities. PLoS One 2008, 3(8):e3042.
7. Beja O, Aravind L, Koonin EV, Suzuki MT, Hadd A, Nguyen LP, Jovanovich SB, Gates CM, Feldman RA, Spudich JL, Spudich EN, DeLong EF: Bacterial rhodopsin: evidence for a new type of phototrophy in the sea. Science 2000, 289(5486):1902-1906.
8. Nicol GW, Schleper C: Ammonia-oxidising Crenarchaeota: important players in the nitrogen cycle? Trends Microbiol 2006, 14(5):207-212.
9. Burke C, Kjelleberg S, Thomas T: Selective extraction of bacterial DNA from the surfaces of macroalgae. Appl Environ Microbiol 2009, 75(1):252-256.
10. Delmont TO, Robe P, Clark I, Simonet P, Vogel TM: Metagenomic comparison of direct and indirect soil DNA extraction approaches. J Microbiol Methods 2011, 86(3):397-400.
11. Knight R, Desai N, Field D, Fierer N, Fuhrman J, Gordon J, Hu B, Hugenholtz P, Jansson J, Meyer F, Stevens R, Bailey M, Kowalchuk G, Gilbert J: Designing Better Metagenomic Surveys: The role of experimental design and metadata capture in making useful metagenomic datasets for ecology and biotechnology. Nature Biotechnology in review
12. Thomas T, Rusch D, DeMaere MZ, Yung PY, Lewis M, Halpern A, Heidelberg KB, Egan S, Steinberg PD, Kjelleberg S: Functional genomic signatures of sponge bacteria reveal unique and shared features of symbiosis. ISME J 2010, 4(12):1557-1567.
13. Palenik B, Ren Q, Tai V, Paulsen IT: Coastal Synechococcus metagenome reveals major roles for horizontal gene transfer and plasmids in population diversity. Environ Microbiol 2009, 11(2):349-359.
14. Angly FE, Felts B, Breitbart M, Salamon P, Edwards RA, Carlson C, Chan AM, Haynes M, Kelley S, Liu H, Mahaffy JM, Mueller JE, Nulton J, Olson R, Parsons R, Rayhawk S, Suttle CA, Rohwer F: The marine viromes of four oceanic regions. PLoS Biol 2006, 4(11):e368.
15. Abbai NS, Govender A, Shaik R, Pillay B: Pyrosequence analysis of unamplified and whole genome amplified DNA from hydrocarbon-contaminated groundwater. Mol Biotechnol 2011.
16. Lasken RS: Genomic DNA amplification by the multiple displacement amplification (MDA) method. Biochem Soc Trans 2009, 37(Pt 2):450-453.
17. Ishoey T, Woyke T, Stepanauskas R, Novotny M, Lasken RS: Genomic sequencing of single microbial cells from environmental samples. Curr Opin Microbiol 2008, 11(3):198-204.
18. Goltsman DS, Denef VJ, Singer SW, VerBerkmoes NC, Lefsrud M, Mueller RS, Dick GJ, Sun CL, Wheeler KE, Zemla A, Baker BJ, Hauser L, Land M, Shah MB, Thelen MP, Hettich RL, Banfield JF: Community genomic and proteomic analyses of chemoautotrophic iron-oxidizing "Leptospirillum rubarum" (Group II) and " Leptospirillum ferrodiazotrophum" (Group III) bacteria in acid mine drainage biofilms. Appl Environ Microbiol 2009, 75(13):4599-4615.

19. Sorek R, Zhu Y, Creevey CJ, Francino MP, Bork P, Rubin EM: Genome-wide experimental determination of barriers to horizontal gene transfer. Science 2007, 318(5855):1449-1452.
20. Metzker ML: Sequencing technologies - the next generation. Nat Rev Genet 2010, 11(1):31-46.
21. Mardis ER: The impact of next-generation sequencing technology on genetics. Trends Genet 2008, 24(3):133-141.
22. Niu B, Fu L, Sun S, Li W: Artificial and natural duplicates in pyrosequencing reads of metagenomic data. BMC Bioinformatics 2010, 11:187.
23. Teal TK, Schmidt TM: Identifying and removing artificial replicates from 454 pyrosequencing data. Cold Spring Harb Protoc 2010, 2010(4):pdb prot5409.
24. Rho M, Tang H, Ye Y: FragGeneScan: predicting genes in short and error-prone reads. Nucleic Acids Res 2010, 38(20):e191.
25. Wommack KE, Bhavsar J, Ravel J: Metagenomics: read length matters. Appl Environ Microbiol 2008, 74(5):1453-1463.
26. White RA, Blainey PC, Fan HC, Quake SR: Digital PCR provides sensitive and absolute calibration for high throughput sequencing. BMC Genomics 2009, 10:116.
27. Adey A, Morrison HG, Asan Xun X, Kitzman JO, Turner EH, Stackhouse B, MacKenzie AP, Caruccio NC, Zhang X, Shendure J: Rapid, low-input, low-bias construction of shotgun fragment libraries by high-density in vitro transposition. Genome Biol 2010, 11(12):R119.
28. Bentley DR, Balasubramanian S, Swerdlow HP, Smith GP, Milton J, Brown CG, Hall KP, Evers DJ, Barnes CL, Bignell HR, Boutell JM, Bryant J, Carter RJ, Keira Cheetham R, Cox AJ, Ellis DJ, Flatbush MR, Gormley NA, Humphray SJ, Irving LJ, Karbelashvili MS, Kirk SM, Li H, Liu X, Maisinger KS, Murray LJ, Obradovic B, Ost T, Parkinson ML, Pratt MR, et al.: Accurate whole human genome sequencing using reversible terminator chemistry. Nature 2008, 456(7218):53-59.
29. Nakamura K, Oshima T, Morimoto T, Ikeda S, Yoshikawa H, Shiwa Y, Ishikawa S, Linak MC, Hirai A, Takahashi H, Altaf-Ul-Amin M, Ogasawara N, Kanaya S: Sequence-specific error profile of Illumina sequencers. Nucleic Acids Res 2011, 39(13):e90.
30. Hess M, Sczyrba A, Egan R, Kim TW, Chokhawala H, Schroth G, Luo S, Clark DS, Chen F, Zhang T, Mackie RI, Pennacchio LA, Tringe SG, Visel A, Woyke T, Wang Z, Rubin EM: Metagenomic discovery of biomass-degrading genes and genomes from cow rumen. Science 2011, 331(6016):463-467.
31. Qin J, Li R, Raes J, Arumugam M, Burgdorf KS, Manichanh C, Nielsen T, Pons N, Levenez F, Yamada T, Mende DR, Li J, Xu J, Li S, Li D, Cao J, Wang B, Liang H, Zheng H, Xie Y, Tap J, Lepage P, Bertalan M, Batto JM, Hansen T, Le Paslier D, Linneberg A, Nielsen HB, Pelletier E, Renault P, et al.: A human gut microbial gene catalogue established by metagenomic sequencing. Nature 2010, 464(7285):59-65.
32. Gulig PA, de Crecy-Lagard V, Wright AC, Walts B, Telonis-Scott M, McIntyre LM: SOLiD sequencing of four Vibrio vulnificus genomes enables comparative genomic analysis and identification of candidate clade-specific virulence genes. BMC Genomics 2010, 11:512.

33. Tyler HL, Roesch LF, Gowda S, Dawson WO, Triplett EW: Confirmation of the sequence of 'Candidatus Liberibacter asiaticus' and assessment of microbial diversity in Huanglongbing-infected citrus phloem using a metagenomic approach. Mol Plant Microbe Interact 2009, 22(12):1624-1634.
34. Kunin V, Raes J, Harris JK, Spear JR, Walker JJ, Ivanova N, von Mering C, Bebout BM, Pace NR, Bork P, Hugenholtz P: Millimeter-scale genetic gradients and community-level molecular convergence in a hypersaline microbial mat. Mol Syst Biol 2008, 4:198.
35. Rasko DA, Webster DR, Sahl JW, Bashir A, Boisen N, Scheutz F, Paxinos EE, Sebra R, Chin CS, Iliopoulos D, Klammer A, Peluso P, Lee L, Kislyuk AO, Bullard J, Kasarskis A, Wang S, Eid J, Rank D, Redman JC, Steyert SR, Frimodt-Moller J, Struve C, Petersen AM, Krogfelt KA, Nataro JP, Schadt EE, Waldor MK: Origins of the E. coli strain causing an outbreak of hemolytic-uremic syndrome in Germany. N Engl J Med 2011, 365(8):709-717.
36. Drmanac R, Sparks AB, Callow MJ, Halpern AL, Burns NL, Kermani BG, Carnevali P, Nazarenko I, Nilsen GB, Yeung G, Dahl F, Fernandez A, Staker B, Pant KP, Baccash J, Borcherding AP, Brownley A, Cedeno R, Chen L, Chernikoff D, Cheung A, Chirita R, Curson B, Ebert JC, Hacker CR, Hartlage R, Hauser B, Huang S, Jiang Y, Karpinchyk V, et al.: Human genome sequencing using unchained base reads on self-assembling DNA nanoarrays. Science 2010, 327(5961):78-81.
37. Chevreux B, Wetter T, Suhai S: Genome Sequence Assembly Using Trace Signals and Additional Sequence Information Computer Science and Biology. Proceedings of the German Conference on Bioinformatics 1999, 99:45-56.
38. Miller JR, Koren S, Sutton G: Assembly algorithms for next-generation sequencing data. Genomics 2010, 95(6):315-327.
39. Pevzner PA, Tang H, Waterman MS: An Eulerian path approach to DNA fragment assembly. Proc Natl Acad Sci USA 2001, 98(17):9748-9753.
40. Zerbino DR, Birney E: Velvet: algorithms for de novo short read assembly using de Bruijn graphs. Genome Res 2008, 18(5):821-829.
41. Li R, Li Y, Kristiansen K, Wang J: SOAP: short oligonucleotide alignment program. Bioinformatics 2008, 24(5):713-714.
42. Peng Y, Leung HC, Yiu SM, Chin FY: Meta-IDBA: a de Novo assembler for metagenomic data. Bioinformatics 2011, 27(13):i94-101.
43. Glass EM, Wilkening J, Wilke A, Antonopoulos D, Meyer F: Using the metagenomics RAST server (MG-RAST) for analyzing shotgun metagenomes. Cold Spring Harb Protoc 2010., 2010(1)
44. McHardy AC, Martin HG, Tsirigos A, Hugenholtz P, Rigoutsos I: Accurate phylogenetic classification of variable-length DNA fragments. Nat Methods 2007, 4(1):63-72.
45. Li W, Godzik A: Cd-hit: a fast program for clustering and comparing large sets of protein or nucleotide sequences. Bioinformatics 2006, 22(13):1658-1659.
46. Edgar RC: Search and clustering orders of magnitude faster than BLAST. Bioinformatics 2010, 26(19):2460-2461.
47. Chan CK, Hsu AL, Halgamuge SK, Tang SL: Binning sequences using very sparse labels within a metagenome. BMC Bioinformatics 2008, 9:215.

48. Zheng H, Wu H: Short prokaryotic DNA fragment binning using a hierarchical classifier based on linear discriminant analysis and principal component analysis. J Bioinform Comput Biol 2010, 8(6):995-1011.
49. Diaz NN, Krause L, Goesmann A, Niehaus K, Nattkemper TW: TACOA: taxonomic classification of environmental genomic fragments using a kernelized nearest neighbor approach. BMC Bioinformatics 2009, 10:56.
50. Markowitz VM, Ivanova NN, Szeto E, Palaniappan K, Chu K, Dalevi D, Chen IM, Grechkin Y, Dubchak I, Anderson I, Lykidis A, Mavromatis K, Hugenholtz P, Kyrpides NC: IMG/M: a data management and analysis system for metagenomes. Nucleic Acids Res 2008, (36 Database):D534-538.
51. Huson DH, Auch AF, Qi J, Schuster SC: MEGAN analysis of metagenomic data. Genome Res 2007, 17(3):377-386.
52. Krause L, Diaz NN, Goesmann A, Kelley S, Nattkemper TW, Rohwer F, Edwards RA, Stoye J: Phylogenetic classification of short environmental DNA fragments. Nucleic Acids Res 2008, 36(7):2230-2239.
53. Monzoorul Haque M, Ghosh TS, Komanduri D, Mande SS: SOrt-ITEMS: Sequence orthology based approach for improved taxonomic estimation of metagenomic sequences. Bioinformatics 2009, 25(14):1722-1730.
54. Liu B, Gibbons T, Ghodsi M, Treangen T, Pop M: Accurate and fast estimation of taxonomic profiles from metagenomic shotgun sequences. BMC Genomics 2011, 12(Suppl 2):S4.
55. Brady A, Salzberg SL: Phymm and PhymmBL: metagenomic phylogenetic classification with interpolated Markov models. Nat Methods 2009, 6(9):673-676.
56. Leung HC, Yiu SM, Yang B, Peng Y, Wang Y, Liu Z, Chen J, Qin J, Li R, Chin FY: A robust and accurate binning algorithm for metagenomic sequences with arbitrary species abundance ratio. Bioinformatics 2011, 27(11):1489-1495.
57. Yung PY, Burke C, Lewis M, Egan S, Kjelleberg S, Thomas T: Phylogenetic screening of a bacterial, metagenomic library using homing endonuclease restriction and marker insertion. Nucleic Acids Res 2009, 37(21):e144.
58. Aziz RK, Bartels D, Best AA, DeJongh M, Disz T, Edwards RA, Formsma K, Gerdes S, Glass EM, Kubal M, Meyer F, Olsen GJ, Olson R, Osterman AL, Overbeek RA, McNeil LK, Paarmann D, Paczian T, Parrello B, Pusch GD, Reich C, Stevens R, Vassieva O, Vonstein V, Wilke A, Zagnitko O: The RAST Server: rapid annotations using subsystems technology. BMC Genomics 2008, 9:75.
59. Markowitz VM, Mavromatis K, Ivanova NN, Chen IM, Chu K, Kyrpides NC: IMG ER: a system for microbial genome annotation expert review and curation. Bioinformatics 2009, 25(17):2271-2278.
60. Lukashin AV, Borodovsky M: GeneMark.hmm: new solutions for gene finding. Nucleic Acids Res 1998, 26(4):1107-1115.
61. Delcher AL, Harmon D, Kasif S, White O, Salzberg SL: Improved microbial gene identification with GLIMMER. Nucleic Acids Res 1999, 27(23):4636-4641.
62. McHardy ACZ, Wenhan Martin HGL, Alexandre Tsirigos A, Hugenholtz P, Rigoutsos IB, Mark : Accurate phylogenetic classification of variable-length DNA fragments. Nat Methods 2007, 4(1):63-72.

63. Noguchi H, Taniguchi T, Itoh T: MetaGeneAnnotator: detecting species-specific patterns of ribosomal binding site for precise gene prediction in anonymous prokaryotic and phage genomes. DNA Res 2008, 15(6):387-396.
64. Hoff KJ, Lingner T, Meinicke P, Tech M: Orphelia: predicting genes in metagenomic sequencing reads. Nucleic Acids Res 2009, (37 Web Server):W101-105.
65. Yok NG, Rosen GL: Combining gene prediction methods to improve metagenomic gene annotation. BMC Bioinformatics 2011, 12:20.
66. Gardner PP, Daub J, Tate JG, Nawrocki EP, Kolbe DL, Lindgreen S, Wilkinson AC, Finn RD, Griffiths-Jones S, Eddy SR, Bateman A: Rfam: updates to the RNA families database. Nucleic Acids Res 2009, (37 Database):D136-140.
67. Lowe TM, Eddy SR: tRNAscan-SE: a program for improved detection of transfer RNA genes in genomic sequence. Nucleic Acids Res 1997, 25(5):955-964.
68. Bendtsen JD, Nielsen H, von Heijne G, Brunak S: Improved prediction of signal peptides: SignalP 3.0. J Molec Biol 2004, 340(4):783-795.
69. Bland C, Ramsey TL, Sabree F, Lowe M, Brown K, Kyrpides NC, Hugenholtz P: CRISPR recognition tool (CRT): a tool for automatic detection of clustered regularly interspaced palindromic repeats. BMC Bioinformatics 2007, 8:209.
70. Grissa I, Vergnaud G, Pourcel C: CRISPRFinder: a web tool to identify clustered regularly interspaced short palindromic repeats. Nucleic Acids Res 2007, (35 Web Server):W52-57.
71. Pruesse E, Quast C, Knittel K, Fuchs BM, Ludwig W, Peplies J, Glöckner FO: SILVA: a comprehensive online resource for quality checked and aligned ribosomal RNA sequence data compatible with ARB. Nucleic Acids Res 2007, 35(21):7188-7196.
72. DeSantis TZ, Hugenholtz P, Larsen N, Rojas M, Brodie EL, Keller K, Huber T, Dalevi D, Hu P, Andersen GL: Greengenes, a chimera-checked 16S rRNA gene database and workbench compatible with ARB. Appl Environ Microbiol 2006, 72(7):5069-5072.
73. Cole JR, Wang Q, Cardenas E, Fish J, Chai B, Farris RJ, Kulam-Syed-Mohideen AS, McGarrell DM, Marsh T, Garrity GM, Tiedje JM: The Ribosomal Database Project: improved alignments and new tools for rRNA analysis. Nucleic Acids Res 2009, (37 Database):D141-145.
74. Sun S, Chen J, Li W, Altintas I, Lin A, Peltier S, Stocks K, Allen EE, Ellisman M, Grethe J, Wooley J: Community cyberinfrastructure for Advanced Microbial Ecology Research and Analysis: the CAMERA resource. Nucleic Acids Res 2011, (39 Database):D546-551.
75. Gilbert JA, Field D, Swift P, Thomas S, Cummings D, Temperton B, Weynberg K, Huse S, Hughes M, Joint I, Somerfield PJ, Muhling M: The taxonomic and functional diversity of microbes at a temperate coastal site: a 'multi-omic' study of seasonal and diel temporal variation. PLoS One 2010, 5(11):e15545.
76. Yooseph S, Sutton G, Rusch DB, Halpern AL, Williamson SJ, Remington K, Eisen JA, Heidelberg KB, Manning G, Li W, Jaroszewski L, Cieplak P, Miller CS, Li H, Mashiyama ST, Joachimiak MP, van Belle C, Chandonia JM, Soergel DA, Zhai Y, Natarajan K, Lee S, Raphael BJ, Bafna V, Friedman R, Brenner SE, Godzik A,

Eisenberg D, Dixon JE, Taylor SS, et al.: The Sorcerer II Global Ocean Sampling expedition: expanding the universe of protein families. PLoS Biol 2007, 5(3):e16.
77. Godzik A: Metagenomics and the protein universe. Curr Opin Struct Biol 2011, 21(3):398-403.
78. Wilkening J, Desai N, Meyer F, A W: Using clouds for metagenomics - case study. IEEE Cluster 2009.
79. Ye Y, Choi JH, Tang H: RAPSearch: a fast protein similarity search tool for short reads. BMC Bioinformatics 2011, 12:159.
80. Kent WJ: BLAT-the BLAST-like alignment tool. Genome Res 2002, 12(4):656-664.
81. Wang W, Zhang P, Liu X: Short read DNA fragment anchoring algorithm. BMC Bioinformatics 2009, 10(Suppl 1):S17.
82. Kanehisa M, Goto S, Kawashima S, Okuno Y, Hattori M: The KEGG resource for deciphering the genome. Nucleic Acids Res 2004, (32 Database):D277-280.
83. Muller J, Szklarczyk D, Julien P, Letunic I, Roth A, Kuhn M, Powell S, von Mering C, Doerks T, Jensen LJ, Bork P: eggNOG v2.0: extending the evolutionary genealogy of genes with enhanced non-supervised orthologous groups, species and functional annotations. Nucleic Acids Res 2010, (38 Database):D190-195.
84. Tatusov RL, Fedorova ND, Jackson JD, Jacobs AR, Kiryutin B, Koonin EV, Krylov DM, Mazumder R, Mekhedov SL, Nikolskaya AN, Rao BS, Smirnov S, Sverdlov AV, Vasudevan S, Wolf YI, Yin JJ, Natale DA: The COG database: an updated version includes eukaryotes. BMC Bioinformatics 2003, 4:41.
85. Finn RD, Mistry J, Tate J, Coggill P, Heger A, Pollington JE, Gavin OL, Gunasekaran P, Ceric G, Forslund K, Holm L, Sonnhammer EL, Eddy SR, Bateman A: The Pfam protein families database. Nucleic Acids Res 2010, (38 Database):D211-222.
86. Selengut JD, Haft DH, Davidsen T, Ganapathy A, Gwinn-Giglio M, Nelson WC, Richter AR, White O: TIGRFAMs and Genome Properties: tools for the assignment of molecular function and biological process in prokaryotic genomes. Nucleic Acids Res 2007, (35 Database):D260-264.
87. Field D, Amaral-Zettler L, Cochrane G, Cole JR, Dawyndt P, Garrity GM, Gilbert J, Glockner FO, Hirschman L, Karsch-Mizrachi I, Klenk HP, Knight R, Kottmann R, Kyrpides N, Meyer F, San Gil I, Sansone SA, Schriml LM, Sterk P, Tatusova T, Ussery DW, White O, Wooley J, Yilmaz P, Gilbert JA, Johnston A, Vaughan R, Hunter C, Park J, Morrison N, et al.: The Genomic Standards Consortium: Minimum information about a marker gene sequence (MIMARKS) and minimum information about any (x) sequence (MIxS) specifications. PLoS Biol 2011, 9(6):e1001088.
88. Prosser JI: Replicate or lie. Environ Microbiol 2010, 12(7):1806-1810.
89. Clarke KR: Non-parametric multivariate analyses of changes in community structure. Australian J Ecology 1993, (18):117-143.
90. White JR, Nagarajan N, Pop M: Statistical methods for detecting differentially abundant features in clinical metagenomic samples. PLoS Comput Biol 2009, 5(4):e1000352.
91. Turnbaugh PJ, Hamady M, Yatsunenko T, Cantarel BL, Duncan A, Ley RE, Sogin ML, Jones WJ, Roe BA, Affourtit JP, Egholm M, Henrissat B, Heath AC, Knight R, Gordon JI: A core gut microbiome in obese and lean twins. Nature 2009, 457(7228):480-484.

92. Kristiansson E, Hugenholtz P, Dalevi D: ShotgunFunctionalizeR: an R-package for functional comparison of metagenomes. Bioinformatics 2009, 25(20):2737-2738.
93. Burke C, Steinberg P, Rusch D, Kjelleberg S, Thomas T: Bacterial community assembly based on functional genes rather than species. Proc Natl Acad Sci USA 2011, 108(34):14288-14293.
94. Mou X, Sun S, Edwards RA, Hodson RE, Moran MA: Bacterial carbon processing by generalist species in the coastal ocean. Nature 2008, 451(7179):708-711.
95. Yilmaz P, Kottmann R, Field D, Knight R, Cole JR, Amaral-Zettler L, Gilbert JA, Karsch-Mizrachi I, Johnston A, Cochrane G, Vaughan R, Hunter C, Park J, Morrison N, Rocca-Serra P, Sterk P, Arumugam M, Bailey M, Baumgartner L, Birren BW, Blaser MJ, Bonazzi V, Booth T, Bork P, Bushman FD, Buttigieg PL, Chain PS, Charlson E, Costello EK, Huot-Creasy H, et al.: Minimum information about a marker gene sequence (MIMARKS) and minimum information about any (x) sequence (MIxS) specifications. Nat Biotechnol 2011, 29(5):415-420.
96. Hsi-Yang Fritz M, Leinonen R, Cochrane G, Birney E: Efficient storage of high throughput DNA sequencing data using reference-based compression. Genome Res 2011, 21(5):734-740.

PART III

SINGLE NUCLEOTIDE POLYMORPHISM IS THE IDEAL MARKER

CHAPTER 14

SNP MARKERS AND THEIR IMPACT ON PLANT BREEDING

JAFAR MAMMADOV, RAJAT AGGARWAL, RAMESH BUYYARAPU, and SIVA KUMPATLA

14.1 INTRODUCTION

Allelic variations within a genome of the same species can be classified into three major groups that include differences in the number of tandem repeats at a particular locus [microsatellites, or simple sequence repeats (SSRs)] [1], segmental insertions/deletions (InDels) [2], and single nucleotide polymorphisms (SNPs) [3]. In order to detect and track these variations in the individuals of a progeny at DNA level, researchers have been developing and using genetic tools called molecular markers [4]. Although SSRs, InDels, and SNPs are the three major allelic variations discovered so far, a plethora of molecular markers were developed to detect the polymorphisms that resulted from these three types of variation [5]. Evolution of molecular markers has been primarily driven by the throughput and cost of detection method and the level of reproducibility [6]. Depending on detection method and throughput, all molecular markers can be divided

This chapter was originally published under the Creative Commons Attribution License. Mammadov J, Aggarwal R, Buyyarapu R, and Kumpatla S. SNP Markers and Their Impact on Plant Breeding. International Journal of Plant Genomics *2012 (2012), 11 pages. doi:10.1155/2012/728398.*

into three major groups: (1) low-throughput, hybridization-based markers such as restriction fragment length polymorphisms (RFLPs) [4]; (2) medium-throughput, PCR-based markers that include random amplification of polymorphic DNA (RAPD) [7], amplified fragment length polymorphism (AFLP) [8], SSRs [9]; (3) high-throughput (HTP) sequence-based markers: SNPs [3]. In late eighties, RFLPs were the most popular molecular markers that were widely used in plant molecular genetics because they were reproducible and codominant [10]. However, the detection of RFLPs was an expensive, labor- and time-consuming process, which made these markers eventually obsolete. Moreover, RFLP markers were not amenable to automation. Invention of PCR technology and the application of this method for the rapid detection of polymorphisms overthrew low-throughput RFLP markers, and new generation of PCR-based markers emerged in the beginning of nineties. RAPD, AFLP, and SSR markers are the major PCR-based markers that research community has been using in various plant systems. RAPDs are able to simultaneously detect polymorphic loci in various regions of a genome [11]. However, they are anonymous and the level of their reproducibility is very low due to the non-specific binding of short, random primers. Although AFLPs are anonymous too, the level of their reproducibility and sensitivity is very high owing to the longer +1 and +3 selective primers and the presence of discriminatory nucleotides at 3′ end of each primer. That is why AFLP markers are still popular in molecular genetics research in crops with little to zero reference genome sequence available [12]. However, AFLP markers did not find widespread application in molecular breeding owing to the lengthy and laborious detection method, which was not amenable to automation either. Therefore, it was not surprising that soon after the discovery of SSR markers in the genome of a plant, they were declared as "markers of choice" [13], because SSRs were able to eliminate all drawbacks of the above-mentioned DNA marker technologies. SSRs were no longer anonymous; they were highly reproducible, highly polymorphic, and amenable to automation. Despite the cost of detection remaining high, SSR markers had pervaded all areas of plant molecular genetics and breeding in late 90s and the beginning of 21st century. However, during the last five years, the hegemony of medium-throughput SSRs was eventually broken by SNP markers. First discovered in human genome, SNPs proved to be universal as well as the most abundant forms of genetic variation among individuals of the same species [14]. Although SNPs are less polymorphic than SSR markers because of their biallelic nature, they easily compensate this drawback by being abundant, ubiquitous, and amenable to high- and ultra-high-throughput automation. However, despite these obvious advantages, there were only a limited number of examples of application of SNP markers in plant breeding by 2009 [15]. In this paper, we tried to summarize the recent progress in the utility of SNP markers in plant breeding.

14.2 SNP DISCOVERY IN COMPLEX PLANT GENOMES

While SNP discovery in crops with simple genomes is a relatively straightforward process, complex genomes pose serious obstacles for the researchers interested in developing SNPs. One of the major problems is the highly repetitive nature of the plant genomes [16]. Prior to the emergence of next-generation sequencing (NGS) technologies, researchers used to rely on different experimental strategies to avoid repetitive portions of the genome. These include discovery of SNPs experimentally by resequencing of unigene-derived amplicons using Sanger's method [17] and in silico SNP discovery through the mining of SNPs within EST databases followed by PCR-based validation [18]. Although these approaches allowed the detection of gene-based SNPs, their frequency is generally low in conserved genic regions, and they were unable to discover SNPs located in low-copy noncoding regions and intergenic spaces. Additionally, amplicon resequencing was an expensive and labor-intensive procedure [15]. As many crops are ancient tetraploids with mosaics of scattered duplicated regions [19], in silico and experimental mining of EST databases resulted in the discovery of a large number of nonallelic SNPs that represented paralogous sequences and were suboptimal for application in molecular breeding [20]. Recent emergence of NGS technologies such as 454 Life Sciences (Roche Applied Science, Indianapolis, IN), HiSeq (Illumina, San Diego, CA), SOLiD and Ion Torrent (Life Technologies Corporation, Carlsbad, CA) has eliminated the problems associated with low throughput and high cost of SNP discovery [21]. Transcriptome resequencing using NGS technologies allows rapid and inexpensive SNP discovery within genes and avoids highly repetitive regions of a genome [22]. This methodology was successfully applied in several plant genomes, including maize [23], canola [24], eucalyptus [25], sugarcane [26], tree species [27], wheat [28], avocado [29], and black currant [30]. Originally developed for human disease diagnostic research, the NimbleGen sequence capture technology (Roche Applied Science, IN) [31] brought the detection of gene-based SNPs in plants into higher throughput and coverage level [32]. This technology consists of exon sequence capture and enrichment by microarray followed by NGS for targeted resequencing. Similar in-solution target capture technologies, such as Agilent SureSelect, are also commercially available for genome/exome mining studies. However, this technology would be efficient only for crops with available reference genome sequence or large transcriptome (EST) datasets, since the design of capture probes requires these reference resources.

Despite the attractiveness of SNP discovery via transcriptome or exome resequencing, this process is targeted, focusing solely on coding regions. It is obvious that the availability of SNPs within coding sequences is a very powerful tool for molecular geneticists to detect a causative mutation [33]. However, often QTL are located in noncoding regulatory sequences such as enhancers or locus control regions, which could be located several megabases away from genes within intergenic spaces [34]. Discovery of SNPs located within those regulatory elements via transcriptome or exon sequencing is limited. In order to discover SNPs in a genome-wide fashion and avoid repetitive and duplicated DNA, it is very important to employ genome complexity reduction techniques coupled with NGS technologies. Several genome complexity reduction techniques have been developed over the years, including High Cot selection [35], methylation filtering [36], and microarray-based genomic selection [37]. These techniques mainly reduce the number of repetitive sequences but lack the power to recognize and eliminate duplicated sequences, which cause the detection of false-positive SNPs. Unlike the above-mentioned techniques, recently developed genome complexity reduction technologies such as Complexity Reduction of Polymorphic Sequences (CRoPS) (Keygene N.V., Wageningen, The Netherlands) [38] and Restriction Site Associated DNA (RAD)

(Floragenics, Eugene, OR, USA) [39] are computationally well equipped and capable of filtering out duplicated SNPs. These systems were successfully applied to discover SNPs in crops with [40] and without reference genome sequences [41].

Although several complexity reduction approaches are being developed to generate data from NGS platforms, it is often challenging to identify candidate SNPs in polyploid crops species such as potato, tobacco, cotton, canola, and wheat. In general, minor allele frequency could be used as a measure to identify candidate SNPs in diploid species [42]. However, in polyploid crops, you often find loci that are polymorphic within a single genotype due to the presence of either homoeologous loci from the individual subgenomes (homoeologous SNPs) or paralogous loci from duplicated regions of the genome. Such false positive SNPs are not useful for genetic mapping purposes and often lead to a lower validation rate during assays. Successful SNP validation in allopolyploids depends upon differentiation of the sequence variation classes [43]. Use of haplotype information beside the allelic frequency would help to identify homologous SNPs (true SNPs) from those of homoeologous loci (false positives). Bioinformatic programs such as HaploSNPer [44] would facilitate identification of candidate loci for assay design purposes in polyploid crops. Elimination of homoeologous loci for the assay design process would improve the validation rate. Such approaches could also be extended to other complex and highly repetitive diploid genomes such as barley. Complexity reduction approaches, combined with sophisticated computational tools, would expedite SNP discovery and validation efforts in polyploids.

Although CRoPS and RAD technologies are powerful tools to detect SNPs in genome-wide fashion, they can hardly be called HTP, because on an average only ~1,000 SNPs pass stringent quality control [40]. While these numbers are enough to generate genetic linkage maps of reasonable saturation and carry out preliminary QTL mapping, they are not adequate to implement genome-wide association studies (GWAS). Depending on the rate of linkage disequilibrium decay, GWAS might require several million genetic landmarks. From this point of view, genotyping-by-sequencing (GBS) technique offers many more opportunities. Discovery of a large number of SNPs using GBS was demonstrated in maize [45] and sorghum [46]. GBS not only increases the sequencing throughput by several orders of magnitude but also has multiplexing capabilities [47]. To eliminate a large portion of repetitive sequences, a type II restriction endonuclease, ApeKI, is applied to digest DNA prior to sequencing to generate reduced representation libraries (genome complexity reduction component), which are further subject to sequencing [47]. In polyploid crops, GBS might be challenging, but the associated complexity reduction methods could be used for SNP discovery. For discovery purposes, the availability of a reference genome is not an absolute requirement to implement GBS approach. However, in organisms that do not have a reference genome, GBS-derived SNPs must be validated using one of the techniques that are described in the following section, which might dramatically increase per marker price. Validation needs to be done primarily to discard paralogous SNPs. For organisms with a reference genome sequence, the validation step is replaced by in silico mapping of the sequenced fragments to the genome. Although GBS has the potential to discover several million SNPs, one of the major drawbacks of this technique is large numbers of missing data. To solve this problem, computational biologists developed data imputation models such as BEAGLE v3.0.2 [48] and IMPUTE v2 [49], to bring imputed data as close as possible to the real data [50, 51].

14.3 SNP VALIDATION AND MODERN GENOTYPING PLATFORMS AND CHEMISTRIES

The availability of reference sequence and sophisticated software does not always guarantee that the discovered SNP can be converted into a valid marker. In order to insure that the discovered SNP is a Mendelian locus, it has to be validated. The validation of a marker is the process of designing an assay based on the discovered polymorphism and then genotyping a panel of diverse germplasm and segregating population. Compared to the collection of unrelated lines, a segregating population is more informative as a validation panel because it allows the inspection of the discriminatory ability and segregation patterns of a marker which helps the researcher to understand whether it is a Mendelian locus or a duplicated/repetitive sequence that escaped the software filter [40].

The most popular HTP assays/chemistries and genotyping platforms that are currently being used for SNP validation are Illumina's BeadArray technology-based Golden Gate (GG) [52] and Infinium assays [53], Life Technologies' TaqMan [54] assay coupled with OpenArray platform (TaqMan OpenArray Genotyping system, Product bulletin), and KBiosciences' Competitive Allele Specific PCR (KASPar) combined with the SNP Line platform (SNP Line XL; http://www.kbioscience.co.uk). These modern genotyping assays and platforms differ from each other in their chemistry, cost, and throughput of samples to genotype and number of SNPs to validate. The choice of chemistry and genotyping platform depends on many factors that include the length of SNP context sequence, overall number of SNPs to genotype, and finally the funds available to the researcher, because most of these chemistries still remain cost intensive. Comparative analyses of these four genotyping assays and platforms were described in Kumpatla et al. [55].

Though all genotyping chemistries and platforms are applicable to generate genotypic data in polyploid crops, analysis of SNP calls is somewhat challenging in polyploids due to multiallele combinations in the genotypes. SNPs in polyploid species can be broadly classified as simple SNPs, hemi-SNPs, and homoeo-SNPs. Here, we describe simple, hemi-, and homoeo-SNPs using an example of allele calls in tetraploid and diploid cotton species (Figure 1). Genomes of tetraploid cotton species, Gossypium hirsutum (AD1) and G. barbadense (AD2), consist of two subgenomes A and D, where A genome was derived from diploid progenitors, such as G. herbaceum (A1) and G. arboreum (A2), and D genome resulted from another diploid progenitor G. raimondii (D5). Simple, or true SNPs are markers that detect allelic variation between homologous loci of the same subgenome of two tetraploid samples. For example, in Figure 1(a), a SNP marker clearly detects polymorphism within A subgenomes of G. hirsutum (AD1) and G. barbadense (AD2) and separates samples into homozygous A (blue) and B (red) clusters. This marker does not discriminate polymorphism in D subgenome, because the D genome allele is absent there (pink dot in G. raimondii). In contrast to simple SNPs, hemi-SNPs detect allelic variation in the homozygous state in one sample and the heterozygous state in the other sample. In Figure 1(b), SNP marker detects both alleles

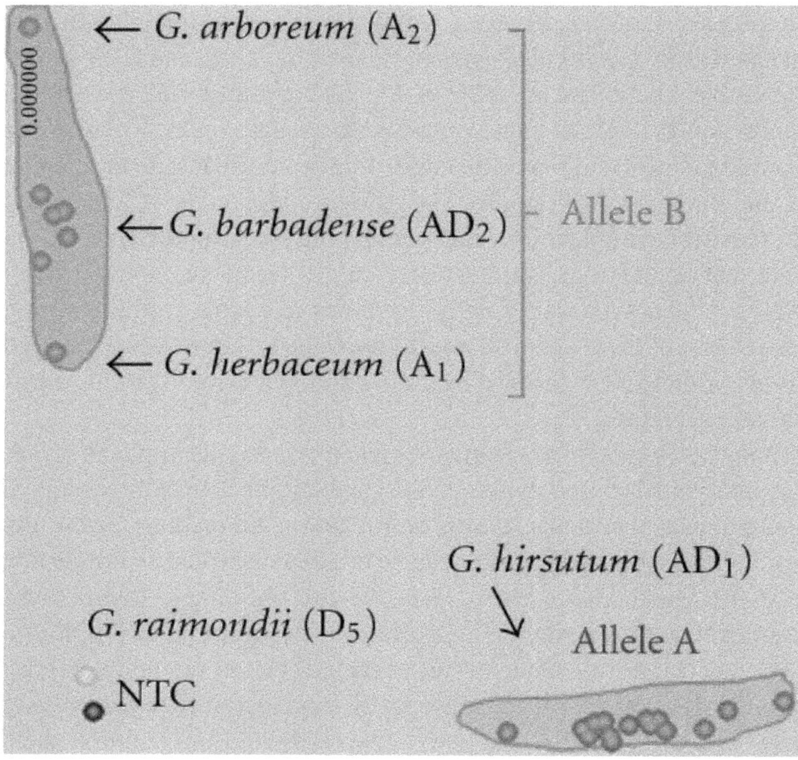

FIGURE 1: Segregation patterns of simple, hemi-, and homoeo-SNPs assayed using KASPar chemistry across tetraploid cotton species [G. hirsutum (AD1) and G. barbadense (AD2)] and their diploid progenitors [G. arboreum, G. herbaceum (A subgenome), and G. raimondii (D-subgenome)]. (a) Simple, or true SNP detects allelic variation between homologous loci of A subgenome of G. hirsutum (AD1) and G. barbadense (AD2). (b) Hemi-SNPs detect allelic variation in homozygous state in G. barbadense (AD2) and heterozygous state in G. hirsutum (AD1). (c) Homoeo-SNP detects homoelogous and/or paralogous loci both in A and D subgenomes, which are monomorphic between G. barbadense (AD2) and G. hirsutum (AD1). Allele calls depicted in blue, red, green, pink, and black represent alleles A, B, and AB, no amplification or missing locus, and no template control (NTC), respectively.

(A and B) in G. hirsutum (heterozygous green cluster) and one allele A in G. barbadense (a homozygous blue cluster) and could be vice versa. Homoeo-SNPs detect homoeologous and possibly paralogous loci both in A and D subgenomes and result in monomorphic loci in tetraploid species (right image). In Figure 1(c) A genome progenitors (G. herbaceum and G. arboreum) had allele A (blue) and D genome progenitor (G. raimondii) had allele B (red), but both tetraploid species (G. hirsutum and G. barbadense) were grouped into heterozygous AB (green) cluster. As homoeo-SNPs can detect paralogous loci, the diploid progenitors both have different alleles.

Simple SNPs as well as hemi-SNPs are useful markers for genetic mapping and diversity screening studies. Simple SNPs segregate like the markers in diploids in most of the mapping populations and would account for approximately 10–30% of total polymorphic SNPs in various polyploid crop species. Hemi-SNPs form a major category (30–60%) of polymorphic SNPs in a polyploid crop species and could be used for genetic mapping purposes in F2, RIL, and DH populations. Homoeo-SNPs are of lesser value for mapping purposes as most of the genotypes result in heterologous loci due to polymorphism between the homoeologous genomes or duplicated loci within each of the polyploid genotypes [56].

14.4 APPLICATION OF SNP MARKERS IN GENE/QTL DISCOVERY

14.4.1 BIPARENTAL APPROACH

Genetic mapping studies involve genetic linkage analysis, which is based on the concept of genetic recombination during meiosis [57]. This encompasses developing genetic linkage maps following genotyping of individuals in segregating populations with DNA markers covering the genome of that organism. Since their discovery in the 1980s, DNA-based markers have been widely used in developing saturated genetic linkage maps as well as for the mapping and discovery of genes/QTL. With the large-scale availability of the sequence information and development of HTP technologies for SNP genotyping, SNP markers have been increasingly used for QTL mapping studies. This is primarily, because SNPs are highly abundant in the genomes and, therefore, they can provide the highest map resolution compared to other marker systems [58, 59]. A review of the selected examples of QTL and gene discovery using SNP markers is presented below.

14.4.1.1 EXAMPLES IN RICE

A recent study on QTL analysis in rice for yield and three-yield-component traits, number of tillers per plant, number of grains per panicle, and grain weight compared a SNP-based map to that of a previous RFLP/SSR-based QTL map generated using the same mapping population [42]. Using the ultra-high-density SNP map, the authors showed that this map had more power and resolution relative to the RFLP/SSR map. This was clearly evident by the analysis of the two main QTL for grain weight, kgw3a (GS3) and kgw5 (GW5/qSW5). Using the SNP bin map, GW5/qSW5 QTL for grain width was accurately narrowed down to a 123 kb region as compared to the 12.4 Mb region based on the RFLP/SSR genetic map. Likewise, GS3 QTL for grain length was mapped to a 197 kb interval in comparison to 6 Mb region with the RFLP/SSR genetic map. Beside the power and the resolution, maps based on high-density SNP markers are also highly suitable for fine mapping and cloning of QTL and at times SNPs on these maps are also functionally associated with the natural variation in the trait. In another QTL mapping project, SNP and InDel markers were used to fine map qSH1 gene, a major QTL of seed shattering trait in rice [60]. The QTL were initially detected using RFLP and RAPD markers on F2 plants. Using large BC4F2 and BC3F2 populations in fine mapping approach with SNP and InDel markers, the authors mapped the functional natural variation to a 612 bp interval between the QTL flanking markers and discovered only one SNP. They further showed that this SNP in the 5′ regulatory region of the qSH1 gene caused loss of seed shattering. Fine mapping approach was also taken to positionally clone the rice bacterial blight resistance gene xa5, by isolating the recombination breakpoints to a pair of SNPs followed by sequencing of the corresponding 5 kb region [61]. Several studies have shown that the SNPs and InDels are highly abundant and present throughout the genome in various species including plants [62–64]. SNP genotyping is a valuable tool for gene mapping, map-based cloning, and marker assisted selection (MAS) in crops [65]. A study was conducted to assess the feasibility of SNPs and InDels as DNA markers in genetic analysis and marker-assisted breeding in rice by analyzing these sequence polymorphisms in the genomic region containing Piz and Piz-t rice blast resistance genes and developing PCR-based SNP markers [65]. The authors discovered that SNPs were abundant in the Piz and Piz-t (averaging one SNP every 248 bp), while InDels were much lower. This dense distribution of SNPs helped in developing SNP markers in the vicinity of these genes. Advancements in rice genomics have led to mapping and cloning of several genes and QTL controlling agronomically important traits, enabled routine use of SNP markers for MAS, gene pyramiding, and marker-assisted breeding (MAB) [66–68].

14.4.1.2. EXAMPLES IN MAIZE

SNP markers have facilitated the dissection of complex traits such as flowering time in maize. Using a set of 5000 RILs, which represent the nested association mapping (NAM) population and genotyping with 1,200 SNP markers, the authors discovered that the genetic architecture of flowering time is controlled by small additive QTL rather than a single large-effect QTL [69]. The same NAM population was used for mapping resistance to northern leaf blight disease [70]. Twenty-nine QTL were discovered and candidate genes were identified with genome-wide NAM approach using 1.6 million SNPs. Proprietary SNP markers developed by companies are being predominantly used in their private breeding programs. A study from Pioneer Hi-Bred International Inc. reported identifying a high-oil QTL (qHO6) affecting maize seed oil and oleic acid contents. This QTL encodes an acyl-CoA:diacylglycerol acyltransferase (DGAT1-2), which catalyzes the final step of oil synthesis [71].

14.4.1.3 EXAMPLES IN WHEAT

Recent advances in wheat genomics have led to the implementation of high-density SNP genotyping in wheat [72–75]. Gene-based SNP markers were developed for Lr34/Yr18/Pm38 locus that confers resistance to leaf rust, stripe rust, and powdery mildew diseases [76]. These markers serve as efficient tools for MAS and MAB of disease resistant wheat lines. Another economically important wheat disease, Fusarium head blight (FHB), has been extensively studied. Several QTL controlling FHB resistance have been identified, with the most important being Fhb1 [77]. Recently, SNP markers were mapped between the known flanking markers for Fhb1 [78]. These new markers would be useful for MAS and fine mapping towards cloning the Fhb1 gene. MAS in wheat has been extensively applied for simple traits that are difficult to score [79].

14.4.1.4 EXAMPLES IN SOYBEAN

In order to improve the effectiveness of MAS and clone soybean aphid resistance gene, Rag1, fine mapping was done to accurately position the gene, which was previously mapped to a 12 cM interval [80]. The authors mapped the gene between two SNP markers that corresponded to a physical distance of 115 kb and identified several candidate genes. Similarly, another aphid resistance gene, Rag2, originally mapped to a 10 cM interval, was fine mapped to a 54 kb interval using SNP markers that were developed by resequencing of target intervals and sequence-tagged sites [81]. In another study that used a similar approach, the authors identified SNP markers tightly linked to a QTL conferring resistance to southern root-knot nematode by developing these SNP markers from the bacterial artificial chromosome (BAC) ends and SSR-containing genomic DNA clones [82]. In all of these examples the main idea behind the identification of closely linked SNP markers was to enhance the efficiency and cost effectiveness through MAS and increase the resolution within the target locus.

14.4.1.5 EXAMPLES IN OTHER CROPS

In a study conducted in canola to map the fad2 and fad3 gene, single nucleotide mutations were identified by sequencing the genomic clones of these genes and subsequently SNP markers were developed [83]. Allele-specific PCR assays were developed to enable direct selection of desirable fad2 and fad3 alleles in marker-assisted trait introgression and breeding. In barley, SNP markers were identified that were linked to a covered smut resistance gene, Ruh.7H, by using high-resolution melting (HRM) technique [84]. In sugar beet, an anchored linkage map based on AFLP, SNP, and RAPD markers was developed to map QTL for Beet necrotic yellow vein virus resistance genes, Rz4 [85] and Rz5 [86]. A consensus genetic map based on EST-derived SNPs was developed for cowpea that would be an important resource for genomic and QTL mapping studies in this crop [87]. In one of the post-genomic era studies in 2002, the fine mapping and map-based cloning approaches were used to clone the VTC2 gene in Arabidopsis [88]. The authors fine mapped the gene interval from ~980 kb region to a 20 kb interval with SNP and InDel markers. Additional nine candidate genes were identified in that interval and subsequently the underlying mutation was discovered. Although only a few examples that demonstrate the application of SNP markers in QTL mapping and genomic studies have been mentioned here, several other studies have been published in this area. Recent advances in HTP genotyping technologies and sequence information will further pave the way for rapid identification of causative variations and cloning of QTL of interest for use in MAB.

14.4.2 GENOME-WIDE ASSOCIATION STUDY APPROACH

GWAS is increasingly becoming a popular tool for dissecting complex traits in plants [89–92]. The idea behind GWAS is to genotype a large number of markers distributed across the genome so that the phenotype or the functional alleles will be in LD with one or few markers that could then be used in the breeding program. However, due to limited extent of LD, a greater number of markers are required for sufficient power to detect linkage between the marker and the underlying phenotypic variation. Several studies on association mapping in plants have been published and reviewed in the past [89, 90, 92, 93]. A few selected examples on the GWAS and candidate gene association (CGA) studies that utilized SNP markers are described below.

The successful use and first time demonstration of the power of GWAS was through the identification of a putative gene associated with a QTL in maize [94]. In that study, a single locus with major effect on oleic acid was mapped to a 4 cM genetic interval by using SNP haplotypes at 8,590 loci. The authors identified a fatty acid desaturase gene, fad2, at ~2 kb from one of the associated markers, and this was considered a likely causative gene. With the discovery of millions of SNPs in maize and the availability of tools such as NAM populations, GWAS was effectively applied to dissect the genetic architecture of leaf traits and it was also shown that variations at the liguleless genes contributed to more upright leaf phenotype [95]. Utility of the GWAS approach was demonstrated in barley through the mapping of a QTL for spot blotch disease resistance [96]. Using the diversity array technology (DArT) and SNP markers, the authors identified several QTL, some of which were not identified for this trait earlier. Another variant of the association mapping method is the CGA, where the association between one or few gene candidate loci and the trait of interest is tested. Using this approach 24 gene candidates were analyzed for association with the field resistance to late blight disease in potato and plant maturity. Nine SNPs were identified to be associated with maturity corrected resistance, explaining 50% of the genetic variance of this trait [97]. Two SNPs at the allene oxide synthase 2 (StAOS2) gene locus were associated with the largest effect on the trait of interest. A GWAS approach was also successfully applied to understand the genetic architecture of complex diseases such as northern and southern corn leaf blights [70, 98]. Although the number of papers dedicated to the application of GWAS to reveal the genetic basis of agronomic traits is growing, the practical utility of minor QTL in molecular breeding is yet to be shown. As GWAS requires large number of molecular markers, the utility of GWAS in dissection of molecular basis of traits in polyploid crops such as canola, wheat, and cotton has been fairly limited due to the insufficient number of polymorphic markers and the absence of reference genome. However, recently developed associative transcriptomics method has a potential to overcome the above-mentioned shortages [99]. Harper et al. [99] leveraged differentially expressed transcriptome sequences to develop molecular markers in tetraploid crop Brassica napus and associated them with glucosinolate content variation in seeds. Due to the precision of this method, scientists were able to correlate specific deletions in canola genome with two QTL controlling the trait. Annotation of deleted regions revealed the orthologs of the transcription factor HAG1, which controlled aliphatic glucosinolate biosynthesis in A. thaliana. This research work gives an optimism on successful application of GWAS in polyploid crops.

14.5 IMPLEMENTATION OF SNP MARKERS IN PLANT BREEDING

Due to the availability of HTP SNP detection and validation technologies, the development of SNP markers becomes a routine process, especially in crops with reference genome. How has that influenced the application of SNP markers in plant breeding? In a review article, Xu and Crouch [100] indicated fairly low number of articles dedicated to the marker assisted selection for the 1986–2005 period. The combination of three key phrases ("marker-assisted selection" AND "SNP" AND "plant breeding"), indeed, shows only 637 articles at Google Scholar for that period. However, similar search for the period, spanning 2006 through 2012, demonstrates almost sevenfold (~4,560) increase in the number of articles indicating the application of SNPs in MAS. A vast majority of those publications are from public sector and primarily describe mapping QTL using SNPs and state the potential usefulness of those markers in MAS without any experimental support for that. For most of those research studies, QTL mapping is the final destination and further application of those markers in actual MAS leading to the development of varieties seldom happens. Fairly low impact of academic research in the MAS-based variety development can be explained by the lack of funding to complete the entire marker development pipeline (MDP), which can be long term and cost intensive. MDP includes several steps such as (1) population development, (2) initial QTL mapping, (3) QTL validation (testing in several locations and years and implementing fine mapping), and (4) marker validation (development of inexpensive but HTP and automation amenable assays) [101]. Every step of the development of markers linked to QTL is associated with numerous constraints, which may take several years and substantial funding to resolve. However, since 2006, there have been a few success stories about the development of varieties using SNPs in publications derived from academic research, including the development of submergence-tolerant rice cultivars [102], rice cultivars with improved eating, cooking, and sensory quality [103], leaf rust resistant wheat variety "Patwin" [104], and maize cultivar with low phytic acid [105]. Although the private sector does not normally release details of its breeding methodologies to the public, several papers published by Monsanto [106, 107], Pioneer Hi-bred [71], Syngenta [108], and Dow AgroSciences [109] indicate that commercial organizations are the main drivers in the application of SNP markers in MAS [110].

Current MAS strategies fit the breeding programs for the traits that are highly heritable and governed by a single gene or one major QTL that explains a large portion of the phenotypic variability. In reality, most of the agronomic traits such as yield, drought and heat tolerance, nitrogen and water use efficiency, and fiber quality in cotton have complex inheritance that is controlled by multiple QTL with minor effect. Use of one of those minor QTL in MAS will be inefficient because of its negligible effect on phenotype.

The MAS scheme using paternity testing has recently been proposed to address challenges associated with selection gains that can be achieved in outbred forage crops [111]. Paternity testing, a nonlinkage-based MAS scheme, improves selection gains by increasing parental control in the selection gain equation. The authors demonstrated paternity testing MAS in three red clover breeding populations by using permutation-based truncation selection for a biomass-persistence index trait and achieved paternity-based selection gains that were greater than double the selection gains based on maternity alone. The paternity was determined by using a small set (11) of SSR markers. SNP markers can also be used for paternity testing, but one would require a relatively larger number of SNP loci [112].

Meuwissen et al. [113] described a new methodology in plant breeding called genomic selection (GS) that was intended to solve problems related to MAS of complex traits. This methodology also applies molecular markers but in a different fashion in both diploid and polyploid crop species. Unlike MAS, in GS markers are not used for tracking a trait. In GS high-density marker coverage is needed to potentially have all QTL in LD with at least one marker. Then the comprehensive information on all possible loci, haplotypes, and marker effects across the entire genome is used to calculate genomic estimated breeding value (GEBV) of a particular line in the breeding population.

GS of superior lines can be carried out within any breeding population. In order to enable successful GS, the experimental population must be identified. The population should not be necessarily derived from biparental cross but must be representative of selection candidates in the breeding program to which GS will be applied [114]. The experimental population must be genotyped with a large number of markers. Taking into account the low cost of sequencing, the best choice is the GBS implementation, which will yield maximum number of polymorphisms. The sequence of the two events, that is, phenotypic and genotypic data collection, is arbitrary and can be done in parallel. When both phenotypic and genotypic data are ready, one can start "training" molecular markers [115]. In order to train the GS model, the effect of each marker is calculated computationally. The effect of a marker is represented by a number with a positive or negative sign that indicates the positive or negative effect, respectively, of a particular locus to the phenotype. When the effects of all markers are known, they are considered "trained" and ready to assess any breeding population different from the experimental one for the same trait. Availability of trained GS model does not require the collection of phenotypic data from new breeding populations. The same set of "trained" markers will be used to genotype a new breeding population. Based on genotypic data, the known effects of each marker will be summed and GEBV of each line will be calculated. The higher the GEBV value of an individual line, the more the chances that this line will be selected and advanced in the breeding cycle. Thus, GS using high-density marker coverage has a potential to capture QTL with major and minor effects and eliminate the need to collect phenotypic data in all breeding cycles. Also, the application of GS was demonstrated to reduce the number of breeding cycles and increase the annual gain [114]. One of the problems of GS is the level of GEBV accuracy. Simulation studies based on simulated and empirical data demonstrated that GEBV accuracy could be within 0.62–0.85. Heffner et al. [114] used previously reported GEBV accuracy of 0.53 and reported three- and twofold annual gain in maize and winter barley, respectively. The obvious advantages of GS over traditional MAS have been successfully proven in animal breeding [116]. Rapid evolution of sequencing technologies and HTP SNP genotyping systems are enabling generation and validation of millions of markers, giving a "cautious optimism" for successful application of GS in breeding for complex traits [117–120].

14.6 CONCLUSION

SNP markers have become extremely popular in plant molecular genetics due to their genome-wide abundance and amenability for high- to ultra-high-throughput detection platforms. Unlike earlier marker systems, SNPs made it possible to create saturated, if not, supersaturated genetic maps, thereby enabling genome-wide tracking, fine mapping of target regions, rapid association of markers with a trait, and accelerated cloning of gene/QTL of interest. On the flip side, there are some challenges that need to be addressed or overcome while using SNPs. For example, the biallelic nature of SNPs needs to be compensated by discovering and using a larger number of SNPs to arrive at the same or higher power as that of earlier-generation molecular markers. This could be cost prohibitive depending on the crop and the sequence resources available for that genome. Working with polyploid crops is another challenge where useful SNPs are only a small percentage of the total available polymorphisms. Creative strategies need to be employed to generate a reasonable number of SNPs in those species. The use of SNP markers in MAB programs has been growing at a faster pace and so is the development of technologies and platforms for the discovery and HTP screening of SNPs in many crops. SNP chips are currently available for several crops; however, one disadvantage is that these readily available chips are made based on SNPs discovered from certain genotypes and, therefore, may not be ideal for projects utilizing unrelated genotypes. This necessitates creation of multiple chips or the usage of technologies that permit design flexibility but are economical. Although GBS creates great opportunities to discover a large number of SNPs at lower per sample cost within the genotypes of interest, the lack of adequate computational capabilities such as reliable data imputation algorithms and powerful computers allowing quick processing and the storage of a large amount of sequencing data becomes a major bottleneck. Despite certain disadvantages or challenges, it is clear that SNP markers, in combination with genomics and other next-generation technologies, have been accelerating the pace and gains of plant breeding.

REFERENCES

1. J. L. Weber and P. E. May, "Abundant class of human DNA polymorphisms which can be typed using the polymerase chain reaction," American Journal of Human Genetics, vol. 44, no. 3, pp. 388–396, 1989.
2. R. Ophir and D. Graur, "Patterns and rates of indel evolution in processed pseudogenes from humans and murids," Gene, vol. 205, no. 1-2, pp. 191–202, 1997.
3. D. G. Wang, J. B. Fan, C. J. Siao et al., "Large-scale identification, mapping, and genotyping of single- nucleotide polymorphisms in the human genome," Science, vol. 280, no. 5366, pp. 1077–1082, 1998.
4. D. Botstein, R. L. White, M. Skolnick, and R. W. Davis, "Construction of a genetic linkage map in man using restriction fragment length polymorphisms," American Journal of Human Genetics, vol. 32, no. 3, pp. 314–331, 1980.
5. P. K. Gupta, R. K. Varshney, P. C. Sharma, and B. Ramesh, "Molecular markers and their applications in wheat breeding," Plant Breeding, vol. 118, no. 5, pp. 369–390, 1999.
6. R. Bernardo, "Molecular markers and selection for complex traits in plants: learning from the last 20 years," Crop Science, vol. 48, no. 5, pp. 1649–1664, 2008.
7. J. Welsh and M. McClelland, "Fingerprinting genomes using PCR with arbitrary primers," Nucleic Acids Research, vol. 18, no. 24, pp. 7213–7218, 1990.
8. P. Vos, R. Hogers, M. Bleeker et al., "AFLP: a new technique for DNA fingerprinting," Nucleic Acids Research, vol. 23, no. 21, pp. 4407–4414, 1995.
9. H. J. Jacob, K. Lindpaintner, S. E. Lincoln et al., "Genetic mapping of a gene causing hypertension in the stroke-prone spontaneously hypertensive rat," Cell, vol. 67, no. 1, pp. 213–224, 1991.
10. E. S. Lander and S. Botstein, "Mapping mendelian factors underlying quantitative traits using RFLP linkage maps," Genetics, vol. 121, no. 1, p. 185, 1989.
11. J. G. K. Williams, A. R. Kubelik, K. J. Livak, J. A. Rafalski, and S. V. Tingey, "DNA polymorphisms amplified by arbitrary primers are useful as genetic markers," Nucleic Acids Research, vol. 18, no. 22, pp. 6531–6535, 1990.

12. Z. Zhang, X. Guo, B. Liu, L. Tang, and F. Chen, "Genetic diversity and genetic relationship of Jatropha curcas between China and Southeast Asian revealed by amplified fragment length polymorphisms," African Journal of Biotechnology, vol. 10, no. 15, pp. 2825–2832, 2011.
13. W. Powell, G. C. Machray, and J. Proven, "Polymorphism revealed by simple sequence repeats," Trends in Plant Science, vol. 1, no. 7, pp. 215–222, 1996.
14. S. Ghosh, P. Malhotra, P. V. Lalitha, S. Guha-Mukherjee, and V. S. Chauhan, "Novel genetic mapping tools in plants: SNPs and LD-based approaches," Plant Science, vol. 162, no. 3, pp. 329–333, 2002.
15. M. W. Ganal, T. Altmann, and M. S. Röder, "SNP identification in crop plants," Current Opinion in Plant Biology, vol. 12, no. 2, pp. 211–217, 2009.
16. B. C. Meyers, S. V. Tingey, and M. Morgante, "Abundance, distribution, and transcriptional activity of repetitive elements in the maize genome," Genome Research, vol. 11, no. 10, pp. 1660–1676, 2001.
17. S. I. Wright, I. V. Bi, S. C. Schroeder et al., "Evolution: the effects of artificial selection on the maize genome," Science, vol. 308, no. 5726, pp. 1310–1314, 2005.
18. J. Batley, G. Barker, H. O'Sullivan, K. J. Edwards, and D. Edwards, "Mining for single nucleotide polymorphisms and insertions/deletions in maize expressed sequence tag data," Plant Physiology, vol. 132, no. 1, pp. 84–91, 2003.
19. A. Pratap, S. Gupta, J. Kumar, and R. Solanki, "Soybean," Technological Innovations in Major World Oil Crops, vol. 1, pp. 293–321, 2012.
20. I. Y. Choi, D. L. Hyten, L. K. Matukumalli et al., "A soybean transcript map: gene distribution, haplotype and single-nucleotide polymorphism analysis," Genetics, vol. 176, no. 1, pp. 685–696, 2007.
21. E. R. Mardis, "The impact of next-generation sequencing technology on genetics," Trends in Genetics, vol. 24, no. 3, pp. 133–141, 2008.
22. O. Morozova and M. A. Marra, "Applications of next-generation sequencing technologies in functional genomics," Genomics, vol. 92, no. 5, pp. 255–264, 2008.
23. W. B. Barbazuk, S. J. Emrich, H. D. Chen, L. Li, and P. S. Schnable, "SNP discovery via 454 transcriptome sequencing," The Plant Journal, vol. 51, no. 5, pp. 910–918, 2007.
24. M. Trick, Y. Long, J. Meng, and I. Bancroft, "Single nucleotide polymorphism (SNP) discovery in the polyploid Brassica napus using Solexa transcriptome sequencing," Plant Biotechnology Journal, vol. 7, no. 4, pp. 334–346, 2009.
25. E. Novaes, D. R. Drost, W. G. Farmerie et al., "High-throughput gene and SNP discovery in Eucalyptus grandis, an uncharacterized genome," BMC Genomics, vol. 9, article 312, 2008.
26. P. C. Bundock, F. G. Eliott, G. Ablett et al., "Targeted single nucleotide polymorphism (SNP) discovery in a highly polyploid plant species using 454 sequencing," Plant Biotechnology Journal, vol. 7, no. 4, pp. 347–354, 2009.
27. T. L. Parchman, K. S. Geist, J. A. Grahnen, C. W. Benkman, and C. A. Buerkle, "Transcriptome sequencing in an ecologically important tree species: assembly, annotation, and marker discovery," BMC Genomics, vol. 11, no. 1, article 180, 2010.
28. K. Lai, C. Duran, P. J. Berkman et al., "Single nucleotide polymorphism discovery from wheat next-generation sequence data," Plant Biotechnology Journal, vol. 10, no. 6, pp. 743–749, 2012.

29. D. Kuhn, "Design of an Illumina Infinium 6k SNPchip for genotyping two large avocado mapping populations," in Proceedings of the 20th Conference on Plant and Animal Genome, San Diego, CA, January 2012.
30. J. R. Russell, M. Bayer, C. Booth et al., "Identification, utilisation and mapping of novel transcriptome-based markers from blackcurrant (Ribes nigrum)," BMC Plant Biology, vol. 11, article 147, 2011.
31. E. Hodges, Z. Xuan, V. Balija et al., "Genome-wide in situ exon capture for selective resequencing," Nature Genetics, vol. 39, no. 12, pp. 1522–1527, 2007.
32. N. M. Springer, K. Ying, Y. Fu et al., "Maize inbreds exhibit high levels of copy number variation (CNV) and presence/absence variation (PAV) in genome content," PLoS Genetics, vol. 5, no. 11, Article ID e1000734, 2009.
33. R. K. Varshney, "Gene-based marker systems in plants: high throughput approaches for marker discovery and genotyping," in Molecular Techniques in Crop Improvement, S. M. Jain and D. S. Brar, Eds., pp. 119–142, 2009.
34. A. Dean, "On a chromosome far, far away: LCRs and gene expression," Trends in Genetics, vol. 22, no. 1, pp. 38–45, 2006.
35. Y. Yuan, P. J. SanMiguel, and J. L. Bennetzen, "High-Cot sequence analysis of the maize genome," The Plant Journal, vol. 34, no. 2, pp. 249–255, 2003.
36. J. Emberton, J. Ma, Y. Yuan, P. SanMiguel, and J. L. Bennetzen, "Gene enrichment in maize with hypomethylated partial restriction (HMPR) libraries," Genome Research, vol. 15, no. 10, pp. 1441–1446, 2005.
37. D. T. Okou, K. M. Steinberg, C. Middle, D. J. Cutler, T. J. Albert, and M. E. Zwick, "Microarray-based genomic selection for high-throughput resequencing," Nature Methods, vol. 4, no. 11, pp. 907–909, 2007.
38. N. J. van Orsouw, R. C. J. Hogers, A. Janssen et al., "Complexity reduction of polymorphic sequences (CRoPS): a novel approach for large-scale polymorphism discovery in complex genomes," PLoS ONE, vol. 2, no. 11, Article ID e1172, 2007.
39. N. A. Baird, P. D. Etter, T. S. Atwood et al., "Rapid SNP discovery and genetic mapping using sequenced RAD markers," PLoS ONE, vol. 3, no. 10, Article ID e3376, 2008.
40. J. A. Mammadov, W. Chen, R. Ren et al., "Development of highly polymorphic SNP markers from the complexity reduced portion of maize (Zea mays L.) genome for use in marker-assisted breeding," Theoretical and Applied Genetics, vol. 121, no. 3, pp. 577–588, 2010.
41. Y. Chutimanitsakun, R. W. Nipper, A. Cuesta-Marcos et al., "Construction and application for QTL analysis of a Restriction Site Associated DNA (RAD) linkage map in barley," BMC Genomics, vol. 12, article 4, 2011.
42. H. Yu, W. Xie, J. Wang et al., "Gains in QTL detection using an ultra-high density SNP map based on population sequencing relative to traditional RFLP/SSR markers," PLoS ONE, vol. 6, no. 3, Article ID e17595, 2011.
43. A. Bus, J. Hecht, B. Huettel, R. Reinhardt, and B. Stich, "High-throughput polymorphism detection and genotyping in Brassica napus using next-generation RAD sequencing," BMC Genomics, vol. 13, no. 1, p. 281, 2012.
44. J. Tang, J. A. M. Leunissen, R. E. Voorrips, C. G. van der Linden, and B. Vosman, "HaploSNPer: a web-based allele and SNP detection tool," BMC Genetics, vol. 9, article 23, 2008.

45. A. Narechania, M. A. Gore, E. S. Buckler, et al., "Large-scale discovery of gene-enriched SNPs," The Plant Genome, vol. 2, no. 2, pp. 121–133, 2009.
46. J. C. Nelson, S. Wang, Y. Wu et al., "Single-nucleotide polymorphism discovery by high-throughput sequencing in sorghum," BMC Genomics, vol. 12, article 352, 2011.
47. R. J. Elshire, J. C. Glaubitz, Q. Sun et al., "A robust, simple genotyping-by-sequencing (GBS) approach for high diversity species," PLoS ONE, vol. 6, no. 5, Article ID e19379, 2011.
48. S. R. Browning and B. L. Browning, "Rapid and accurate haplotype phasing and missing-data inference for whole-genome association studies by use of localized haplotype clustering," American Journal of Human Genetics, vol. 81, no. 5, pp. 1084–1097, 2007.
49. B. N. Howie, P. Donnelly, and J. Marchini, "A flexible and accurate genotype imputation method for the next generation of genome-wide association studies," PLoS Genetics, vol. 5, no. 6, Article ID e1000529, 2009.
50. X. Huang, X. Wei, T. Sang et al., "Genome-wide asociation studies of 14 agronomic traits in rice landraces," Nature Genetics, vol. 42, no. 11, pp. 961–967, 2010.
51. J. Marchini and B. Howie, "Genotype imputation for genome-wide association studies," Nature Reviews Genetics, vol. 11, no. 7, pp. 499–511, 2010.
52. J. B. Fan, A. Oliphant, R. Shen et al., "Highly parallel SNP genotyping," Cold Spring Harbor Symposia on Quantitative Biology, vol. 68, pp. 69–78, 2003.
53. F. J. Steemers and K. L. Gunderson, "Whole genome genotyping technologies on the BeadArray™ platform," Biotechnology Journal, vol. 2, no. 1, pp. 41–49, 2007.
54. K. J. Livak, S. J. A. Flood, J. Marmaro, W. Giusti, and K. Deetz, "Oligonucleotides with fluorescent dyes at opposite ends provide a quenched probe system useful for detecting PCR product and nucleic acid hybridization," Genome Research, vol. 4, no. 6, pp. 357–362, 1995.
55. S. P. Kumpatla, R. Buyyarapu, I. Y. Abdurakhmonov, and J. A. Mammadov, "Genomics-assisted plant breeding in the 21st century: technological advances and progress," in Plant Breeding, I. Y. Abdurakhmonov, Ed., pp. 131–184.
56. R. Buyyarapu, R. Ren, S. Kumpatla et al., "In silico discovery and validation of SNP markers for molecular breeding in cotton," in Proceedings of the 19th Conference on Plant & Animal Genome, San Diego, Calif, USA, January 2011.
57. S. D. Tanksley, "Mapping polygenes," Annual Review of Genetics, vol. 27, pp. 205–233, 1993.
58. D. Bhattramakki, M. Dolan, M. Hanafey et al., "Insertion-deletion polymorphisms in 3′ regions of maize genes occur frequently and can be used as highly informative genetic markers," Plant Molecular Biology, vol. 48, no. 5-6, pp. 539–547, 2002.
59. E. S. Jones, H. Sullivan, D. Bhattramakki, and J. S. C. Smith, "A comparison of simple sequence repeat and single nucleotide polymorphism marker technologies for the genotypic analysis of maize (Zea mays L.)," Theoretical and Applied Genetics, vol. 115, no. 3, pp. 361–371, 2007.
60. S. Konishi, T. Izawa, S. Y. Lin et al., "An SNP caused loss of seed shattering during rice domestication," Science, vol. 312, no. 5778, pp. 1392–1396, 2006.
61. A. S. Iyer and S. R. McCouch, "The rice bacterial blight resistance gene xa5 encodes a novel form of disease resistance," Molecular Plant-Microbe Interactions, vol. 17, no. 12, pp. 1348–1354, 2004.

62. E. Drenkard, B. G. Richter, S. Rozen et al., "A simple procedure for the analysis of single nucleotide polymorphism facilitates map-based cloning in Arabidopsis," Plant Physiology, vol. 124, no. 4, pp. 1483–1492, 2000.
63. K. Garg, P. Green, and D. A. Nickerson, "Identification of candidate coding region single nucleotide polymorphisms in 165 human genes using assembled expressed sequence tags," Genome Research, vol. 9, no. 11, pp. 1087–1092, 1999.
64. S. Nasu, J. Suzuki, R. Ohta et al., "Search for and analysis of single nucleotide polymorphisms (SNPS) in rice (Oryza sativa, Oryza rufipogon) and establishment of SNP markers," DNA Research, vol. 9, no. 5, pp. 163–171, 2002.
65. K. Hayashi, N. Hashimoto, M. Daigen, and I. Ashikawa, "Development of PCR-based SNP markers for rice blast resistance genes at the Piz locus," Theoretical and Applied Genetics, vol. 108, no. 7, pp. 1212–1220, 2004.
66. M. Ashikari and M. Matsuoka, "Identification, isolation and pyramiding of quantitative trait loci for rice breeding," Trends in Plant Science, vol. 11, no. 7, pp. 344–350, 2006.
67. K. K. Jena and D. J. Mackill, "Molecular markers and their use in marker-assisted selection in rice," Crop Science, vol. 48, no. 4, pp. 1266–1276, 2008.
68. R. K. Varshney, D. A. Hoisington, and A. K. Tyagi, "Advances in cereal genomics and applications in crop breeding," Trends in Biotechnology, vol. 24, no. 11, pp. 490–499, 2006.
69. E. S. Buckler, J. B. Holland, P. J. Bradbury et al., "The genetic architecture of maize flowering time," Science, vol. 325, no. 5941, pp. 714–718, 2009.
70. J. A. Poland, P. J. Bradbury, E. S. Buckler, and R. J. Nelson, "Genome-wide nested association mapping of quantitative resistance to northern leaf blight in maize," Proceedings of the National Academy of Sciences of the United States of America, vol. 108, no. 17, pp. 6893–6898, 2011.
71. P. Zheng, W. B. Allen, K. Roesler et al., "A phenylalanine in DGAT is a key determinant of oil content and composition in maize," Nature Genetics, vol. 40, no. 3, pp. 367–372, 2008.
72. E. Akhunov, C. Nicolet, and J. Dvorak, "Single nucleotide polymorphism genotyping in polyploid wheat with the Illumina GoldenGate assay," Theoretical and Applied Genetics, vol. 119, no. 3, pp. 507–517, 2009.
73. A. M. Allen, G. L. Barker, S. T. Berry et al., "Transcript-specific, single-nucleotide polymorphism discovery and linkage analysis in hexaploid bread wheat (Triticum aestivum L.)," Plant Biotechnology Journal, vol. 9, no. 9, pp. 1086–1099, 2011.
74. A. Bérard, M. C. Le Paslier, M. Dardevet et al., "High-throughput single nucleotide polymorphism genotyping in wheat (Triticum spp.)," Plant Biotechnology Journal, vol. 7, no. 4, pp. 364–374, 2009.
75. M. O. Winfield, P. A. Wilkinson, A. M. Allen et al., "Targeted re-sequencing of the allohexaploid wheat exome," Plant Biotechnology Journal, vol. 10, no. 6, pp. 733–742, 2012.
76. E. S. Lagudah, S. G. Krattinger, S. Herrera-Foessel et al., "Gene-specific markers for the wheat gene Lr34/Yr18/Pm38 which confers resistance to multiple fungal pathogens," Theoretical and Applied Genetics, vol. 119, no. 5, pp. 889–898, 2009.

77. H. Buerstmayr, T. Ban, and J. A. Anderson, "QTL mapping and marker-assisted selection for Fusarium head blight resistance in wheat: a review," Plant Breeding, vol. 128, no. 1, pp. 1–26, 2009.
78. A. N. Bernardo, H. Ma, D. Zhang, and G. Bai, "Single nucleotide polymorphism in wheat chromosome region harboring Fhb1 for Fusarium head blight resistance," Molecular Breeding, vol. 29, no. 2, pp. 477–488, 2012.
79. P. K. Gupta, P. Langridge, and R. R. Mir, "Marker-assisted wheat breeding: present status and future possibilities," Molecular Breeding, vol. 26, no. 2, pp. 145–161, 2010.
80. K. S. Kim, S. Bellendir, K. A. Hudson et al., "Fine mapping the soybean aphid resistance gene Rag1 in soybean," Theoretical and Applied Genetics, vol. 120, no. 5, pp. 1063–1071, 2010.
81. K. S. Kim, C. B. Hill, G. L. Hartman, D. L. Hyten, M. E. Hudson, and B. W. Diers, "Fine mapping of the soybean aphid-resistance gene Rag2 in soybean PI 200538," Theoretical and Applied Genetics, vol. 121, no. 3, pp. 599–610, 2010.
82. B. K. Ha, R. S. Hussey, and H. R. Boerma, "Development of SNP assays for marker-assisted selection of two southern root-knot nematode resistance QTL in soybean," Crop Science, vol. 47, no. 2, pp. S73–S82, 2007.
83. X. Hu, M. Sullivan-Gilbert, M. Gupta, and S. A. Thompson, "Mapping of the loci controlling oleic and linolenic acid contents and development of fad2 and fad3 allele-specific markers in canola (Brassica napus L.)," Theoretical and Applied Genetics, vol. 113, no. 3, pp. 497–507, 2006.
84. A. Lehmensiek, M. W. Sutherland, and R. B. McNamara, "The use of high resolution melting (HRM) to map single nucleotide polymorphism markers linked to a covered smut resistance gene in barley," Theoretical and Applied Genetics, vol. 117, no. 5, pp. 721–728, 2008.
85. M. K. Grimmer, S. Trybush, S. Hanley, S. A. Francis, A. Karp, and M. J. C. Asher, "An anchored linkage map for sugar beet based on AFLP, SNP and RAPD markers and QTL mapping of a new source of resistance to Beet necrotic yellow vein virus," Theoretical and Applied Genetics, vol. 114, no. 7, pp. 1151–1160, 2007.
86. M. K. Grimmer, T. Kraft, S. A. Francis, and M. J. C. Asher, "QTL mapping of BNYVV resistance from the WB258 source in sugar beet," Plant Breeding, vol. 127, no. 6, pp. 650–652, 2008.
87. W. Muchero, N. N. Diop, P. R. Bhat et al., "A consensus genetic map of cowpea [Vigna unguiculata (L) Walp.] and synteny based on EST-derived SNPs," Proceedings of the National Academy of Sciences of the United States of America, vol. 106, no. 43, pp. 18159–18164, 2009.
88. G. Jander, S. R. Norris, S. D. Rounsley, D. F. Bush, I. M. Levin, and R. L. Last, "Arabidopsis map-based cloning in the post-genome era," Plant Physiology, vol. 129, no. 2, pp. 440–450, 2002.
89. I. Y. Abdurakhmonov and A. Abdukarimov, "Application of association mapping to understanding the genetic diversity of plant germplasm resources," International Journal of Plant Genomics, vol. 2008, Article ID 574927, 2008.

90. D. Hall, C. Tegström, and P. K. Ingvarsson, "Using association mapping to dissect the genetic basis of complex traits in plants," Briefings in Functional Genomics and Proteomics, vol. 9, no. 2, pp. 157–165, 2010.
91. S. Myles, J. Peiffer, P. J. Brown et al., "Association mapping: critical considerations shift from genotyping to experimental design," Plant Cell, vol. 21, no. 8, pp. 2194–2202, 2009.
92. M. Gore, E. S. Buckler, J. Yu, and C. Zhu, "Status and prospects of association mapping in plants," The Plant Genome, vol. 1, no. 1, pp. 5–20, 2008.
93. J. A. Rafalski, "Association genetics in crop improvement," Current Opinion in Plant Biology, vol. 13, no. 2, pp. 174–180, 2010.
94. A. Beló, P. Zheng, S. Luck et al., "Whole genome scan detects an allelic variant of fad2 associated with increased oleic acid levels in maize," Molecular Genetics and Genomics, vol. 279, no. 1, pp. 1–10, 2008.
95. F. Tian, P. J. Bradbury, P. J. Brown et al., "Genome-wide association study of leaf architecture in the maize nested association mapping population," Nature Genetics, vol. 43, no. 2, pp. 159–162, 2011.
96. J. K. Roy, K. P. Smith, G. J. Muehlbauer, S. Chao, T. J. Close, and B. J. Steffenson, "Association mapping of spot blotch resistance in wild barley," Molecular Breeding, vol. 26, no. 2, pp. 243–256, 2010.
97. K. Pajerowska-Mukhtar, B. Stich, U. Achenbach et al., "Single nucleotide polymorphisms in the Allene Oxide Synthase 2 gene are associated with field resistance to late blight in populations of tetraploid potato cultivars," Genetics, vol. 181, no. 3, pp. 1115–1127, 2009.
98. K. L. Kump, P. J. Bradbury, R. J. Wisser et al., "Genome-wide association study of quantitative resistance to southern leaf blight in the maize nested association mapping population," Nature Genetics, vol. 43, no. 2, pp. 163–168, 2011.
99. A. L. Harper, M. Trick, J. Higgins et al., "Associative transcriptomics of traits in the polyploid crop species Brassica napus," Nature Biotechnology, vol. 30, no. 8, pp. 798–802, 2012.
100. Y. Xu and J. H. Crouch, "Marker-assisted selection in plant breeding: from publications to practice," Crop Science, vol. 48, no. 2, pp. 391–407, 2008.
101. B. C. Y. Collard and D. J. Mackill, "Marker-assisted selection: an approach for precision plant breeding in the twenty-first century," Philosophical Transactions of the Royal Society B, vol. 363, no. 1491, pp. 557–572, 2008.
102. E. M. Septiningsih, A. M. Pamplona, D. L. Sanchez et al., "Development of submergence-tolerant rice cultivars: the Sub1 locus and beyond," Annals of Botany, vol. 103, no. 2, pp. 151–160, 2009.
103. L. Jin, Y. Lu, Y. Shao et al., "Molecular marker assisted selection for improvement of the eating, cooking and sensory quality of rice (Oryza sativa L.)," Journal of Cereal Science, vol. 51, no. 1, pp. 159–164, 2010.
104. M. Asif, T. Shaheen, N. Tabbasam, Y. Zafar, and A. H. Paterson, "Marker-assisted breeding in higher plants," Alternative Farming Systems, Biotechnology, Drought Stress and Ecological Fertilisation, vol. 6, pp. 39–76, 2011.

105. R. Naidoo, G. M. F. Watson, J. Derera, P. Tongoona, and M. Laing, "Marker-assisted selection for low phytic acid (lpa1-1) with single nucleotide polymorphism marker and amplified fragment length polymorphisms for background selection in a maize backcross breeding programme," Molecular Breeding, vol. 30, pp. 1207–1217, 2012.
106. S. R. Eathington, T. M. Crosbie, M. D. Edwards, R. S. Reiter, and J. K. Bull, "Molecular markers in a commercial breeding program," Crop Science, vol. 47, supplement 3, pp. S154–S163, 2007.
107. M. L. Rosso, S. A. Burleson, L. M. Maupin, and K. M. Rainey, "Development of breeder-friendly markers for selection of MIPS1 mutations in soybean," Molecular Breeding, vol. 28, no. 1, pp. 127–132, 2011.
108. J. M. Ribaut and M. Ragot, "Marker-assisted selection to improve drought adaptation in maize: the backcross approach, perspectives, limitations, and alternatives," Journal of Experimental Botany, vol. 58, no. 2, pp. 351–360, 2007.
109. R. Ren, B. A. Nagel, S. P. Kumpatla et al., "Maize Cytoplasmic Male Sterility (Cms) C-Type Restorer Rf4 Gene, Molecular Markers And Their Use," Google Patents, 2011.
110. M. Ragot, M. Lee, E. Guimarães et al., "Marker-assisted selection in maize: current status, potential, limitations and perspertives from the private and public sectors," Marker-Assisted Selection, Current Status and Future Perspectives in Crops, Livestock, Forestry and Fish, pp. 117–150, 2007.
111. H. Riday, "Paternity testing: a non-linkage based marker-assisted selection scheme for outbred forage species," Crop Science, vol. 51, no. 2, pp. 631–641, 2011.
112. D. W. Gjertson, C. H. Brenner, M. P. Baur et al., "ISFG: recommendations on biostatistics in paternity testing," Forensic Science International, vol. 1, no. 3-4, pp. 223–231, 2007.
113. T. H. E. Meuwissen, B. J. Hayes, and M. E. Goddard, "Prediction of total genetic value using genome-wide dense marker maps," Genetics, vol. 157, no. 4, pp. 1819–1829, 2001.
114. E. L. Heffner, M. E. Sorrells, and J. L. Jannink, "Genomic selection for crop improvement," Crop Science, vol. 49, no. 1, pp. 1–12, 2009.
115. Z. Shengqiang, J. C. M. Dekkers, R. L. Fernando, and J. L. Jannink, "Factors affecting accuracy from genomic selection in populations derived from multiple inbred lines: a barley case study," Genetics, vol. 182, no. 1, pp. 355–364, 2009.
116. B. Hayes and M. Goddard, "Genome-wide association and genomic selection in animal breeding," Genome, vol. 53, no. 11, pp. 876–883, 2010.
117. J. L. Jannink, A. J. Lorenz, and H. Iwata, "Genomic selection in plant breeding: from theory to practice," Briefings in Functional Genomics and Proteomics, vol. 9, no. 2, pp. 166–177, 2010.
118. A. M. Mastrangelo, E. Mazzucotelli, D. Guerra, P. Vita, and L. Cattivelli, "Improvement of drought resistance in crops: from conventional breeding to genomic selection," Crop Stress and Its Management, pp. 225–259, 2012.

119. M. D. V. Resende, M. F. R. Resende Jr., C. P. Sansaloni et al., "Genomic selection for growth and wood quality in Eucalyptus: capturing the missing heritability and accelerating breeding for complex traits in forest trees," New Phytologist, vol. 194, no. 1, pp. 116–128, 2012.
120. Y. Zhao, M. Gowda, W. Liu et al., "Accuracy of genomic selection in European maize elite breeding populations," Theoretical and Applied Genetics, vol. 124, no. 4, pp. 769–776, 2012.

CHAPTER 15

GENOTYPING-BY-SEQUENCING IN PLANTS

STÉPHANE DESCHAMPS, VICTOR LLACA,
and GREGORY D. MAY

15.1 INTRODUCTION

The analysis of genomic variation is an essential part of plant genetics and crop improvement programs. DNA polymorphisms can be directly related to phenotype differences, be genetically linked to its causative factor, or indicate relationships between individuals in populations [1]. Over the last 30 years, the use of genotyping has enabled the characterization and mapping of genes and metabolic pathways in plants as well as the study of species diversity and evolution, marker-assisted selection (MAS), germplasm characterization and seed purity. Single Nucleotide Polymorphisms (SNPs) have emerged as the most widely used genotyping markers due to their abundance in the genome and the relative ease in determining their frequency in a cost-effective and parallel manner in a given panel of individuals.

The field of agricultural genomics is in the midst of a technological revolution caused by the relatively sudden emergence of "next-generation" DNA sequencing technologies, driven in part by the completion of the human genome and the desire to apply the benefits of genomics to a better understanding of diseases and a more personalized view of medicine. By greatly reducing limitations in generating sequence information, these technological advances have facilitated the characterization of genes and genomes, and started to provide a more comprehensive view of diversity and gene function in plants. The increased ability to sequence in a cost-effective manner large numbers of individuals within the same species has altered the concept of variant discovery and genotyping in mapping studies, especially in plant species with complex genomes or limited public resources available. A new concept, namely genotyping-by-sequencing (GBS), has emerged, where the detection of sequence differences (namely SNPs) in a large segregating or mutant population is combined with scoring, thus allowing a rapid and direct study of its diversity targeted towards the mapping of a trait or a mutation of interest. This review will summarize the current state of genotyping and next-generation DNA sequencing technologies then provide some examples of studies where next-generation DNA sequencing has been used in plant species for genotyping applications.

15.2 GENOTYPING APPLICATIONS

The first plant DNA markers were based on restriction fragment length polymorphisms (RFLPs) [2]. Early hybridization-based, radioactive RFLP techniques were inherently challenging and time consuming, and were eventually replaced by less complex, more cost-effective PCR-based markers. Among them, simple-sequence repeats (SSR) [3] were particularly useful as genetic markers. They were relatively inexpensive, abundant in plant genomes and more informative than bi-allelic markers. Additional marker strategies were developed using different combinations of PCR, restriction digestion and gel electrophoresis techniques. Major marker techniques included random amplification of polymorphic DNAs (RAPDs) [4]; sequence characterized amplified region (SCARs) [5]; cleaved amplified polymorphic sequences (CAPS) [6]; Intersimple Sequence Repeats (ISSRs) [7]; amplified fragment length polymorphisms (AFLPs) [8]; and direct amplification of length polymorphism (DALP)

[9]. The improvement of Sanger sequencing throughput in the 1990's in combination with the start of genome and expressed sequence tag (EST) sequencing programs in model plant species, led to the acceleration in the identification of variation at the single base pair resolution [10]. The use of single-nucleotide polymorphisms (SNPs) as markers for genotyping using direct re-sequencing increased the potential to score variation in specific targets. More importantly, the increase in information about potentially millions of genome-wide SNPs or small insertion-deletions and their surrounding sequence context set the foundation of high-throughput genotyping.

For the past 15 years, automation and miniaturization in SNP-based marker technologies has increased marker density and reduced genotyping costs and time by orders of magnitude in relation to earlier approaches. Some of the most commonly used systems are based on fluorescent detection of SNP-specific hybridization probes on PCR products such as Taqman, Molecular Beacons and Invader [11–16]. Other strategies such as Sequenom homogeneous Mass Extend (hME) and iPLEX genotyping systems involve MALDI-TOF mass spectrophotometry of SNP-specific PCR primer extension products. These technologies originally allowed data collection of hundreds to up to a few thousand SNP/samples per day/machine. With the increasing need for higher throughput, end-point fluorescent assays such as Taqman and Invader have been significantly enhanced by the use of array tape technology in place of 96, 384 or 1,536-well microtiter plates, reducing cost per assay to a fraction of the original reaction and increasing throughput in a format that allows the PCR processing of the equivalent of hundreds of microtiter plates on one roll of tape [17].

The emergence of massively parallel array systems has further enabled parallel scoring of up to hundreds of thousands of markers in plants [18]. These ultra-high throughput technologies are wide-ranging and researchers can now select methods based on application, assay simplicity, cost, throughput and accuracy. Among the most widely used array-based systems in plants are the GoldenGate and Infinium assays, which consist on multistep protocols based on Illumina's BeadArray/BeadChip technology. The GoldenGate system allows screening of a large number of samples using a single multiplexed assay that can include as many as 3,072 SNPs. The Illumina's Infinium provides considerably higher throughput, of up to four million SNPs on a single sample, or up to several hundred thousand on multiple samples in the same array. In Infinium, samples are incubated on bead chips where they anneal to locus-specific 50-mers covalently linked to beads. After hybridization, oligos are subject to allele-specific single-base extension; followed by fluorescent staining, signal amplification, scanning in a dual-color channel reader, and analysis. A major advantage of Infinium is the availability of commercially available validated chips in selected species, such as the MaizeSNP50, which includes more than 56,000 SNP markers derived from the comparison of the B73 maize reference genome sequence to multiple lines. The use of pre-made arrays reduces cost considerably although the actual number of markers derived from this array will be considerably lower, depending on the relationship to the reference and gene representation in the interrogated plants. Other arrays provide comparable levels of throughput. Beckman Coulter's GenomeLAb SNPstream allows the processing of up to three million genotypes in 384 samples per day per instrument. The widely used Affimetrix GeneChip system cannot only detect hundreds of thousands of SNPs in a single array but it can also be used for SNP discovery by sequencing by hybridization (SbH).

15.3 ULTRA-HIGH THROUGHPUT GENOTYPING APPLICATIONS

In plant genetics, not all marker-related applications require massively parallel, genome-wide genotyping. Plant phylogenetic and diversity studies have successfully exploited relatively low marker densities or regional markers to determine relationship in plants at the interspecific and intraspecific levels [19,20]. The ability of highly polymorphic SSRs and AFLPs to differentiate individuals in a population has made them markers of choice for pedigree analyses and cultivar identification. In linkage mapping, relatively low marker density has been sufficient to enable the mapping and characterization of simple traits and quantitative trait loci (QTL) with large effects in the total genetic variance. Recent linkage mapping studies in maize have identified QTL with relatively large effects in oil content [21] and root architecture [22]. However, in most cases QTL characterization by linkage mapping can be problematic as intervals may encompass large genetic and physical distances and require walking through several megabase-pairs of sequence, with a large number of potential candidates [23–25]. Only a small fraction of mapped plant QTL has been cloned by linkage mapping due to the low resolution of available mapping strategies [26,27]. Increasing marker density may not provide additional benefits as map resolution can be limited by the relatively few recombinants generated from two original parents in a limited number of generations and progeny as polymorphisms are identified between two parents and then followed in a segregating population. Finally, medium to high marker densities may be required in marker assisted selection (MAS) to allow early testing of specific traits using linked markers to reduce breeding time and number of plants and space needed. In cases of selection for specific traits to reduce linkage drag or pyramiding genes for the same trait, the use of low density or regional markers may be sufficient.

The development of ultra-high-throughput genotyping technologies and later the emergence of new sequencing platforms has enabled the development of high-density applications in QTL characterization and plant breeding that had been difficult to accomplish or not feasible before. Major marker-intensive applications include genome-wide association studies (GWAS) and bulked segregant analysis (BSA) [28,29]. Unlike linkage mapping, GWAS exploits the natural diversity generated by multi-generational recombination events in a population or panel [30–33]. This strategy can result in increased resolution compared to linkage mapping populations, as long as enough markers are provided. Only the markers that are in linkage disequilibrium (LD) with the trait of interest will show association to such trait. Without enough genome-wide marker coverage, association mapping studies need to focus on polymorphisms in candidate genes that are suspected to have roles in controlling phenotypic variation for one specific trait of interest [34]. The concept of GWAS predates high throughput genotyping technologies and the first genome-wide association study in plants was conducted more than 10 years ago in sea beet (Beta vulgaris ssp. maritima) [35]. However, the effective implementation of genome-wide association mapping in plants and the determination of optimal marker density have been problematic because of lack of knowledge regarding the degree and structure of LD distribution in specific target populations. LD can be affected by multiple factors at the species or population level, including the degree of selfing, epistasis, admixture and population bottlenecks followed by genetic drift. Complex breeding history and limited gene flow are common factors in plants generating stratification and uneven distribution of alleles in populations, which can lead to false associations [36–38]. Depending on the size and LD characteristics of the population under analysis, tens of thousands or even millions of independent genetic markers may be needed to correct the effects of population structure and achieve optimal resolution in a genome-wide scan. In plant populations with low LD, genotyping costs have been a serious limiting factor thus far, deeming candidate-gene association analysis the method of choice. In plants, the ability to create lines of individuals with identical or near identical background offer the potential to create public GWAS resources that can be accessed by multiple groups and rapidly resolve complex traits. Plant GWAS can be performed in large numbers of samples in replicated trials using inbreds, double haploid (DH) lines and recombinant inbred lines (RILs) [39]. The determination of large numbers of genome variants in combination with transcription profiling can be used to determine expression quantitative loci (eQTLs) [40], mapping regions with cis- and trans-effects [41–44]. Bulked segregant analysis can be used as a time- and cost-effective way to identify markers associated to specific phenotypes without the need of having a linkage map or sampling large numbers of samples in a population [45]. BSA can be used for extreme mapping, where plants from extreme ends of the phenotype range in a population derived from a single cross are bulked, or pooled, and the genotype differences correlated to the trait of interest. This method allows the easy detection of QTL in large populations as long as the number of markers available is large and widely distributed along the genome [46].

Finally, the increase in availability and cost reduction of markers has made feasible the concept of Genomic Selection for plant breeding. Genomic, or Genome-wide, Selection (GS) [47] has been proposed as an effective method to breed for traits involving multiple QTL with low heritability. In GS, unlike GWAS, arrays of markers are selected without establishing association with traits, and are used to predict phenotypes. Only genotypic data need to be used in a breeding population as predictor to select individuals with the best breeding values. In GS, an initial training population is used to capture both phenotypic and genotyping data from a very large number of markers, to capture all additive genetic variance for specific traits. Breeding values are then estimated in a breeding population solely based on genotype and the estimated marker effects.

With all their potential to increase SNP density and resolution in large samples for GWAS, BSA and GS, current array-based technologies have clear limitations. They require prior generation of sequence information, identification of polymorphisms, validation and array production. The value of Sanger or NGS-driven massive polymorphisms discovery and ultra-high throughput platforms can be seriously restricted by cost and time limitations in the design, validation and deployment of molecular markers. Furthermore, the significant sequence diversity and the high structural polymorphism observed in important plant models such as maize imposes a challenge to these knowledge-based platforms. Structural genome differences, including translocations, copy number variation and presence-absence variation are observed in landraces and lines in the same species and correspond to differences in repetitive, non-coding DNA and gene content [48,49]. There is an inherent bias towards cultivars used as reference in genome projects. Molecular markers may be absent near or within the gene space located in larger structural variations (i.e., CNVs, PAVs, and large indels). With the falling cost of NGS there is an increased interest in genotyping-by-sequencing (GBS), where obtained sequence differences are used directly as markers for analysis. We will describe here a number of GBS strategies applied to populations or panels in plants genetics and breeding.

15.4 ULTRA HIGH THROUGHPUT DNA SEQUENCING

The field of DNA sequencing recently has been marked by dramatic increases in throughput combined with a significant decrease in cost per base of raw sequence. For over two decades, the advent of the modern genomics era has been characterized by major prokaryotic and eukaryotic genome sequencing projects achieved using the Sanger method of sequencing [50,51]. Sanger sequencing still is the gold standard in terms of generating high quality sequencing information as many finished-grade whole genome sequencing drafts were achieved using that method [52–59]. Sanger sequencing also can be used to discover genetic variations (including SNPs) within a set of individuals in a population. One particular method employs the PCR amplification of genomic DNA in multiple individuals as a mean to generate homologous DNA fragments that are end-sequenced and compared to reveal particular sequence variations [60,61]. Since its inception more than 35 years ago, the Sanger sequencing method has gone through several iterations of improvements, including automated sequencers [62,63] and the emergence of fluorescent dye terminators to capture nucleotide incorporation events [64,65]. However, the high costs and labor generally associated with the Sanger sequencing technique fundamentally limit its reach and use in large multi-genome comprehensive studies, both for medical and agricultural applications. These limitations have contributed to the emergence in the past eight years of "next generation sequencing" (NGS) technologies that rely on massively parallel sequencing and imaging techniques to yield several hundreds of millions to several hundreds of billions of bases per run [66]. Current NGS platforms can be divided into two categories, labeled as second-generation and third-generation depending, mostly, on whether DNA templates are amplified on an immobilized support prior to sequencing, and the subsequent generation of sequencing data either from clustered copies originating from the same DNA strand, or directly from single DNA molecules [67].

Second-generation DNA sequencing strategies (Table 1) all follow a similar pattern for DNA template preparation, where universal adapters are ligated at both ends of randomly sheared DNA fragments. They all also rely on the cyclic interrogation of millions of clonally amplified DNA molecules immobilized on a synthetic surface to generate up to several billions of sequences in a massively parallel fashion. Sequencing is performed in an iterative manner, where the incorporation of one or more nucleotides is followed by the emission of a signal and its detection by the sequencer [68].

TABLE 1: Comparison of representative next-generation sequencing technologies.

Sequencing Platform	Sequencing Chemistry	Detection Chemistry	Run Time[a]	Read Length (bp)	Reads per Run (million)	Through-put per Run (Gbp)
Roche 454 FLX Titanium	Sequencing by Synthesis	Light	23 hours	~800	~1	~0.7
Illumina MiSeq	Sequencing by Synthesis	Fluores-cence	39 hours	2 × 250[b]	~1	~8
Illumina HiSeq2500	Sequencing by Synthesis	Fluores-cence	11 days (high output)/27 hours (rapid run)	2 × 100[b]	~3,000	~600 (high out-put)/~120 (rapid run)
Life Technolo-gies 5500xl	Sequencing by Ligation	Fluores-cence	8 days	75 + 35[b]	~5,000	~310
Ion Torrent PGM	Sequencing by Synthesis	pH	4 hours	100	1	~0.1

a Not including library construction; b Paired-end read sequencing.

The first NGS platform to become commercially available was the Roche 454 GS20 sequencer, which later was replaced by the 454 GS FLX Titanium sequencer [69].The 454 technology combines clonal amplification of a single DNA molecule by PCR in a water-oil emulsion [70] with a sequencing-by-synthesis approach known as pyrosequencing [71].Here, the sequential release of single nucleotides, followed by the detection of pyrophosphate release during nucleotide incorporation, generates series of chemiluminescent signals whose intensities are used to determine the number of bases being incorporated to the elongating DNA strand. The 454 FLX Titanium sequencer currently is capable of generating approximately 450Mbp of sequences per 10-hour run with a read length up to 600bp and 99.99% accuracy [72]. The NGS technology commercialized by Illumina [73] generates shorter reads, ranging from 50 to 150bp, with sequencing throughputs ranging from ~1.5Gbp to ~600Gbp depending on the platform being used. Several instruments are commercialized by Illumina, ranging from the bench top MiSeq sequencer to the high-throughput HiSeq2500 sequencer. The Illumina sequencing technology combines clonal amplification of a single DNA molecule with a cyclical sequencing-by-synthesis approach. The PCR amplification is performed using a solid phase amplification protocol, also known as "bridge amplification" [74], to generate up to 1,000 copies of an original molecule of DNA, grouped together into a cluster. Sequencing is performed with proprietary reversible fluorescent terminator deoxyribonucleotides, in a series of cycles consisting of single base extension, fluorescence detection (where the nature of the signal is used to determine the identity of the base being incorporated) and cleavage of both the fluorescent label and of chemical moieties at the 3' hydroxyl position to allow for the next cycle to occur. Another NGS manufacturer, Life Technologies, currently offers two series of NGS instruments: the large-scale 5500 series, whose yields and read lengths are up to >20Gbp per day and 75bp, respectively [72], and the small-scale Ion Torrent series, yielding up to 10Gbp per run in less than a day [75]. The Ion Torrent series of instruments (PGM and Ion Proton) are smaller instruments that use semi-conductor chip technology to capture a signal after incorporation of a single base to the elongating strand of DNA. Similarly to the 454 sequencing technology, DNA fragments flanked by universal adapters are clonally amplified by emulsion PCR prior to being sequenced, and sequencing is performed by releasing single nucleotide types sequentially and detecting the release of one or more protons, and the subsequent local change in pH, during nucleotide incorporation.

Applications of second-generation sequencing technologies are numerous and include de novo assemblies of prokaryotic and eukaryotics genomes [76,77], alignment and comparison or targeted regions for variant discovery [78–81], profiling of transcripts [82–84] and small RNAs [85–87], profiling of epigenetics patterns [88,89] and chromatin structure [90,91], and species classification via metagenomics studies [92].

15.5 APPLICATIONS OF NGS TECHNOLOGIES TO GENOTYPING-BY-SEQUENCING

As described in a previous paragraph, the development of markers, as well as their scoring across populations, traditionally has been a high-cost process with many labor-intensive and time-consuming steps. The emergence of SNP arrays has reduced the time and efforts spent on scoring but the development of new markers still requires significant investments. These markers also are specific to the population in which they are developed, and the resulting allelic bias can be problematic in some divergent populations and species. Preliminary sequence information of regions flanking a SNP of interest also is required to develop marker assays, and only a few SNPs derived from sequencing data generally can be considered suitable for marker development, due to several factors, including proximity to repetitive regions, to known markers or to other regions of interest. By contrast, as chemistry and software improvements are leading to significant decreases in the overall cost of NGS, resequencing extended to entire populations, rather than to a few parental individuals for the sole purpose of discovering variants, enables the simultaneous genome-wide detection and scoring of hundreds of thousands of markers [93]. This "genotyping-by-sequencing" (GBS) approach also uses data directly from the populations being genotyped, thus removing ascertainment bias towards a particular population. A typical GBS procedure is shown in Figure 1. Genetic maps generated using GBS-based sequencing information then can be used subsequently for identifying loci of interest from different sets of individuals, including segregating populations or mutant pools.

GBS can be performed either through a reduced-representation or a whole-genome resequencing approach. The presence of repetitive elements in plants [94] can represent a significant challenge for de novo assembly, alignment to a reference sequence and sequence comparison for variant discovery. The choice of whether to sequence the entire genome or a reduced portion of it is generally dictated by several factors, including repetitive content, ploidy, and presence or absence of homeologs [95,96]. Whole genome resequencing has been performed in Arabidopsis [97] and rice [98]. In larger and more complex genomes, such as maize [93] or wheat [99], where much of the sequence is repetitive, the use of reduced-representation resequencing is generally preferred. Several strategies are available for reducing the complexity of a genome. The "mRNA-Seq" strategy, where cDNA molecules are chemically cleaved and the resulting fragments are end-sequenced, is an effective way of targeting

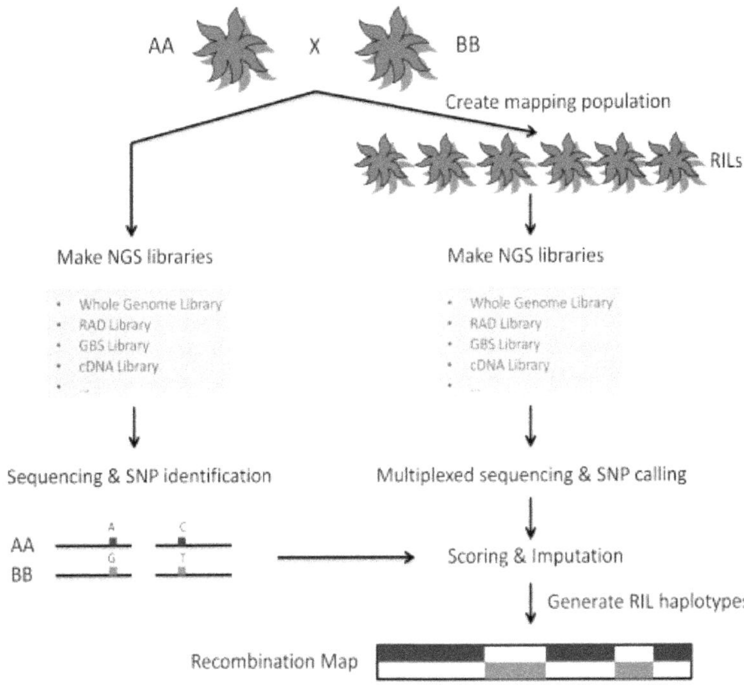

FIGURE 1: Schematic diagram of a representative GBS procedure. Two parents (AA and BB) are selected to create a mapping population. The parents are deeply sequenced using NGS technologies. SNPs and other variations between them are identified. The RILs are prepared using the same library construction strategy as the two parents (see text for details) and sequenced at lower coverage using NGS technologies. The resulting sequences are used to determine allelic diversity for each individual. Genotypes are assigned based on parental information. Haplotypes and recombination maps are created for each RIL. Blocks of haplotypes can be used directly as markers for mapping applications.

coding regions of the genome [100]. Other genome reduction approaches are based on the distinct methylation pattern of plant genomes [101,102] and include the use of methylation-sensitive restriction endonucleases to enrich for low-copy hypomethylated regions of the genome [103,104]. Other strategies for genome reduction such as multiplexed amplification of target sequences [105], molecular inversion probes (MIPs) [106] or the use of probes to capture DNA fragments by direct hybridization prior to sequencing [107,108] are available but can be labor intensive and rely heavily on existing sequence information, thus potentially limiting their value in large and highly divergent populations or species.

15.6 POLYMORPHISM DETECTION FROM NGS DATA

SNP calling and genotyping of NGS data must take into accounts characteristics inherent to NGS technologies. NGS data typically have a higher error rate than traditional Sanger sequencing or SNP genotyping methods. This traditionally has been addressed via deeper sequencing, thus increasing the confidence that a particular SNP call is correct. However, recent GBS studies, where whole genomic samples from large mapping populations are sequenced, have mostly relied on low sequencing coverage [98]. Li et al. [109] also have developed models showing that sequencing many individuals at low depth (2–4×) was a powerful alternative strategy to sequencing few individuals at high depth (30×) for complex trait association studies (it must be noted that such a low coverage model assumes a diploid genome and therefore needs to be evaluated in plants, where genome duplication and polyploidy are prevalent). NGS technologies also generate shorter reads than Sanger sequencing, thus increasing the risk of aligning a particular sequence to the wrong region. For all these reasons, calling SNPs within a population remain challenging. Several algorithms have been developed to address these issues and improve confidence for SNP calls or genotypes in relation to a particular read aligned to a reference sequence. While some methods are simply based on filtering out low quality data and counting alleles for a given locus, others are more probabilistic in nature, incorporating errors introduced during basecalling, alignment and assembly and coupling them with existing data such as allele frequencies, reference haplotypes or linkage disequilibrium information [110]. Table 2 provides several examples of software that have been developed for SNP detection from NGS data.

While SNP calls can be performed independently at each locus, linkage disequilibrium (LD) often can be used to impute missing data in SNP datasets for a particular population. LD is defined in population genetics as the non-random association of linked alleles at two or more loci, and typically exists within regions located on the same chromosome. As a consequence, entire haplotypes often can be defined by a few SNPs located at the boundaries of such regions in LD, and individuals within a recombinant population sharing alleles flanking a region of interest are expected to share the same haplotype within this region. As a consequence, several statistical methods have been developed for genotype imputation in panels of related individuals (using identity-by-descent information in linkage and association studies) and unrelated individuals (by comparing markers to a reference panel of haplotypes) (for review, see [111]). It must be noted that the use of imputation to estimate the genotype of many individuals that are not directly sequenced in specific areas is central to the concept of GBS, as biological and technical bias during sample preparation and sequencing generally lead to variable sequencing coverage at a particular locus between individuals.

TABLE 2: Non-exhaustive list of available SNP calling software for NGS data.

Software	Link	Input Format	Primary Function
MAQ	http://maq.sourceforge.net/	FASTA, FASTQ	Mapping and Assembly
SAMtools	http://samtools.sourceforge.net/	SAM, BAM	Alignments
GATK	http://www.broadinstitute.org/gatk/	SAM	Alignments
SOAPsnp	http://soap.genomics.org.cn/soapsnp.html	SOAP	Mapping and Assembly
SNIP-Seq	http://polymorphism.scripps.edu/~vbansal/software/SNIP-Seq/	Pileup	Alignments
MapNext	http://evolution.sysu.edu.cn/english/software/mapnext.htm	FASTA, FASTQ	Alignments

15.7 GENOTYPING-BY-SEQUENCING IN PLANTS

Many traits in plants, such as yield, are quantitative, resulting from the combinatorial effect of many genes [112]. The mapping of underlying quantitative trait loci (QTL) has been made possible by the emergence of molecular markers, genotyping technologies and related statistical methodologies [1]. Initially, the identification of QTL was mostly based on linkage mapping strategies, where polymorphisms between two parents are detected in a segregating population, and the linkage of a particular region to a given phenotype can be determined by genotyping recombinants exhibiting phenotypic variations for a trait of interest [21,113]. However, the relatively small number of recombinants generated from two parents in a limited number of generations means that linkage mapping generally has low resolution, encompassing very large genetic and physical distance, with many possible candidate genes for a trait of interest. This has led to the emergence of association mapping studies, which utilize the natural diversity present in a multi-generational population and provides higher resolution than linkage mapping populations to map traits of interest [28,114,115]. Larger genome-wide association studies (GWAS) require hundreds of thousands to millions of markers to generate sufficient information and coverage, and getting such resolution has been greatly enhanced by the emergence of NGS technologies [116–118]. More recently, NGS technologies have been used to resequence collections of recombinant inbred lines (RILs) to analyze, in correlation to appropriate phenotypic values, and map various traits of interest in specific environments. Such resources have been generated in maize [31,119] where a collection of 5,000 RILs derived from a nested association mapping (NAM) population have been resequenced using a restriction endonuclease-based reduced-representation approach and the Illumina sequencing technology, generating a total of 1.4 million SNPs and 200,000 indels. However, such resources have limited value beyond their population and a number of related sequencing protocols have been applied to other population for linkage mapping, association mapping, or bulked segregant analysis studies. In rice (Oryza sativa), Huang et al. [98] re-sequenced with the Illumina technology 150 F11 RILs derived from a cross between Indica and Japonica rice cultivars. The resulting sequences, generated after whole-genome re-sequencing of each RILs to an average 0.02X coverage, resulted in the discovery of 1,226,791 SNPs, separated by an average of 40Kbp. Haplotypes and recombination breakpoints could be determined for each RIL, using the parental origins of SNPs in discrete regions of the genomes, and a recombination bin map made of 2,334 bins for the 150 RILs was constructed from the haplotypes. Using each bin as a genetic marker, 49 QTLs linked to various phenotypes in rice could be detected, including 5 QTLs physically located at positions overlapping with the location of candidate genes described in previous studies [120].

Construction of a low-density GBS linkage map using the reduced-representation sequence-based marker discovery technique known as restriction site associated DNA sequencing (RAD) [121] has been reported in barley in barley (*Hordeum vulgare*) [122]. The RAD approach does not require any prior knowledge of the genome of the species being investigated and produces two types of markers: (1) co-dominant markers from sequence variations present in short targeted regions of the genomes immediately adjacent to selected restriction endonuclease cutting sites and (2) dominant markers from sequence variations present within the selected restriction endonuclease cutting sites. RAD sequencing was used to generate a set of 530 fixed SNP markers from the Oregon Wolfe Barley (OWB) parental inbred lines. These markers were classified as codominant, selected from approximately 10,000 clusters of RAD sequences obtained on the Illumina Genome Analyzer, and compared between lines using a k-mer algorithm allowing 0, 1 or 2 mismatches for every 28bp of sequence. After having excluded 94 markers from the analysis due to missing data and absence of linkage, the remaining 436 markers then were used to score RAD sequences obtained from a set of 93 individuals from a double haploid (DH) OWB mapping population and assist in the construction of a linkage map with an average marker density of 5cM. The RAD map and a higher density map generated by combining RAD markers with 2,383 markers previously reported by Szücs et al. [123] both allowed for the detection of the same large-effect QTLs for reproductive fitness traits, confirming the value of RAD markers for developing linkage maps and QTL mapping. RAD sequencing also was used in perennial ryegrass (*Lolium perenne*) to construct a linkage map and detect QTLs associated with resistance to stem rust caused by the pathogen *Puccinia graminis* subsp. *graminicola* [124]. The obligate outcrossing mating system of *Lolium*, resulting in high levels of heterozygosity and population heterogeneity, makes any attempt at marker development a challenging endeavor. A pseudo-testcross approach [125] combined with sequence-based marker development was tested for the identification of markers associated with stem rust resistance. RAD sequencing was performed on 188 F1 individuals, following the development of 1,733 RAD markers, characterized by 1 or 2 bi-allelic SNPs or small indels, from the resistant (male) and susceptible (female) parental lines. The analysis of the F1 RAD data led to the identification of 329 RAD markers for the female map and 305 RAD markers for the male map, with an average distance between markers of 2.3cM and 2.6cM for the female and male maps, respectively. Three QTL for stem rust resistance were subsequently identified in this population from linkage maps generated from the selected RAD markers, combined with SSR and STS markers [126–129] and tested against parental DNA and a random panel of six F1 progeny DNA. Finally, the RAD approach was used in narrow-leafed lupine (*Lupinus angustifolius*) to discover new markers closely associated with a single dominant gene, known as Lanr1, conferring resistance to anthracnose caused by the pathogen *Colletotrichum lupini* [130]. RAD sequencing first was performed on 20 F8 individuals and the two resistant and susceptible parents of a mapping population. A total of 38 co-dominant RAD markers were selected as candidate markers linked to the Lanr1 gene, from an initial pool of 8,207 putative SNP markers. Resistant and susceptible alleles for all 38 markers were confirmed from both the parental and progeny RAD marker data. A subset of 5 RAD markers then were converted into PCR-based markers exhibiting co-dominant polymorphic bands on SSCP gels and tested on 186 F8 individuals. Linkage analysis using the PCR-based marker genotyping score data and the anthracnose phenotyping data confirmed that all five newly developed markers were linked to the Lanr1 gene, including two markers flanking the gene within 0.9cM, thus enabling a very simple and efficient assay for marker-assisted selection in lupine breeding programs.

Another important application derived from RAD markers has been the development of SSR markers for mapping purposes. Barchi et al. [131] generated RAD sequences in eggplant from a pair of mapping parents, enabling the discovery of ~10,000 SNPs, out of which a representative subset of 384 was used for fingerprinting a panel of eggplant germplasm using an Illumina GoldenGate assay. In the same study, RAD sequences also led to the identification of ~2,000 putative SSR markers that have been applied for genetic mapping and diversity analysis. Readers are directed towards a review article by Zalapa et al. [132] for more information on SSR marker identification in plants using NGS technologies.

Data shown above clearly confirm the value of reduced-representation sequencing approaches such as RAD sequencing for variant discovery and genotyping by sequencing, including in species with very limited public resources. Elshire et al. [93] demonstrated the feasibility of another reduced-representation highly-multiplexed GBS strategy in the complex genomes of maize (*Zea mays*) and barley (*Hordeum vulgare)* using a simple procedure targeting regions flanking restriction endonuclease sites. The approach included digestion of genomic DNA with a methylation-sensitive restriction endonuclease followed by ligation to barcoded adapters, pooling, PCR-based amplification, and sequencing of the amplified pool on a single lane of an Illumina flow cell. In maize, two parents and 276 RILs from the maize IBM (B73 × Mo17) mapping population were sequenced on six lanes of a single Illumina flow cell at 48-plex. A total of 809,651 sequences occurring at least five times and aligning uniquely to the reference genome were selected and generated a total of 25,185 bi-allelic SNPs that were added to a reference map. No alternate allele was found for 584,119 sequences and, by treating these as dominant data, an additional 167,494 markers were added to the map, out of which 133,129 uniquely aligned to the reference genome. In barley, two parents and 43 double haploid lines from the OWB mapping population were sequenced on one lane of an Illumina flow cell. A total of 2.1 million reads present in at least 20% of the RILs were selected and mapped to the OWB framework map by considering sequences as dominant markers. Prior to mapping, the genetic map was collapsed to retain 436 bi-allelic markers containing unique linkage information in the 43 lines. A total of 24,186 sequences then were mapped and, for 4,596 of them present in one of the lines, 99% agreed on parental origin with the reference markers. A modified version of the protocol described in Elshire et al. [93] was successfully applied to the complex genomes of barley and wheat [133]. SNP detection in wheat, as in barley, is a challenging endeavor for multiple reasons. First, the very large genome sizes (~16Gbp for hexaploid wheat vs. ~0.13Gbp for *Arabidopsis*) warrant using a reduced-representation strategy for sequencing. Second, the polyploid nature of wheat and the existence of homeologous sub-genomes sharing ~96%–98% identities in tetraploid or hexaploid wheat easily confound SNP detection, due to the existence of polymorphisms between them, known as "inter-homeologue polymorphisms" (IHP). Here, two restriction endonucleases (MspI and PstI) were used to generate digested fragments whose ligation to universal adapters, including "Y-shaped" adapters for the more common MspI overhang (due to the presence of MspI-MspI fragments), allowed for the specific amplification of PstI-MspI digested DNA fragments. The presence of a short 4-9bp barcode on the PstI adapter enabled multiplexed sequencing of the amplified DNA fragments on the Illumina sequencers. A total of 82 double haploid (DH) lines from the OWB mapping population in barley and 164 DH lines from the SynPoDH mapping population in wheat (a cross between the cultivar Opata85 and the hexaploid W9784 line) were sequenced, along with their respective parental lines. Bi-allelic SNPs were detected in both populations and a Fisher's exact test of independence enabled the detection of putatively paralogous SNPs, as they are expected to segregate independently. The resulting bi-allelic SNPs then were added to the reference maps and placed on recombination bins if the parental information for the SNP of interest matched that of the bin markers for all lines present in that interval. A total of 34,396 bi-allelic SNPs, along with 241,159 sequence tags (treated as dominant markers) were added to the OWB map. In wheat, AntMap [134] was used to first create a GBS linkage map, where 1,485 SNP markers were assembled into 21 linkage groups representing the 21 wheat chromosomes. A total of 19,720 SNP markers and 367,423 sequence tags then were mapped on this newly created GBS map.

In another study, Harper et al. [135] developed a new concept, labeled as "associative transcriptomics" in the complex polyploid genome of rapeseed (Brassica napus) where they used transcriptome sequencing (mRNA-Seq) for association studies. First, a pre-existing *B. napus* SNP linkage map [Bancroft] was used to improve the order and orientation of genome sequence scaffolds of diploid ancestors *B. rapa* (which contributed to the *B. napus* A genome) and *B. oleracea* (which contributed to the *B. napus* C genome), creating pseudomolecules representative of the polyploid *B. napus* chromosomes. The pseudomolecules were then used to infer gene order for a set of reference unigenes assembled de novo from a previous *B. napus* mRNA-Seq dataset [80]. *B. napus* mRNA-Seq data generated from an 84-line diversity panel were subsequently aligned to the reference unigenes, leading to the detection of 101,644 SNPs within 11,743 unigenes, out of which 62,980 were kept for further analysis, following the removal of SNPs with minor allele frequencies of less than 5%. These data were used in conjunction with the putative gene order on the B. napus pseudomolecules to study the genetic basis of two traits of interest (erucic acid content of seed oil and seed glucosinolate content). Diversity analysis on 53 of the B. napus accessions showed strong associations over previously identified QTLs for both traits. In addition, mRNA-Seq data also can be used to profile transcript abundance (thus enabling association studies with gene expression markers, or GEMs), and profiling data for the A and C genome copies of each unigene were used to detect unigenes (in one or both genomes) showing significant association between transcript abundance and glucosinolate content of seeds. Positioning SNP and GEM markers on the pseudomolecules identified two QTL regions containing orthologs of a transcription factor, known to control aliphatic glucosinolate biosynthesis in *A. thaliana,* whose loss by deletion causes a reduced seed glucosinolate phenotype in selected *B. napus* accessions.

In a separate study, Maughan et al. [136] re-sequenced two *Arabidopsis thaliana* parents and 58 RILs on the Roche 454 and Illumina platforms. Prior to sequencing, the genomic DNA from each individual first was digested with the restriction endonucleases EcoRI and BfaI, and the resulting DNA fragments were ligated to specific barcoded adapters, PCR amplified, pooled and size-selected. A total of 6,159 SNPs and 701 SNPs were discovered from the Roche 454 and Illumina data sets, respectively. 1,712 Roche 454 SNPs (selected after applying a 20% threshold for maximum missing data) were used for linkage mapping analysis. After removing a subset of SNPs showing either significant segregation distortion or linkage disequilibrium (LD) with other SNPs, pairwise linkage analysis grouped the remaining 1,555 SNPs into five distinct linkage groups (corresponding to all 5 chromosomes). The linkage order of the SNPs on the genetic map also was shown to be highly related to the order of the SNPs on the physical map.

In addition to resequencing segregating populations, GBS also has been used to sequence pools of mutants in bulked segregant analysis studies. In *Arabidopsis*, Schneeberger et al. [137] sequenced, via whole genome shotgun sequencing on the Illumina platform, a pool of 500 F2 plants generated by crossing a recessive ethane methyl sulfonate (EMS)-induced

Col-0 mutant characterized by slow growth and light green leaves, with a wild type Ler (*Landsberg erecta*) line. Interval analysis of the relative parental allele frequencies using the newly developed software package SHOREmap [138] revealed a narrow candidate region on chromosome 4. A mutation leading to a non-synonymous codon change in a putative gene of interest distant by only 4Kbp from the peak then was suggested as being the causal mutation for this particular phenotype. In a similar study, Austin et al. [139] used an Illumina whole genome shotgun sequencing approach to resequence three pools of 80 F2 cell wall-related mutants generated by crossing individuals corresponding to three separate EMS-induced Col-0 lines with a wild type Ler mapping line. At least 230,000 SNPS were discovered for each mutant pool by aligning to the *A. thaliana* Col-0 reference genome. Regions of the genome lacking SNPs were discovered, corresponding to non-recombinant haplotypic blocks linked to the recessive mutations. A modification of the Illumina "chastity" statistic, which is normally used by the basecalling software to measure cross-talk between dyes during the sequencing process (and thus the "purity" of a specific base call at a given sequencing cycle) was then used to measure the proportion of reads that are completely homozygous for bases that differ from the reference genome, thus further narrowing the search window for a putative causative SNP of interest within these blocks. Finally, density interval analysis measuring the frequency of SNPs with discordant "chastity" values returned several non-synonymous SNPs for all three mutants, including three likely candidates located in putative genes of interest with roles ranging from actin cytoskeleton organization to sugar transport. Finally, Trick et al. [99] used a reduced-representation sequencing approach in wheat to identify SNPs between two parental lines in wheat and examine their frequency in two bulks of 28 homozygous recombinant lines differentiated by high and low grain protein contents, respectively. SNP detection first was performed by re-sequencing with the Illumina sequencing technology the transcriptome "mRNA-Seq" of two parental lines, LDN and RSL65, segregating for the trait of interest. Individual sequences from each parental then were mapped to the NCBI wheat unigene dataset separately, thus creating two SNP sets, and a custom Perl script was used to determine differences between the two parental SNP sets. IHPs representing a consensus sequence with the same ambiguity code in each parental line and common between the two parental sets were removed, leading to the identification of 3,963 putative SNPs between LDN and RSL65. This dataset was later reduced to 3,427 SNPs, after examining SNP frequency distribution for each mapped unigene, and the possibility of mapping putative SNPs to closely related paralogues. mRNA-seq data then were generated for each bulk "high" and "low" protein contents) in order to compare the allelic frequencies for the parental SNPs between the two of them. Relative SNP frequency measurements led to the characterization of two tightly linked unigenes located on wheat chromosome arm 6BS.

15.8 CONCLUSIONS

High-throughput variant discovery has been made possible in multiple species by the recent advent of next-generation DNA sequencing technologies. Continuous increase in sequencing throughput and the accompanying decrease in consumable cost per Gbp has allowed researchers to switch focus from resequencing small panels of parental individuals for the sole purpose of discovering variants to resequencing much larger pools of individuals within a population, where the sequenced differences are used directly as genotypic markers. This genotyping-by-sequencing (GBS) approach has several advantages, including the facts that no preliminary sequence information is required and that all newly discovered markers originate from the population being genotyped. On the other hand, due to several biological and technical factors, such as PCR amplification bias during the library construction step, not all sequenced regions of interest are evenly covered in all individuals within a population, reaffirming the need for imputing missing data using pedigree or parental information when available.

Because DNA fragments are more readily prepared using a genome-wide approach (as opposed to a targeted approach where only a small region of the genome is sequenced), the advent of GBS is expected to have a more profound impact on mapping strategies benefiting from a dense genome-wide distribution of markers. Such strategies include Genome-Wide Association Studies (GWAS), Bulked Segregant Analysis (BSA) and Genomic Selection (GS).

Successive improvements of the sequencing chemistries and basecalling software are allowing NGS technologies to deliver higher sequencing throughputs per run, which in turn enables deeper multiplexing for a fixed average sequencing depth per sample. Although the cost of sample preparation and bioinformatics analysis are not decreasing as rapidly as the cost of sequencing, such a trend is already enabling GBS to be a cost-competitive alternative to other whole-genome genotyping platforms. It is expected that, as the amount and quality of sequencing information generated per run keeps increasing, thus allowing even higher plexing and lower costs per samples, plant breeders soon may be able to sequence even larger populations, allowing genomic selection or the determination of a population structure without prior knowledge of the diversity present in the species [93].

REFERENCES AND NOTES

1. Rafalski, A. Applications of single nucleotide polymorphisms in crop genetics. Curr. Opin. Plant Biol. 2002, 5, 94–100.
2. Bernatsky, R.; Tanksley, S. Toward a saturated linkage map in tomato based on isozymes and random cDNA sequences. Genetics 1986, 112, 887–898.
3. Litt, M.; Luty, J.A. A hypervariable microsatellite revealed by in vitro amplification of a dinucleotide repeat within the cardiac muscle actin gene. Am. J. Hum. Genet. 1986, 44, 397–401.
4. Williams, J.G.; Kubelik, A.R.; Livak, K.J.; Rafalski, J.A.; Tingey, S.V. DNA polymorphisms amplified by arbitrary primers are useful as genetic markers. Nucleic Acids Res. 1990, 18, 6531–6535.
5. Paran, I.; Michelmore, R.W. Development of reliable PCR-based markers linked to downy mildew resistance genes in lettuce. Theor. Appl. Genet. 1993, 85, 985–993.
6. Konieczny, A.; Ausubel, F.M. A procedure for mapping Arabidopsis mutations using co-dominant ecotype-specific PCR-based markers. Plant J. 1993, 4, 403–410.
7. Salimath, S.S.; De Oliveira, A.C.; Bennetzen, J.; Godwin, I.D. Assessment of genomic origin and genetic diversity in the genus Eleusine with DNA markers. Genome 1995, 38, 757–763.
8. Vos, P.; Hogers, R.; Bleeker, M.; Reijans, M.; van de Lee, T.; Hornes, M.; Frijters, A.; Pot, J.; Peleman, J.; Kuiper, M. AFLP: A new technique for DNA fingerprinting. Nucleic Acids Res. 1995, 23, 4407–4414.
9. Desmarais, E.; Lanneluc, I.; Lagnel, J. Direct amplification of length polymorphisms (DALP), or how to get and characterize new genetic markers in many species. Nucleic Acids Res. 1998, 26, 1458–1465.

10. Wang, D.G.; Fan, J-B.; Siao, C.J.; Berno, A.; Young, P.; Sapolsky, R.; Ghandour, G.; Perkins, N.; Winchester, E.; Spencer, J.; et al. Large-scale identification, mapping, and genotyping of single-nucleotide polymorphisms in the human genome. Science 1998, 280, 1077–1082.
11. Täpp, I.; Malmberg, L.; Rennel, E.; Wik, M.; Syvanen, A.C. Homogeneous scoring of single-nucleotide polymorphisms: comparison of the 5'-nuclease TaqMan assay and molecular beacon probes. Biotechniques 2000, 28, 732–738.
12. Prince, J.A.; Feuk, L.; Howell, W.M.; Jobs, M.; Emahazion, T.; Blennow, K.; Brookes, A.J. Robust and accurate single nucleotide polymorphism genotyping by dynamic allele-specific hybridization (DASH): design criteria and assay validation. Genome Res. 2001, 11,152–162.
13. Storm, N.; Darnhofer-Patel, B.; van den Boom, D.; Rodi, C.P. MALDI-TOF mass spectrometry-based SNP genotyping. Methods Mol. Biol. 2003, 212, 241–262.
14. Livak, K.J. SNP genotyping by the 50-nuclease reaction. Methods Mol. Biol. 2003, 212, 129–147.
15. Olivier, M. The Invader assay for SNP genotyping. Mutat. Res. 2005, 573, 103–110.
16. Ragoussis, J. Genotyping technologies for all. Drug Discov. Today Technol. 2006, 3, 115–122.
17. Procunier, J.D.; Prashar, S.; Chen, G.; Wolfe, D.; Fox, S.; Ali, M.L.; Gray, M.; Zhou, Y.; Shillinglaw, M.; Roeven, R.; Ron DePauw, R. Rapid ID technology (RIDT) in plants: High-speed DNA fingerprinting in grain seeds for the identification, segregation, purity, and traceability of varieties using lab automation robotics. J. Lab. Automat. 2009, 14, 221–231.
18. Gupta, P.K.; Rustgi, S.; Mir, R.R. Array-based high-throughput DNA markers for crop improvement. Heredity 2008, 101, 5–18.
19. Nybon, H. Comparison of different nuclear DNA markers for estimating intraspecific genetic diversity in plants. Mol. Ecol. 2004, 13, 1143–1155.
20. Arif, I.A.; Bakir, M.A.; Khan, H.A.; Al Farhan, A.H.; Al Homaidan, A.A.; Bahkali, A.H.; Al Sadoon, M.; Shobrak, M. A brief review of molecular techniques to assess plant diversity. Int. J. Mol. Sci. 2010, 11, 2079–2096.
21. Zheng, P.; Allen, W.B.; Roesler, K.; Williams, M.E.; Zhang, S.; Li, J.; Glassman, K.; Ranch, J.; Nubel, D.; Solawetz, W.; Bhattramakki, D.; et al. A phenylalanine in DGAT is a key determinant of oil content and composition in maize. Nat. Genet. 2008, 40, 367–372.
22. Ruta, N.; Liedgens, M.; Fracheboud, Y.; Stamp, P.; Hund, A. QTLs for the elongation of axile and lateral roots of maize in response to low water potential. Theor. Appl. Genet. 2010, 120, 621–631.
23. Yano, M.; Harushima, Y.; Nagamura, Y.; Kurata, N.; Minobe, Y.; Sasaki, T. Identification of quantitative trait loci controlling heading date in rice using a high-density linkage map. Theor. Appl. Genet. 1997, 95, 1025–1032.
24. El-Din El-Assal, S.; Alonso-Blanco, C.; Peeters, A.J.; Raz, V.; Koornneef, M. The cloning of a flowering time QTL reveals a novel allele of CRY2. Nat. Genet. 2001, 29, 435–440.
25. Liu, J.; van Eck, J.; Cong, B.; Tanksley, S.D. A new class of regulatory genes underlying the cause of pear-shaped tomato fruit. Proc. Natl. Acad. Sci. USA 2002, 99, 13302–13306.

26. Salvi, S.; Tuberosa, R. To clone or not to clone plant QTLs: Present and future challenges. Trends Plant Sci. 2005, 10, 297–304.
27. Frary, A.; Nesbitt, T.C.; Grandillo, S.; Knaap, E.; Cong, B.; Liu, J.; Meller, J.; Elber, R.; Alpert, K.B.; Tanksley, S.D. fw2.2: A quantitative trait locus key to the evolution of tomato fruit size. Science 2000, 289, 85–88.
28. Rafalski, A. Association genetics in crop improvement. Curr. Opin. Plant Biol. 2010, 13, 174–180.
29. Schneeberger, K.; Weigel, D. Fast-forward genetics enabled by new sequencing technologies. Trends Plant Sci. 2011, 16, 282–288.
30. Risch, N.; Merikangas, K. The future of genetic studies of complex human diseases. Science 1996, 273, 1516–1517.
31. Yu, J.; Buckler, E.S. Genetic association mapping and genome organization of maize. Curr. Opin. Biotech. 2006, 17, 155–160.
32. Beló, A.; Zheng, P.; Luck, S.; Shen, B.; Meyer, D.J.; Li, B.; Tingey, S.; Rafalski, A. Whole genome scan detects an allelic variant of fad2 associated with increased oleic acid levels in maize. Mol. Genet. Genomics 2008, 279, 1–10.
33. Nordborg, M.; Weigel, D. Next-generation genetics in plants. Nature 2008, 456, 720–723.
34. Thornsberry, J.M.; Goodman, M.M.; Doebley, J.; Kresovich, S.; Nielsen, D.; Buckler, E.S. Dwarf8 polymorphisms associate with variation in flowering time. Nat. Genet. 2001, 28, 286–289.
35. Hansen, M.; Kraft, T.; Ganestam, S.; Säll, T.; Nilsson, N.O. Linkage disequilibrium mapping of the bolting gene in sea beet using AFLP markers. Genet. Res. 2001, 77, 61–66.
36. Flint-Garcia, S.A.; Thornsberry, J.M.; Buckler, E.S. Structure of linkage disequilibrium in plants. Ann. Rev. Plant Biol. 2003, 54, 357–374.
37. Flint-Garcia, S.A.; Thuillet, A.C.; Yu, J.; Pressoir, G.; Romero, S.M.; Mitchell, S.E.; Doebley, J.; Kresovich, S.; Goodman, M.M.; Buckler, E.S. Maize association population: A high-resolution platform for quantitative tgrait locus dissection. Plant J. 2005, 44, 1054–1064.
38. Sharbel, T.F.; Haubold, B.; Mitchell-Olds, T. Genetic isolation by distance in Arabidopsis thaliana: biogeography and postglacial colonization of Europe. Mol. Ecol. 2000, 9, 2109–2118.
39. Zhu, C.; Gore, M.; Buckler, E.S.; Yu, J. Status and prospects of association mapping in plants. Plant Gen. 2008, 1, 5–20.
40. Damerval, C.; Maurice, A.; Josse, J.M.; de Vienne, D. Quantitative trait loci underlying gene product variation: A novel perspective for analyzing genome expression. Genetics 1994, 137, 289–301.
41. Holloway, B.; Li, B. Expression QTLs: Applications for crop improvement. Mol. Breeding 2010, 26, 381–391.
42. Holloway, B.; Luck, S.; Beatty, M.; Rafalski, J.A.; Li, B. Genome-wide expression quantitative trait loci (eQTL) analysis in maize. BMC Genomics 2011, 12, 1–14.
43. West, M.A.; Kim, K.; Kliebenstein, D.J.; van Leeuwen, H.; Michelmore, R.W.; Doerge, R.W.; St Clair, D.A. Global eQTL mapping reveals the complex genetic architecture of transcript-level variation in Arabidopsis. Genetics 2007, 175, 1441–1450.

44. Swanson-Wagner, R.A.; DeCook, R.; Jia, Y.; Bancroft, T.; Ji, T.; Zhao, X.; Nettleton, D.; Schnable, P.S. Paternal dominance of trans-eQTL influences gene expression patterns in maize hybrids. Science 2009, 326, 1118–1120.
45. Becker, A.; Chao, D.Y.; Zhang, X.; Salt, D.E.; Baxter, I. Bulk Segregant Analysis Using Single Nucleotide Polymorphism Microarrays. PLoS One 2011, 6, e15993.
46. Wolyn, D.J.; Borevitz, J.O.; Loudet, O.; Schwartz, C.; Maloof, J.; Ecker, J.R.; Berry, C.C.; Chory, J. Light-response quantitative trait loci identified with composite interval and eXtreme array mapping in Arabidopsis thaliana. Genetics 2004, 167, 907–917.
47. Meuwissen, T.H.E.; Hayes, B.J.; Goddard, M.E. Prediction of total genetic value using genome-wide dense marker maps. Genetics 2001, 157, 1819–1829.
48. Morgante, M.; de Paoli, E.; Radovic, S. Transposable elements and the plant pangenomes. Curr. Opin. Plant Biol. 2007, 10, 149–155.
49. Llaca, V.; Campbell, M.; Deschamps, S. Genome diversity in maize. J. Botany 2011, 104172, 1–10.
50. Sanger, F.; Nicklen, S.; Coulson, A.R. DNA sequencing with chain-terminating inhibitors. Proc. Natl. Acad. Sci. USA 1977, 74, 5463–5467.
51. Hunkapiller, T.; Kaiser, R.J.; Koop, B.F.; Hood, L. Large-scale and automated DNA sequence determination. Science 1991, 254, 59–67.
52. Lander, E.S.; Linton, L.M.; Birren, B.; Nusbaum, C.; Zody, M.C.; Baldwin, J.; Devon, K.; Dewar, K.; Doyle, M.; FitzHugh, W.; et al. Initial sequencing and analysis of the human genome. Nature 2011, 409, 860–921.
53. Goff, S.A.; Ricke, D.; Lan, T.H.; Presting, G.; Wang, R.; Dunn, M.; Glazebrook, J.; Sessions, A.; Oeller, P.; Varma, H.; et al. A draft sequence of the rice genome (Oryza sativa L. ssp. japonica). Science 2002, 296, 92–100.
54. Yu, J.; Hu, S.; Wang, J.; Wong, G.K.; Li, S.; Liu, B.; Deng, Y.; Dai, L.; Zhou, Y.; Zhang, X.; et al. A draft sequence of the rice genome (Oryza sativa L. ssp. indica). Science 2002, 296, 79–92.
55. Tuskan, G.A.; Difazio, S.; Jansson, S.; Bohlmann, J.; Grigoriev, I.; Hellsten, U.; Putnam, N.; Ralph, S.; Rombauts, S.; Salamov, A.; et al. The genome of black cottonwood, Populus trichocarpa (Torr. & Gray). Science 2006, 313, 1596–1604.
56. Jaillon, O.; Aury, J.M.; Noel, B.; Policriti, A.; Clepet, C.; Casagrande, A.; Choisne, N.; Aubourg, S.; Vitulo, N.; Jubin, C.; et al. The grapevine genome sequence suggests ancestral hexaploidization in major angiosperm phyla. Nature 2007, 449, 463–467.
57. Paterson, A.H.; Bowers, J.E.; Bruggmann, R.; Dubchak, I.; Grimwood, J.; Gundlach, H.; Haberer, G.; Hellsten, U.; Mitros, T.; Poliakov, A.; et al. The Sorghum bicolor genome and the diversification of grasses. Nature 2009, 457, 551–556.
58. Schnable, P.S.; Ware, D.; Fulton, R.S.; Stein, J.C.; Wei, F.; Pasternak, S.; Liang, C.; Zhang, J.; Fulton, L.; Graves, T.A.; et al. The B73 maize genome: complexity, diversity, and dynamics. Science 2009, 326, 1112–1115.
59. Young, N.D.; Debellé, F.; Oldroyd, G.E.; Geurts, R.; Cannon, S.B.; Udvardi, M.K.; Benedito, V.A.; Mayer, K.F.; Gouzy, J.; Schoof, H.; et al. The Medicago genome provides insight into the evolution of rhizobial symbioses. Nature 2011, 480, 520–524.
60. Rostoks, N.; Mudie, S.; Cardle, L.; Russell, J.; Ramsay, L.; Booth, A.; Svensson, J.T.; Wanamaker, S.I.; Walia, H.; Rodriguez, E.M.; et al. Genome-wide SNP discov-

ery and linkage analysis in barley based on genes responsive to abiotic stress. Mol. Genet. Genomics 2005, 274, 515–527.
61. Choi, I.Y.; Hyten, D.L.; Matukumalli, L.K.; Song, Q.; Chaky, J.M.; Quigley, C.V.; Chase, K.; Lark, K.G.; Reiter, R.S.; Yoon, M.S.; et al. A soybean transcript map: Gene distribution, haplotype and single nucleotide polymorphism analysis. Genetics 2007, 176, 685–696.
62. Luckey, J.A.; Drossman, H.; Kostichka, A.J.; Mead, D.A.; D'Cunha J.; Norris, T.B.; Smith, L.M. High speed DNA sequencing by capillary electrophoresis. Nucleic Acids Res. 1990, 18, 4417–4421.
63. Swerdlow, H.; Gesteland, R. Capillary gel electrophoresis for rapid, high resolution DNA sequencing. Nucleic Acids Res. 1990, 18, 1415–1419.
64. Smith, L.M.; Sanders, J.Z.; Kaiser, R.J.; Hughes, P.; Dodd, C.; Connell, C.R.; Heiner, C.; Kent, S.B.; Hood, L.E. Fluorescence detection in automated DNA sequence analysis. Nature 1986, 321, 674–679.
65. Prober, J.M.; Trainor, G.L.; Dam, R.J.; Hobbs, F.W.; Robertson, C.W.; Zagursky, R.J.; Cocuzza, A.J.; Jensen, M.A.; Baumeister, K. A system for rapid DNA sequencing with fluorescent chain-terminating dideoxynucleotides. Science 1987, 238, 336–341.
66. Shendure, J.; Ji, H. Next-generation DNA sequencing. Nat. Biotechnol. 2008, 26, 1135–1145.
67. Niedringhaus, T.P.; Milanova, D.; Kerby, M.B.; Snyder, M.P.; Barron, A.E. Landscape of next-generation sequencing technologies. Anal. Chem. 2011, 83, 4327–4341.
68. Metzker, M.L. Sequencing technologies – the next generation. Nat. Rev. Genet. 2010, 11, 31–46.
69. Margulies, M.; Egholm, M.; Altman, W.E.; et al. Genome sequencing in microfabricated high-density picolitre reactors. Nature 2005, 437, 376–380.
70. Dressman, D.; Yan, H.; Traverso, G.; Kinzler, K.W.; Vogelstein, B. Transforming single DNA molecules into fluorescent magnetic particles for detection and enumeration of genetic variations. Proc. Natl. Acad. Sci. USA 2003, 100, 8817–8822.
71. Ronaghi, M. Pyrosequencing sheds light on DNA sequencing. Genome Res. 2001, 11, 3–11.
72. Thudi, M.; Li, Y.; Jackson, S.A.; May, G.D.; Varshney, R.K. Current state-of-art of sequencing technologies for plant genomics research. Brief. Funct. Genomics 2012, 11, 3–11.
73. Bentley, D.R.; Balasubramanian, S.; Swerdlow, H.P.; Smith, G.P.; Milton, J.; Brown, C.G.; Hall, K.P.; Evers, D.J.; Barnes, C.L.; Bignell, H.R.; et al. Accurate whole human genome sequencing using reversible terminator chemistry. Nature 2008, 456, 53–59.
74. Fedurco, M.; Romieu, A.; Williams, S.; Lawrence, I.; Turcatti, G. BTA, a novel reagent for DNA attachment on glass and efficient generation of solid-phase amplified DNA colonies. Nucleic Acids Res. 2006, 34, e22.
75. Rothberg, J.M.; Hinz, W.; Rearick, T.M.; Schultz, J.; Mileski, W.; Davey, M.; Leamon, J.H.; Johnson, K.; Milgrew, M.J.; Edwards, M.; et al. An integrated semiconductor device enabling non-optical genome sequencing. Nature 2011, 475, 348–352.

76. Velasco, R.; Zharkikh, A.; Troggio, M.; Cartwright, D.A.; Cestaro, A.; Pruss, D.; Pindo, M.; Fitzgerald, L.M.; Vezzulli, S.; Reid, J.; Malacarne, G.; et al. A high quality draft consensus sequence of the genome of a heterozygous grapevine variety. PLoS One 2007, 2, e1326.
77. Wheeler, D.A.; Srinivasan, M.; Egholm, M.; Shen, Y.; Chen, L.; McGuire, A.; He, W.; Chen, Y.J.; Makhijani, V.; Roth, G.T.; et al. The complete genome of an individual by massively parallel DNA sequencing. Nature 2008, 452, 872–876.
78. Novaes, E.; Drost, D.R.; Farmerie, W.G.; Pappas, G.J., Jr.; Grattapaglia, D.; Sederoff, R.R.; Kirst, M. High-throughput gene and SNP discovery in Eucalyptus grandis, an uncharacterized genome. BMC Genomics 2008, 9, 312.
79. Van Tassell, C.P.; Smith, T.P.; Matukumalli, L.K.; Taylor, J.F.; Schnabel, R.D.; Lawley, C.T.; Haudenschild, C.D.; Moore, S.S.; Warren, W.C.; Sonstegard, T.S. SNP discovery and allele frequency estimation by deep sequencing of reduced representation libraries. Nat. Methods 2008, 5, 247–252.
80. Trick, M.; Long, Y.; Meng, J.; Bancroft, I. Single nucleotide polymorphism (SNP) discovery in the polyploid Brassica napus using Solexa transcriptome sequencing. Plant Biotechnol. J. 2009, 7, 334–346.
81. Deschamps, S.; la Rota, M.; Ratashak, J.P.; Biddle, P.; Thureen, D.; Farmer, A.; Luck, S.; Beatty, M.; Nagasawa, N.; Michael, L.; et al. Rapid genome-wide single nucleotide polymorphism discovery in soybean and rice via deep resequencing of reduced representation libraries with the Illumina Genome Analyzer. Plant Genome 2010, 3, 53–68.
82. Mortazavi, A.; Williams, B.; McCue, K.; Schaeffer, L.; Wold, B. Mapping and quantifying mammalian transcriptomes by RNA-Seq. Nat. Methods 2008, 5, 621–628.
83. Sultan, M.; Schulz, M.H.; Richard, H.; Magen, A.; Klingenhoff, A.; Scherf, M.; Seifert, M.; Borodina, T.; Soldatov, A.; Parkhomchuk, D.; et al. A global view of gene activity and alternative splicing by deep sequencing of the human transcriptome. Science 2008, 321, 956–960.
84. Wilhelm, B.; Marguerat, S.; Watt, S.; Schubert, F.; Wood, V.; Goodhead, I.; Penkett, C.; Rogers, J.; Bähler, J. Dynamic repertoire of a eukaryotic transcriptome surveyed at single nucleotide resolution. Nature 2008, 453, 1239–1243.
85. Fahlgren, N.; Howell, M.D.; Kasschau, K.D.; Chapman, E.J.; Sullivan, C.M.; Cumbie, J.S.; Givan, S.A.; Law, T.F.; Grant, S.R.; Dangl, J.L.; Carrington, J.C. High-throughput sequencing of Arabidopsis microRNAs: evidence for frequent birth and death of MIRNA genes. PLoS One 2007, 2, e219.
86. Kasschau, K.D.; Fahlgren, N.; Chapman, E.J.; Sullivan, C.M.; Cumbie, J.S.; Givan, S.A.; Carrington, J.C. Genome-wide profiling and analysis of Arabidopsis siRNAs. PLoS Biol. 2007, 5, e57.
87. Sunkar, R.; Zhou, X.; Zheng, Y.; Zhang, W.; Zhu, J.K. Identification of novel and candidate miRNAs in rice by high throughput sequencing. BMC Plant Biol. 2008, 8, 25.
88. Cokus, S.J.; Feng, S.; Zhang, X.; Chen, Z.; Merriman, B.; Haudenschild, C.D.; Pradhan, S.; Nelson, S.F.; Pellegrini, M.; Jacobsen, S.E. Shotgun bisulphate sequencing of the Arabidopsis genome reveals DNA methylation patterning. Nature 2008, 452, 215–219.

89. Lister, R.; O'Malley R.C.; Tonti-Filippini, J.; Gregory, B.D.; Berry, C.C.; Millar, A.H.; Ecker, J.R. Highly integrated single-base resolution maps of the epigenome in Arabidopsis. Cell 2008, 133, 523–536.
90. Barski, A.; Cuddapah, S.; Cui, K.; Roh, T.Y.; Schones, D.E.; Wang, Z.; Wei, G.; Chepelev, I.; Zhao, K. High-resolution profiling of histone methylations in the human genome. Cell 2007, 129, 823–837.
91. Johnson, D.S.; Mortazavi, A.; Myers, R.M.; Wold, B. Genome-wide mapping of in vitro protein-DNA interactions. Science 2007, 316, 1497–1502.
92. Petrosino, J.F.; Highlander, S.; Luna, R.A.; Gibbs, R.A.; Versalovic, J. Metagenomic pyrosequencing and microbial identification. Clin. Chem. 2009, 55, 856–866.
93. Elshire, R.J.; Glaubitz, J.C.; Sun, Q.; Poland, J.A.; Kawamoto, K.; Buckler, E.S.; Mitchell, S.E. A robust, simple genotyping-by-sequencing (GBS) approach for high diversity species. PLoS One 2011, 6, e19379.
94. Flavell, A.J.; Pearce, S.R.; Kumar, A. Plant transposable elements and the genome. Curr. Opin. Genet. Dev. 1994, 4, 838–844.
95. SanMiguel, P.; Tikhonov, A.; Jin, Y.K.; Motchoulskaia, N.; Zakharov, D.; Melake-Berhan, A.; Springer, P.S.; Edwards, K.J.; Lee, M.; Avramova, Z.; Bennetzen, J.L. Nested retrotransposons in the intergenic regions of the maize genome. Science 1996, 274, 765–768.
96. Bennetzen, J.L.; Ma, J.; Devos, K.M. Mechanisms of recent genome size variation in flowering plants. Ann. Bot. 2005, 95, 127–135.
97. Ossowski, S.; Schneeberger, K.; Clark, R.M.; Lanz, C.; Warthmann, N.; Weigel, D. Sequencing of natural strains of Arabidopsis thaliana with short reads. Genome Res. 2008, 18, 2024–2033.
98. Huang, X.; Feng, Q.; Qian, Q.; Zhao, Q.; Wang, L.; Wang, A.; Guan, J.; Fan, D.; Weng, Q.; Huang, T.; Dong, G.; Sang, T.; Han, B. High-throughput genotyping by whole-genome resequencing. Genome Res. 2009, 19, 1068–1076.
99. Trick, M.; Adamski, N.M.; Mugford, S.G.; Jiang, C.-C.; Febrer, M.; Uauy, C. Combining SNP discovery from next-generation sequencing data with bulked segregant analysis (BSA) to fine-map genes in polyploidy wheat. BMC Plant Biol. 2012, 12, 14.
100. Barbazuk, W.B.; Emrich, S.J.; Chen, H.D.; Li, L.; Schnable, P.S. SNP discovery via 454 transcriptome sequencing. Plant J. 2007, 51, 910–918.
101. Rabinowicz, P.; McCombie, W.R.; Martienssen, R.A. Gene enrichment in plant genomic shotgun libraries. Curr. Opin. Plant Biol. 2003, 6, 150–156.
102. Rabinowicz, P.D.; Citek, R.; Budiman, M.A.; Nunberg, A.; Bedell, J.A.; Lakey, N.; O'Shaughnessy, A.L.; Nascimento, L.U.; McCombie, W.R.; Martienssen, R.A. Differential methylation of genes and repeats in land plants. Genome Res. 2005, 15, 1431–1440.
103. Fellers, J.P. Genome filtering using methylation-sensitive restriction enzymes with six base pair recognition sites. The Plant Genome 2008, 1, 146–152.
104. Gore, M.A.; Wright, M.H.; Ersoz, E.S.; Bouffard, P.; Szekeres, E.S.; Jarvie, T.P.; Hurwitz, B.L.; Narechania, A.; Harkins, T.T.; Grills, G.S.; et al. Large-scale discovery of gene-enriched SNPs. The Plant Genome 2008, 2, 121–133.
105. Tewhey, R.; Warner, J.B.; Nakano, M.; Libby, B.; Medkova, M.; David, P.H.; Kotsopoulos, S.K.; Samuels, M.L.; Hutchison, J.B.; Larson, J.W.; et al. Microdroplet-

based PCR enrichment for large-scale targeted sequencing. Nat. Biotechnol. 2009, 27, 1025–1031.
106. Hardenbol, P.; Yu, F.; Belmont, J.; Mackenzie, J.; Bruckner, C.; Brundage, T.; Boudreau, A.; Chow, S.; Eberle, J.; Erbilgin, A.; et al. Highly multiplexed molecular inversion probe genotyping: over 10,000 targeted SNPs genotyped in a single tube assay. Genome Res. 2005, 15, 269–275.
107. Hodges, E.; Xuan, Z.; Balija, V.; Kramer, M.; Molla, M.N.; Smith, S.W.; Middle, C.M.; Rodesch, M.J.; Albert, T.J.; Hannon, G.J.; McCombie, W.R. Genome-wide in situ exon capture for selective resequencing. Nat. Genet. 2007, 39, 1522–1527.
108. Gnirke, A.; Melnikov, A.; Maguire, J.; Rogov, P.; LeProust, E.M.; Brockman, W.; Fennell, T.; Giannoukos, G.; Fisher, S.; Russ, C.; et al. Solution hybrid selection with ultra-long oligonucleotides for massively parallel targeted sequencing. Nat. Biotechnol. 2009, 27, 182–189.
109. Li, Y.; Sidore, C.; Kang, H.M.; Boehnke, M.; Abecasis, G.R. Low-coverage sequencing: implications for design of complex trait association studies. Genome Res. 2011, 21, 940–951.
110. Nielsen, R.; Paul, J.S.; Albrechtsen, A.; Song, Y.S. Genotype and SNP calling from next-generation sequencing data. Nat. Rev. Genet. 2011, 12, 443–451.
111. Li, Y.; Willer, C.; Sanna, S.; Abecasis, G. Genotype imputation. Annu. Rev. Genomics Hum. Genet. 2009, 10, 387–406.
112. Paran, I.; Zamir, D. Quantitative traits in plants: beyond QTL. Trends Genet. 2003, 19, 303–306.
113. Rahman, H.; Pekic, S.; Lazic-Jancic, V.; Quarrie, S.A.; Shah, S.M.; Pervez, A.; Shah, M.M. Molecular mapping of quantitative trait loci for drought tolerance in maize plants. Genet. Mol. Res. 2011, 10, 889–901.
114. Kump, K.L.; Bradbury, P.J.; Wisser, R.J.; Buckler, E.S.; Belcher, A.R.; Oropeza-Rosas, M.A.; Zwonitzer, J.C.; Kresovich, S.; McMullen, M.D.; Ware, D.; Balint-Kurti, P.J.; Holland, J.B. Genome-wide association study of quantitative resistance to southern leaf blight in the maize nested association mapping population. Nat. Genet. 2011, 43, 163–168.
115. Tian, F.; Bradbury, P.J.; Brown, P.J.; Hung, H.; Sun, Q.; Flint-Garcia, S.; Rocheford, T.R.; McMullen, M.D.; Holland, J.B.; Buckler, E.S. Genome-wide association study of leaf architecture in the maize nested association-mapping population. Nat. Genet. 2011, 43, 159–162.
116. Edwards, D.; Batley, J. Plant genome sequencing: applications for crop improvement. Plant Biotechnol. J. 2010, 8, 2–9.
117. Morrell, P.L.; Buckler, E.S.; Ross-Ibarra, J. Crop genomics: Advances and applications. Nat. Rev. Genet. 2011, 13, 85–96.
118. Schneeberger, K.; Weigel, D. Fast-forward genetics enabled by new sequencing technologies. Trends Plant Sci. 2011, 16, 282–288.
119. Gore, M.A.; Chia, J.M.; Elshire, R.J.; Sun, Q.; Ersoz, E.S.; Hurwitz, B.L.; Peiffer, J.A.; McMullen, M.D.; Grills, G.S.; Ross-Ibarra, J.; Ware, D.H.; Buckler, E.S. A first-generation haplotype map of maize. Science 2009, 326, 1115–1117.
120. Wang, L.; Wang, A.; Huang, X.; Zhao, Q.; Dong, G.; Qian, Q.; Sang, T.; Han, B. Mapping 49 quantitative trait loci at high resolution through sequencing-based

genotyping of rice recombinant inbred lines. Theor. Appl. Genet. 2011, 122, 327–340.
121. Baird, N.A.; Etter, P.D.; Atwood, T.S.; Currey, M.C.; Shiver, A.L.; Lewis, Z.A.; Selker, E.U.; Cresko, W.A.; Johnson, E.A. Rapid SNP discovery and genetic mapping using sequenced RAD markers. PLoS One 2008, 3, e3376.
122. Chutimanitsakun, Y.; Nipper, R.W.; Cuesta-Marcos, A.; Cistué, L.; Corey, A.; Filichkina, T.; Johnson, E.A.; Hayes, P.M. Construction and application for QTL analysis of a Restriction Site Associated DNA (RAD) linkage map in barley. BMC Genomics 2011, 12, 4.
123. Szucs, P.; Blake, V.C.; Bhat, P.R.; Chao, S.; Close, T.J.; Cuesta-Marcos, A.; Muehlbauer, G.J.; Ramsay, L.V.; Waugh, R.; Hayes, P.M. An integrated resource for barley linkage map and malting quality QTL alignment. Plant Genome 2009, 2, 134–140.
124. Pfender, W.F.; Saha, M.C.; Johnson, E.A.; Slabaugh, M.B. Mapping with RAD (restriction-site associated DNA) markers to rapidly identify QTL for stem rust resistance in Lolium perenne. Theor. Appl. Genet. 2011, 122, 1467–1480.
125. Grattapaglia, D.; Sederoff, R. Genetic linkage maps of Eucalyptus grandis and Eucalyptus urophylla using a pseudo-testcross mapping strategy and RAPD markers. Genetics 1994, 137, 1121–1137.
126. Saha, M.C.; Mian, R.; Eujayl, I.; Zwonitzer, J.C.; Wang, L.; May, G.D. Tall fescue EST-SSR markers with transferability across several grass species. Theor. Appl. Genet. 2004, 109, 783–791.
127. Saha, M.C.; Mian, R.; Zwonitzer, J.C.; Chekhovskiy, K.; Hopkins, A.S. An SSR- and AFLP-based genetic linkae map of tall fescue (Festuca arundinacea Schreb.). Theor. Appl. Genet. 2005, 110, 323–336.
128. Kantety, R.V.; Rota, M.L.; Matthews, D.E.; Sorrells, M.E. Data mining for simple sequence repeats in expressed sequence tags from barley, maize, rice, sorghum and wheat. Plant Mol. Biol. 2002, 48, 501–510.
129. Lauvergeat, V.; Barre, P.; Bonnet, M.; Ghesquiere, M. Sizty simple sequence repeat markers for use in the Festuca-Lolium complex of grasses. Mol. Ecol. 2005, 5, 401–405.
130. Yang, H.; Tao, Y.; Zheng, Z.; Li, C.; Sweetingham, M.; Howieson, J. Application of nextgeneration sequencing for rapid marker development in molecular plant breeding: A case study on antrachnose disease resistance in Lupinus angustifolius L. BMC Genomics 2012, 13, 318.
131. Barchi, L.; Lanteri, S.; Portis, E.; Acquadro, A.; Valè, G.; Toppino, L.; Rotino, G.L. Identification of SNP and SSR markers in eggplant using RAD tag sequencing. BMC Genomics 2011, 12, 304.
132. Zalapa, J.E.; Cuevas, H.; Zhu, H.; Steffan, S.; Senalik, D.; Zeldin, E.; McCown, B.; Harbut, R.; Simon, P. Using next-generation sequencing approaches to isolate simple sequence repeat (SSR) loci in the plant sciences. Am. J. Bot. 2012, 99, 193–208.
133. Poland, J.A.; Brown, P.J.; Sorrells, M.E.; Jannink, J.L. Development of high-density genetic maps for barley and wheat using a novel two-enzyme genotyping-by-sequencing approach. PLoS One 2012, 7, e32253.
134. Iwata, H.; Ninomiya, S. AntMap: Constructing genetic linkage maps using an ant colony optimization algorithm. Breed. Sci. 2006, 56, 371–377.

135. Harper, A.L.; Trick, M.; Higgins, J.; Fraser, F.; Clissold, L.; Wells, R.; Hattori, C.; Werner, P.; Bancroft, I. Associative transcriptomics of traits in the polyploidy crop species Brassica napus. Nat. Biotechnol. 2012, 30, 798–802.
136. Maughan, P.J.; Yourstone, S.M.; Byers, R.L.; Smith, S.M.; Udall, J.A. Single-nucleotide polymorphism genotyping in mapping populations via genomic reduction and next-generation sequencing: proof of concept. Plant Genome 2010, 3, 166–178.
137. Schneeberger, K.; Ossowski, S.; Lanz, C.; Juul, T.; Petersen, A.H.; Nielsen, K.L.; Jørgensen, J.E.; Weigel, D.; Andersen, S.U. SHOREmap: Simultaneous mapping and mutation identification by deep sequencing. Nat. Methods 2009, 6, 550–551.
138. 1001 Genomes. A Catalog of Arabidopsis thaliana Genetic Variation. Available online: http://1001genomes.org/downloads/shore.html (accessed on 6 September 2012).
139. Austin, R.S.; Vidaurre, D.; Stamatiou, G.; Breit, R.; Provart, N.J.; Bonetta, D.; Zhang, J.; Fung, P.; Gong, Y.; Wang, P.W.; McCourt, P.; Guttman, D.S. Next-generation mapping of Arabidopsis genes. Plant J. 2011, 67, 715–725.

AUTHOR NOTES

CHAPTER 2

Acknowledgments

The authors would like to acknowledge the computer resources of the Plataforma Andaluza deBioinformática of the University of Málaga, Spain. This study was funded by Spanish MICINN (BIO2009-07490) and Junta de Andalucía (P10-CVI-6075 and BIO-114).

CHAPTER 3

Acknowledgments

Work partly supported by the Italian Ministry of Instruction University and Research: PRIN 20087ATS57 "Food allergens". C. Leoni was recipient of a fellowship from the Italian Consortium for Biotechnologies. Authors wish to thank Graziano Pesole (Department of Biosciences, Biotechnology and Pharmacological Sciences, University of Bari) for his helpful comments.

CHAPTER 4

Acknowledgments

This research was funded in part by the Ministry of University and Science, Project PRIN 2008: "Investigations of the genetic and epigenetic changes in induced polysomic polyploids of crop plants", PI Fabio Veronesi. This paper received the contribution number 272 by the Department of Soil, Plant, Environmental and Animal Production Sciences.

CHAPTER 5

Acknowledgments

The authors would like to thank the Indian National Science Academy (INSA) for award of an INSA Honorary Scientist Position to PKG, Department of Biotechnology (DBT) for providing financial support to R. R. Mir, A. Mohan, J. Kumar, and also to the Head of Department of Genetics and Plant Breeding, Chaudhary Charan Singh University, Meerut, for providing the facilities.

CHAPTER 6

Acknowledgments

The authors thank all the participants of the Rice Genome Research Program (RGP) and the International Rice Genome Sequencing Project (IRGSP). This work was supported by grants from the Ministry of Agriculture, Forestry, and Fisheries of Japan (MAFF) through the Rice Genome Project, Green Technology Project, and GD 2007.

CHAPTER 7

Acknowledgments

The authors would like to acknowledge funding support from the Grains Research and Development Corporation (Project DAN00117) and the Australian Research Council (Projects LP0882095, LP0883462 and LP110100200). Support from the Australian Genome Research Facility (AGRF), the Queensland Cyber Infrastructure Foundation (QCIF) and the Australian Partnership for Advanced Computing (APAC) is gratefully acknowledged.

CHAPTER 8

Acknowledgment

We are grateful to the anonymous referees whose valuable comments helped to substantially improve the paper. This work was supported by the CNR-Bioinformatics Project.

CHAPTER 9

Competing interests

The authors declare that they have no competing interests.

Authors' contributions

All authors contributed to the manuscript. JAR, SEQ and SJS performed the statistical and bioinformatic analysis. CJB developed the data simulation algorithm. CJB, SRW and JMT conceived and designed the study. All authors read and approved the final manuscript.

Acknowledgments

JAR is funded by the OCE Science team. This project is supported by Australian Research Council Discovery Grant DP1094699. We used [45] as our Bibtex references manager. We used the TexMed database [46] as our bibtex references source. Additional file 2: Figure S4 was generated using the "VennDiagram" R-based package by Chen and Boutros [47].

CHAPTER 10

Acknowledgments

This work was supported by a grant from the NSF Plant Genome Program (DBI-0922526).

CHAPTER 11

Author Contributions

Conceived and designed the experiments: CNH NdL SMK CRB. Performed the experiments: CNH BV RSS. Analyzed the data: CNH BV. Contributed reagents/materials/analysis tools: NdL SMK CRB. Wrote the paper: CNH CRB SMK NdL BV RSS.

CHAPTER 12

Acknowledgments

This work was supported in part by a grant-in-aid for Scientific Research on Innovative Areas (24113509) to R. Fujimoto.

CHAPTER 13

Competing interests

The authors declare that they have no competing interests.

Authors' contributions

All authors contributed to the conception and writing of the review article. All authors have read and approved the final manuscript.

Acknowledgments

This work was supported by the Australian Research Council and the U.S. Dept. of Energy under Contract DE-AC02-06CH11357. The submitted manuscript has been created by UChicago Argonne, LLC, Operator of Argonne National Laboratory ("Argonne"). Argonne, a U.S. Department of Energy Office of Science laboratory, is operated under Contract No. DE-AC02-06CH11357. The U.S. Government retains for itself, and others acting on its behalf, a paid-up nonexclusive, irrevocable worldwide license in said article to reproduce, prepare derivative works, distribute copies to the public, and perform publicly and display publicly, by or on behalf of the Government.

CHAPTER 14

Acknowledgments

The authors would like to thank Drs. Shunxue Tang and Peizhong Zheng of the Trait Genetics and Technologies Department of Dow AgroSciences (DAS) and Raghav Ram of the IP Portfolio Development Department of DAS for careful review of the paper and the DAS Seeds and Traits R & D leaders Drs. David Meyer and Steve Thompson for general support and help.

INDEX

α-haemolysin (αHL), 18

A

Aegilops tauschii, 103, 108, 136, 142, 152, 156, 159
amplified fragment length polymorphisms (AFLPs), 103–104, 107, 133, 193, 388, 406, 412, 416, 418
angiosperm, xx, 27, 29, 36, 51, 73–74, 76, 80, 91, 93–94, 188, 440
Applied Biosystems, 2–3, 6, 29, 209, 245, 363
Arabidopsis lyrata, 28, 33, 94, 283, 337, 342–348, 353, 355
Arabidopsis thaliana, xxiv, 28, 31, 33, 48, 52, 75, 83–88, 90, 93–94, 96, 119, 121, 177, 184, 186–187, 281, 283, 294–295, 300, 323, 328–332, 334, 336–355, 401, 434–435, 439–440, 443, 446
asexual, 34

B

bacterial artificial chromosome (BAC), xxii, 28, 35, 108–109, 112–113, 116–117, 122, 133, 140–141, 143–145, 147–148, 169, 172, 174, 179–183, 187–188, 398
binning, xxvi, 358, 365–368, 379–380
biocatalysts, 357
biodiversity, xi, xv–xvii, 167, 358, 374
biogenesis, 62, 133, 282, 284–285, 288–290, 293, 295
biological replication, xxiv, 250, 252, 259, 265
biological variability, 265
bioluminescence, 220
biotinylated primers, 57, 60–62
Brassica oleracea, 80, 93, 433

C

cancer, 4, 19, 22, 213, 242–243
C. elegans, 244, 281, 294
candidate gene association (CGA), 400
Cap Analysis of Gene Expression (CAGE), 212, 246
Carica papaya (papaya), 28, 36–37, 48
centromeres, 38, 168–169, 174–175, 177–178, 183, 186
cereal, x, xxii, 73–74, 115, 121, 123, 131, 133–134, 137, 139, 143, 146–149, 151, 160, 163–168, 184, 189, 196, 409, 411
charge-coupled device (CCD), 8, 362
chloroplast, 35, 41, 62, 127, 132–134, 165, 181, 187
  chloroplast genome, 132, 187
chromatin immunoprecipitation (ChIP), xxiii, 7, 10, 13, 15, 21–22, 88, 110, 186, 211, 246, 328, 332, 424
cleavage, 2, 6, 18, 86, 219, 284–285, 295, 424

cluster analysis, 306
colinearity, 121–122, 141, 323
computation, xv, 39, 44, 246
copy number variation (CNV), xxiii, 15, 25, 48–49, 94, 211, 300, 309, 313, 315, 322, 407, 421
coverage variability, 42, 44
crop yields, 19
Cyclic Reversible Termination (CRT), 219, 240, 381

D

data sharing, xxvi, 375
de novo sequencing, xx, 19–20, 48
deletion, 11–13, 100, 105–107, 113–114, 120, 132, 134, 138–139, 163, 194, 202, 221, 224, 227, 362, 365, 408, 434
deoxyribonucleoside monophosphate (dNMP), 18
detection algorithms, 264, 370
differential expression (DE), viii, xxiii–xxiv, 86, 209, 228, 230–231, 235–237, 240, 246–247, 249, 251, 253–255, 257, 259, 261, 263, 265, 267, 269, 271, 273, 275–279
Digestion Ligation Amplification (DLA), 58, 69
diversity, x, xix, xxiv, xxvi, 28, 34, 48, 54, 79, 87–88, 92, 97, 105, 111, 120, 127–128, 130, 138, 143, 151, 160, 183, 187, 193–196, 198, 208, 238, 281–282, 284, 293–294, 299–301, 303–309, 311–313, 315–317, 319, 321–323, 325, 334–335, 350, 357, 360–361, 366, 373, 376–377, 379, 381, 392, 395, 400, 406, 408, 410, 415–416, 418–419, 421, 426, 429, 431, 433–434, 437–438, 440, 443
    diversity array technologies (DArT), 79, 92, 102, 105, 130, 133, 138, 193, 400
DNA
    cDNA, 31, 83–84, 87, 101, 106, 110, 124, 139, 185, 193, 212, 215–218, 243, 269, 271, 284, 316, 320, 425, 437
    centromeric satellite DNA, 175
    contamination, 41
    DNA short fragment library, 20
    elimination, 76
    extraction, 358, 360, 377
    junk DNA, 208
dNTP, 5, 220
domestication syndrome factors (DSFs), 128
doubled haploid (DH), 104, 118, 131, 137, 155, 157, 161, 165, 380, 382, 395, 420, 430, 433

E

edgeR, xxiv, 247, 254, 257–259, 264–266, 268–271, 275, 278
embryogenesis, 332
epiallele, 328, 330, 332, 334–337, 348
epigenomics, 19
Escherichia coli, 10, 12, 20, 24, 53, 379
Eucaliptus regnans, 27
eukaryotes, 36, 50, 56, 61–62, 66, 68, 73, 113–115, 144, 175, 177, 187, 208, 241, 245–246, 300, 327–328, 382, 421, 442
experimental design, xxiii–xxiv, xxvi, 239, 249–254, 258, 262, 265, 268, 275, 358, 372–374, 377, 411
expressed sequence tags (ESTs), xxi, 101, 106–107, 110–111, 122, 133,

135, 141, 143, 146, 169, 172, 201, 222, 224, 409, 445

F

fingerprint contigs (FPCs), 108
flow cytometry, 35, 81, 116, 360
fold-changes, 262, 264–265, 274–275
food security, 167, 184
Fragaria vesca, 35, 51, 81, 94
fragmentation, 59, 93, 218, 240, 250
functional markers (FMs), 105, 133–134, 163

G

gel electrophoresis, 87, 217, 416, 441
gene
  gene rich regions (GRRs), 100–101, 106, 108, 113–115, 128, 131, 138–139, 144,196
  gene-poor regions (GPRs), xxii, 113–114
  redundancy, 76, 119
  retention, 32, 35
genetic mapping, 77–79, 89, 92, 101, 117, 135, 141, 146, 152, 156, 158, 391, 395, 405–407, 431, 444
genome
  Genome Analyzer (GA), xx, 2–3, 8, 20–21, 53, 182, 209, 243, 317, 430, 442
  Genome Walking (GW), ix, xx–xxi, 55–69, 71, 276, 376–377
  genome-wide association studies (GWAS), 391, 399–40, 408, 419–421, 429, 436
  genome-wide duplication (WGD), 74, 80
  reference, xix, 168, 183, 220, 222, 224, 227, 229, 245, 278, 290, 300, 309–310, 312, 364, 388, 390–392, 401, 418, 432, 435
genomics
  comparative genomics, ix, xxii, 30, 50, 80, 93–94, 101, 106, 110, 121–122, 134, 183–184, 192, 200, 322, 371
  complete genomics, 15, 364
  databases, vii, 189–191, 193, 195, 197, 199, 201, 203
  Genomic Diversity and Phenotype Data Model (GDPDM), 193
  genomic estimated breeding value (GEBV), 403–404
genotyping-by-sequencing, viii, x, xxvii, 66, 70, 88, 97, 304, 318, 323, 391, 408, 415–417, 419, 421, 423–425, 427–429, 431, 433, 435–437, 439, 441, 443, 445
germplasm, xxv, 83, 120, 131, 144, 150, 155, 165, 193, 195, 300, 303, 306, 313, 315–316, 392, 410, 415, 431
Gilbert, Walter, 2
Green Revolution, 168
GS RunProcessor, 6

H

haplotype, 34–35, 51–52, 92, 127–128, 130, 147, 151, 194, 247, 305, 322, 340, 391, 400, 403, 406, 408, 426–429, 440, 444
heterochromatic, 38, 174, 177–178, 186, 327, 355
heterosis, xxi, xxv, 50, 76, 309, 315–316, 323–324

heterozygosity, 33–34, 81, 304, 430
High Resolution Melting (HRM), 81–82, 95, 399, 410
high-throughput (HTP), xxiii, xxvi, 20, 23–24, 49, 52–53, 59, 68–70, 81, 87–88, 105, 138, 146, 169, 171, 198, 244–246, 276, 278, 294, 296, 328, 388–389, 391, 393, 395, 399, 401–402, 404, 406–409, 417, 419, 424, 436, 438, 442–443
high-throughput insertion tracking by deep sequencing (HITS), 59, 234, 302
HiSeq control system (HCS), 8–9
HIV, 56, 68
human Microbiome project (HMP), 22
hybridization, xxiii, 15, 20, 68, 74, 77, 83, 86, 91, 94, 96, 108, 116, 141–142, 145, 148–149, 169, 177, 181, 209, 212–214, 218–219, 231, 309, 311, 344, 346, 388, 408, 416–418, 426, 438

I

in situ hybridization, 77, 108, 141–142, 145, 148, 177
    fluorescent in situ hybridization (FISH), 40, 77–78, 90, 108–109, 137, 141–143, 177, 287, 381, 412
    genomic in situ hybridization (GISH), 77–78, 108, 142, 185, 324
insertional mutagenesis, 56, 68
International Crop Information System (ICIS), 190, 195, 199
International Nucleotide Sequence Database Collaboration (INSDC), 190–191, 199
International Rice Genome Sequencing Project (IRGSP), 168–169, 172, 174, 184–185, 448
International Triticeae Mapping Initiative (ITMI), 102–103, 116, 191, 196

L

linkage disequilibrium (LD), 129–130, 134, 321, 391, 400, 403, 406, 419–420, 427–428, 434, 439

M

marker assisted selection (MAS), xxii, 52, 101–102, 105, 111, 118, 130–132, 134, 155, 165, 193, 397–398, 401–404, 409–412, 415, 419, 431
mass spectrometry (MS), 88, 378–379, 438
massively parallel signature sequencing (MPSS), 118
mate-pair, 19–20, 211, 218
maximum likelihood estimate (MLE), 272
Medicago sativa (alfalfa), 74–75, 78–79, 82, 92, 95
megasporogenesis, 75
meiosis, 74, 76, 90–91, 100, 351, 395
metabolism, 36, 76, 293, 376
metagenomics, x, xxv–xxvi, 19, 357–383, 424, 443
methylation, xxv, 7, 21–22, 24–25, 30, 69, 76, 79–80, 88, 92, 122–124, 148–149, 283–284, 295, 327–356, 390, 426, 432, 442–443
    Methylation-Sensitive Amplified Polymorphism (MSAP), 79, 122, 124, 350
    hypo-methylation, 331–333
microarrays, 25, 84–85, 118, 212, 214, 225, 231, 236, 242, 250, 277, 439

microsatellite (SSR), 79, 92, 102–103, 106, 110, 135–139, 143, 152–163, 191, 193–194, 200, 202, 388–389, 396, 398, 402, 407, 416, 431, 437, 445
Microsatellite Allele Dose and Configuration Establishment (MADCE), 79, 92
mid-parent value (MPV), 85, 88
MiSeq, xx, 8–9, 13–14, 363, 423–424
molecular cytogenetics, 77, 100, 141
multidimensional scaling (MDS), 373
multiple displacement amplification (MDA), 360, 377

N

nanopore sequencing, 18, 24
nested association mapping (NAM), 299, 321–322, 397, 400, 409, 411, 429, 444
next-generation sequencing (NGS), vx–xvi, xx–xxi, xxiii, xxvi, 1–3, 5, 7, 9, 11, 13, 15, 17–19, 21, 23–25, 28–33, 37, 40–47, 49–50, 53, 56, 58–59, 61–62, 65–68, 70–71, 83, 86, 88, 95, 189, 191, 209, 211, 213, 215, 218–222, 226, 238, 241, 243, 245–246, 250–251, 253, 277, 284, 296, 358, 361, 363, 378–379, 389–391, 406, 421–429, 431, 436, 441, 443–445
Nipponbare, 168–169, 171, 174–176, 179–180, 182
nonrestrictive Linear Amplification-Mediated PCR (nrLAM-PCR), 58, 60, 68
northern blot, 83
nucleotide barcode, 58

O

oligonucleotide, 25, 55, 58–59, 116, 214, 245, 379
open reading frames (ORFs), 62, 132, 216
Oryza sativa (rice), ix, xxii, 25, 28, 33, 36, 38, 48–49, 52, 92, 101, 111, 114–115, 117, 121–122, 132–133, 141, 143, 147–149, 152, 160, 167–169, 171–175, 177–179, 181–187, 189–192, 196, 198, 200, 282–283, 291–294, 306, 312–313, 320, 323, 325, 328, 334, 336, 349–351, 353, 396–397, 402, 408–409, 411, 425, 429, 438, 440, 442, 444–445, 448
overdispersion, 265, 272

P

Panzea, 190, 193, 200
paramutation, 328, 336, 339, 341, 350
personal genome machines (PGM), xx, 2, 9–10, 14, 66, 70, 423–424
phenotype, xxi, xxv, 34, 69, 76, 91, 119–120, 137, 193, 198, 241, 329–331, 333, 336–337, 340, 348, 351, 355–356, 399–400, 402–403, 415, 420, 429, 434–435
photoheterotrophy, 358
phylogeny, 30, 51, 150, 289, 357
physical map, xxi, 34–35, 94, 105–108, 113, 133, 138–139, 169, 174, 179–181, 187–188, 434
plant genome assembly, ix, 37, 39, 45
plasticity, x, 34, 89, 124, 164, 241, 348
Poisson distribution, 232, 234–236
polymerase chain reaction (PCR), xx, xxv, 5, 16, 44, 53, 55–69, 71, 83, 86, 116, 118, 120, 151, 156–158, 161, 193–194, 217–218, 250, 284, 311–312, 316, 320, 345, 361–362, 378, 388–389, 393, 397, 399, 405, 408–409, 416–417, 422–424, 431–432, 434, 436–437, 443

emulsion polymerase chain reaction (ePCR), 361–362
polyploid, ix, xxi, 34–36, 51, 73–97, 111, 114, 119, 127, 138, 140–141, 146, 148, 150, 164, 166, 391–393, 395, 401, 403–404, 406, 409, 411, 427, 432–433, 442–443, 445, 447
allopolyploids, 34, 74, 78–79, 83–87, 91–92, 95–97, 124, 149, 294, 391
autopolyploids, 74, 78, 83, 85, 88, 92, 97
polyploidization, xxi, 34, 52, 73–78, 88–89, 96, 124, 147
presence/absence variation (PAV), xxv, 49, 300, 308–309, 311, 313, 315, 320, 322, 407
processing load, 44
protein quantitative loci (PQL), 124–125
pseudomolecules, 302–304, 306, 309–311, 314, 317–320, 433–434
pyrophosphate (PPi), 5, 220, 361, 423
pyrosequencing, 3, 5, 24, 49, 53, 56–60, 66, 68–69, 87, 181–183, 188, 220, 240, 361–362, 364, 378, 423, 441, 443

Q

quality value (QV), 10, 12, 43, 221, 363
quantitative trait loci (QTL), xxii, xxvi, 78, 92, 101, 104, 106, 109–110, 113, 116, 118, 124–125, 128–131, 134, 136–137, 141, 143, 149–150, 153–155, 157–158, 160–164, 182, 390–391, 395–404, 407, 409, 418–420, 428, 430–431, 434, 438–440, 444–445

R

radiation hybrid mapping (RH), 106, 108, 140
read length, 2–3, 5–7, 9–14, 16, 18, 29, 39–40, 42–43, 209, 319, 361–364, 378, 423
real-time analyzer (RTA), 8–9
receiver operator characteristic (ROC), 22
recombination, 52, 69, 80, 95, 100, 102, 105, 114–115, 128, 132, 138, 144, 152, 154, 177, 340, 395–396, 419, 426, 429, 433
replication, xxiv, 20, 81, 94, 132, 250, 252, 254, 259–260, 262, 264–265, 275, 373–374
restriction fragment length polymorphisms (RFLPs), 103–104, 107, 133, 193, 388, 405, 416
retrovirus, 68
Rice Genome Research Program (RGP), 115, 169, 172, 174, 178–179, 185, 448
RNA, x, xxii–xxv, 7, 19–21, 25, 30, 37, 52, 54, 59, 83, 85–87, 96–97, 101, 118–119, 133, 139, 146, 178, 207–209, 211–219, 221–247, 249–255, 257, 259, 261, 263–269, 271, 273, 275–279, 281–282, 284, 287–293, 295–297, 299–301, 303–305, 307, 309, 311, 313, 315–321, 323–325, 330, 337, 339, 341, 346, 349, 351–352, 354–355, 381, 442
interference, xxii, 101, 118, 146, 330
miRNA, xxiv, 20, 37, 85–87, 119, 208, 215–216, 227, 277–278, 281–296, 442
mRNA, 21, 85, 118, 207–208, 214–217, 233, 243–246, 267, 271,

276, 285–286, 293, 295, 316, 319, 324, 349, 425, 433–436
non-coding RNAs (ncRNAs), 37, 52, 293, 296
RNAi, xxii, 118–119, 134, 295, 330, 352–353
RNA-seq, ix–ix, xxiii–xxv, 19, 21, 25, 52, 54, 86–87, 96–97, 207, 209, 211–215, 217, 219, 221–247, 249–255, 257, 259, 261, 263–269, 271, 273, 275–279, 300, 303–305, 309, 311, 313, 315–320, 324, 442
transcripts, 212, 215
Roche 454, 5, 7, 9, 56, 423, 434

S

Salmonella enterica, 60–61
sample processing, x, xxvi, 358
Sanger method, 2, 28, 182, 238, 421
Sanger, Frederick, 2–3, 28–30, 40, 43, 46, 49, 54, 181–183, 189, 209, 211, 238, 241, 312, 320, 361, 389, 417, 421–422, 427, 440
self-organising maps (SOMs), 367
sequencing
sequence variation, xxiv, 84, 94, 187, 194, 313, 391, 422, 430
Sequencing by Oligo Ligation Detection (SOLiD), xx, 2–9, 15, 20–22, 24, 29, 56, 65–67, 117, 211, 215, 217–221, 245, 260, 262, 362–363, 378, 389, 424, 441
sequencing by synthesis (SBS), 3, 8, 13–14, 56, 59, 61, 63–64, 66, 423
sequencing depth, xxiii–xxiv, 11, 13, 251–254, 259–262, 264–267, 275–276, 289, 292, 366, 437
sequencing error, 40, 42, 44, 221, 224–225, 227, 250, 369
serial analysis of gene expression (SAGE), 117–118, 212, 242
short interspersed nuclear element (SINE), 332, 340–343, 345, 347, 355
simple sequence repeats (SSRs), 38, 103–104, 106–107, 110–111, 118, 133, 135–136, 139, 143, 186–187, 193–194, 387–388, 387, 406, 418, 445
single nucleotide polymorphism (SNP), x, xxv–xxvi, 15, 23, 82, 86, 94–95, 102, 105, 111, 141, 146, 179, 183, 187, 194–196, 198, 202, 226, 301–307, 315, 317–318, 321, 385, 387–413, 417–418, 421, 425, 427–428, 430–436, 438–440, 442–445
single nucleotide variations (SNV), 219, 243
single-molecule real-time (SMRT), 16, 18
single-strand conformational polymorphism (SSCP), 84, 431
software, 2, 6–9, 15, 41, 44, 47, 50, 52, 101, 169, 173, 182, 194, 221–223, 227, 229, 239–240, 245, 255–256, 288, 296, 363, 365, 367, 392, 425, 427–428, 435–436
soybean, 81, 93–94, 97, 197, 199, 201, 398, 406, 410, 412, 440, 442
Staphylococcus aureus, 18
sterility, 74, 333, 412
Streptococcus pneumoniae, 62
sustainable development, 167

T

tandem duplications, 20, 38, 343, 346–348
targeted sequence capture, 88, 97
taxonomic, 80, 197, 199, 366, 368–370, 372–373, 380–381

taxonomy, 191, 194, 200
telomere, 177–179, 186–187
telomeres, 38, 52, 168, 174, 177–179, 183, 186
  Telomeric sequences, 38
The Million Eco-System Genomes Project, 19
third-generation sequencing, 2, 16, 47
transcriptome, xix, xxiii, 4, 7, 19–21, 25, 31–32, 39, 41, 45–46, 49–52, 54, 81, 83–85, 87, 94–97, 118, 146, 207–209, 211, 213–217, 221–223, 225, 227, 229, 231–233, 235, 237, 239, 241–247, 249–251, 266–267, 272, 275–277, 281, 283, 285–287, 289, 291, 293, 295–297, 300, 304, 306–308, 310, 313, 315–316, 318–319, 323, 349, 353, 390, 401, 406–407, 433, 435, 442–443
transposable insertion, 337
transposon, 33, 35, 37, 51, 58–63, 65–67, 69, 76, 87, 328, 330, 335–337, 348–349, 351–354

U

unigene-derived microsatellites (UGMS), 106

V

variation
  epigenetic, xxv, 322, 327–329, 331, 333–335, 337, 339, 341, 343, 345, 347, 349–355
  phenotypic, xxv, 75, 90, 303, 327, 329, 333–334, 400, 419

W

wheat, ix, xxi–xxii, 32–33, 35, 66, 78–85, 87–88, 93–97, 99–168, 185, 190–191, 193, 196, 201–203, 390–391, 398, 401–402, 405–406, 409–410, 425, 432–433, 435–436, 443, 445
  hexaploid, 35, 82, 87–88, 95–97, 108, 113, 117–118, 128, 134–135, 137–138, 140–141, 144, 146–150, 152, 154, 158, 163, 432
  wheat breeding, 105, 131, 135, 150, 161, 165, 405, 410
  Wheat Genome Sequencing Consortium (IWGSC), 101, 108, 117, 131, 135, 185
  wheat production, 99, 135
whole genome sequencing (WGS), 10, 20, 24, 49, 116–117, 169, 171, 244, 304, 306, 323, 422

Z

Zea mays (maize), x, xxiv, 28, 33, 48, 51–52, 58, 62, 66, 69, 82–83, 87–88, 94–97, 101, 111, 115, 132–133, 143, 164–165, 167, 186, 189–190, 192–193, 196, 198, 200, 202, 299–301, 303–305, 307–313, 315–317, 319–323, 325, 328, 334, 336, 349–350, 354, 390–391, 397, 400, 402, 404, 406–409, 411–413, 418, 421, 425, 429, 432, 438–440, 443–445
zero-mode waveguides (ZMWs), 16

For Product Safety Concerns and Information please contact our EU
representative GPSR@taylorandfrancis.com
Taylor & Francis Verlag GmbH, Kaufingerstraße 24, 80331 München, Germany

www.ingramcontent.com/pod-product-compliance
Ingram Content Group UK Ltd.
Pitfield, Milton Keynes, MK11 3LW, UK
UKHW022208190926
475702UK00003B/256

* 9 7 8 1 7 7 1 8 8 0 0 3 9 *